Blue Carbon Dynamics of the Indian Ocean

Abhra Chanda · Sourav Das · Tuhin Ghosh
Editors

Blue Carbon Dynamics of the Indian Ocean

The Present State of the Art

Springer

Editors
Abhra Chanda
School of Oceanographic Studies
Jadavpur University
Kolkata, West Bengal, India

Sourav Das
School of Oceanographic Studies
Jadavpur University
Kolkata, West Bengal, India

Tuhin Ghosh
School of Oceanographic Studies
Jadavpur University
Kolkata, West Bengal, India

ISBN 978-3-030-96557-0 ISBN 978-3-030-96558-7 (eBook)
https://doi.org/10.1007/978-3-030-96558-7

This Springer imprint is published by the registered company Springer Nature Switzerland AG
The registered company address is: Gewerbestrasse 11, 6330 Cham, Switzerland

Dedicated to the millions of coastal people who are incessantly struggling with climatic disasters

Preface

In the name of development and modernization, human beings have unearthed several compounds and elements from the earth's crust, degraded the natural landscape largely, and emitted several gases and liquids that can significantly pollute the environment. Among the various anthropogenic activities that have taken a toll on the present-day environment, carbon emission is perhaps the most widely discussed issue. Carbon-containing gaseous compounds like carbon dioxide (CO_2) and methane (CH_4) have been long recognized as potent greenhouse gases, and human beings through fossil fuel combustion, forest degradation, and what continually emit these more. Ever since the role of these greenhouse gases in changing the climate has been realized, environmental scientists have been desperately looking for options to combat this necessary evil. In the present era of climate change, the need to understand our environment and the role of several ecosystems in mitigating climate change has become an utmost priority.

In this regard, the concept of blue carbon has lately emerged as one of the most important topics that emphasize the role of oceans and coastal ecosystems in combating greenhouse gas emissions. Oceans are known to act as one of the most important carbon repositories of the world. Besides the oceanic water volume, the carbon sink potential of several marine ecosystems like mangroves, seagrass, salt-marshes, and tidal flats has drawn the attention of the present scientific community. Many researchers and scholars have argued to have found an avenue to mitigate climate change through proper restoration, conservation, and management of these blue carbon ecosystems. Though blue carbon ecosystems are found throughout the world's oceans' periphery, the tropics and the subtropics are comparatively much more carbon enriched than the higher latitudes. Among the five oceans of the world, the Indian Ocean covers one of the most extensive blue carbon ecosystems in its periphery, with more than 50 % of the global mangrove cover and a large proportion of seagrass beds. A substantial part of the Indian Ocean lies within the 30°N and 30°S encompassing a shoreline of about 66,500 km and experiences a typical tropical to subtropical climate. This is why the productivity of the Indian Ocean blue carbon ecosystems is much higher than that found in the other parts of the world. The following countries bordering the Indian Ocean is considered in this book—Iran,

Pakistan, India, Bangladesh, Myanmar, Thailand, Singapore, Indonesia, Malaysia, Yemen, Oman, United Arab Emirates (UAE), Djibouti, Somalia, Kenya, Tanzania, Mozambique, South Africa, Sri Lanka, Maldives, Seychelles, Madagascar, Comoros, Mauritius, and Australia. The present book has collated and discussed the present status of the Indian Ocean blue carbon repositories in the coastline and their country-wise scenario, along with the carbon sequestration potential of the open oceanic part. Many of the world's largest estuaries drain into the Indian Ocean. This book has dealt in detail with the information gained so far on the carbon dynamics of such estuarine systems. The carbon biogeochemistry in the aquatic and land compartments in and around the mangrove systems is discussed in detail along with the factors regulating the carbon cycles. Similar treatments are also done on the seagrass, saltmarshes, and tidal flats found in the periphery of the Indian Ocean. Though corals are debated not to be an integral part of the blue carbon repositories, their role in regulating the carbon dynamics in the Indian Ocean is also explicitly discussed in the present book. Besides the biogeochemical aspects of these ecosystems, the livelihood opportunities and ecosystem services of these systems toward human beings and the importance of their conservation and management from the incessant threats these ecosystems are facing at present, are also highlighted in this book. On the whole, the present book has covered every aspect of blue carbon from the perspective of the Indian Ocean with the intent to provide all the possible information related to the blue carbon dynamics of the Indian Ocean under one umbrella.

Kolkata India

Abhra Chanda
Sourav Das
Tuhin Ghosh

Acknowledgments

We thank Dr. Anirban Akhand and Dr. Somnath Hazra for reviewing many of the chapters of this book. We express our deepest regards to Dr. Rabindranath Bhattacharya for his constant support and encouragement. We also thank the editor of Springer Nature, Ms. Doris Bleier, for her kind help and encouragement. We also extend our gratitude to Mrs. Jayanthi Krishnamoorthi, Project Coordinator, Books Production, Springer Nature, for making the publishing process hassle-free and easy for us. We especially thank Prof. Andy Large and Dr. Andrew C. G. Henderson (School of Geography, Politics and Sociology, Newcastle University, UK) for their kind support and encouragement throughout this book project.

Kolkata, India

Abhra Chanda
Sourav Das
Tuhin Ghosh

Contents

Editors and Contributors

About the Editors

Dr. Abhra Chanda is an Assistant Professor in the School of Oceanographic Studies at Jadavpur University, India. His research interest encompasses the biogeochemistry of lentic and lotic aquatic systems and pollutant dynamics. Blue carbon dynamics is one of the priorities of his research. He has eleven years of research experience and has more than 70 research articles in reputed peer-reviewed journals, 10 chapters in edited book volume, and one book to his credit.

Dr. Sourav Das is a Post-Doctoral Research Associate of the UKRI-GCRF Living Deltas Hub in the School of Oceanographic Studies at Jadavpur University, India. His research interests include biogeochemistry of marine and coastal water, ocean remote sensing, optical properties of water (inherent and apparent optical properties), and air quality monitoring. He has 12 years of research experience and has more than 60 research articles in reputed peer-reviewed journals including two books.

Prof. Tuhin Ghosh is Professor and Director of the School of Oceanographic Studies at Jadavpur University, India. He has more than two decades of research experience in coastal geomorphology, disaster management, climate change impacts, adaptation strategies, and human migration. He has more than 80 research articles in reputed peer-reviewed journals to his credit.

Contributors

Anirban Akhand Coastal and Estuarine Environment Research Group, Port and Airport Research Institute, Nagase, Kanagawa, Yokosuka, Japan

Abhra Chanda School of Oceanographic Studies, Jadavpur University, Kolkata, West Bengal, India

Sourav Das School of Oceanographic Studies, Jadavpur University, Kolkata, West Bengal, India

Tuhin Ghosh School of Oceanographic Studies, Jadavpur University, Kolkata, West Bengal, India

List of Figures

List of Tables

Chapter 1
A General Introduction to the Concept of Blue Carbon

Abhra Chanda, Sourav Das, and Tuhin Ghosh

Abstract Human beings have been on earth for almost six million years. Through this long journey, we have evolved physiologically and mentally and brought ourselves up as social animals. We learned to utilize the earth's valuable resources to meet daily life activities. However, in the past three centuries, the consumption of fossil fuels to meet the energy demand has pushed the environment to such an extent that it has started affecting all of us through various natural disasters. At present, phenomena like climate change, global warming, and sea-level rise have bothered all of us, directly or indirectly. Retracting the course of human development is nigh impossible. Thus, it is high time we looked for options to heal the environment instead of only using it to our benefit. Since the last century, we have been desperately looking for options to enhance the carbon sink potential of mother earth, given the excessive amount of CO_2 we have emitted, and continue to do so in the name of development. We recognized the importance of terrestrial forests in sequestering CO_2 long ago. We have lately discovered the potential of marine ecosystems, especially the angiosperm-dominated ones that thrive in the periphery of the coastal margins and store a substantial amount of carbon. These ecosystems include mangrove forests, seagrass meadows, salt marshes, and tidal flats. Together they are referred to as the 'blue carbon ecosystems. Some evidence shows the promising potential of these ecosystems to combat the evil of climate change. The concept of blue carbon is merely a decade old, and already it has become quite popular among academicians and researchers. In this chapter, which introduces this book, we have given a detailed overview of the very concept of blue carbon. We tried to provide a holistic understanding of the importance of these ecosystems in mitigating climate change. We have also explicitly discussed the carbon dynamics of these ecosystems and the emerging dimensions in this arena. We also included a brief review of the present state of awareness and efforts for conservation and management practiced throughout the world. Lastly, we introduced the backdrop of the Indian Ocean, on which we have focused this entire book.

A. Chanda (✉) · S. Das · T. Ghosh
School of Oceanographic Studies, Jadavpur University, 188 Raja S.C. Mullick Road, Kolkata, West Bengal 700084, India
e-mail: abhrachanda.sos.rs@jadavpuruniversity.in

A. Chanda et al. (eds.), *Blue Carbon Dynamics of the Indian Ocean*,
https://doi.org/10.1007/978-3-030-96558-7_1

Keywords Blue carbon · Mangrove · Seagrass · Saltmarshes · Corals · Macroalgae · Estuaries · Open ocean · Indian Ocean countries

1.1 The Changing Environment and Greenhouse Gases

The history of humankind took a critical turn when human beings invented fire. According to several anthropologists, this invention was perhaps one of the most important that marked the beginning of development, as far as the human race is concerned. This invention marked the beginning of an era that has witnessed the burning of the earth's resources to meet the infinite demand of the human race. Scarcely the human beings realized in the then time that evil inevitably accompanies every boon. As the course of materialistic development went forward, 1760 marked the beginning of a crucial phase with the onset of the first industrial revolution. The world we know today would not have been possible without the commencement of this event. The industrial revolution indeed changed the very definition of human life, and we stepped into the modern era of human civilization, where machines started deciding the present and the future of humankind. While galloping over time accompanied by such growth and development, we forgot that it is only us, the human beings, who witnessed development through our course of action, and the environment wherein we live has degraded beyond measures. To meet the ever-increasing demand of human beings, we have indiscriminately used the earth's resources leaving behind some byproducts, which took a disastrous toll on environmental health. Since the last century, the ill effects of this so-called development started hitting us back in the form of various disasters like climate change, global warming, environmental pollution, and more. Such phenomena made us realize that it is time we looked at the well-being of the environment. Or else, in the name of development and modernization, we are very likely on the verge of inviting a dangerous future for ourselves.

The global atmosphere, especially the thin blanket of air mass close to the earth's surface named troposphere, where all the life forms mainly thrive, is composed of a mixture of gases, and it took almost 4.6 billion years to reach the present gaseous composition (Mohanakumar 2008). The troposphere is mainly composed of 78% nitrogen gas (N_2), followed by 21% oxygen gas (O_2), 1% argon gas (Ar), 0.036% of carbon dioxide (CO_2), several other trace gases, and varying concentrations of water vapor (Monks 2005). Oxygen is integral to the sustenance of life on earth. Nitrogen also takes part in many biochemical processes. The atmospheric concentration of both these gases has not changed over the past millennia. However, in the last two to three centuries, the trace gas concentrations in the atmosphere have witnessed a radical change due to a wide range of anthropogenic activities (Overpeck et al. 1997). Among the various gases that exist in the atmosphere, some of the gases can act as greenhouses gases, which include CO_2, water vapor, Ozone (O_3), and methane (CH_4). The presence of these radiatively active gases makes this planet habitable for life forms to flourish, as these gases raise the ambient temperature in the troposphere by almost 30 K (Mitchell 1989). Earth, like all the other planets of this solar system, receive the incoming solar radiation in the visible range of the spectrum. However, the energy radiated by the earth's surface mostly escapes to space in the form of longer-wavelength infrared radiation. The greenhouse gases are capable of absorbing these outgoing radiations and retaining the heat energy, which, in turn, enhances the atmospheric temperature. According to estimates made by several researchers, the global mean temperature of the earth's Surface would have been −18 °C in the absence of any atmosphere, which is at present close to 15 °C (MacCracken and Luther 1985).

Though available in trace quantities, the umbrella of greenhouse gases encompasses a large number of gases. These greenhouse gases comprise two broad categories: (i) gases that occur in the atmosphere naturally (water vapor, CO_2, CH_4, and N_2O) and (ii) gases that are manufactured and introduced into the atmosphere through anthropogenic activities [chlorofluorocarbons (CFCs)]. Except for a few gases, all the other greenhouse gases contain carbon. Researches show that the anthropogenically manufactured CFCs have increased significantly over the last century due to their extensive use as a coolant, refrigerant, and several metal-processing plants (Ming et al. 2016; Rigby et al. 2019). However, the point of concern remains to be the conventional greenhouse gases like CO_2, CH_4, and N_2O. According to the estimates for the year 2017, the global mean atmospheric CO_2 concentration was 407.58 ppm, followed by CH_4 (1.849 ppm) and N_2O (0.330 ppm) (European Environment Agency 2019). The concentration of these three greenhouse gases that occur naturally in the atmosphere has steeply increased in the atmosphere since the industrial revolution (Meinshausen et al. 2017) (Fig. 1.1). Though the concentrations of CH_4 and N_2O are much less than that of CO_2, the high global warming potential of these two gases (CH_4—25; N_2O—298) makes these gases equally important (IPCC 2001). A wide range of anthropogenic activities like the use of inorganic nitrogen-based fertilizer in the food production sector, excreta processing of animal manure, and agricultural soil tillage have disturbed the natural nitrogen cycle and led to considerable N_2O emissions (Syakila and Kroeze 2011). However, in terms of increment in emission, N_2O still ranks to be the third most important greenhouse gas (Forster et al. 2007), with CO_2 and CH_4 holding the first and second rank, respectively. The principal chunk of greenhouse gas emissions to the atmosphere in the form of CO_2 and CH_4 originates due to fossil fuel combustion, forest degradation, land-use change, agriculture, aquaculture, livestock farming, and coal, cement, and lime production (Olivier et al. 2017).

1.2 Carbonaceous Green Gases and the Manifestations of Climate Change

As per the recent estimates, CO_2 and CH_4 contribute 72% and 19% of the total greenhouse gas emission, respectively (Olivier et al. 2019). In the year 2018, the

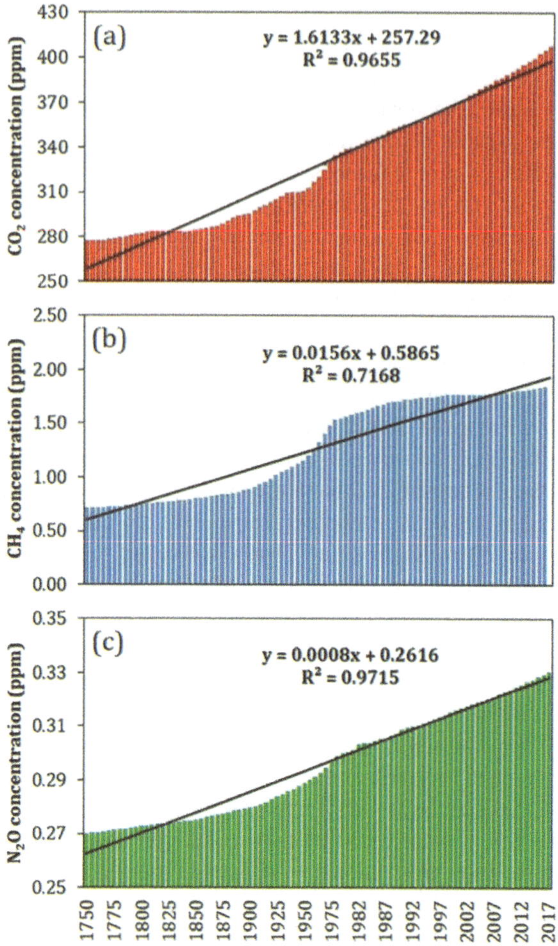

Fig. 1.1 The atmospheric concentration of **a** CO_2, **b** CH_4, and **c** N_2O from the year 1750 to 2017. (*Data source* CO_2—National Oceanic and Atmospheric Administration (NOAA); CH_4 and N_2O—Advanced Global Atmospheric Gases Experiment (AGAGE))

global greenhouse gas emission was around 51.8 Gigatonne CO_2 equivalent, which marked a continued annual increment of about ±1% since the year 2012 (UNEP 2019). The year 2018 also witnessed one of the highest increases in the global annual mean atmospheric CO_2 concentration (2.86 ppm) (NOAA/ESRL 2019). During the years 1970 to 2003, the CO_2 emission kept on increasing at a rate of 1.6% per year (on average). We observed an acceleration in this rate during the years 2003 to 2011 (3.2%), followed by a reduction during the years 2012 to 2015 (1.3%). However, since the year 2016, we have noticed a significant increase in CO_2 emission, which crossed the previous estimates (BP 2018; IEA 2019). We observed similar scenarios in the case of CH_4 emission also. During the years 2004 to 2014, CH_4 emission increased at a mean rate of 1.6% per year. However, it was interesting to observe that the production of fossil fuels like coal, oil, and natural gas and their corresponding transmission contributed to most of the CH_4 emission, but not fossil fuel combustion

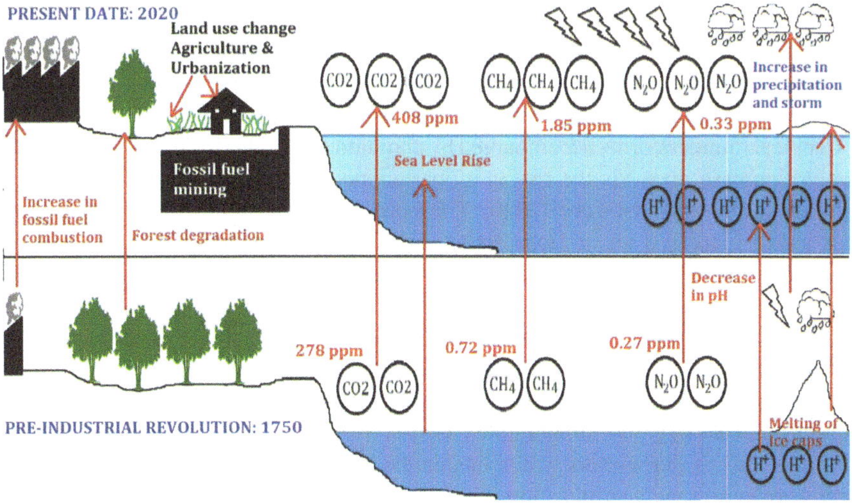

Fig. 1.2 A diagram showing the principal changes that took place since the pre-industrial revolution era

(Olivier et al. 2019). Most of us have a common perception that fossil fuel combustion in the less oxygenated environment leads to the production of CH_4, which is true, but CH_4 produced in this way contributes negligibly to the total CH_4 emission. On the contrary, livestock farming, especially cattle breeding, followed by domestic and industrial waste, contributes to a significant share of the global CH_4 emission.

The manifestations of such an increase in global CO_2 and CH_4 emission occurred throughout the world, especially in the last few decades (Fig. 1.2). The global atmospheric temperature increased at an alarming rate. Out of the ten warmest years recorded since the monitoring began in the year 1880, 9 occurred since the year 2005 (NOAA 2019; Olivier et al. 2019). Arnell et al. (2019) estimated that the intensity and occurrence of heatwaves are quite likely to increase by 28% by the next 50 years if the temperature increases by 1.5 °C. The chances of heat waves can increase by 92% if the temperature increases by 4 °C. Because of such an increase in heat waves, agricultural droughts, river flooding, and crop degradation are likely to increase manifold soon (Zhao et al. 2017; Arnell et al. 2019). Several researchers have argued that the ongoing increase in global temperature is also responsible for the increment in the frequency of forest fires observed throughout the world (Seidl et al. 2017). Melting of polar ice caps and deglaciation-induced rise in sea level has perhaps been the topic of most severe concern in the last decades (Wong et al. 2014; Arnell et al. 2016; Le Bars et al. 2017).

In the present date, restricting the global mean temperature rise to below 2 °C above the pre-industrial levels and striving hard to delimit the increase to 1.5 °C happens to the main aim of all the 196 parties of the United Nations as an increase beyond this threshold might lead to cataclysmic climate change risks (UNFCCC 2015). Jevrejeva et al. (2018) estimated a median global sea-level rise of 52 cm and

63 cm corresponding to a mean temperature rise of 1.5 °C and 2 °C, respectively, by the end of the twenty-first century. Sea level rise can prove to be catastrophic soon, as 600 million people all over the world live within 10 m from the sea level (McGranahan et al. 2007). There is also an overwhelming consensus that the frequency and intensity of tropical thunderstorms accompanied by anomalously high and erratic precipitation would increase vis-à-vis the rise in global temperature (Wasko and Sharma 2017; Zhao et al. 2018; Knutson et al. 2020). The increasing atmospheric CO_2 concentration has led to increased absorption of CO_2 by the global oceans, which in turn, has decreased the pH of the oceanic waters by almost 0.1 in the last 150 years. Such a decrease in pH is equivalent to an increase of 26% of free hydrogen ions (Hoegh-Guldberg et al. 2017). This phenomenon, named ocean acidification, is biologically significant, as it hampers the ability of the calcareous marine organisms to build structures made up of inorganic carbonates, like corals (Kroeker et al. 2013). Besides this, an increase in global temperature can directly lead to an outbreak of vector-borne diseases, as warmer winters would enable the viruses and pathogens to stay active, and warmer summers would shorten the time needed to infect human beings (Chan and Johansson 2012; Gloria-Soria et al. 2017; Wilke et al. 2019). Indirectly, the temperature increases induced heavy rainfall and consequent flooding, especially in the urban areas, would proliferate the breeding sites of several arthropods (Jia et al. 2017). All these observations indicate that it is high time we gave priority to reduce the greenhouse gases and stabilize the global temperature to prevent ourselves from consequences that are unknown but far more disastrous than what we have observed so far.

1.3 Natural Sinks for Carbon

Though we recognize the ill effects of fossil fuel combustion, land-use change, and forest degradation, ceasing all these processes forever at an instant to safeguard the environment is not at all possible. In the present day, the entire world population is very reliant on these processes to carry out their daily life activities. Under these circumstances, the environmentalists were desperately looking for alternatives to combat the evil of growing carbon emission, and such endeavors led to some fruitful results. There are mainly two ways to tackle the acute problem of increasing carbon emission. The first way is to reduce the present rate of emission by sustainable use of non-renewable fossils and by regulating a wide range of anthropogenic activities that indirectly lead to carbon emissions, like forest degradation and deforestation. The second way is to come up with new technologies that can artificially trap and sequester GHGs from the atmosphere and help reduce the carbon load of the atmosphere. Specialists belonging to the disciplines of environmental engineering, biotechnology, and many others are incessantly working hard to synthesize various materials and develop newer technologies to curb carbon emissions and artificially sequester substantial quantities of carbon (NASEM 2018; Singh and Dhar 2019; Zhao et al. 2020). However, several researchers unanimously believe that nothing can beat the option of strengthening nature's own carbon sequestration potential to combat this evil. This realization urged several researchers all over the world to understand, characterize, and identify avenues through which we can increase the carbon stock of the several natural reservoirs of this planet, except the atmosphere.

The terrestrial forests, the pedosphere, and the oceans are the largest carbon stockholders of this planet, and these three reservoirs maintain a dynamic equilibrium of carbon by exchanging carbon in several chemical forms (Ciais et al. 2013) (Fig. 1.3). The terrestrial vegetation absorbs a substantial amount of CO_2 by carrying out photosynthesis during the daytime, and it can potentially offset the anthropogenic CO_2 emissions (Running 2012; Zhang et al. 2019).

Terrestrial gross primary production (GPP) is the highest global CO_2 flux (~123 ± 8 Pg C yr^{-1}) that governs several ecosystem functions (Beer et al. 2010). However, the global mean GPP is one of the most variable carbon fluxes that depends on the changing environment and the biogeochemical cycles of the earth (Campbell et al. 2017). The soil surface up to a depth of 1 m is the largest known carbon reservoir in the world, with an estimated carbon stock ranging from 2500 to 3400 Pg C (Tifafi et al. 2018). The soils encompass almost 3.3 times the carbon present in the atmosphere and 4.5 times the carbon present in the biotic communities (Lal 2004; Trumbore 2009). However, this enormous reservoir of carbon is very much susceptible to anthropogenic disturbances, and there is substantial uncertainty related to the carbon dynamics in the topmost part of the pedosphere (Köchy et al. 2015). Tang et al. (2020) recently estimated the global mean soil heterotrophic respiration rate of 57.2 ± 0.6 Pg C during the years 1980 to 2016. All the parties of the United Nations Framework Convention on Climate Change (UNFCCC) have formulated a

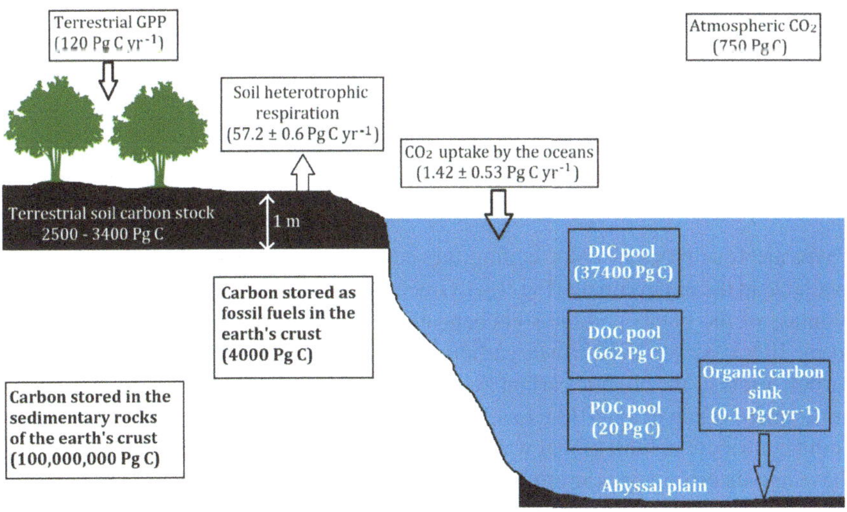

Fig. 1.3 The carbon pool in some of the major carbon reservoirs of the planet and the exchanges among the reservoirs that take place per year

joint international collaboration to reduce the emissions by preventing the degradation of forests and terrestrial vegetation and enhance the forest carbon sequestration, named as "Reducing emissions from deforestation and forest degradation (REDD+)" (Huang et al. 2020, and the references therein). The broader intentions of mechanisms as REDD+ is not only to prevent deforestation but also to conserve, sustainably manage, and enhance the forest carbon stock (Lund et al. 2017). No country at present can afford to deny the essential need to comply with such initiatives to bolster climate governance (Tateishi et al. 2020).

By the turn of the previous century, it was evident that the oceans have played an extremely crucial role in the carbon cycling of the earth. The biological pump of the oceans refers to the process through which carbon near the sea surface undergoes conversion from the inorganic to organic form by autotrophic activities and sinks to the depths of the ocean, where the recalcitrant carbon remains for a long time as a strong net sink of carbon (Jiao et al. 2010). A substantial portion of the CO_2 is thus fixed by the biological pump as organic carbon undergoes re-mineralization back to CO_2, which eventually finds its way back to the atmosphere; however, the recalcitrant dissolved organic matter (RDOM) remains in the abyssal depths for more than a million years as long-term carbon sinks. According to the latest estimates, oceans uptake CO_2 at about -1.42 ± 0.53 Pg C yr^{-1} (Landschützer et al. 2014). Though the open ocean comprises a large surface area (3.33×10^8 km^2) and volume of water mass, the total net carbon sequestration only in the abyss of the open ocean is minimal (6 Tg C yr^{-1}). However, the coastal vegetated ecosystems situated in the periphery of the oceans sequester disproportionately large quantities of carbon compared to that of the Open Ocean (Nellemann et al. 2009). The scientific community realized the capacity of these coastal vegetated ecosystems to sequester and store anomalously high quantities of carbon compared to the rest of the ocean long ago. However, the scientific attention that these ecosystems were receiving increased manifold once the term "blue carbon" came into operation.

1.4 The Term "Blue Carbon"

There are four types of color-coding tagged with the various categories of carbon available in the environment (Fig. 1.4). Green carbon refers to the carbon stored in the biomass of the plants and the soils beneath those plants, exclusively in the organic form. Black carbon and brown carbon are the other two types of carbon. Brown carbon refers to the organic carbon available in the form of aerosols suspended in the atmosphere (Feng et al. 2013). Lately, we realized that this particular type of carbon could contribute to both global warming and cooling, depending on the fact whether these are absorbing or scattering solar radiation (Zhang et al. 2017). Black carbon emission refers to the carbon emitted to the atmosphere as a by-product of fossil fuel combustion. The emission of these carbons holds the second rank in contributing to global warming after the gaseous CO_2 increase in the atmosphere (Ramanathan and Carmichael 2008).

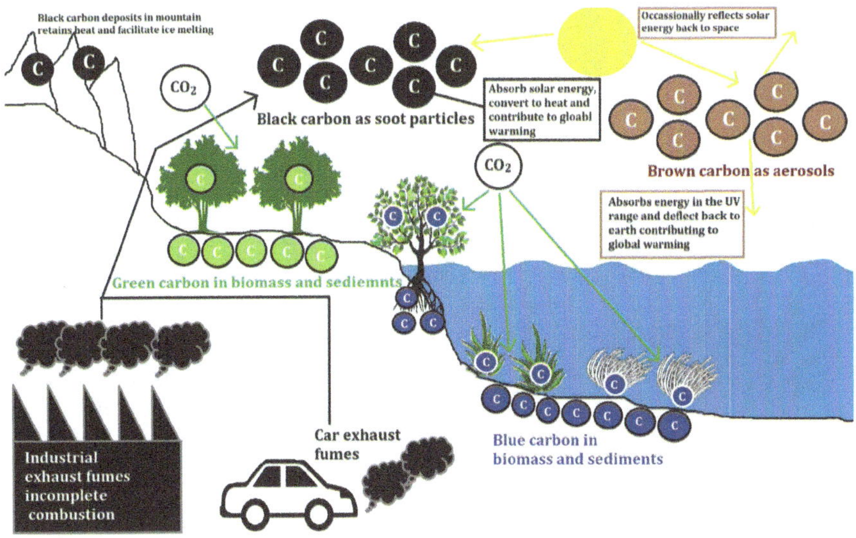

Fig. 1.4 A diagram showing the presence and function of the green, blue, black, and brown carbon in the environment

In the league of these colored carbons, blue carbon is perhaps the newest inclusion. Nellemann et al. (2009) officially gave birth to the term blue carbon in their publication, wherein they advocated the need to emphasize the share of green carbon stored by the oceans to mitigate climate change. According to them, the carbon sequestered by the global ocean holds more than 55% of the global total green carbon.

The copybook definition of blue carbon is the carbon dioxide removed from the atmosphere and stored by the coastal ocean ecosystems, mostly mangroves, salt marshes, seagrasses, and potentially macroalgae, through plant growth and the accumulation and burial of organic matter in the soil. The concept of blue carbon is comparatively new, and it is still evolving with each passing day. Besides the conventional blue carbon ecosystems, there are arguments, doubts, and concerns about several other marine ecosystems, which fall outside the ambit of the blue carbon framework, like fishes, bivalves, coral reefs, and phytoplankton (Kuwae and Hori 2018). However, once we can clear the doubts and minimize the uncertainties related to the carbon processing aspects of these crucial ecosystems, these ecosystems can officially come under the envelope of blue carbon (Lovelock and Duarte 2019). Keeping the technicalities aside, one thing is for sure that coining the term "blue carbon" had a noble intention to aware the present civilization about the fact that marine ecosystems are exceedingly efficient in the long-term conservation of carbon and that their adequate management can usher us to a world rid of the overwhelming problem of carbon emissions. However, the process through which the formation of new organic carbon takes place in the live compartments of the blue carbon ecosystems at the cost of depletion in atmospheric gaseous CO_2 levels is not as simple as observed in the case of the terrestrial plants.

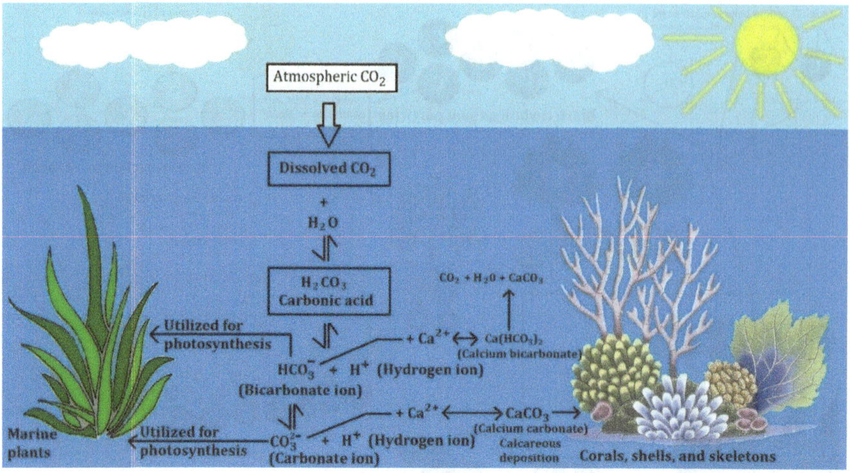

Fig. 1.5 A diagram showing the carbonate-bicarbonate equilibrium reactions and the association of marine plants and calcareous organisms

Terrestrial plants can directly absorb atmospheric CO_2 through the stomatal openings of their leaves. Vegetation submerged in the water cannot absorb the dissolved CO_2 in the aquatic phase (Fig. 1.5). Dissolution of CO_2 in water forms a weak compound known as carbonic acid (H_2CO_3), which in turn, depending upon the pH of the water, dissociates into bicarbonate ions (HCO_3^-) and hydrogen ions (H^+). At a higher pH (in waters having fewer hydrogen ions), the bicarbonate ions dissociate to carbonate ions (CO_3^{2-}) and hydrogen ions (Al-Gousous et al. 2019). The aqueous CO_2, the dissolved bicarbonate, and the carbonate ions comprise the dissolved inorganic carbon (DIC). Instead of gaseous CO_2, the aquatic plants can intake the dissolved CO_2 (mainly in the form of HCO_3^-), convert this inorganic carbon to organic carbon, and store it in their biomass (Poschenrieder et al. 2018). The dissociation of dissolved CO_2 gives rise to free H^+ ions, which decreases the pH of the water column, and in the case of oceans, this phenomenon leads to ocean acidification. However, according to the basic principle of Henry Louis le Chatelier, the more the intake of bicarbonates and carbonates, the lesser would be the abundance of H^+ ions in the water column (Middelburg et al. 2020). Thus, the higher the removal of DIC from the water column, the lesser would be the ocean acidification, and the higher would be the CO_2 sink potential of the water column under consideration. Now, questions may arise as to what is the source of CO_2 in the water column that the marine plants utilize during photosynthesis. The answer lies in the fact that all the surficial aquatic bodies maintain a steady gas exchange with the adjacent atmosphere. In the water column, CO_2 mostly comes from the atmosphere, which reduces the atmospheric CO_2 load. CO_2 being a sparingly soluble gas, can enter the aquatic phase from the atmosphere as well as leave the same and return to the atmosphere. The difference in the partial pressure of CO_2 (pCO_2) between the atmosphere and the hydrosphere governs the direction of the CO_2 fluxes. When the pCO_2 is higher in the water compared to that in the immediate atmosphere, CO_2 molecules tend to diffuse out of the water bodies and move toward the atmosphere, and vice-versa (Laruelle et al. 2018). At present, the pCO_2 level in the oceans is much lower than the pCO_2 level in the atmosphere.

Due to this difference in pCO_2 levels, the oceans act as a sink for CO_2. However, the more the oceans absorb the CO_2 from the atmosphere, the less becomes the buffering capacity of the oceans (the potential to absorb more CO_2). Given the volume of seawater present in the earth, it would take a substantial amount of CO_2 emissions to put an end to the sink character of the oceans. However, this would lead to an irreversible climate change risk that we cannot even imagine. If the oceanic biological pump along with the blue carbon ecosystems consistently remove an adequate amount of dissolved carbon, convert it to the organic form, and store this carbon for high residence time, it would enable the oceans to maintain high buffering capacity and keep on absorbing CO_2. The importance of blue carbon ecosystems lies here, and this is why the concept of blue carbon has received increasing attention.

1.5 Why Is "Blue Carbon" so Crucial?

Several coastal vegetated ecosystems like mangroves, seagrass, and saltmarshes accumulate and sequester in their live and dead biomass along with the soil substratum disproportionately high quantity of organic carbon compared to their global areal cover (Pendleton et al. 2012). The coastal vegetated ecosystems comprise an area of 4.9×10^5 km^2 (Barbier et al. 2011), which is only 0.00068 times the area of the open ocean (3.33×10^8 km^2). Yet these vegetated systems altogether sequester carbon at a rate (114 – 131 Tg C yr^{-1}) (Kuwae and Hori 2018), which is 19to 22 times higher than that observed in the open ocean (6 Tg C yr^{-1}).

Moreover, the substratum on which the blue carbon vegetation grows is usually rich in organic matter and is capable of storing organic carbon to the depths of several meters. There are several factors, including oxygen deficiency in these depths that facilitate these ecosystems to store substantial quantities of carbon (Kristensen et al. 2008). The rate of carbon storage surpasses many of the terrestrial forest ecosystems in terms of carbon locked per unit area (Donato et al. 2011, 2012). The role of blue carbon in climate change mitigation has two facets of equal importance. Firstly, these natural ecosystems have the potential to absorb and store CO_2 from the atmosphere for a long time when left undisturbed (Sasmito et al. 2020). Hence, an approach to enhance the areal cover of these blue carbon systems by proper restoration and afforestation can help reduce the atmospheric CO_2 load (Wylie et al. 2016). Secondly, the blue carbon already stored in these ecosystems can lead to potential CO_2 emission if degradation and deterioration of these ecosystems take place due to anthropogenic activities and human negligence (Pendleton et al. 2012; Lovelock et al. 2017) (Fig. 1.6). Such observations indicate that proper conservation, management, and restoration of blue carbon ecosystems vis-à-vis enhancement and afforestation of

Global Extent (M ha)	13.8 - 15.2	17.7 - 60.0	2.2 - 40.0
Current conversion rate (%/year)	0.7 - 3.0	0.4 - 2.6	1.0 - 2.0
Carbon emission due to degrdation (Pg CO_2/year)	0.09 - 0.45	0.05 - 0.33	0.02 - 0.24
Economic cost (Billion US $/year)	3.6 - 18.5	1.9 - 13.7	0.64 - 9.7

Fig. 1.6 The rate of degradation of the blue carbon ecosystems and the potential CO2 emission (modified after Pendleton et al. 2012)

these ecosystems both are required to prevent the emission of carbon due to degradation and habitat conversion of the blue carbon ecosystems and to remove more CO_2 from the atmosphere.

Among the various factors, the one that makes the blue carbon ecosystems stand out is their higher rate of carbon burial compared to all the other natural carbon sinks (McLeod et al. 2011). The blue carbon science, being comparatively new, is still evolving, which leaves several questions that are unanswered at present. Unlike the terrestrial ecosystems, a suite of physical and biogeochemical processes coupled with natural and anthropogenic disturbances govern the carbon burial dynamics in the blue carbon ecosystems, which in turn regulates the amount of carbon production, the amount of carbon available for burial, the rate of sedimentation, and sequestration (Macreadie et al. 2019). Blue carbon ecosystems seldom act or function as separate entities. These ecosystems carry out life activities in the presence and association of many other living systems. Several researchers argue that in the case of blue carbon ecosystems, the phytoplankton productivity in the water column lying above, the presence of macrophytes, the role of marine fauna on the ecological food chain, and the efficiency of the microbial community can play a decisive role in the carbon burial (Atwood et al. 2015; Kellewayet al. 2017; Liu et al. 2017). We could understand the role of physical factors like hydrodynamics and vegetation density on the carbon burial to some extent (Gacia and Durate 2001; Horstman et al. 2014); however, we cannot explain the variability of the organic carbon stock in the substratum of some of the ecosystems like seagrass (Ricart et al. 2015). A complex interaction between several biological as well as physical factors determines the carbon burial efficiency of these ecosystems. Such unaccounted variations in carbon stock often lead to uncertainties in the estimate of blue carbon. We should try to understand the controls of other factors, and this might help us to enhance the carbon stock of these ecosystems by proper human intervention. These observations indicate the importance of blue carbon ecosystems in carbon sequestration and, at the same time, emphasize the necessity for advanced studies to gain insights into the complex processes that take place within the ecosystems for better carbon conservation as well as augment their sink potential.

1.6 Blue Carbon Ecosystems

Mangroves, seagrasses, and saltmarshes are the three conventional blue carbon ecosystems recognized and unanimously accepted by the global scientific community (Barbier et al. 2011). Mangroves exclusively thrive in the tropics and subtropics; however, the increasing global temperature has widened the distribution of mangroves as the areal cover of mangroves is steadily increasing in the temperate regions (Bulmer et al. 2016; Suyadi et al. 2018). Seagrasses and saltmarshes thrive throughout the world's coastal margins from the equator to the poles (except Antarctica), but saltmarshes are most common in the temperate regions (Pendleton et al. 2012). Besides acting as a strong sink of organic carbon, these ecosystems furnish a wide range of ecosystem services directly or indirectly to human beings as well as to each other (when situated in a continuum). The ecosystem services provided by mangroves vary from provisioning (timber and fuelwood) to regulating (protection from the storm, flood, erosion, and salt-water intrusion) to habitat (nursing ground of fishes) to cultural services (aesthetic and recreation purposes) (UNEP 2006). According to the purchase power parity-based valuation of the international dollar for the year 2007, mangroves have an estimated valuation for its ecosystem service of >50,000 Int$ ha^{-1} yr^{-1} (Mukherjee et al. 2014). Seagrasses and salt marshes also provide several ecosystem services like providing a habitat for fish and another biodiversity, coastal protection, sediment accretion and stabilization, water purification, and recreation (MtwanaNordlund et al. 2016; Duarte et al. 2018; Himes-Cornell et al. 2018). However, the economic quantification of these ecosystem services is a cumbersome process. Carbon sequestration is also one of the most important regulating ecosystem services that all these ecosystems provide.

1.6.1 Mangroves

Mangroves are a group of tidal halophytic plants that occupy the sheltered shorelines, deltaic regions, estuaries, and coastal stretches throughout the tropic and the subtropics (Tomlinson 1986), with increasing occurrence in the temperate regions, also. Globally, the mangrove floral community consists of almost seventy species, which includes several taxonomically diverse shrubs and ferns along with trees (Ellison et al. 1999). These seventy species belong to twenty-seven families under twenty families and nine orders (Fig. 1.7). Among all the blue carbon ecosystems, mangroves are the ecotones in the strictest sense, as this community exhibits characteristics of both terrestrial and marine biomes (Alongi 2012). Mangroves typically flourish under saline and anoxic conditions, which compelled these species to develop a suite of morphological and physiological adaptions (Farnsworth and Farrant 1998). These plants have developed specialized aerial root systems, known as pneumatophores, to sustain under waterlogged conditions. Viviparous germination is also a unique characteristic feature of the mangroves, which allows these plants to germinate their seeds while remaining attached to the parent plant. The leaves of these plants have become thick, succulent, and sclerophyllous over time to prevent the high rate of photo-transpiration (Ball 1988). The ambient temperature mainly plays a regulating role behind the spatial extent of mangroves in the world.

However, on the regional scale, the spatial distribution, as well as the species assemblages, depend on a set of physical and atmospheric factors like the hydrogeomorphology of the substratum, the precipitation characteristics, the wave dynamics, and the tidal nature and amplitude (Mazda and Wolanski 2009). There are several coastal peripheries where no mangroves are available despite having all the characteristics required for mangrove growth. There are indeed no set rules for the natural occurrence of mangroves; however, the development of mangroves usually takes place in coastal regions having a flat topography coinciding with the sea level (Ellison 2009). Initiatives of artificial restoration and afforestation of mangroves have shown that the mangroves are capable of surviving in coastal regions, which were not originally mangrove habitats (Erftemeijer and Lewis 1999). Such observations indicate that mangroves are quite robust and can acclimatize themselves under unfavorable and unknown conditions. Shreds of evidence are there that show the mangroves through different ecological successional stages show significant resilience to both natural as well as anthropogenic disturbances, as the mangrove community as a whole can change with the changing ambiance and environmental setting (Berger et al. 2006). Despite such unique characteristics and so much importance, this ecosystem is at present under immense threat due to uncountable anthropogenic disturbances.

Fig. 1.7 Images of some of the most dominant mangrove species from the world's largest mangrove forest of Sundarban, India

1.6.2 Seagrasses

Seagrasses are flowering plants (angiosperms) that thrive in various places along the coastal periphery of the world (Larkum et al. 2006). These flowering plants usually prefer to remain submerged in the shallow coastal regions and thrive throughout the world except Antarctica. The seagrass community does not represent a taxonomic group. They form an ecological group. Seagrasses all over the world mainly fall under three families, namely the Cymodoceaceae, the Zosteraceae, and the Posidoniaceae (Larkum et al. 2017). According to Short et al. (2011), there are seventy-two species found all over the world. Like mangroves, seagrasses have adapted to survive under varying degrees of salinity and temperature. Their thick stems known as rhizomes enable the seagrasses to cope with the stress generated by waves and water current (Gambi et al. 1990; Widdows et al. 2008). The rhizome blades are capable of growing straight toward the sunlight, and at the same time, these are flexible enough to bend in concordance with the water current. The roots of the seagrasses go deep down the seabed, which helps them to colonize and stabilize the sediment substratum (Hemminga and Durate 2000; Wilkie et al. 2012). The habitat types where seagrass grows vary from open and exposed areas to sheltered bays and estuaries. The depth at which seagrasses thrive also ranges from the very intertidal zones to 40 m depth (Carruthers et al. 2007). Depending upon the type of habitat, the geophysical characteristics and the environmental setting vary in terms of light penetrability and transparency of the water column, temperature, and hydrodynamics. The species diversity in the seagrass meadows along with the meadow properties like density, productivity, and biomass depends on the environmental setting (Alcoverro et al. 2001). Unlike the mangroves, restoration and artificial afforestation of seagrass is a tedious endeavor, and reports of such nature are scarce. Being submerged in vegetation, implementing conservation measures is difficult in the case of seagrass; however, pieces of evidence show that these species are at immense threat due to various natural and anthropogenic disturbances. Research shows that the behavior and functioning of this crucial blue carbon ecosystem are most likely going to experience a paradigm shift due to the ongoing climate change, sea-level rise, and ocean acidification (Mazarrasa et al. 2018).

1.6.3 Saltmarshes

Unlike mangroves and seagrasses, which essentially represent a type of floral community, the term "saltmarshes" refers to a particular type of landscape found throughout the world's coastline. These are coastal wetlands made up of a deep layer of mud and peat, which generates the marshy look of these ecosystems. The floral diversity of the minimal vegetation cover of this ecosystem is surprisingly high; nearing 500 species reported so far (Silliman 2014). Though distributed all over the world, this ecosystem is more common in the middle to higher latitudes and usually prefers to thrive in sheltered regions with minimal disturbances from wave action (Mcowen et al. 2017). The floral community of the saltmarshes encompasses several halophytic types of grass, herbs, and shrubs. The species diversity of the saltmarshes exhibits significant spatial variability throughout the world (Fischer et al. 2000). The geomorphological characteristics of the terrain principally govern the extent and the species composition of the saltmarshes, and the presence of these species can regulate the geomorphology of a particular region (Adam 1990). Saltmarshes vegetation generally prefers the soft coasts for colonizing, and they thrive within a very shallow altitudinal margin ranging between the highest high tide line and mean high water neap tide line (Adam 2002). However, in terms of tidal nature, saltmarshes thrive under both microtidal and macrotidal conditions, with regular to occasional flooding regimes (Hodgkin and Hamilton 1998). Merely a century ago, people used to consider saltmarshes as wastelands of no worth. Several pieces of research and intervention from many scholars established the key role of this ecosystem in shoreline protection, wildlife conservation, nutrient generation, and many more immeasurable ecosystem services (Boorman 1999). However, like all the other blue carbon ecosystems, we have not spared this ecosystem from bearing the ill effects of human disturbances. At present, the future of this vegetation type is at stake due to multiple factors like the rising sea level, decline in sediment supply due to construction of dams and barrages, conversion to aquaculture, and a suite of aquatic pollution (Adam 2002).

1.6.4 Other Less Conventional Blue Carbon Ecosystems

Coral reefs are calcareous structures that occupy the shallow regions (0 m to 30 m depth) of the ocean. Four decades back, this ecosystem encompassed an area of 617,000 km^2on the ocean floor (Smith 1978). Later on, Spalding and Grenfell (1997) challenged the 1978 estimate of Smith (1978), and according to the former's conservative estimation, coral reefs stretch over 255,000 km^2. Corals are marine invertebrates under the class Anthozoa. They secrete an exoskeleton made of hard carbonate deposits, which gives rise to the myriad of colorful rocky structures that we call the reefs. Despite being a crucial marine ecosystem, from the viewpoint of providing a bunch of ecosystem services like supporting biodiversity, coastal protection, fisheries, medicine, tourism, and recreation (Moberg and Folke 1999), these ecosystems remain outside the ambit of the conventional blue carbon ecosystems. Two biological processes mainly drive the growth and functioning in the reef ecosystems: (i) the metabolism of organic carbon through photosynthesis and respiration and (ii) the processing of inorganic carbon through precipitation and calcification (Frankignoulle et al. 1996). The building process of these reefs depends on marine calcification, and there is no evidence of net organic carbon accumulation within the reefs (Watanabe and Nakamura 2018). The biogeochemical reaction that leads to marine calcification produces CO_2, and hence coral reefs act as sources of CO_2. These qualities of the corals bar the scientific community to include this ecosystem under the umbrella of blue carbon ecosystems. However, several researchers have addressed concerns and the need to tag this ecosystem as a blue carbon ecosystem because of its close association with mangroves and seagrasses (Guannel et al. 2016). Coral reefs often overlap with the mangroves and seagrasses, and the carbon dynamics of such a continuum becomes so intertwined with each other that it sometimes becomes impossible to distinguish the carbon pathways between these ecosystems. Contrasting observations show that coral reef waters can act as both sources (Gattuso et al. 1996) and sinks of CO_2 (Kayanne et al. 1995). Such observations have lately compelled many researchers to rethink this issue. They have expressed concerns about the fact that our knowledge about the net organic carbon assimilation and the CO_2 dynamics in the corals is still limited. Before ruling out the possibility of their inclusion in the arena of blue carbon, we should carry out more exhaustive and in-depth studies (Lovelock and Duarte 2019).

Another such controversial ecosystem that has lately come under the lenses is the macroalgae, commonly known as seaweeds. In terms of color, there are three categories of seaweeds, known as green algae (*Chlorophyta*), brown algae (*Phaeophyceae*), and red algae (*Rhodophyta*). There are almost 9800 macroalgal species available all around the world. All these macroalgae consist of chlorophyll in their tissues and are capable of carrying out photosynthesis (Clements et al. 2020). The macroalgae vary in size from few millimeters to almost 50 m. Seaweeds grow over a varied terrain but mostly prefer rocky and hard substratum. Smith (1981) was perhaps the first one to point out the capabilities of these macroalgae to act as a sink for carbon. As the blue carbon concept started gaining momentum, several authors have argued in the recent past and proposed for its rightful inclusion in the blue carbon arena (Krause-Jensen and Duarte 2016). The scientific community has reservations to consider macroalgae as a blue carbon ecosystem because this ecosystem thrives on rocky bases and cannot store carbon in their sediments. However, the export of photosynthetically captured carbon by the seaweeds takes place to the other nearby blue carbon ecosystems and the deep oceans, where the carbon remains locked for a long time (Barrón and Duarte 2015; Almahasheer et al. 2017). Thus, unlike the other marine ecosystems, the contribution of macroalgae in removing the atmospheric CO_2 and sequestering it does take place in areas beyond the macroalgal habitat (Hill et al. 2015). According to the latest estimates, macroalgal cover extends for 3,400,000 km^2 all over the world and absorbs carbon at a net rate of 1.5 PgC yr^{-1} (Krause-Jensen and Duarte 2016). Organic carbon export at a rate of 2.4 Pg C yr^{-1} takes place from the macroalgal beds toward the ocean, and the other angiosperm-dominated blue carbon ecosystems and deep benthic sediments together can sequester 11% (173 Tg C yr^{-1}) of this export (Krause-Jensen et al. 2018). Besides the corals and the macroalgae, Lovelock and Duarte (2019) have explicitly discussed the criteria in favor and against in reaching a consensus on whether to include other ecosystems like fish, bivalves, and phytoplankton in the blue carbon framework. However, studies concerning these possibilities have only begun.

1.7 How Do the Blue Carbon Ecosystems Store Such High Quantities of Carbon?

The mangrove forests are architecturally much simpler than many of the tropical forests with almost no lower canopy or understory, and yet their carbon potential is equivalent to that of the tropical rainforests, one of the most biodiverse and complex forests of the world (Alongi 2009). The net primary productivity varies in the mangrove forests depending upon the location, species composition, age, and height of the trees with a global mean rate of 11.1 t dry weight ha^{-1} yr^{-1}, and these are capable of storing carbon for a century if left undisturbed (Alongi 2012). As icebergs consist of the larger fraction of ice below the water surface, the key behind the carbon sink potential of the mangroves lies in their belowground carbon stock, which we cannot make an idea of when we behold the aboveground forest structure. The belowground carbon stock in the extensive root system and the soil organic carbon pool (mainly driven by a high rate of litterfall) is almost equivalent to the aboveground carbon stock, if not more (Kauffman et al. 2011). The remarkable ability of sediment accretion makes the mangroves active sinks of carbon, which remains adhered to the fine particles, and the process of steady deposition of such particulates enhances the soil carbon pool of the mangroves forest floor up to depths of several meters. The global mean sediment accretion rate for the mangroves is 5 mm yr^{-1} (Alongi 2012), and the best possible estimate for global mean organic carbon burial is 24 Tg C yr^{-1}, equivalent to 174 g C m^{-2} yr^{-1} (Bouillon et al. 2008; 2009). However, all these characteristic features of mangroves that are critical to the substantially high rate of carbon sequestration go boomerang when these forests are disturbed, leading to high emissions 0.02–0.12 PgC yr^{-1}, which is equivalent to 2–10% of the CO_2emissions that arises due to global deforestation (van der Werf et al. 2009; Donato et al. 2011).

Similar to mangroves, several specialized ecosystem functions in the seagrass meadows and saltmarshes lead to a high rate of organic carbon burial. Macroalgae often accompany seagrass, and together these vegetated covers exhibit a very high net primary productivity (Miyajima et al. 2017). The high nutrient dynamics coupled with a steady allochthonous carbon input lead to a very high net primary productivity for the saltmarshes as well (Bertness et al. 2008). Both of the seagrasses and saltmarshes are efficient in slowing down the flow velocity of the ambient water column, thereby traps a significant amount of organic carbon (especially in the particulate form) and buries the carbon for a long term in the seagrass beds (Bridgham et al. 2006; Duarte et al. 2013). Seagrasses and saltmarshes mostly flourish under anoxic conditions. The seagrass beds being almost devoid of oxygen reduces the rate of mineralization of the organic carbon trapped by this ecosystem. There are pieces of evidence that show the efficiency of these two ecosystems in trapping allochthonous carbon of terrestrial origin and from the dead phytoplankton. Stable isotopic analyses indicated that almost 50% of the carbon locked in the seagrass beds are allochthonous (Miyajimaet al. 2015). Of course, the fraction of allochthonous carbon in the seagrass beds depends on several factors like the geomorphological setting of the meadows and their species diversity (Serrano et al. 2014). Besides accumulating carbon in their habitat, seagrasses also export a substantial quantity of organic carbon toward the outer ocean. During the transit, there is a high possibility of re-mineralization of this organic carbon, but a significant portion of this carbon can find a long-term sink in the nearby continental shelves and the deep oceans under suitable conditions (Sugimatsu et al. 2015). In this way, seagrasses and saltmarshes store a substantial amount of organic carbon both within their habitat and beyond; however, like the mangroves, disturbances posed in this ecosystem can also lead to a significant amount of CO_2 emissions (Lovelock et al. 2017).

1.8 Carbon Speciation in Blue Carbon Ecosystems

Carbon exists in various forms within the blue carbon ecosystems. Carbon in the particulate and dissolved form is prevalent in these ecosystems. In terms of chemical states, carbon is present in the inorganic and organic states as well. Thus, there are mainly four types of carbonaceous compounds within these ecosystems, namely dissolved inorganic carbon (DIC), dissolved organic carbon (DOC), particulate inorganic carbon (PIC), and particulate organic carbon (POC). These states of carbon interchange among one another via a suite of biogeochemical processes (Fig. 1.8). We can consider all of these states of carbon in the marine systems as carbon sinks because the removal of atmospheric CO_2 eventually through a series of biochemical reactions gives rise to all these forms. All these states of carbon can also return to the

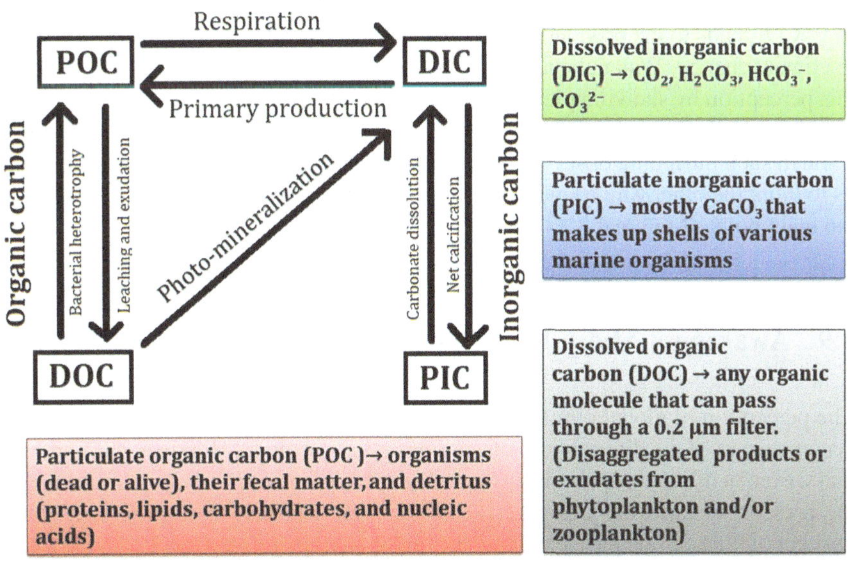

Fig. 1.8 A schematic diagram showing the pathways through which the different forms of carbon interchange among one another

atmosphere as CO_2. However, the residence time (the time for which the carbon in the atmospheric CO_2 remains in the other form within the marine ecosystems) varies significantly.

On a quantitative basis, a large fraction of carbon in the seawater exists in the form of DIC. However, the chemical components that constitute the DIC can readily change to aqueous CO_2 depending on the pH-based chemical equilibrium of the water and, thus, DIC has a short residence time within the marine systems. The several life forms that thrive in the marine ecosystems store substantial quantities of carbon in the organic form locked in their biomass, but their death and eventual decay mineralizes the carbon and finds its way back to the atmosphere as CO_2 over a comparatively shorter time (days to decades). The two main carbon pools in the marine system that shows a slow decomposition rate is the sedimentary organic carbon (SOC) pool and the refractory dissolved organic carbon (RDOC) pool, where carbon in the organic carbon is least susceptible to conversion to atmospheric CO_2 and remains sequestered for a long time (century to millennia) (Kuwae et al. 2018).

The change of state that is of fundamental interest to us is the utilization of DIC through photosynthetic activities and its conversion to POC, as the carbon sequestered in the form of organic carbon (especially in sediments) can remain so for a long time compared to the other forms of carbon. This conversion takes place at a high rate in the blue carbon ecosystems, and human beings can play a decisive role in conserving these ecosystems and provide negative feedback to climate change. The knowledge gained so far indicates that human intervention cannot regulate the RDOC sink in the deep oceans. Recently, Saderne et al. (2019) estimated that the seagrass and the mangrove sites of the world bury inorganic carbon in the particulate form (as $CaCO_3$) at a rate of 15–62 $TgC_{inorg}\ yr^{-1}$ and 0.8 $TgC_{inorg}\ yr^{-1}$, respectively. As precipitation of $CaCO_3$ leads to the production of CO_2, PIC sequestration gives positive feedback to climate change (Watanabe and Nakamura 2018). Saderne et al. (2019) challenged this perception by showing pieces of evidence that the blue carbon sites deposit PIC, which is of allochthonous origin and overall acts as sites of net $CaCO_3$ dissolution. Though such intriguing observations about the other forms of carbon are compelling us to think the other way round, the organic carbon in the solid form continues to be the main point of interest.

1.9 Awareness About Blue Carbon

The perception of global climate change and environmental degradation has indeed become a part of the dining room discussions of a household. However, the common mass, even in this date, appreciates the significance of terrestrial forests in combating the present scenario. It is an undeniable fact that scholarly articles related to the concept of blue carbon have exponentially increased in the last decade, but the refection of the same in the newspaper articles, blogs, and television programs, which serves as efficient tools for mass propaganda, is still scarce (Kuwae and Hori 2018). For any ordinary person, the importance of marine ecosystems mostly lies in the fact that these ecosystems furnish many usable products and foods. However, people hardly understand their immeasurable contribution to mitigating climate change. With a noble intent to modify this perception and draw the attention of the global population toward the efficiency of marine ecosystems in removing atmospheric CO_2, the United Nations Environment Program Planning Unit (UNEP), the United Nations Educational Science and Culture Organization (UNESCO), and the United Nations Food and Agriculture Organization (FAO) jointly published a report (Nellemann et al. 2009). This report not only coined the term blue carbon for the first but also served as the base document to emphasize the angiosperm-dominated marine ecosystems for their carbon sequestration potential. At the conference of parties, organized in Paris in December 2015, all the countries under the United Nations Framework Convention on Climate Change (UNFCCC) pledged to reduce carbon emissions and combat climate change. At present, 197 countries have ratified this agreement, commonly referred to as the Paris Agreement. Article 5 of this agreement encourages all the nations to explore their natural sinks and reservoirs of greenhouse gases and conserve as well as enhance their potential. A school of scholars has already proposed that countries with oceanic borders should optimally utilize their coastal blue carbon repository to strengthen their goals in maintaining the nationally determined contribution (NDC) of the Paris Agreement (Herr and Landis 2016; Taillardat et al. 2018). Taillardat et al. (2018) advocated that countries with strong mangrove vegetation cover should conserve their mangrove ecosystems to negate their emissions, and where mangrove clearing took place indiscriminately in the past decades, should strive hard toward restoration and afforestation of mangroves to meet the goals of the Paris Agreement in the true essence. Several countries have already started exploring the avenues of conserving and enhancing their blue carbon strength by carrying out research projects (Crooks et al. 2014). International organizations have also inclined toward the development of effective strategies to corroborate the blue carbon initiatives within UNFCCC mechanisms like REDD+ (Herr et al. 2012). Various scholars believe that an efficient amalgamation of carbon conservation projects and proper implementation of carbon finance mechanisms can lead to some meaningful outcomes.

1.10 Management of Blue Carbon: The Need of the Hour

Despite being one of the most productive ecosystems, providing a range of ecosystem services, the blue carbon vegetative covers are under enormous threat. These ecosystems are undergoing rapid destruction and degradation due to a suite of land-use changes and other associated activities. In the recent past, the rate of degradation has accelerated manifold. McLeod et al. (2011) estimated that in the last century, the blue carbon ecosystems lost almost 25 to 50% of their global areal cover. Annually the rate of degradation ranges between 0.5 and 3% depending on the ecosystem type (Valiela et al. 2001; Waycott et al. 2009; Pendleton et al. 2012). Several researchers projected that if we do not check and restrict the present rate of degradation, almost 40% of the seagrass meadows and saltmarshes stretch along with 100% of the mangrove cover could be lost by the end of the twenty-first century (Duke et al. 2007; IPCC 2007). Disturbances to the natural ecosystems account for almost 30% of the global anthropogenic CO_2 emissions (Houghton 2003), and marine ecosystems hold a fair share in this estimate. Implementing the technique of uncertainty-propagation Pendleton et al. (2012) estimated that the degradation of blue carbon ecosystems leads to ~ 0.15–1.02 Pg C of CO_2 emissions, worth 6 to 42 billion US dollars (according to the social cost of carbon in the year 2007). The potential emission of CO_2 arising out of the destruction of blue carbon ecosystems constitutes almost 3 to 19% of the global CO_2 emission that takes place only due to deforestation. The estimates indicate that our negligence toward the marine vegetation is much higher than that we have shown toward the terrestrial vegetation. The coastal margins of the world have witnessed an increase in the degree of human interference due to the ongoing population explosion. Anthropocene has been a challenging epoch for all the natural ecosystems on earth, and it has been exceedingly harsh toward marine ecosystems (Aswani 2019).

Change of land use in the mangrove environment through human intervention leads to substantial CO_2 emissions (Kauffman et al. 2017). In recent times, widespread clearing of mangroves for aquaculture practice occurred all through the world. The carbon locked for more than centuries in the mangrove soils goes back to the atmosphere due to such conversion (Arifanti et al. 2019). The transformation from mangroves to agricultural land also took place in several parts of the world. However, recent studies registered a global decline in mangrove forest loss, and the increased conservation efforts are mainly accountable for such a decrease (Goldberg et al. 2020). These observations indicate that prioritizing conservation endeavors can lead to fruitful results, but there is still a scope of humongous work for the future. Industrial effluents, heavy metal pollution, over-exploitation of forest resources including woods, eutrophication, and many other factors are responsible for the global decline in not only mangroves but also the seagrass and saltmarshes. The conservation practitioners often do not consider many of these factors while optimizing the measures required for management, but all these factors are critical to the well-being of these ecosystems. Unfortunately, the geographical location of the blue carbon ecosystems in many of the developing and developed countries is associated with the rural people who live below the poverty line (Alongi 2018). A large number of poor people are directly or indirectly dependent on the blue carbon usufructs. Without their involvement and participation, the effective formulation of management strategies is nigh impossible. In the present date, the poverty-stricken section of the society in the coastal belt sustains their daily life at the cost of indiscriminate degradation of these ecosystems. As an example, the coastal poor people perceive shrimp culture as a respite for their struggle, which in turn leads to the degradation of several marine ecosystems (Ahmed et al. 2017). The priority of hunger comes ahead of all conservation measures, and we can never afford to overlook the desperate situation of millions of people if we achieve a sustainable future for these ecosystems. Thus, the policymakers should look out for options through which we would be able to achieve a symbiotic conjugation between local people and these ecosystems. The conservation practitioners should cultivate avenues through which the millions of poor can manage a modest livelihood through the conservation of these ecosystems. We can utilize the infinite resource in the form of ecosystem services that these ecosystems are capable of providing as a measure of poverty alleviation in many of the developing countries. Several studies show that we can confidently envisage the blue carbon ecosystems as not only a carbon sink but as a platform that can eradicate poverty for a large number of people (Ahmed et al. 2017; Herr et al. 2019).

1.11 The Blue Carbon Economics: An Emerging Area of Interest

Human beings in the modern world are incessantly running after money and comfort for the betterment of their respective lives. We cannot afford to ignore this undeniable truth while looking for options to safeguard the environment. Our realizations about environmental degradation continue to remain in pen and paper unless we can formulate adequate strategies to implement the course of action necessary to conserve any and every ecosystem, and such formulation needs the involvement of money. Without a proper finance mechanism, it is impossible to carry out any physical effort, be it restoration, conservation, or management of these ecosystems, which requires manual labor and, at the same time, an expectation of profit for those indulging in this business. The monetary aspect comes in the form of two ways in the blue carbon arena, one being the finance and the other the market. The finance, in this case, refers to the investments that could be from the public as well as private sectors, which would aim to develop blue carbon projects to conserve these ecosystems, whereas, market involves the trade and exchange of money centering the ecosystem services furnished by these ecosystems (Thomas 2014). In simpler words, there should be an investment of money (which at present mostly comes from the government sector) to uplift the status of these ecosystems. There are ways by which we may attract the private sectors as well to invest in the restoration of these ecosystems. However, the involvement of the private sector without any profit for the organization seems irrational, as no one would be willing to pay money for no return in any form. Leaving the fate of the blue carbon ecosystems at the mercy of generous donations would never enable us to achieve the degree of conservation that we intend to see shortly. The economists and environmentalists have been trying to amalgamate their knowledge to develop quantitative estimates for all the plausible ecosystem services that these ecosystems fetch. At present, two terms, namely the payment for ecosystem services (PES) and the markets for ecosystem services (MES), have become relevant in this field. PES refers to a wide range of provisions through which the beneficiaries of environmental services reward the stakeholders or landowners (whose lands and property deliver the sustenance of the ecosystem services) through incentives and subsidies. MES involves a suite of interactions between the businessperson and stakeholders in the form of exchange and trading of money to fulfill the ultimate motive of nature conservation. Several non-market schemes are in practice in many of the developing nations, whereby the national government contributes by subsidizing the management of blue carbon in ways other than the cap-and-trade mechanism (Ullman et al. 2013). There are debates on the execution of many of these schemes, and several aspects of such endeavors are still not very clear, but the inclusion of the blue carbon in financial market-oriented climate policy is imperative to conserve these ecosystems.

1.12 The Blue Carbon of the Indian Ocean

Though the blue carbon ecosystems exist throughout the world's oceans' periphery, the tropics and the subtropics are comparatively much more carbon enriched than the higher latitudes. Among the five oceans of the world, the Indian Ocean covers one of the most extensive blue carbon ecosystems in its periphery, with more than 50% of the global mangrove cover and a large proportion of seagrass beds (Vanderklift et al. 2019). A substantial part of the Indian Ocean lies within the 30°N and 30°S, encompassing a shoreline of about 66,500 km, and experiences a typical tropical to subtropical climate which, makes the productivity of the Indian Ocean blue carbon ecosystems much higher than that found in the other parts of the world.

Several aspects make the Indian Ocean a unique region and different from the other major oceans in the world. The importance of the Indian Ocean from the perspective of blue carbon ecosystems not only lies in the fact that this region shelters one of the largest blue carbon vegetation in the world but also because the rate of degradation of these ecosystems is very high in this region. Almost 35% of the global population lives in the countries within and surrounding the Indian Ocean having a large number of poor people (Vanderklift et al. 2019). Many of these countries are underdeveloped according to the list of United Nations (https://unctad.org/topic/vulnerable-economies/least-developed-countries/list) (Table 1.1). Though there are a few developing countries with emerging economies, these countries have a substantially high population with a disproportionate distribution of wealth, making a large number of people vulnerable to poverty. The Indian Ocean experiences natural calamities in the form of cyclones and tsunamis more than any other ocean of the world, and the disaster management potential of most of the countries lying in its periphery is not up to the mark. Every year these natural disasters lead to loss of life and property in almost all these countries. In India and Bangladesh, the fury of nature leaves such a disastrous signature that it strongly affects the country's economy.

The blue carbon stands, especially that of mangroves, plays a very crucial role in safeguarding the life of millions of people from these natural disasters and pays the price through natural deforestation due to associated effects like storm surge and over-salinization of soil.

As per the best possible recent estimate, the mangroves and the seagrass meadows in the Indian Ocean countries stretch for about 76,275 km^2 and 1,17,177 km^2, respectively (Vanderklift et al. 2019). It is troublesome to map the seagrass meadows,

Table 1.1 The list of Indian Ocean countries considered in this book along with the length of their coastline and present human population

Country	Present human population[a] (in million)	Length of the coastline (km)[b]
India (AS)	~1380	~7516
Bangladesh (AS)	~164.7	~580
Myanmar (AS)	~54.4	~1930
Thailand (AS)	~69.8	~3219
Singapore (AS)	~5.85	~193
Indonesia (AS)	~273.5	~99,083
Malaysia (AS)	~32.3	~4675
Yemen (AS)	~29.8	~1906
Oman (AS)	~5.10	~3165
UAE (AS)	~9.89	~1318
Iran (AS)	~83.9	~3294
Pakistan (AS)	~220.8	~1046
South Africa (AF)	~59.3	~2798
Mozambique (AF)	~31.2	~2470
Tanzania (AF)	~59.7	~1424
Kenya (AF)	~53.7	~536
Somalia (AF)	~15.8	~3333
Djibouti (AF)	~0.98	~314
Sri Lanka (AS)	~21.4	~1340
Maldives (AS)	~0.54	~644
Seychelles (AF)	~0.098	~491
Madagascar (AF)	~27.6	~4828
Comoros (AF)	~0.86	~340
Mauritius (AF)	~1.27	~177

[a] https://www.worldometers.info/world-population/population-by-country/
[b] CIA World Fact book: Coastline

as they remain submerged under the water. Even with the most advanced techniques of remote sensing, identifying seagrass cover in deeper waters remains a challenge. Thus, several authors have argued that the actual seagrass cover in the Indian Ocean largely remains underestimated. Despite such a lack of proper estimates, one can unequivocally claim that the blue carbon ecosystems of the Indian Ocean have promising potential to combat climate change and needs attention. The Paris Agreement motivated several countries in the periphery of the Indian Ocean to include efforts for the conservation of these ecosystems as an integral part of their respective NDCs. India, Bangladesh, Sri Lanka, Myanmar, and Madagascar have already shown their commitment to the restoration and conservation of mangrove forests. Bangladesh went on a step forward to go for widespread afforestation in the mangrove forest of Sundarban, the world's largest mangrove forest. The intention behind such an afforestation program, where the Bangladesh Government planted almost 16,000,000 mangrove seedlings covering an area of 1950 km^2 of mangrove forest, was to build a defense mechanism against the recurrent tropical cyclones and furnish a suitable habitat for the fishes to safeguard the livelihood of the poor. Kenya and Madagascar have also carried out such restoration of mangroves. Despite having other intentions, such initiatives enhance the carbon sequestration potential nonetheless. These observations indicate that any mechanism, thinking, or approach that ultimately safeguards these crucial ecosystems are welcome, given the present scenario of climate change. United Arab Emirates (UAE), Seychelles, and Sri Lanka have also taken similar initiatives to safeguard their seagrass meadows. There are at present 15 Indian Ocean countries that have considered their respective blue carbon ecosystems to meet the goals of NDC. Despite such initiatives already taken so far, Vanderklift et al. (2019), in their latest review on the blue carbon of the Indian Ocean, confidently emphasized that the attention paid so far remains insufficient, given the potential of blue carbon stock in the Indian Ocean. They also advocated that if the nations surrounding the Indian Ocean join hands and work together, they could lead from the front and set an example for the other countries of the world.

1.13 The Scope of the Present Book

Going through the literature on blue carbon, the authors observed that there are extensive blue carbon ecosystems all along the periphery of the Indian Oceans and in the island margins situated in the same. Compared to the vast expanse of these ecosystems already noted in this region, the attention provided toward the carbon dynamics of these ecosystems from the Indian Ocean region is insufficient.

The authors immediately felt the need to assimilate and collate all the finds made so far in this regard to develop an idea about the entire Indian Ocean collectively, as well as the present state of the art of the respective countries bordering the Indian Ocean. We considered the following countries bordering the Indian Ocean in this book—Iran, Pakistan, India, Bangladesh, Myanmar, Thailand, Singapore, Indonesia, Malaysia, Yemen, Oman, United Arab Emirates (UAE), Djibouti, Somalia, Kenya, Tanzania, Mozambique, South Africa, Sri Lanka, Maldives, Seychelles, Madagascar, Comoros, and Mauritius (Fig. 1.9). We did not consider the countries exclusively facing the Red Sea and the Persian Gulf, as these water bodies are semi-closed and technically do not form a part of the open Indian Ocean. We devoted a separate chapter to the Australian Blue Carbon ecosystem as most of the blue carbon ecosystems in the Australian periphery faces the Pacific Ocean. The countries considered in this book belong to the Asian and African continents, and as mentioned above, many of these countries are underdeveloped. Some of these countries belong to the developing category. The present book has collated and discussed the status of the Indian Ocean blue carbon repositories in the coastline and their country-wise scenario, along with

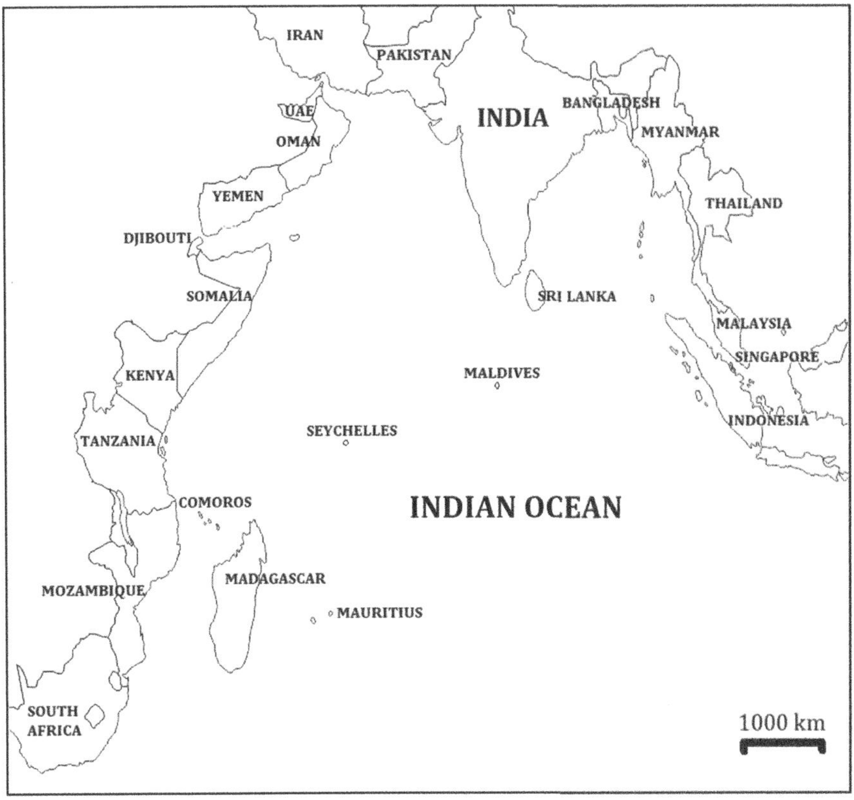

Fig. 1.9 A map showing the Indian Ocean countries considered in this book

the carbon sequestration potential of the open oceanic part. Some of the world's largest estuaries drain into the Indian Ocean. This book has dealt with details about the information gained so far on the carbon dynamics of such estuarine systems. We discussed in detail the carbon biogeochemistry in the aquatic and land compartments in and around the mangrove systems along with the factors regulating the carbon cycles. We paid the same attention to the seagrass, saltmarshes, and tidal flats found in the periphery of the Indian Ocean. Though corals in the present date do not qualify as a conventional blue carbon ecosystem, we discussed their role in regulating the carbon dynamics in the Indian Ocean in this book. Besides the biogeochemical aspects of these ecosystems, we also emphasized the livelihood opportunities and ecosystem services of these systems toward human beings. We also discussed the threats that these ecosystems are facing at present. We devoted a separate section to the importance of their conservation and management. Overall, the present book has covered every aspect of blue carbon from the perspective of the Indian Ocean with the intent to provide all the possible information related to the blue carbon dynamics of the Indian Ocean under one umbrella. Through this book, we have tried our level best to pinpoint the areas where research is lagging and needs more attention from the viewpoint of all the countries considered in this book. We honestly believe that this topic will become more significant in the days to come, and this will lead to the influx of several scholars in this domain who would be interested in research with the ultimate motive to conserve and restore these crucial blue carbon ecosystems. This book can serve as a ready reference for scholars, academicians, and policy managers.

References

Adam P (1990) The saltmarsh biota. In: Saltmarsh ecology. Cambridge Studies in Ecology, Cambridge University Press, pp 72–145. https://doi.org/10.1017/cbo9780511565328.003

Adam P (2002) Saltmarshes in a time of change. Environ Conserv 29(1):39–61. https://doi.org/10.1017/S0376892902000048

Ahmed N, Cheung WW, Thompson S, Glaser M (2017) Solutions to blue carbon emissions: Shrimp cultivation, mangrove deforestation and climate change in coastal Bangladesh. Mar Pol 82:68–75. https://doi.org/10.1016/j.marpol.2017.05.007

Alcoverro T, Cerbiãn E, Ballesteros E (2001) The photosynthetic capacity of the seagrass Posidoniaoceanica: influence of nitrogen and light. J Exp Mar Biol Ecol 261(1):107–120. https://doi.org/10.1016/S0022-0981(01)00267-2

Al-Gousous J, Salehi N, Amidon GE, Ziff RM, Langguth P, Amidon GL (2019) Mass transport analysis of bicarbonate buffer: effect of the CO_2–H_2CO_3 hydration-dehydration kinetics in the fluid boundary layer and the apparent effective pKa controlling dissolution of acids and bases. Mol Pharm 16(6):2626–2635. https://doi.org/10.1021/acs.molpharmaceut.9b00187

Almahasheer H, Serrano O, Duarte CM, Arias-Ortiz A, Masque P, Irigoien X (2017) Low carbon sink capacity of Red Sea mangroves. Sci Rep 7:9700. https://doi.org/10.1038/s41598-017-10424-9

Alongi DM (2009) The energetics of mangrove forests. Springer, Amsterdam, The Netherlands

Alongi DM (2012) Carbon sequestration in mangrove forests. Carbon Manag 3(3):313–322. https://doi.org/10.4155/cmt.12.20

Alongi DM (2018) The blue economy: mitigation and adaptation. In: Blue carbon. Springer Briefs in Climate Studies. Springer, Cham. https://doi.org/10.1007/978-3-319-91698-9_6

Arifanti VB, Kauffman JB, Hadriyanto D, Murdiyarso D, Diana R (2019) Carbon dynamics and land use carbon footprints in mangrove-converted aquaculture: the case of the Mahakam Delta, Indonesia. Forest Ecol Manag 432:17–29. https://doi.org/10.1016/j.foreco.2018.08.047

Arnell NW, Brown S, Gosling SN, Hinkel J, Huntingford C, Lloyd-Hughes B, Lowe JA, Osborn T, Nicholls RJ, Zelazowski P (2016) Global-scale climate impact functions: the relationship between climate forcing and impact. Clim Chang 134:475–487. https://doi.org/10.1007/s10584-013-1034-7

Arnell NW, Lowe JA, Challinor AJ, Osborn TJ (2019) Global and regional impacts of climate change at different levels of global temperature increase. Clim Chang 155(3):377–391. https://doi.org/10.1007/s10584-019-02464-z

Aswani S (2019) Perspectives in coastal human ecology (CHE) for marine conservation. Biol Conserv 236:223–235. https://doi.org/10.1016/j.biocon.2019.05.047

Atwood T, Connolly R, Ritchie E, Lovelock CE, Heithaus MR, Hays GC, Fourqurean JW, Macreadie PI (2015) Predators help protect carbon stocks in blue carbon ecosystems. Nat Clim Chang 5:1038–1045. https://doi.org/10.1038/nclimate2763

Ball MC (1988) Ecophysiology of mangroves. Trees 2:129–142. https://doi.org/10.1007/BF00196018

Barbier EB, Hacker SD, Kennedy C, Koch EW, Stier AC et al (2011) The value of estuarine and coastal ecosystem services. Ecol Monogr 81:169–193. https://doi.org/10.1890/10-1510.1

Barrón C, Duarte CM (2015) Dissolved organic carbon pools and export from the coastal ocean. Glob Biogeochem Cycl 29:1725–1738. https://doi.org/10.1002/2014GB005056

Beer C, Reichstein M, Tomelleri E, Ciais P, Jung M, Carvalhais N, Rödenbeck C, AltafArain M, Baldocchi D, Bonan GB, Bondeau A, Cescatti A, Lasslop G, Lindroth A, Lomas M, Luyssaert S, Margolis H, Oleson KW, Roupsard O, Veenendaal E, Viovy N, Williams C, Ian Woodward F, Papale D (2010) Terrestrial gross carbon dioxide uptake: global distribution and covariation with climate. Science 329(5993):834–838. https://doi.org/10.1126/science.1184984

Berger U, Adams M, Grimm V, Hildenbrandt H (2006) Modelling secondary succession of subtropical mangroves: causes and consequences of growth reduction in pioneer species. Persp Plant Ecol Evol Syst 7:243–252. https://doi.org/10.1016/j.ppees.2005.08.001

Bertness MD, Crain C, Holdredge C, Sala N (2008) Eutrophication and consumer control of New England salt marsh primary productivity. Conserv Biol 22(1):131–139. https://doi.org/10.1111/j.1523-1739.2007.00801.x

Boorman LA (1999) Salt marshes–present functioning and future change. Mangrove Salt Marsh 3(4):227–241. https://doi.org/10.1023/A:1009998812838

Bouillon S, Rivera-Monroy VH, Twilley RR et al (2009) Mangroves. In: Laffoley D, Grimsditch G (eds) The management of natural coastal carbon sinks. International Union for Conservation of Nature, Gland, Switzerland, pp 13–20

Bouillon S, Borges AV, Castaneda-Moya E et al (2008) Mangrove production and carbon sinks: a revision of global budget estimates. Glob Biogeochem Cycles 22, GB2013. https://doi.org/10.1029/2007GB003052

BP (2018) Statistical review of world energy 2019. https://www.bp.com/en/global/corporate/energy-economics/statistical-review-ofworldenergy.html (Accessed on 30 Sept 2020)

Bridgham SD, Megonigal JP, Keller JK, Bliss NB, Trettin C (2006) The carbon balance of North American wetlands. Wetlands 26:889–916. https://doi.org/10.1672/0277-5212(2006)26[889:TCBONA]2.0.CO;2

Bulmer RH, Schwendenmann L, Lundquist CJ (2016) Carbon and nitrogen stocks and below-groundallometry in temperate mangroves. Front Mar Sci 3:150. https://doi.org/10.3389/fmars.2016.00150

Campbell JE, Berry JA, Seibt U, Smith SJ, Montzka SA, Launois T et al (2017) Large historical growth in global terrestrial gross primary production. Nature 544(7648):84–87. https://doi.org/10.1038/nature22030

Carruthers TJB, Dennison WC, Kendrick GA, Waycott M, Walker DI, Cambridge ML (2007) Seagrasses of south–west Australia: a conceptual synthesis of the world's most diverse and extensive seagrass meadows. J Exp Mar Biol Ecol 350(1–2):21–45. https://doi.org/10.1016/j.jembe.2007.05.036

Chan M, Johansson MA (2012) The incubation periods of dengue viruses. PLoS ONE 7:1–7. https://doi.org/10.1371/journal.pone.0050972

Ciais P, Sabine C, Bala G, Bopp L, Brovkin V, Canadell J et al (2013) Carbon and other biogeochemical cycles. In: Stocker TF et al (eds) Climate change 2013: the physical science basis. Contribution of working group I to the fifth assessment report of the intergovernmental panel on climate change. Cambridge University Press, Cambridge, UK and New York, pp 465–570. https://doi.org/10.1017/CBO9781107415324.015

Clements CS, Burns AS, Stewart FJ, Hay ME (2020) Seaweed-coral competition in the field: effects on coral growth, photosynthesis and microbiomes require direct contact. Proc Royal Soc B 287(1927):20200366. https://doi.org/10.1098/rspb.2020.0366

Crooks S, Orr M, Emmer I, von Unger M, Brown B, Murdiyarso D (2014) Guiding principles for delivering coastal wetland carbon projects. Center for International Forestry Research. http://www.jstor.com/stable/resrep01961

Donato DC, Kauffman JB, Murdiyarso D, Kurnianto S, Stidham M et al (2011) Mangroves among the most carbon-rich forests in the tropics. Nat Geosci 4:293–297. https://doi.org/10.1038/ngeo1123

Donato DC, Kauffman JB, Mackenzie RA, Ainsworth A, Pfleeger AZ (2012) Whole-island carbon stocks in the tropical Pacific: implications for mangrove conservation and upland restoration. J Environ Manag 97:89–96. https://doi.org/10.1016/j.jenvman.2011.12.004

Duarte CM, Kennedy H, Marba N, Hendriks I (2013) Assessing the capacity of seagrass meadows for carbon burial: current limitations and future strategies. Ocean Coast Manag 83:32–38. https://doi.org/10.1016/j.ocecoaman.2011.09.001

Duarte B, Martins I, Rosa R, Matos AR, Roleda MY, Reusch TBH, Engelen AH, Serrão EA, Pearson GA et al (2018) Climate change impacts on seagrass meadows and macroalgal forests: an integrative perspective on acclimation and adaptation potential. Front Mar Sci 5:190. https://doi.org/10.3389/fmars.2018.00190

Duke NC, Meynecke JO, Dittmann S, Ellison AM, Anger K et al (2007) A world without mangroves? Science 317(5834):41–42. https://doi.org/10.1126/science.317.5834.41b

Ellison AM, Farnsworth EJ, Merkt RE (1999) Origins of mangrove ecosystems and the mangrove biodiversity anomaly. Glob Ecol Biogeogr 8(2):95–115. https://doi.org/10.1046/j.1466-822X.1999.00126.x

Ellison JC (2009) Geomorphology and sedimentology of mangroves. In: Perillo GME, Wolanski E, Cahoon DR, Brinson MM (eds) Coastal Wetlands: an integrated ecosystem approach, Elsevier, Amsterdam, The Netherlands, pp 565–591

Erftemeijer PL, Lewis RR (1999). Planting mangroves on intertidal mudflats: habitat restoration or habitat conversion. In: Proceedings of the ECOTONE VIII seminar enhancing coastal ecosystems restoration for the 21st century, Ranong, Thailand, pp 23–28

European Environment Agency (2019) https://www.eea.europa.eu/data-and-maps/daviz/atmospheric-concentration-of-carbon-dioxide-5#tab-chart_5 (Accessed on 30 Sept 2020)

Farnsworth EJ, Farrant JM (1998) Reductions in abscisic acid are linked with viviparous reproduction in mangroves. Am J Bot 85:760–769. https://doi.org/10.2307/2446410

Feng Y, Ramanathan V, Kotamarthi VR (2013) Brown carbon: a significant atmospheric absorber of solar radiation? Atmos Chem Phys Discuss 13(1). https://doi.org/10.5194/acp-13-8607-2013

Fischer J, Reed-Andersen T, Klug J, Chalmers A (2000) Spatial pattern of localized disturbance along a southeastern salt marsh tidal creek. Estuaries 23(4):565. https://doi.org/10.2307/1353146

Forster P, Ramaswamy V, Artaxo P, Bernsten T, Betts R, Fahey DW, Haywood J, Lean J, Lowe DC, Myhre G, Nganga J, Prinn R, Raga G, Schulz M, Van Dorland R (2007) Changes in atmospheric constituents and in radiative forcing In: Solomon S, Qin D, Manning M, Chen Z, Marquis M, Averyt KB, Tignor M, Miller BR (eds) Climate change 2007: the physical science basis. Contribution of working group I to the fourth assessment report of the intergovernmental panel on climate change. Cambridge University Press, Cambridge, UK and New York, US pp 500–587

Frankignoulle M, Gattuso JP, Biondo R, Bourge I, Copin-Montégut G, Pichon M (1996) Carbon fluxes in coral reefs. II. Eulerian study of inorganic carbon dynamics and measurement of air-sea CO_2 exchanges. Mar Ecol Prog Ser 145:123–132. https://doi.org/10.3354/meps145123

Gacia E, Duarte CM (2001) Sediment retention by a mediterranean Posidonia oceanica meadow: the balance between deposition and resuspension. Estuar Coast Shelf Sci 52:505–514. https://doi.org/10.1006/ecss.2000.0753

Gambi MC, Nowell AR, Jumars PA (1990) Flume observations on flow dynamics in Zostera marina (eelgrass) beds. Mar Ecol Prog Ser 61(1/2):159–169. https://www.jstor.org/stable/24842256

Gattuso J-P, Pichon M, Delesalle B, Canon C, Frankignoulle M (1996) Carbon fluxes in coral reefs. I. Lagrangian measurement of community metabolism and resulting air–sea CO_2 disequilibrium. Mar Ecol Prog Ser 145:109–121. https://doi.org/10.3354/meps145109

Gloria-Soria A, Armstrong PM, Turner PE, Turner PE (2017) Infection rate of Aedesaegypti mosquitoes with dengue virus depends on the interaction between temperature and mosquito genotype. Proc R Soc B Biol Sci 284:20171506. https://doi.org/10.1098/rspb.2017.1506

Goldberg L, Lagomasino D, Thomas N, Fatoyinbo T (2020) Global declines in human-driven mangrove loss. Glob Chang Biol 26(10):5844–5855. https://doi.org/10.1111/gcb.15275
Guannel G, Arkema K, Ruggiero P et al (2016) The power of three: coral reefs, seagrasses and mangroves protect coastal regions and increase their resilience. PLoS One 11(7):e0158094. https://doi.org/10.1371/journal.pone.0158094
Hemminga MA, Duarte CM (2000) Seagrass ecology. Cambridge University Press, Cambridge
Herr D, Blum J, Himes-Cornell A, Sutton-Grier A (2019) An analysis of the potential positive and negative livelihood impacts of coastal carbon offset projects. J Environ Manag 235:463–479. https://doi.org/10.1016/j.jenvman.2019.01.067
Herr D, Landis E (2016) Coastal blue carbon ecosystems. Opportunities for nationally determined contributions. Policy brief. Gland, Switzerland: IUCN, Washington, DC: TNC
Herr D, Pidgeon E, Laffoley DDA (2012) Blue carbon policy framework 2.0: based on the discussion of the International Blue Carbon Policy Working Group. IUCN
Hill R, Bellgrove A, Macreadie PI, Petrou K, Beardall J, Steven A, Ralph PJ (2015) Can macroalgae contribute to blue carbon? an Australian perspective. Limnol Oceanogr 60:1689–1706. https://doi.org/10.1002/lno.10128
Himes-Cornell A, Pendleton L, Atiyah P (2018) Valuing ecosystem services from blue forests: a systematic review of the valuation of salt marshes, sea grass beds and mangrove forests. Ecosyst Serv 30:36–48. https://doi.org/10.1016/j.ecoser.2018.01.006
Hodgkin EP, Hamilton B (1998) Changing estuarine wetlands: a long term perspective for management. In: Davis JA (ed) Wetlands for the future McComb AJ. Gleneagles Publishing, Adelaide, Australia, pp 243–255
Hoegh-Guldberg O, Poloczanska ES, Skirving W, Dove S (2017) Coral reef ecosystems under climate change and ocean acidification. Front Mar Sci 4:158. https://doi.org/10.3389/fmars.2017.00158
Horstman EM, Dohmen-Janssen CM, Narra PMF, van den Berg NJF, Siemerink M, Hulscher SJMH (2014) Wave attenuation in mangroves: a quantitative approach to field observations. Coast Eng 94:47–62. https://doi.org/10.1016/j.coastaleng.2014.08.005
Houghton RA (2003) Revised estimates of the annual net flux of carbon to the atmosphere from changes in land use and land management 1850–2000. Tellus 55B:378–390. https://doi.org/10.3402/tellusb.v55i2.16764
Huang L, Zhou M, Lv J, Chen K (2020) Trends in global research in forest carbon sequestration: A bibliometric analysis. J Clean Prod 252:119908. https://doi.org/10.1016/j.jclepro.2019.119908
IEA (2019) World energy balances, 1970–2016. International Energy Agency, Paris
IPCC (2007) Intergovernmental panel on climate change. Climate change 2007: Assessment Report. IPCC, Valencia
IPCC (2001) Climate change 2001: the scientific basis. In: Houghton JT, Ding Y, Griggs DJ, Noguer M, van der Linden PJ, Dai X, Maskell K, Johnson CA (eds) Contributions of working group I to the third assessment of the intergovernmental panel on climate change. Cambridge, UK, p 881
Jevrejeva S, Jackson LP, Grinsted A, Lincke D, Marzeion B (2018) Flood damage costs under the sea level rise with warming of 1.5 C and 2 C. Environ Res Lett 13(7):074014. https://doi.org/10.1088/1748-9326/aacc76
Jia P, Chen X, Chen J, Lu L, Liu Q, Tan X (2017) How does the dengue vector mosquito Aedesalbopictus respond to global warming? Parasit Vector 10:140. https://doi.org/10.1186/s13071-017-2071-2
Jiao N, Herndl GJ, Hansell DA, Benner R, Kattner G, Wilhelm SW et al (2010) Microbial production of recalcitrant dissolved organic matter: long-term carbon storage in the global ocean. Nat Rev Microbiol 8(8):593–599. https://doi.org/10.1038/nrmicro2386
Kauffman JB, Heider C, Cole TG et al (2011) Ecosystem carbon stocks of Micronesian mangrove forests. Wetlands 31:343–352. https://doi.org/10.1007/s13157-011-0148-9
Kauffman JB, Arifanti VB, Hernández Trejo H, del Carmen JM, Norfolk J, Cifuentes M et al (2017) The jumbo carbon footprint of a shrimp: carbon losses from mangrove deforestation. Front Ecol Environ 15(4):183–188. https://doi.org/10.1002/fee.1482

Kayanne H, Suzuki A, Saito H (1995) Diurnal changes in the partial pressure of carbon dioxide in coral reef water. Science 269:214–216. https://doi.org/10.1126/science.269.5221.214

Kelleway JJ, Saintilan N, Macreadie PI, Baldock JA, Ralph PJ (2017) Sediment and carbon deposition vary among vegetation assemblages in a coastal salt marsh. Biogeosciences 14:3763–3779. https://doi.org/10.5194/bg-14-3763-2017

Knutson T, Camargo SJ, Chan JC, Emanuel K, Ho CH, Kossin J et al (2020) Tropical cyclones and climate change assessment: Part II: projected response to anthropogenic warming. Bull Am Met Soc 101(3):E303–E322. https://doi.org/10.1175/BAMS-D-18-0194.1

Köchy M, Hiederer R, Freibauer A (2015) Global distribution of soil organic carbon—Part 1: Masses and frequency distributions of SOC stocks for the tropics, permafrost regions, wetlands, and the world. Soil 1(1):351–365. https://doi.org/10.5194/soil-1-351-2015

Krause-Jensen D, Duarte CM (2016) Substantial role of macroalgae in marine carbon sequestration. Nat Geosci 9:737–742. https://doi.org/10.1038/ngeo2790

Krause-Jensen D, Lavery P, Serrano O, Marba N, Masque P, Duarte CM (2018) Sequestration of macroalgal carbon: the elephant in the Blue Carbon room. Biol Lett 14(6):20180236. https://doi.org/10.1098/rsbl.2018.0236

Kristensen E, Bouillon S, Dittmar T, Marchand C (2008) Organic carbon dynamics in mangrove ecosystems. Aquat Bot 89:201–219. https://doi.org/10.1016/j.aquabot.2007.12.005

Kroeker KJ, Kordas RL, Crim R, Hendriks IE, Ramajo L, Singh GS et al (2013) Impacts of ocean acidification on marine organisms: quantifying sensitivities and interaction with warming. Glob Chang Biol 19:1884–1896. https://doi.org/10.1111/gcb.12179

Kuwae T, Kanda J, Kubo A, Nakajima F, Ogawa H, Sohma A, Suzumura M (2018) CO_2 uptake in the shallow coastal ecosystems affected by anthropogenic impacts. In: Kuwae T, Hori M (eds) Blue carbon in shallow coastal ecosystems: carbon dynamics, policy, and implementation. Springer, Singapore, pp 295–319. https://doi.org/10.1007/978-981-13-1295-3_11

Kuwae T, Hori M (2018) Blue carbon in shallow coastal ecosystems. Springer, Singapore

Lal R (2004) Soil carbon sequestration impacts on global climate change and food security. Science 304(5677):1623–1627. https://doi.org/10.1126/science.1097396

Landschützer P, Gruber N, Bakker DC, Schuster U (2014) Recent variability of the global ocean carbon sink. Glob Biogeochem Cycl 28(9):927–949. https://doi.org/10.1002/2014GB004853

Larkum AW, Orth RJ, Duarte CM (2006) Seagrasses. Springer, Dordrecht, The Netherlands, p 691

Larkum AWD, Davey PA, Kuo J, Ralph PJ, Raven JA (2017) Carbon-concentrating mechanisms in seagrasses. J Exp Bot 68(14):3773–3784. https://doi.org/10.1093/jxb/erx206

Laruelle GG, Cai WJ, Hu X, Gruber N, Mackenzie FT, Regnier P (2018) Continental shelves as a variable but increasing global sink for atmospheric carbon dioxide. Nat Commun 9(1):454. https://doi.org/10.1038/s41467-017-02738-z

Le Bars D, Drijfhout S, de Vries H (2017) A high-end sea level rise probabilistic projection including rapid Antarctic ice sheet mass loss. Environ Res Lett 12(4):044013. https://doi.org/10.1088/1748-9326/aa6512

Liu S, Jiang Z, Zhang J, Wu Y, Huang X, Macreadie PI (2017) Sediment microbes mediate the impact of nutrient loading on blue carbon sequestration by mixed seagrass meadows. Sci Total Environ 599:1479–1484. https://doi.org/10.1016/j.scitotenv.2017.05.129

Lovelock CE, Duarte CM (2019) Dimensions of blue carbon and emerging perspectives. Biol Lett 15(3):20180781. https://doi.org/10.1098/rsbl.2018.0781

Lovelock CE, Atwood T, Baldock J, Duarte CM, Hickey S, Lavery PS et al (2017) Assessing the risk of carbon dioxide emissions from blue carbon ecosystems. Front Ecol Environ 15(5):257–265. https://doi.org/10.1002/fee.1491

Lund JF, Sungusia E, Mabele MB, Scheba A (2017) Promising change, delivering continuity: REDD+ as conservation fad. World Dev 89:124–139. https://doi.org/10.1016/j.worlddev.2016.08.005

MacCracken MC, Luther FM (eds) (1985) Detecting the climatic effects of increasing carbon dioxide, Rep. DOE/ER0235, U.S. Dep. of Energy, Washington, D.C

Macreadie PI, Anton A, Raven JA, Beaumont N, Connolly RM, Friess DA, Kelleway JJ et al (2019) The future of blue carbon science. Nat Commun 10(1):1–13. https://doi.org/10.1038/s41467-019-11693-w

Mazarrasa I, Samper-Villarreal J, Serrano O, Lavery PS, Lovelock CE, Marbà N et al (2018) Habitat characteristics provide insights of carbon storage in seagrass meadows. Mar Poll Bull 134:106–117. https://doi.org/10.1016/j.marpolbul.2018.01.059

Mazda Y, Wolanski E (2009) Hydrodynamics and modeling of water flow in mangrove areas. In: Perillo GME, Wolanski E, Cahoon DR, Brinson MM (eds) Coastal Wetlands: an integrated ecosystem approach. Elsevier, Amsterdam, The Netherlands, pp 231–262

McGranahan G, Balk D, Anderson B (2007) The rising tide: assessing the risks of climate change and human settlements in low elevation coastal zones. Environ Urban 19:17–37. https://doi.org/10.1177/0956247807076960

McLeod E et al (2011) A blueprint for blue carbon: toward an improved understanding of the role of vegetated coastal habitats in sequestering CO_2. Front Ecol Environ 9:552–560. https://doi.org/10.1890/110004

Mcowen CJ, Weatherdon LV, Van Bochove JW, Sullivan E, Blyth S, Zockler C et al (2017) A global map of saltmarshes. Biodivers Data J (5). https://doi.org/10.3897/BDJ.5.e11764

Meinshausen M, Vogel E, Nauels A, Lorbacher K, Meinshausen N, Etheridge DM et al (2017) Historical greenhouse gas concentrations for climate modelling (CMIP6). Geosci Model Dev 10:2057–2116. https://doi.org/10.5194/gmd-10-2057-2017

Middelburg JJ, Soetaert K, Hagens M (2020) Ocean alkalinity, buffering and biogeochemical processes. Rev Geophys 58(3):e2019RG000681. https://doi.org/10.1029/2019RG000681

Ming T, Caillol S, Liu W (2016) Fighting global warming by GHG removal: destroying CFCs and HCFCs in solar-wind power plant hybrids producing renewable energy with no-intermittency. Int J Greenh Gas Control 49:449–472. https://doi.org/10.1016/j.ijggc.2016.02.027

Mitchell JF (1989) The "greenhouse" effect and climate change. Rev Geophys 27(1):115–139. https://doi.org/10.1029/RG027i001p00115

Miyajima T, Hori M, Hamaguchi M, Shimabukuro H, Adachi H, Yamano H, Nakaoka M (2015) Geographic variability in organic carbon stock and accumulation rate in sediments of East and Southeast Asian seagrass meadows. Glob Biogeochem Cycl 29:397–415. https://doi.org/10.1002/2014GB004979

Miyajima T, Hori M, Hamaguchi M, Shimabukuro H, Yoshida G (2017) Geophysical constraints for organic carbon sequestration capacity of Zostera marina seagrass meadows and surrounding habitats. Limnol Oceanogr 62(3):954–972. https://doi.org/10.1002/lno.10478

Moberg F, Folke C (1999) Ecological goods and services of coral reef ecosystems. Ecol Econ 29(2):215–233. https://doi.org/10.1016/S0921-8009(99)00009-9

Mohanakumar K (2008) Stratosphere troposphere interactions: an introduction. Springer Science & Business Media

Monks PS (2005) Gas-phase radical chemistry in the troposphere. Chem Soc Rev 34(5):376–395. https://doi.org/10.1039/B307982C

MtwanaNordlund L, Koch EW, Barbier EB, Creed JC (2016) Seagrass ecosystem services and their variability across genera and geographical regions. PLoS One 11(10):e0163091. https://doi.org/10.1371/journal.pone.0163091

Mukherjee N, Sutherland WJ, Dicks L, Huge J, Koedam N, Dahdouh-Guebas F (2014) Ecosystem service valuations of mangrove ecosystems to inform decision making and future valuation exercises. PloS One 9(9):e107706. https://doi.org/10.1371/journal.pone.0107706

NASEM (2018) National ACADEMIES OF SCIENCES, ENGINEERING, AND Medicine. Negative emissions technologies and reliable sequestration: a research agenda. National Academies Press

Nellemann C, Corcoran E, Duarte CM et al (2009) Blue carbon: a rapid response assessment. United Nations Environmental Programme, GRID-Arendal, BirkelandTrykkeri AS, Birkeland

NOAA (2019) Global climate report, annual 2018. US national oceanic and atmospheric administration's National Centers for Environmental Information (NCEI). https://www.ncdc.noaa.gov/sotc/global/201813

NOAA/ESRL (2019) Trends in Atmospheric Carbon Dioxide. NOAA/ESRL's Global Monitoring Division (GMD). https://www.esrl.noaa.gov/gmd/ccgg/trends/data.html (Accessed on 30 Sept 2020)

Olivier JGJ, Schure KM, Peters JAHW (2017) Trends in global CO_2 and total GHG emissions. 2017 Report. PBL Netherlands Environmental Assessment Agency, The Hague

Olivier JGJ, Schure KM, Peters JAHW (2019) Trends in global CO_2 and total greenhouse gas emissions. 2019 Report. PBL Netherlands Environmental Assessment Agency, The Hague

Overpeck J, Hughen K, Hardy D, Bradley R, Case R, Douglas M et al (1997) Arctic environmental change of the last four centuries. Science 278(5341):1251–1256. https://doi.org/10.1126/science.278.5341.1251

Pendleton L, Donato DC, Murray BC, Crooks S, Jenkins WA, Sifleet S et al (2012) Estimating global "blue carbon" emissions from conversion and degradation of vegetated coastal ecosystems. PloS One 7(9):e43542. https://doi.org/10.1371/journal.pone.0043542

Poschenrieder C, Fernández JA, Rubio L, Pérez L, Terés J, Barceló J (2018) Transport and use of bicarbonate in plants: current knowledge and challenges ahead. Int J Mol Sci 19(5):1352. https://doi.org/10.3390/ijms19051352

Ramanathan V, Carmichael G (2008) Global and regional climate changes due to black carbon. Nat Geosci 1(4):221–227. https://doi.org/10.1038/ngeo156

Ricart AM, York PH, Rasheed MA, Pérez M, Romero J, Bryant CV, Macreadie PI (2015) Variability of sedimentary organic carbon in patchy seagrass landscapes. Mar Pollut Bull 100:476–482. https://doi.org/10.1016/j.marpolbul.2015.09.032

Rigby M, Park S, Saito T, Western LM, Redington AL, Fang X et al (2019) Increase in CFC-11 emissions from eastern China based on atmospheric observations. Nature 569(7757):546–550. https://doi.org/10.1038/s41586-019-1193-4

Running SW (2012) A measurable planetary boundary for the biosphere. Science 337(6101):1458–1459. https://doi.org/10.1126/science.1227620

Saderne V, Geraldi NR, Macreadie PI, Maher DT, Middelburg JJ, Serrano O, Fourqurean JW (2019) Role of carbonate burial in blue carbon budgets. Nat Commun 10(1):1–9. https://doi.org/10.1038/s41467-019-08842-6

Sasmito SD, Sillanpää M, Hayes MA, Bachri S, Saragi-Sasmito MF, Sidik F et al (2020) Mangrove blue carbon stocks and dynamics are controlled by hydrogeomorphic settings and land-use change. Glob Chang Biol 26(5):3028–3039. https://doi.org/10.1111/gcb.15056

Seidl R, Thom D, Kautz M, Martin-Benito D, Peltoniemi M, Vacchiano G et al (2017) Forest disturbances under climate change. Nat Clim Chang 7:395–402. https://doi.org/10.1038/nclimate3303

Serrano O, Lavery PS, Rozaimi M, Mateo M (2014) Influence of water depth on the carbon sequestration capacity of seagrasses. Glob Biogeochem Cycl 28:950–961. https://doi.org/10.1002/2014GB004872

Short FT, Polidoro B, Livingstone SR, Carpenter KE, Bandeira S, Bujang JS et al (2011) Extinction risk assessment of the world's seagrass species. Biol Conserv 144(7):1961–1971. https://doi.org/10.1016/j.biocon.2011.04.010

Silliman BR (2014) Salt marshes. Curr Biol 24(9):R348–R350

Singh J, Dhar DW (2019) Overview of carbon capture technology: microalgalbiorefinery concept and state-of-the-art. Front Mar Sci 6:29. https://doi.org/10.3389/fmars.2019.00029

Smith SV (1978) Coral reef area and the contributions of reefs to processes and resources of the world's ocean. Nature 273:225–232. https://doi.org/10.1038/273225a0

Smith SV (1981) Marine macrophytes as a global carbon sink. Science 211:838–840. https://doi.org/10.1126/science.211.4484.838

Spalding MD, Grenfell AM (1997) New estimates of global and regional coral reef areas. Coral Reef 16(4):225–230. https://doi.org/10.1007/s003380050078

Sugimatsu K, Yagi H, Abo K, Tarutani K, Hori M, Yoshida G, Shimabukuro H, Nakayama A (2015) A coupled particle tracking-carbon cycle modeling system for sedimentary organic carbon derived from drifting seagrass in Seto Inland Sea. J Jpn Soc Civil Eng B2(71), https://doi.org/10.2208/kaigan.71.I_1387 (in Japanese with English abstract)
Suyadi GJ, Lundquist CJ, Schwendenmann L (2018) Sources of uncertainty in mapping temperate mangroves and their minimization using innovative methods. Int J Remote Sens 39(1):17–36. https://doi.org/10.1080/01431161.2017.1378455
Syakila A, Kroeze C (2011) The global nitrous oxide budget revisited. Greenh Gas Measur Manag 1(1):17–26. https://doi.org/10.3763/ghgmm.2010.0007
Taillardat P, Friess DA, Lupascu M (2018) Mangrove blue carbon strategies for climate change mitigation are most effective at the national scale. Biol Lett 14(10):20180251. https://doi.org/10.1098/rsbl.2018.0251
Tang X, Fan S, Du M, Zhang W, Gao S, Liu S et al (2020) Spatial and temporal patterns of global soil heterotrophic respiration in terrestrial ecosystems. Earth Syst Sci Data 12(2):1037–1051. https://doi.org/10.5194/essd-12-1037-2020
Tateishi HR, Bragagnolo C, de Faria RN (2020) Economic and environmental efficiencies of greenhouse gases' emissions under institutional influence. Technol Forecast Soc Chang 161:120321. https://doi.org/10.1016/j.techfore.2020.120321
Thomas S (2014) Blue carbon: knowledge gaps, critical issues, and novel approaches. Ecol Econ 107:22–38. https://doi.org/10.1016/j.ecolecon.2014.07.028
Tifafi M, Guenet B, Hatté C (2018) Large differences in global and regional total soil carbon stock estimates based on SoilGrids, HWSD, and NCSCD: Intercomparison and evaluation based on field data from USA, England, Wales, and France. Glob Biogeochem Cycl 32(1):42–56. https://doi.org/10.1002/2017GB005678
Tomlinson PB (1986) The botany of mangroves. Cambridge University Press, Cambridge
Trumbore S (2009) Radiocarbon and soil carbon dynamics. Annu Rev Earth Planet Sci 37:47–66. https://doi.org/10.1146/annurev.earth.36.031207.124300
Ullman R, Bilbao-Bastida V, Grimsditch G (2013) Including blue carbon in climate market mechanisms. Ocean Coast Manag 83:15–18. https://doi.org/10.1016/j.ocecoaman.2012.02.009
UNEP (United Nations Environment Programme) (2019) The emissions gap report 2019. United Nations Environment Program (UNEP), Nairobi. Internet: https://www.unenvironment.org/resources/emissions-gap-report-2019 (Accessed on 30 Sept 2020)
UNEP (2006) Marine and coastal ecosystems and human well-being: a synthesis report based on the findings of the millennium ecosystem assessment. UN Environment Programme, Nairobi
UNFCCC (2015) Decision 1/CP.21. The Paris Agreement
Valiela I, Bowen JL, York JK (2001) Mangrove forests: one of the world's threatened major tropical environments: at least 35% of the area of mangrove forests has been lost in the past two decades, losses that exceed those for tropical rain forests and coral reefs, two other well-known threatened environments. Bioscience 51:807. https://doi.org/10.1641/0006-3568(2001)051[0807:MFOTW]2.0.CO;2
van der Werf GR, Morton DC, DeFries RS et al (2009) CO_2 emissions from forest loss. Nat Geosci 2:737–738. https://doi.org/10.1038/ngeo671
Vanderklift MA, Gorman D, Steven AD (2019) Blue carbon in the Indian Ocean: a review and research agenda. J India Ocean Reg 15(2):129–138. https://doi.org/10.1080/19480881.2019.1625209
Wasko C, Sharma A (2017) Global assessment of flood and storm extremes with increased temperatures. Sci Rep 7(1):1–8. https://doi.org/10.1038/s41598-017-08481-1
Watanabe A, Nakamura T (2018) Carbon dynamics in coral reefs. In: Kuwae T, Hori M (eds) Blue carbon in shallow coastal ecosystems: carbon dynamics, policy, and implementation. Springer, Singapore, pp 273–293
Waycott M, Duarte CM, Carruthers TJB, Orth RJ, Dennison WC et al (2009) Accelerating loss of seagrasses across the globe threatens coastal ecosystems. Proc Natl Acad Sci USA 106:12377–12381. https://doi.org/10.1073/pnas.0905620106

Widdows J, Pope ND, Brinsley MD, Asmus H, Asmus RM (2008) Effects of seagrass beds (Zostera noltii and Z. marina) on near-bed hydrodynamics and sediment resuspension. Mar Ecol Prog Ser 358:125–136. https://doi.org/10.3354/meps07338

Wilke AB, Beier JC, Benelli G (2019) Complexity of the relationship between global warming and urbanization–an obscure future for predicting increases in vector-borne infectious diseases. Curr Opin Insect Sci 35:1–9. https://doi.org/10.1016/j.cois.2019.06.002

Wilkie L, O'Hare MT, Davidson I, Dudley B, Paterson DM (2012) Particle trapping and retention by Zostera noltii: a flume and field study. Aquat Bot 102:15–22. https://doi.org/10.1016/j.aquabot.2012.04.004

Wong PP, Losada I, Gattuso J-P, Hinkel J, Khattabi A, McInnes K, Saito Y, Sallenger A (2014) Climate change 2014: impacts, adaptation, and vulnerability. Part A: Global and sectoral aspects. Contribution of working group II to the fifth assessment report of the intergovernmental panel on climate change on coastal systems and low-lying areas, Cambridge University Press, Cambridge, pp 361–409

Wylie L, Sutton-Grier AE, Moore A (2016) Keys to successful blue carbon projects: lessons learned from global case studies. Mar Pol 65:76–84. https://doi.org/10.1016/j.marpol.2015.12.020

Zhang Y, Forrister H, Liu J, Dibb J, Anderson B, Schwarz JP et al (2017) Top-of-atmosphere radiative forcing affected by brown carbon in the upper troposphere. Nat Geosci 10(7):486–489. https://doi.org/10.1038/ngeo2960

Zhang Y, Song C, Band LE, Sun G (2019) No proportional increase of terrestrial gross carbon sequestration from the greening Earth. J Geophys Res Biogeosci 124(8):2540–2553. https://doi.org/10.1029/2018JG004917

Zhao C, Liu B, Piao S, Wang X, Lobell DB, Huang Y et al (2017) Temperature increase reduces global yields of major crops in four independent estimates. Proc Natl Acad Sci 114(35):9326–9331. https://doi.org/10.1073/pnas.1701762114

Zhao J, Zhan R, Wang Y (2018) Global warming hiatus contributed to the increased occurrence of intense tropical cyclones in the coastal regions along East Asia. Sci Rep 8(1):1–9. https://doi.org/10.1038/s41598-018-24402-2

Zhao Y, Wang J, Ji Z, Liu J, Guo X, Yuan J (2020) A novel technology of carbon dioxide adsorption and mineralization via seawater decalcification by bipolar membrane electrodialysis system with a crystallizer. Chem Eng J 381:122542. https://doi.org/10.1016/j.cej.2019.122542

Chapter 2
Carbon Dynamics in the Open Indian Ocean

Abhra Chanda

Abstract Oceans have long been realized as a sink for CO_2. Though this nature of oceans has changed many times in the glacial and interstadial phases that this planet has witnessed in the past. These vast water reservoirs play a crucial role in regulating the biogeochemical cycles of several important elements and gases that are integral to life, and carbon is no exception in this regard. Many eminent oceanographers pointed out that among all the oceans of the world, the Indian Ocean is perhaps the most undersampled and understudied. However, several studies have already been carried out in this ocean basin. The present chapter collated all the crucial observations on multifarious aspects related to the Indian Ocean carbon dynamics, like the spatiotemporal distribution of the partial pressure of CO_2 in seawater, biological production, dissolved organic and inorganic carbon in the sea surface and water column along with the air-sea CO_2 exchange. The anthropogenic inventory of CO_2 in this ocean basin and the role of the autotrophic community are also discussed. Several carbon-related characteristic features of the Indian Ocean and their variability in the geological time scales are also kept within the purview of this chapter. Besides developing a fundamental understanding of the open Indian Ocean's biochemical traits concerning carbon dynamics, this chapter has also identified the aspects on which further studies should be framed and executed.

Keyword Partial pressure of CO_2 · Air-sea CO_2 exchange · Dissolved inorganic carbon · Dissolved organic carbon · Anthropogenic carbon inventory · Deep ocean burial · Physical forcing · Upwelling · Downwelling · Nutrient · Primary production

2.1 Global Open Ocean Carbon Sink Potential

The global scientific community has recognized the potential of oceans in capturing carbon long ago. However, this issue became relevant and important, when we could

A. Chanda (✉)
School of Oceanographic Studies, Jadavpur University, 188 Raja S.C. Mullick Road, Kolkata, West Bengal 700084, India
e-mail: abhrachanda.sos.rs@jadavpuruniversity.in

A. Chanda et al. (eds.), *Blue Carbon Dynamics of the Indian Ocean*,
https://doi.org/10.1007/978-3-030-96558-7_2

understand the ill effects of anthropogenic CO_2 emission. Carbon-containing gaseous compounds, which are greenhouse gases by nature, existed on this planet even before human beings started burning fossil fuels. However, since the industrial revolution, anthropogenic activities like fuel combustion and forest clearing have enhanced the rate of CO_2 emission at an alarming rate, which this planet has never witnessed earlier. Once we started emitting CO_2 and CH_4 to meet our daily life requirements and pushed the global atmosphere to such an extent that it started hitting us back, in the form of natural calamities induced by climate change, the crucial question that became most relevant is 'where does all the carbon go?'. In the twentieth century, several pieces of research carried out throughout the world, unambiguously established that the terrestrial biosphere and the oceans absorb a substantial amount of anthropogenic carbon emitted towards the atmosphere as CO_2. However, the CO_2 absorbing potential of the oceans and the dynamics of carbon accumulation remain an area of active research.

Fundamental research has proven that carbon enters the ocean in the form of CO_2 absorbed from the atmosphere and laterally through the rivers as dissolved carbonates, bicarbonates, and organic matter. Worldwide estimates made in the last decade showed that the global oceans have absorbed almost 25% of the total anthropogenic CO_2 emission that took place in the last two decades, with rates varying between 2 and 2.5 Pg C yr^{-1} (Wanninkhof et al. 2013; Le Quéré et al. 2014). However, the carbonates laterally passed on through the rivers to the oceans, degasses out and return to the atmosphere from the oceans, in the form of CO_2, reducing the net CO_2absorption magnitude to 1.5 ± 0.5 Pg C yr^{-1} (Jacobson et al. 2007; Gruber et al. 2009; Takahashi et al. 2009). Apparently, from these estimates, it might seem that we can keep on emitting CO_2 as much as possible, as the oceans can take the responsibility of absorbing all of it. However, the rate at which we emit and the rate at which oceans absorb are not in parity. At the present date, we are emitting CO_2 at a much faster rate than the oceanic uptake rate. Moreover, the substantial amount of CO_2 that the oceans absorb does not remain in the ocean for a long period. In other words, we cannot consider this oceanic CO_2 uptake an avenue of long-term carbon sequestration. The CO_2 absorbed by the oceans undergoes various transformations, as it actively enters the biological life cycle through trophic transfer from the autotrophs to the higher marine lifeforms. Carbon in living organisms can remain stored in the organic form from few hours to a century, depending upon the life span of that particular organism. However, as the organism dies, a substantial part of this organic carbon undergoes mineralization and transforms to inorganic form, which can re-transform to aqueous CO_2, and eventually escape to the atmosphere. A part of this ocean-sequestered CO_2 travels to the abyssal plain of the oceans in the form of organic matter and remains there for millennia. However, according to the latest estimates, the rate of such organic carbon sequestration by the open ocean comes around 6 Tg C yr^{-1} (Nellemann et al. 2009). This magnitude is much lower than that observed in the blue carbon ecosystems in the oceanic periphery, like mangroves, seagrasses, and saltmarshes (Kuwae and Hori 2018). The blue carbon ecosystems, despite covering only a small area compared to the vast oceans, sequester disproportionately high quantities of organic carbon. Due to this reason, the organic carbon sequestered by the open ocean has less importance, though it fulfills the definition of blue carbon. The low long-term organic carbon burial rate of the open oceans can trigger a lack of interest among the blue carbon researchers; however, the carbon-absorbing power of the open oceans is unquestionable. Ample scope of further research on the carbon dynamics of these vast stretches of water mass exists in the present date. The character and biogeochemical behaviour of the oceans are constantly changing, which can significantly affect the functioning of the ecosystems situated in the oceanic periphery. In the pre-industrial era rivers used to drain a substantial amount of carbon in the organic as well as inorganic form. Contrary to the present scenario, Regnier et al. (2013) argued that the remineralization of organic carbon to the inorganic form, and its subsequent removal from the ocean as degassing of CO_2 to the atmosphere used to take place in the open oceanic regions, and the nearshore waters were neutral (neither sink nor source). Besides, the magnitude of organic matter export from the rivers has substantially increased over the years. During the transit from the rivers to the oceans, most of the carbon degasses as CO_2 in the estuarine phase and buries in the sediments of the coastal vegetated ecosystems, and the remaining goes to the ocean. With an increase in the quantity of organic matter transported from the rivers, the likelihood of benthic burial also increases. Due to this and many other factors, the carbon dynamics of the open ocean deserve more attention from the scientific community.

Estimations and drawing global ocean carbon budgets have always been a daunting task owing to the vast stretch of the water mass, lack of field data representing all the nooks and corners, and unfeasibility of carrying out field measurements covering such a huge area. However, oceanographers, mathematicians, and statistical modelers have developed several global ocean biogeochemical models (GOBMs) to overcome such difficulties. Comparing the simulated outputs from such models and data products based on field measurements of partial pressure of CO_2 in seawater [pCO_2(water)] leads to varying estimates of air-sea CO_2 fluxes (Hauck et al. 2020). A suite of climatological and hydrodynamic factors, like the sea-surface temperature (SST), precipitation, river run-off, and oceanic currents can be accounted for or not accounted for while running such simulations. Moreover, every model has some biases and drifts intrinsic to the formulation and logic behind such models, which needs to be minimized to derive the most realistic output (Carroll et al. 2020). The last decade has witnessed many such attempts with the prime intent to quantify the surface ocean pCO_2(water) and the global air-sea CO_2 fluxes (Landschützer et al. 2017; Li and Ilyina 2018; Denvil-Sommer et al. 2019; DeVries et al. 2019; Gruber et al. 2019). The area-weighted global oceanic carbon flux estimation requires a pre-fixed global oceanic areal cover, which incidentally varies from model to model. At the present date, the global ocean surface ranges from 352,050,000 km^2 to 365,980,000 km^2 among the existing ocean models (Hauck et al. 2020). The difference in these estimates arises due to the exclusion/inclusion of the extreme north and southern waters or the coastal periphery. Since the coastal ocean's biogeochemical behaviour varies significantly from that of the open ocean, many eminent scholars opined that it is safe to exclude the coastal waters while drawing any global oceanic budgets. In this regard, the global ocean surface area of 361,900,000 km^2 furnished by the ETOPO1 model is preferred by many (Amante and Eakins 2009; Eakins and Sharman 2010).

However, the differences in the global ocean area considered in the respective models often lead to minimal variation in the modelled global air-sea CO_2 exchange. Hauck et al. (2020) observed that the corrections required to address such considerations ranged from -0.02 PgC yr^{-1} to $+0.05$ PgC yr^{-1}. However, the criticality in the realistic estimation arises due to the inter-annual to decadal variability in the fluxes and their relationship with the ocean biogeochemical dynamics and the changing climate, as well as anthropogenic CO_2 load in the atmosphere (Table 2.1). DeVries et al. (2019) recently observed that the rise in atmospheric CO_2 concentration does not concomitantly take place with the rising anthropogenic carbon emissions. They meticulously observed that the decade 1990–1999 witnessed substantial CO_2 accumulation in the global atmosphere when compared to the global carbon emission, compared to that in the following decade 2000–2009. They inferred that these differences arise due to the varying carbon uptake rate of the oceans and the terrestrial biosphere, which essentially does not maintain a directly proportional relationship with the atmospheric CO_2 concentration.

Besides the temporal variability of the climatic variables and the rate of anthropogenic carbon emissions, another reason that leads to the variation in the global ocean carbon sink rates is the spatial variability of the fluxes. Several studies indicated that the air-sea CO_2 fluxes vary significantly along the latitudinal gradient (Landschützer et al. 2014, 2016). Landschützer et al. (2016) emphasized that the higher latitude oceanic regions are mostly responsible for the decadal variability, whereas, the tropical waters contribute significantly to the inter-annual variability, especially, the equatorial pacific ocean region. On the contrary, the Indian Ocean exhibited the minimum most inter-annual variability when compared to all other ocean basins of the world (Landschützer et al. 2014). The higher latitude waters near the north and the south pole experience typical variation in the atmospheric forcing, which in turn, regulate the total alkalinity and dissolved inorganic carbon in these waters. These variations are primarily held responsible for the decadal variations in the global oceanic sink rates, whereas, in the subtropical regions, the changes in SST and the associated changes in the physical solubility of CO_2 lead to the inter-annual variability of air-sea CO_2 fluxes (Landschützer et al. 2016). The ocean–atmosphere coupled physical forcings like that of the ElNiño-Southern Oscillation (ENSO) cycles also a major role behind such variations. Regional Carbon Cycle Assessment and Processes (RECCAP) projects segregated the global ocean surface into eleven distinct regions (Mikaloff Fletcher et al. 2006; Gurney et al. 2008). Landschützer et al. (2014) computed the air-sea CO_2 flux in these eleven basin-scale regions (from 1998 to 2011) and compared it with the observations of Gruber et al. (2009) and Takahashi et al. (2009) (Fig. 2.1). The pacific ocean, especially the region lying in the north of the 44° S latitudes, has shown remarkable inter-annual variability between -0.11 ± 0.25 Pg C yr^{-1} and -0.59 ± 0.23 Pg C yr^{-1}, which in turn, contribute significantly to the global variation in air-sea CO_2 fluxes. These variations were in good conformity with the ENSO-induced fluctuations (Feely et al. 2006) of the SST, especially in the equatorial pacific, and the El Niño Modoki oscillations (Ashok et al. 2007). The annual variability in the Atlantic was comparatively lower than that

Table 2.1 The decadal mean global ocean CO_2 sink reported by the recent most studies (in Pg C yr^{-1}; 1 Pg = 10^{15} g)

1960–1969	1970–1979	1980–1989	1990–1999	2000–2009	2009–2018	Authors	Remarks
1.0±0.6	1.3±0.6	1.7±0.6	2.0±0.6	2.2±0.6	2.5±0.6	Friedlingstein et al. (2019)	Methodology of Denman et al. (2007); approved by Ciais et al. (2013)
			2.03±0.20	2.21±0.21	2.55±0.24	Hauck et al. (2020)	The simulation that took into account climate and atmospheric CO_2 variability (A)
			1.96±0.27	2.15±0.28	2.50±0.31	Hauck et al. (2020)	A and bias correction applied
			1.97±0.30	2.17±0.31	2.52±0.34	Hauck et al. (2020)	A and bias and area-weighted corrections applied
			1.99±0.25	2.17±0.26	2.52±0.29	Hauck et al. (2020)	A and bias and area-weighted corrections applied + FESOM-REcoM biogeochemical attributes
			2.32±0.18	2.44±0.14	3.09±0.10	Hauck et al. (2020_	Area-weighted data products

Note All the estimates provided in this table include thsse river adjustments of 0.78 Pg C yr^{-1} (Resplandy et al. 2018)

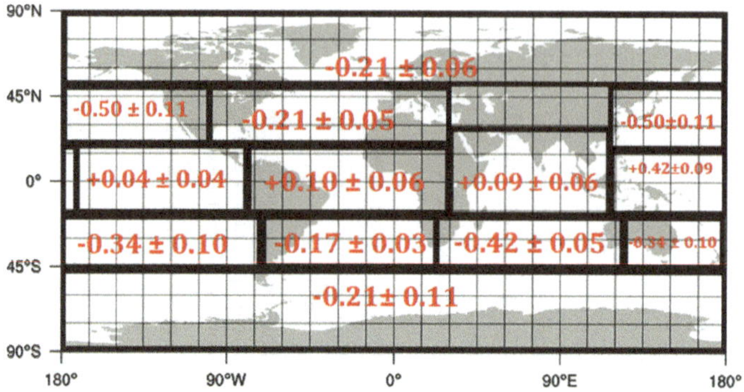

Fig. 2.1 The mean air-sea CO_2 fluxes observed in the major oceanic basins by Landschützer et al. (2014) during the years 1998 to 2011

observed in the pacific. However, the mean sink strength of the Atlantic often superseded the pacific with peak annual mean magnitudes like -0.71 ± 0.16 Pg C yr^{-1} (Landschützer et al. 2013). Though the North and the South Atlantic exhibit almost similar inter-annual variability, overall, the net sink strength of the North Atlantic was observed to be higher than that of the South Atlantic (Landschützer et al. 2014). Since the onset of the twenty-first century, the Southern ocean has exhibited significant net sink strength varying between -0.02 ± 0.10 Pg C yr^{-1} and -0.57 ± 0.13 Pg C yr^{-1}, with very little inter-annual variability (± 0.06 Pg C yr^{-1}). Though the Southern ocean is the most data-deficient ocean in the world, these observations were in parity with the observations recorded by Lenton et al. (2013). Lastly, the Indian Ocean with the least surface area among the other oceans exhibited a modest sink strength varying between -0.25 ± 0.07 Pg C yr^{-1} and -0.41 ± 0.09 Pg C yr^{-1}, with the least inter-annual variability ($^{\pm}0.02$ Pg C yr^{-1}). However, on the temporal scale, the Indian Ocean has so far exhibited the weakest increase in the carbon sink strength in comparison to the other oceanic regions (Landschützer et al. 2014).

2.2 Paleo-Oceanographic and Earlier Studies on the Carbon Dynamics of the Indian Ocean

Though the Indian Ocean received substantial attention from the perspective of characterization of its carbon dynamics, studies focusing on its role during the glacial-interglacial era are scarce. In the mid of the twentieth century, initial measurements and depth profiling in the Indian Ocean strongly indicated that the northern regions of this oceanic basin emitted CO_2 towards the atmosphere and the southern counterparts acted as moderate sinks of CO_2. In one of the earliest studies conducted by Ivanenkov (1964), the surface pCO_2(water) in the Indian Ocean was found to vary between 310 and 430 μatm. The depth profiles of Ivanenkov (1964) along a north–south transect in the 71° E longitudes indicated higher pCO_2(water)at intermediate depths, and near the Arabian Sea, the pCO_2(water) magnitudes at these intermediate depths reached 1000 μatm. Keeling and Waterman (1968) while characterizing the latitudinal gradient in surface pCO_2(water)in the Indian Ocean observed that the region north of 10° S acted as a source of CO_2 and the other half in the south of 10° S acted as a sink for CO_2. Depth-wise the most prominent increase in pCO_2(water) was observed within the thermocline layer and the highest difference between pCO_2(water) and atmospheric CO_2 concentration [pCO_2(air)] was observed in the regions where the monsoon gyres were active (Batrakov et al. 1981). The stagnation of the water layer and exceedingly high rate of production of organic carbon in these intermediate layers due to the high rate of photosynthesis were held accountable for this enhanced pCO_2(water) (Naqvi and Reddy 1975). A paleo-oceanographic study by Socci (1986) indicated that before the last glacial maxima, the primary productivity rate in the Indian Ocean was substantially low, which started increasing along with the surface circulation about 9000 years ago. Socci (1986) also concluded that the productivity levels during the last glacial maximum have remained more or less the same in the Holocene phase. Sengupta and Naqvi (1984) pointed out that the pCO_2(water) maxima inevitably coincided with the oxygen minima in the northern Indian Ocean. Anoxia is an indicator of remarkable oxygen utilization and enrichment of organic matter. Sarkar et al. (1993), through their oxygen and uranium isotopic studies, inferred that the Arabian Sea was significantly devoid of oxygen at the intermediate depths during the entire last glacial period. This indicates that a substantial part of the Indian Ocean lying near the equator has long been acting as a source of CO_2. In due course of time, these anticipations were confirmed by Bentaleb and Fontugne (1998).

They observed that during the Holocene period, the surface water pCO_2(water) was near-neutral in comparison to the then (pre-industrial revolution) atmospheric CO_2 concentration of 270 μatm. However, during the glacial and deglacial phases for the last 40,000 years before the Holocene, the surface pCO_2(water) surpassed the pCO_2(air) by about 50 to 100 μatm, indicating a strong CO_2 source character of the Indian Ocean during that era (Fig. 2.2). They also inferred from their study that the biological pump was not strong enough in the Indian Ocean to reduce the atmospheric CO_2 concentration. Thus, overall, it can be concluded that these studies that the character of the Indian Ocean has not changed much in the last 50,000 years from the perspective of CO_2 dynamics, though the paleoceanographic observations are biased for the northern Indian Ocean. These observations led Naqvi et al. (1994) to infer that though the indo-pacific regions witnessed a significant shift in biogeochemistry during the glacial to Holocene transition, it had little role to play in contributing to the changes observed in the atmospheric CO_2 concentration observed in that era.

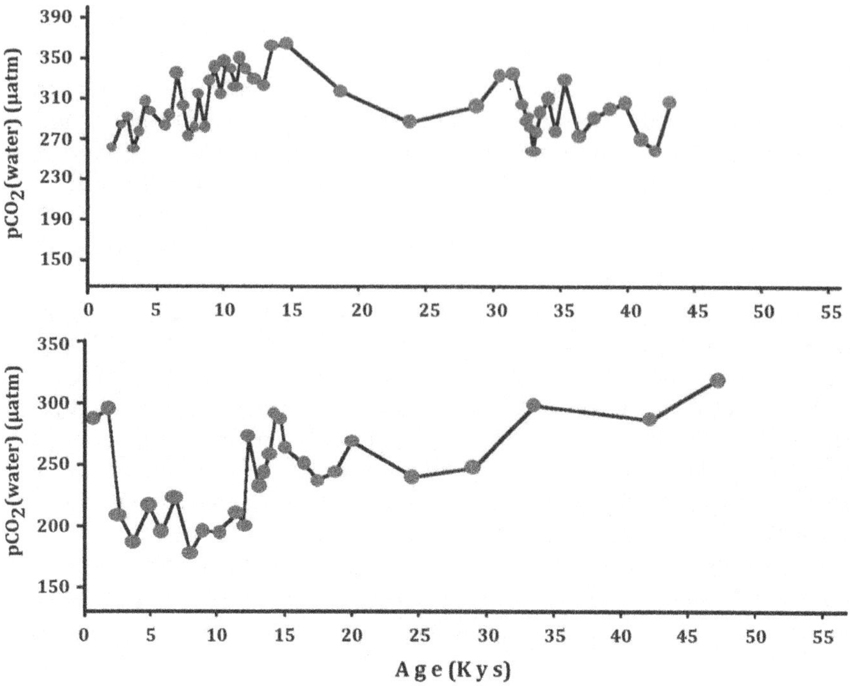

Fig. 2.2 The variation of pCO_2(water) in the sea surface against calibrated ages observed from the marine cores of the southern Indian Ocean (recreated from Bentaleb and Fontugne 1998)

2.3 Anthropogenic Carbon Storage in the Indian Ocean

The major significance of the global oceanic bodies from the perspective of the ongoing climate change is their capability to absorb CO_2 (He et al. 2018). Robust estimates indicate that approximately one-third of the anthropogenic CO_2 emission is absorbed by the global oceans; however, reducing the uncertainties along with the drifts and biases of the models that derive such estimates continues to be a challenging endeavour (Khatiwala et al. 2013). The southern ocean has perhaps received the most attention from this perspective (Langlais et al. 2017; Terhaar et al. 2021). However, a few studies in the last two decades tried to characterize and understand the role of the Indian Ocean in absorbing anthropogenically-emitted CO_2. Gruber et al. (1996) engineered a model perhaps, the first of its kind to estimate the anthropogenic carbon inventory in the oceanic waters. The simplified form of the model measures anthropogenic CO_2 dissolved in ocean water according to the formula: C_{ant} (μmol kg^{-1}) $= Ct - \Delta C_b - C_{atm} - \Delta C_{dis}$, where Cant denotes the anthropogenic carbon concentration in the oceanic water column, Ct denotes the total carbon concentration in the water column, ΔC_b denotes the changes in the total carbon dioxide levels (T_{CO2}) due to the biological processes that take place in the oceans, ΔC_{dis} denotes the difference in the CO_2 concentration between the sea surface and the immediately adjacent atmosphere. Using this method, Sabine et al. (1999) characterized the anthropogenic CO_2 inventory at the end of the twentieth century. They observed that the subtropical convergence zone (30° S to 40° S) in the Indian Ocean had the highest anthropogenic CO_2 concentrations. The depth profiles in these zones showed that the anthropogenic CO_2 went deep down the water column and such deep penetration was absent in the other regions. Whereas, the regions lying to the south of the 50° S latitudes exhibited the poorest accumulation of anthropogenic CO_2. Combining their results for the different regions, Sabine et al. (1999) estimated anthropogenic CO_2 storage of 20.3±3.0 Pg C till the year 1995 (Fig. 2.3). However, Sabine et al. (1999) also laid the stress to the fact that these estimates might incorporate considerable uncertainties, as model derive estimates indicated that the sink potential of the Indian Ocean to the north of 35 °C could be one-half of the estimates furnished by this study, and the southern counterpart can store almost 2.6 times more carbon than that observed by Sabine et al. (1999). However, Sabine and Feely (2001) compared the methodology adopted by Sabine et al. (1999) and the classical approach of Chen and Millero (1979) and observed that the latter has some significant advantages over the former.

In the following years, Hall et al. (2004) used tracers like chlorofluorocarbon to remove the data constraints under varying mixing scenarios and inferred that the Indian Ocean could have stored anthropogenic CO_2 in the range of 14.3–20.5 Pg C till the year 2000. The upper limit of this range was in parity with the earlier observations; however, Hall et al. (2004) emphasized that restricted mixing in the water

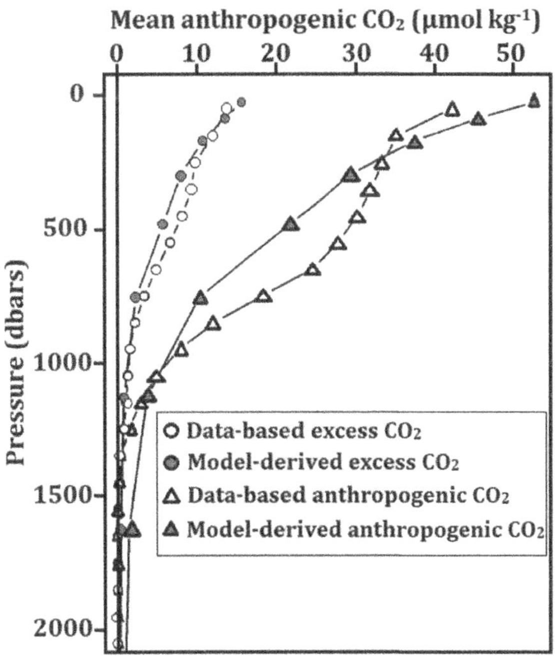

Fig. 2.3 The depth profile of data-based (open marker) and model-derived (solid marker) area-weighted anthropogenic CO_2 concentration in the Indian Ocean region lying north of the 35° S latitudes. The triangles denote the total increase since the pre-industrial revolution levels and the circles denote the increase since the GEOSECS measurement that ended in 1995 (recreated from Sabine et al. 1999)

column could substantially lower the estimates, which was not reported earlier. In the subsequent years, Álvarez et al. (2009) compared five different methods to quantify the anthropogenic CO_2 inventory in the Indian Ocean. Though they did not furnish any basin-wide upscaled estimates, in Terms of CO_2 concentration per unit area, their results ranged between 23.9±2.0 mol C m^{-2} to 30.8±2.5 mol C m^{-2}. Goyet and Tourantier (2009) also appreciated the challenges in drawing realistic estimates of the anthropogenic CO_2 inventory in the Indian Ocean, given the complex hydrodynamics of this ocean basin. They inferred that the two flanks of the northern Indian Ocean, namely the Arabian Sea and the Bay of Bengal, exhibited significant differences in anthropogenic CO_2 absorption. However, they agreed with many earlier studies that the southern counterpart of the Indian Ocean is capable of storing anthropogenic carbon up to great depths. Valsala et al. (2012) identified two potential windows; one lying between the 15° S and 35° S and the other between 35° S and 50° S latitudes in the Indian Ocean, where active CO_2 absorption was found to take place. They also inferred that the solubility pump played a greater role than the biological pump in these windows. More recent information is yet to be generated from the Indian Ocean as the principal focus of the scientific community since the last decade has shifted exclusively to the southern ocean.

2.4 Inorganic Carbon Dynamics in the Indian Ocean

Dissolved inorganic carbon (DIC), which includes the dissolved CO_2, carbonate, and bicarbonate ions, plays a crucial role in regulating the carbon-absorbing capacity of any oceanic water column and phenomena like ocean acidification (Xue and Cai 2020). The DIC dynamics, especially the vertical flux along the water column, exhibits significant decadal variability, throughout the global oceans, which makes it a crucial parameter, as most of the CO_2-oriented estimates include DIC (along with total alkalinity (TAlk), and pH) as the chief parameter (Doi et al. 2015). Recently, Broullón et al. (2020) have furnished a monthly climatological data repository for the total DIC in the global oceans having a spatial resolution of 1° × 1°. Their model-derived estimates were in parity with the other climatologies like that of Takahashi et al. (2014) and Lauvset et al. (2016). They recorded that unlike most of the other oceans of the world, the Indian Ocean and some parts of the pacific as well exhibited a lower concentration of DIC in the higher latitudes. Prominent endeavours of exclusive measurements of DIC in the Indian Ocean dates back to the end of the last century (Johnson et al. 1998), which ended up measuring almost 19,000 samples. In the following years, Sabine et al. (2002) conducted almost 25,000 measurements throughout the length and breadth of the Indian Ocean.

They inferred that the variability of DIC with depth all along the water column did not follow the temperature or salinity gradients, as the biological processes were significantly different across the locations sampled, and it successfully interfered with the circulation-induced fluctuations in DIC. They also recorded that the anthropogenic CO_2 emission enhanced the DIC levels near the water surface, whereas the benthic DIC levels were essentially found to increase in the higher latitudes. Bates et al. (2006) extrapolated the DIC values throughout the Indian Ocean and observed that in the northern Indian Ocean, the Arabian Sea showed fairly high DIC (>2000 μmol kg^{-1}) in the surface layers, whereas the Bay of the Bengal (<1900 μmol kg^{-1}) and the region lying between 0° and 15° S (<2000 μmol kg^{-1}). Down south, near the south equatorial current and subtropical front ranging between 15° S and 20° S, DIC concentrations occasionally surpassed 2100 2000 μmol kg^{-1}. However, the seasonal maximum was observed in the regions lying between 30° S and 45° S, especially during the southwest monsoon and the retreating monsoon (Bates et al. 2006) (Fig. 2.4). It is interesting to note that though the Bay of Bengal and the Arabian Sea exhibits lower DIC than that observed in the southern latitudes, the coastal waters of these two northern flanks of the Indian Ocean receive almost 10.3 Tg C yr^{-1}, as DIC (Krishna et al. 2019). Out of this magnitude, almost three-fourths is received by the Bay of Bengal and the rest by the Arabian Sea. Sarma et al. (2018) observed that the East India Coastal Current plays a crucial role in regulating the export fluxes of DIC to the Bay of Bengal.

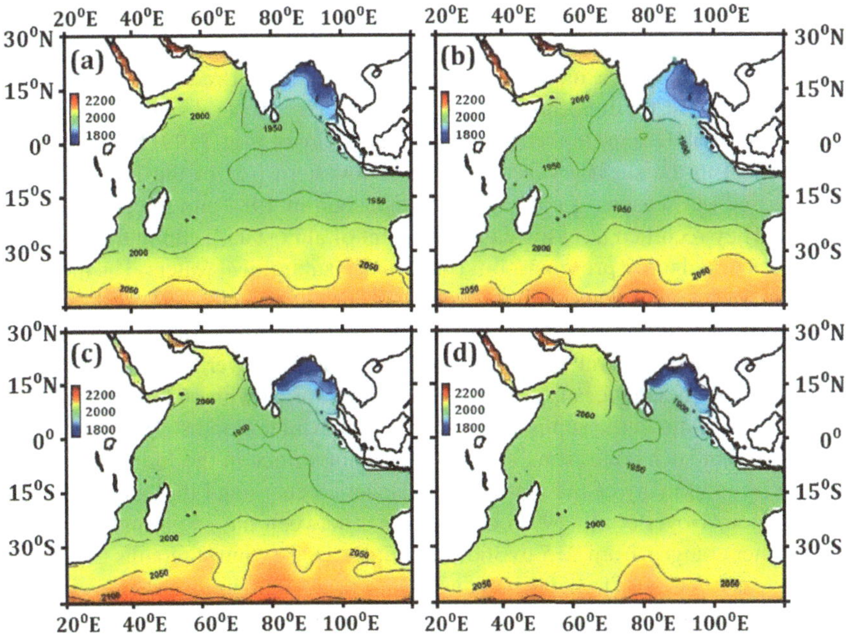

Fig. 2.4 The spatial variation of DIC (μmol kg-1) in the entire Indian Ocean during **a** during the northeastern monsoon (December-February), **b** spring inter-monsoon (March–May), **c** southwest monsoon (June–August), and **d** fall inter-monsoon (September–November) (recreated from Bates et al. 2006)

2.5 pCO_2(Water) Variability: Monthly to Seasonal to Inter-Annual

pCO_2(water), especially on the sea surface, is ultimately the most crucial parameter which principally governs the source/sink nature of the oceans. This parameter depends on a suite of biogeochemical processes including the physical forcings as well as the biological phenomena that incessantly takes place in the oceans with varying rate and magnitude from seconds to millennia (Gloege 2020; Krishna et al. 2020; Chen et al. 2021; Valsala et al. 2021). The sea-surface temperature and salinity exhibit significant spatiotemporal variability in the Indian Ocean, both on the seasonal scale and the multi-decadal scale (Ma et al. 2020; Mohapatra et al. 2020). Moreover, the drastic meteorological shifts and the hydrographic traits of this region, including the ocean currents, which undergo significant seasonal changes make this ocean basin one of the most biogeochemically dynamic regions of the world (Francis et al. 2020; Sen et al. 2020; Vianello et al. 2020). Several pieces of literature documented the efforts of characterizing the spatiotemporal variability of pCO_2(water) in different pockets of the Indian Ocean. The Geochemical Ocean Section Study(GEOSECS) in the 1970s and 80s was perhaps the first worldwide endeavour to document the variability of a suite of physicochemical and hydrographic parameters throughout the world oceans, which included the Indian Ocean also. Bradshaw et al. (1981) reported the total CO_2 concentration and alkalinity from several stations monitored within the Indian Ocean. In the following decade, Goyet et al. (1991) measured the pCO_2(water) in the surface waters of the southwestern part of the Indian Ocean. They carried out seasonal sampling in the monsoon of 1984 and the pre-monsoon of 1985. They recorded a seasonal enhancement of almost 45 μatm in the monsoon season compared to the pre-monsoon season, in the subtropical waters. The monsoon season, in this part of the world, comprises the summer months, whereas, the pre-monsoon season includes the winter months. Thus, they attributed the increase in SST to the rise in the pCO_2(water) in the summer months. However, in the sub-antarctic portions of the Indian Ocean, a reverse scenario prevailed with pCO_2(water) higher by at least 25 μatm in the summer compared to the winter months. The authors also mentioned that such contrasting behaviour was not unique in the Indian Ocean but also observed in many of the northern hemispheric oceans and that this is a global phenomenon. Goyet et al. (1991) observed that between the 23° S and 53° S, the pCO_2(water) ranged from 315 to 360 μatm during the summer months (minimum at 33° S), and from 310 to 350 μatm during the winter months (minimum at 40° S). In the following years, George et al. (1994) measured the pCO_2(water) along the central Arabian Sea and the central Bay of Bengal in the pre-monsoon season of 1991_92. They observed that the mean pCO_2(water) in the Arabian Sea was higher than that observed in the Bay of Bengal, though the maximum magnitudes were observed in the latter. They observed that the subsurface pCO_2(water) showed an increase near the surface and gradually decreased with depth; however, in the Bay of Bengal the highest pCO_2(water) was found to coincide with the highest apparent oxygen utilization, whereas in the Arabian Sea there was no such scenario. This indicates that the Bay of Bengal pCO_2(water) dynamics is to some extent governed by the biological pump, whereas, in the Arabian Sea, the physical pump played a major role. In the subsequent years, Louanchi et al. (1996) attempted to model the monthly fugacity of carbon dioxide (fCO_2)in four locations of the Indian Ocean, with special emphasis on the 20° S to 50° S, taking into account all the physical and biological factors like SST, productivity, salinity, mixed-layer depth, and wind circulation. Their model-derived extrapolation for the entire Indian Ocean was in parity with the earlier observations. Louanchi et al. (1996) observed that in the northern Indian Ocean, the fCO_2 peaked from March onwards till May, and this peak shifted to June near the equator. Whereas, in the southern latitudes, the peak was observed at the end of the monsoon (October–November) or in the winter months of December-January. Jabaud-Jan et al. (2004) observed the variations in fCO_2(water) between the latitudes 20° S and the 60° S in the southern parts of the Indian Ocean. They observed significant inter-annual variability in fCO_2(water). They documented the surface fCO_2(water) variability in the permanent open ocean zone (50° S to 57° S) and the seasonally ice-covered zone (58° S to 60° S) and observed a mean difference of −14.6 μatm between the January data for the years 1998 and 2000, in the former, and a mean difference of +37.9 μatm in the later. This data shows the challenges in predicting and forecasting the variability of surface ocean μatm fCO_2(water) and the degree of its spatiotemporal variability in the Indian Ocean. Between the years 1991 and 2007, Metzl (2009) observed a significant increase in fCO_2(water) in the southern Indian Ocean, which varied between 0.62 and 4.72 μatm yr^{-1}. The 40° S to 42° S region exhibited the highest rate of increase in fCO_2(water), whereas the immediately adjacent 35° S to 40° S region exhibited the lowest increase.

Overall, Metzl (2009) observed an increase in the rate of enhancement of fCO_2(water) from the north towards the south in the Indian Ocean. In the last decade, Xue et al. (2014) observed that in the eastern equatorial Indian Ocean, the summer surface pCO_2(water) increased from 307 to 385 μatm from 1963 to 2012. The rate of increase of surface pCO_2(water) (1.56 μatm yr^{-1}) was comparable with the rate of increase in atmospheric CO_2 concentration (1.46 μatm yr^{-1}) during the same span. Shetye et al. (2015) observed the depth profiles as well as surface pCO_2(water) in the southern Indian Ocean during the austral summer-to-winter transition phase in January, February, and March in the years 2010 to 2012. They observed that mean pCO_2(water) was lowest during January and reached a peak in March. The surface pCO_2(water) values also followed the same trend (Table 2.2). They observed a significant negative relationship between the pCO_2(water) and the dissolved oxygen and further inferred that the pCO_2(water) was mostly governed by biological processes during January and February; however, in March, the physical processes like SST variability and changes in mixed-layer depth played an equally significant role in association with the biological processes. Despite many such observations, the pCO_2(water) measured data from the Indian Ocean remains insufficient to draw a holistic scenario about the entire basin (Wang et al. 2016). However, efforts to establish permanent data logging stations in the most cost-effective way to cover the entire Indian Ocean are on their way (Valsala et al. 2021), and we expect that

Table 2.2 A list of observed pCO_2(water) in the various regions of the Indian Ocean in the last four decades

Oceanic region	Time	Measurement type	pCO_2(water)	Authors
Southwest Indian Ocean 23° S to 53° S; 50° E to 76° E	July 1984	Surface waters	315–360	Goyet et al. (1991)
	February–March 1985		310–350	
Arabian Sea	February 1992	Depth profiles	Mean: 383	George et al. (1994)
Bay of Bengal	March–April 1991	Depth profiles	Mean: 352	
Southern Indian Ocean 50° S to 57° S	January 1998	Surface waters	341±6.7	Jabaud-Jan et al. (2004)
	January 2000	Surface waters	355.6±4.5	
Southern Indian Ocean 58° S to 60° S	January 1998	Surface waters	379.3±6.1	
	January 2000	Surface waters	341.4±8.8	
Equatorial Indian Ocean 2° N to 2° S; 90° E to 95° E	April 1963	Surface waters	Mean: 307	Xue et al. (2014)
	May 1999	Surface waters	Mean: 373	
	April 2007	Surface waters	Mean: 381	
	May 2012	Surface waters	Mean: 385	
Southern Indian Ocean 30° S to 66° S	March 2009	Surface waters	Mean: 337	Shetye et al. (2015)
	February 2010	Surface waters	Mean: 297	
	January 2012	Surface waters	Mean: 286	

very soon the problem regarding data insufficiency of this region to model the overall pCO_2 dynamics would be resolved.

2.6 Air-Sea CO_2 Exchange Dynamics

Quantifying air–water CO_2 fluxes in any aquatic body in a dynamic system remains a challenging task since the exchange depends upon several physical and biological parameters, which vary significantly in a short time. Assessing this exchange of CO_2 through the air-sea interface includes various methods; however, the two most promising and the most used methods are the bulk formula method (Wanninkhof and McGillis 1999) and the eddy-covariance technique (Miller et al. 2010). While the latter has become popular in the last decades, the former has been implemented in several studies and still finds its application. The estimation of air-sea fluxes requires the measurement of pCO_2(water) and pCO_2(air) to compute the ΔpCO_2 (Lerner et al. 2021). However, the trickiest and debated part of this estimation is assigning a suitable gas transfer coefficient (also known as gas transfer velocity) which when multiplied with the ΔpCO_2 leads to the rate of exchange of CO_2 per unit area per unit time (Liss and Merlivat 1986; Wanninkhof 1992; 2014; Ho et al. 2006). Thus, indirect measurement techniques like that of the Bulk formula method incurs substantial uncertainties in the flux estimation. However, in the present date, the uses of gas tracers have substantially minimized the uncertainties (Nightingale et al. 2000; Sweeney et al. 2007; Ho et al. 2011; Ho and Wanninkhof 2016).

Though pCO_2 measurement in the Indian Ocean dates back to the GEOSECS expedition in the 1970s, measurements of fluxes were mostly reported post-1990s. In the year 1991, Poisson et al. (1993) carried out several measurements between the 23° S and 35° S latitudes and reported the air-sea CO_2 fluxes for January, February, April, and May. They reported a steady increase in the CO_2 intake rate from January 1991 to July 1991 (Table 2.3). They inferred that a combination of physical as well as biological pumps governs the exchange dynamics of CO_2 across the air-sea interface in this region. The study conducted by George et al. (1994) was one of the first to characterize the air-sea CO_2 exchange rates in the northern Indian Ocean and they observed significantly high mean CO_2 effluxes from the Arabian Sea (2.90 mmol m^{-2} d^{-1}) compared to the Bay of Bengal (0.33 mmol m^{-2} d^{-1})based on their observed wind speed measurements and measured CO_2 concentration in the air. However, based on climatological wind speed and global average CO_2 concentrations the fluxes varied as 2.52 mmol m^{-2} $d^{-1}$for the Arabian Sea and 0.66 mmol m^{-2} d^{-1} for the Bay of Bengal. Combining the areas of the Bay of Bengal and the Arabian Sea (encompassing around 10,312,000 km^2), the area-weighted mean CO_2 flux from the northern Indian Ocean was computed to range from 1.090 mmol m^{-2} d^{-1} (based on climatological wind speed and air CO_2 concentration) to 1.381 mmol m^{-2} d^{-1} (based on measured wind speed and air CO_2 concentration) (George et al. 1994). During the years, 1991 to 1993, Metzl et al. (1995) carried out several measurements in the 20° S to 50° S meridional ranges of the Indian Ocean basin. They observed an increasing trend in the CO_2 sink rate from the north to the south. According to Metzl et al. (1995), the 20–35° S exhibited a mean CO_2 sink rate of −0.068 to −0.139 Pg C yr^{-1}, and the 35– 50° S exhibited a mean CO_2 sink rate of −0.128 to −0.229 Pg C yr^{-1}. They recorded the highest sink strengths during April–May, whereas the January–February and July–August months exhibited weaker CO_2 intake. Later on, Louanchi et al. (1996) modelled the air-sea CO_2 exchange rates for the entire Indian Ocean basin based on some measured data during the MINERVE ocean expedition programme for measuring pCO_2(water) in the Indian Ocean (Louanchi 1995). They observed that the Indian Ocean altogether acted as a sink for CO_2 at a rate of 0.5 Pg C yr^{-1}. However, they showed a clear latitudinal variability of CO_2 fluxes in the ocean basin. The northern latitudes up to 10° S acted as a source followed by a near-neutral scenario in the 10–20° S region. Further south the sink strength was found to increase considerably up to 50° S latitudes. Their observation was in parity with the earlier study of Metzl et al. (1995); however, the magnitudes of fluxes in the same meridional ranges were almost two times higher than the previous study.

Sabine et al. (2000) meticulously studied the seasonal fluxes in the entire region lying north of 36° S following the methodology of Takahashi et al. (1997; 1999) and their approach in the year 1995. They observed that despite a strong seasonal variability throughout the entire ocean basin, the Arabian Sea has consistently acted as a

Table 2.3 A list of air-sea CO_2 flux observations in various places of the open Indian Ocean

Oceanic region	Time	Measurement type	CO_2 flux	Units	Authors
Indian Ocean 20° S to 35° S	January 1991	Bulk formula	Mean: –0.51	mmol m^{-2} d^{-1}	Poisson et al. (1993)
	February 1991		Mean: –0.62		
	April 1991		Mean: –1.89		
	May 1991		Mean: –3.39		
	July 1991		Mean: –4.50		
Arabian Sea	February 1992	Bulk formula	Mean: 2.52–2.90	mmol m^{-2} d^{-1}	George et al. (1994)
Bay of Bengal	March–April 1991		Mean: 0.33–0.66		
Indian Ocean 20° S to 35° S	1991, 1992, 1993	Bulk formula	Mean: -0.068 to -0.139	Pg C yr^{-1}	Metzl et al. (1995)
Indian Ocean 35° S to 50° S			Mean: –0.128 to –0.229		
Indian Ocean 18° N to 10° S	1993	Remote sensing-in situ coupled	Mean: 0.169	Pg C yr^{-1}	Louanchi et al. (1996)
Indian Ocean 10° S to 20° S			Mean: 0.000		
Indian Ocean 20° S to 35° S			Mean: –0.147		
Indian Ocean 35° S to 50° S			Mean: –0.507		
Indian Ocean North of 36° S	1995	Bulk formula	Mean: –0.15	Pg C yr^{-1}	Sabine et al. (2000)
Western Arabian Sea	1994–1996	Bulk formula	Mean: 5.59	Tg C yr^{-1}	Sarma (2003)
Eastern Arabian Sea			Mean: 8.21		
Northern Arabian Sea			Mean: 76.26		
Southern Indian Ocean 50° S to 57° S	January 1998	Bulk formula	–3.2	mmol m^{-2} d^{-1}	Jabaud-Jan et al. (2004)
	January 2000		–1.3		
Southern Indian Ocean 58° S to 60° S	January 1998		3.2		
	January 2000		–3.8		

(continued)

Table 2.3 (continued)

Oceanic region	Time	Measurement type	CO_2 flux	Units	Authors
Indian Ocean 20° N to 60° S	1780	Bulk formula	−0.2	Pg C yr^{-1}	Hall and Primeau (2004)
	2000		−0.5		
Indian Ocean 30° N to 60° S	1995	Bulk formula	0.24	Pg C yr^{-1}	Bates et al. (2006)
Indian Ocean 30° N to 44° S	1990–2009	Bulk formula	Median of annual mean: −0.37±0.06	Pg C yr^{-1}	Sarma et al. (2013)

strong source of CO_2 and the equatorial Indian Ocean seasonally switches between source and sink. These results prompted exclusive studies in the Arabian Sea. Sarma (2003) assessed several datasets acquired from 1994 to 1996 and reported the spatial distribution of the CO_2 fluxes from the Arabain Sea. He observed that the Arabian Sea in totality emitted around 90 Tg C yr^{-1}. The CO_2 effluxes from the northern Arabian Sea were an order of magnitude higher than that observed in the eastern and the western Arabian Sea. Seasonally the southwest monsoon phase emitted almost 75% of the yearly CO_2emissions from all the regions. Jabaud-Jan et al. (2004) observed significant inter-annual CO_2 fluxes in permanent open oceanic zones and the seasonally ice-capped zones of the southern Indian Ocean. They inferred that SST, varying rates of primary productivity, the lateral flux of nutrients, and occasional phytoplankton bloom formations regulate the CO_2 fluxes in this region. Hall and Primeau (2004) estimated the CO2 fluxes for the year 1780, i.e. during the pre-industrial revolution era, and for the year 2000. They estimated that during the 1780s, the Indian Ocean altogether acted as a sink for CO_2; however, the sink rate was only 0.2 Pg C yr^{-1}. In the year 2000, the enhanced sink rate was estimated to be 0.5 Pg C yr^{-1}. They inferred that due to the anthropogenic emission the emission increased by almost 0.26 Pg C yr^{-1} between 1780 and 2000 (Fig. 2.5). While separating the natural and the anthropogenic CO_2 fluxes, they further observed that the natural fluxes in the present date are critically sensitive to the CO_2 disequilibrium in the present-day environment, whereas the anthropogenic fluxes are scarcely bothered. However, a contemporary study by Bates et al. (2006) reported the entire Indian Ocean as a net source of CO_2 having a mean CO_2 efflux rate of 0.24 Pg C yr^{-1}. The latitudinal variability observed by them was in parity with the observations of Sabine et al. (2000) that the northern latitudes acted as a source throughout the year; however, near the equator the source-sink character switched with the seasons. However, the net efflux magnitude of Bates et al. (2006) was in stark contrast with that of Sabine et al. (2000).

In the last two decades, several advanced technologies were being implemented to measure the CO_2 fluxes in the Indian Ocean. Kondo and Tsukamoto (2007) directly measured the CO_2 fluxes in equatorial India with the eddy-covariance technique along

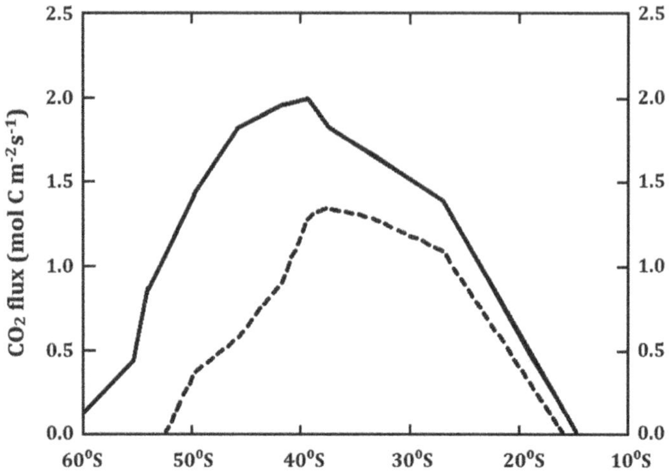

Fig. 2.5 The latitudinal variation of air–water CO_2 exchange along the Indian Ocean was estimated for the years 1780 (dashed line) and 2000 (dark black line) (recreated from Hall and Primeau (2004))

with concomitant bulk formula measurements. They observed that between the two stations they sampled, one station exhibited similarity between the eddy-covariance-based flux estimates and the bulk formula-based fluxes, and at the other station, there was no similarity at all. They inferred that the implementation of Webb correction as a result of changes in latent and sensible heat flux with varying wind velocity might have led to such changes and they advocated the need for measuring the verticle profile of atmospheric CO_2 in air above the sea level to remove the ambiguities and uncertainties in flux estimations through eddy-covariance technique. Chen et al. (2011) coupled field observations and remotely sensed images to draw a spatial scenario of the air-sea CO_2 fluxes in the southern Atlantic and the Indian Ocean. They utilized the empirical relationships obtained between pCO_2(water), SST, and chlorophyll-a observed in situ to compute the fluxes and prepare spatial maps with the help of GIS tools. They characterized significant differences between the south Atlantic and the Indian Ocean from their study. Sarma et al. (2013) compiled an array of models and inversions to compute the air-sea CO_2 fluxes for the entire basin of the Indian Ocean from 1990 to 2009. They computed a median annual mean CO_2 sink rate of -0.37 ± 0.06 Pg C yr^{-1}. They further pinpointed the lacunae in the data like underestimates of fluxes in the Somalian and Oman offshore waters due to overruling the effect of upwelling and in the subtropical convergence zone and overestimates in the Bay of Bengal. They also indicated that the sink strength enhanced marginally in the Indian Ocean and their observations were in parity with the earlier inferences that the Indian Ocean CO_2 concentration has increased at a faster rate than the increase in anthropogenic CO_2 emission. However, at the same time, they mentioned that compared to the degree of seasonal and inter-annual variability that this ocean basin has exhibited in the last decades, the Indian Ocean remains data deficient, and more studies are required in the future to gain confidence about the flux estimates.

2.7 Dissolved Organic Carbon

Pieces of evidence suggest that there could be as much carbon in the oceans in dissolved organic form (dissolved organic carbon; DOC) as there is in the atmosphere as CO_2 (Hedges 1992). However, their participation in the global carbon cycle and the transformation between different carbon species within the marine water mass is poorly constrained (Bercovici et al. 2018). Considerable uncertainties also exist in delineating the residence time of DOC in the various oceanic basins (Matsumoto 2007). As per the latest estimates, the global oceanic waters might hold 662 Pg C as DOC and substantial advancement has already been made regarding characterizing the chemical composition of this collective sub-head of carbon species (Hansell and Carlson 2002). Still, this carbon parameter remains the least studied of all, and the Indian Ocean is perhaps the most data deficient of all the other oceans in the world (Hansell 2009). One of the oldest DOC data from the Indian Ocean dates back to the year 1963. During October–November 1963, the fourth cruise expedition of Anton Frederik Bruun collected DOC data in the depths of 0 m to 2000 m and all along a north-to-south transect in the western Indian Ocean (Menzel 1964) (Table 2.4). Menzel (1964) noticed that there was no significant relationship between DOC and the rate of primary productivity. However, a north-to-south and depth-wise variability in DOC with the varying salinity made him infer that DOC could be strongly conservative in the Indian Ocean. He could differentiate primarily three distinct water masses during the survey through the transect, namely Indian Ocean Central Water, Arabian Seawater, and Equatorial water having characteristic

Table 2.4 A list of observations of dissolved organic carbon (DOC) across the Indian Ocean

Oceanic region/water mass	Time	DOC	Units	Authors
Indian Ocean Central Water	1963	1.5 to 2.0	mg l^{-1}	Menzel (1964)
Arabian Seawater	1963	0.2 to 0.4	mg l^{-1}	Menzel (1964)
Equatorial water	1963	0.6 to 0.8	mg l^{-1}	Menzel (1964)
Northwestern Indian Ocean	1986	49 to 309	μmol dm$-^{3}$	Kumar et al. (1990)
Central Indian Ocean (25–200 m)	1997	52 to 191	μM	Sardessai and de Sousa (2001)
Central Indian Ocean (500–800 m)	1997	41 to 57	μM	Sardessai and de Sousa (2001)
Central Indian Ocean (3500–5300 m)	1997	41 to 64	μM	Sardessai and de Sousa (2001)
Bay of Bengal	1995	73 to 75	μmol kg^{-1}	Hansell (2009)
Arabian Sea	1995	55 to 90	μmol kg^{-1}	Hansell (2009)
Agulhas current system	1995	40.8±0.3 to 71.0±4.5	μmol kg^{-1}	Hansell (2009)
Bay of Bengal	2013–14	75 to 100	μM	Shah et al. (2018)

different DOC concentration ranges. Overall, Menzel (1964) inferred that carbon in the form of DOC is substantially stable in the Indian Ocean, especially in the Arabian Sea, and it is going to be least disturbed by biological activities. Almost two decades later, Kumar et al. (1990) focused on the northwestern Indian Ocean (mainly encompassing the Arabain Sea) and reported a non-conservative nature of DOC, in contrast to the observations of Menzel (1964). They also reported that the overall concentrations observed by them (49 to 309 μmol dm^{-3}) were three to four times higher than those reported by the earlier studies. They observed a direct relationship between the oxidative decomposition of DOC and the depletion in oxygen levels and nitrate concentrations. They observed that the microbial community played a governing role in regulating the DOC concentrations as nitrogen cycling was closely associated with the DOC dynamics. Alike Menzel (1964), they also observed that the DOC concentrations were higher in the equatorial Indian Ocean. Later on, Hansell (2009) modelled the surface concentration variability of DOC, taking into account the autochthonous DOC production, lateral fluxes from the continental margins, and overturning phenomena. They validated their model with measured data from three sectors of the Indian Ocean, namely, the Bay of Bengal, the Arabian Sea, and the Agulhas current system. They observed that the mean DOC in the Bay of Bengal was significantly higher than that observed in the Arabian Sea. The perennial discharge from the Ganges–Brahmaputra system along with other Indian east coast rivers is principally responsible for these elevated DOC levels. They critically observed that six different water mass regulates the DOC dynamics in the Agulhas current system. They further reported that the photosynthetic activity produced substantial DOC and hence in the vertical profile, the highest concentrations were observed in the euphotic depth. The vertical profile of DOC was exclusively focused by Sardessai and de Sousa (2001) in the central Indian Ocean. The maximum concentration, as well as the maximum variability, was observed in the upper euphotic layers, whereas, in the deeper waters the DOC concentration showed negligible variation. Hansell and Carlson (1998) also observed similar observations in the year 1995 in several places of the Indian Ocean throughout the varying depths below 500 m.

This observation reaffirmed that the DOC is mostly refractory and is not easily perturbed by any biochemical activities throughout the water column. Sardessai and de Sousa (2001) further observed that though the minimum DOC coincided with the oxygen minimum zone in the depth profile, there was no significant relationship between the DOC concentrations and apparent oxygen utilization. This observation shows that apart from the oxidation of DOC other biochemical processes also simultaneously utilized dissolved oxygen to oxidize other forms of carbon. The last decade has witnessed several advanced research on this topic in the Indian Ocean. Bercovici et al. (2018) through radioisotope ^{14}C analysis, estimated the age of DOC in the various water mass of the southern Indian Ocean. They observed that the DOC in the polar front (4400 ^{14}C years) was older than the subtropical waters (2000 ^{14}C years). However, the DOC in the circumpolar deep waters was the oldest (5600 ^{14}C years) among all others. Shah et al. (2018) recently quantified the DOC concentrations along a north–south transect in the Bay of Bengal and reported substantially higher magnitudes than that documented in the earlier studies on this region. They held the perennial discharge from several Indian rivers responsible for these enhanced DOC levels and further observed that vertical mixing is substantially inhibited in the Bay of Bengal which enables the surface waters of this region to show considerably high DOC concentrations. Usually, DOC increases towards the equator; however, along the Bay of Bengal, a decrease was observed, most probably due to the steady diminishing of the riverine influence. They also concluded in line with Sardessai and de Sousa (2001) that apparent oxygen utilization had little role to play in the oxidation of DOC (only 18%). Krishna et al. (2015) quantified the lateral fluxes of DOC from twenty-seven Indian estuaries and estimated that these estuaries transfer 2.37±0.47 Tg C as DOC to the northern Indian Ocean, out of which roughly 75% is received by the Bay of Bengal, and the rest goes to the Arabian Sea.

2.8 Deep Oceanic Carbon Burial

Several oceanographers have pointed out that the deep oceanic burial of organic carbon, though a slow process compared to the other biogeochemically dynamic fluxes that take place at various interfaces could act as a long-term sink for carbon (Galy et al. 2007; Owens et al. 2018). However, measurement of vertical organic carbon fluxes into the deep benthic oceans is a tedious endeavour and requires sediment traps and preferably a long-term measurement schedule, which needs to be maintained and monitored (Haake et al. 1993; Rixen et al. 1996). Attempts of quantifying the deep oceanic organic carbon burial rate are very few. Sarma et al. (2007) observed that the carbon burial rate in the northern Indian Ocean, which many of the studies have recognized to mainly act as a source of CO_2 towards the atmosphere far exceeds the rates observed in the pacific ocean, where it is expected to be the highest in the world owing to the oldest age of the water mass. This crucial observation adds another feather to the uniqueness of this oceanic basin. They further mentioned that the sinking fluxes of DIC, particulate organic carbon, and calcium carbonate ($CaCO_3$) were substantially high in both the Arabian Sea and the Bay of Bengal compared to that observed in the Atlantic and Pacific. They measured that the sinking rate of POC (below the depth of 2500 m) varied from 3.0 to 13.3 mg C m^{-2} d^{-1} and 5.6 to 7.3 mg C m^{-2} d^{-1} in the Arabian Sea and the Bay of Bengal, respectively. Whereas, the same varied between 0.07 and 6.5 mg C m^{-2} d^{-1}, and between 0.43 and 5.2 mg C m^{-2} d^{-1} in the Atlantic and Pacific, respectively. Sarma et al. (2007) also emphasized that besides these high sinking rates, the rate of remineralization and bacterial respiration are also substantially high in this region. Overall, this region has a strong biological pump, which can facilitate the storage of substantial quantities of carbon in the deeper layers. Between the two flanks of the northern Indian Ocean, Sarma et al. (2007) stressed the Arabian Sea which stores substantial carbon in the form of DIC (between 6 and 395 $^{\mu mol\ kg-1}$)in the deeper layers. In line with these observations, Rixen et al. (2019a) made some interesting observations from the sediment trap data acquired from 1987 to 2002 in the northern Indian Ocean. They inferred that the land-derived ballast effect of the lithogenic matter played a

Table 2.5 The composite mean organic carbon fluxes (OC Flux; in g C $m^{-2\,yr-1}$) observed at various sediment trap stations based on the data from the years 1986 to 2002

Location name	Lat (°N)	Long (°E)	Depth (m)	OC flux
Western Arabian Sea trap station	16.26	60.58	4032	3.1±13.9
Central Arabian Sea trap station	14.51	64.72	3920	2.3±8.8
Eastern Arabian Sea trap station	15.57	68.73	3791	2.1±15.7
Northern Arabian Sea trap station	19.98	65.73	3147	
East Pakistan trap station	24.77	65.82	1093	11.9
West Pakistan trap station	24.60	65.59	1900	
North-eastern Arabian Sea trap station	16.93	67.84	3545	
Southern Arabian Sea trap station	11.60	66.08	4243	2.3
Equatorial Indian Ocean trap station	3.56	77.78	3400	1.4±20.6
Northern Bay of Bengal trap station—North	17.42	89.65	2267	2.9±12.2
Northern Bay of Bengal trap station—South	15.48	89.45	2709	2.2±0.4
Central Bay of Bengal trap station—North	13.14	84.41	3266	2.9±11.6
Central Bay of Bengal trap station—South	11.03	84.43	3462	2.0
Southern Bay of Bengal trap station	5.09	87.26	3995	2.3±12.7
Java mooring	−8.28	108.02	3250	4.4±7.3

crucial role in enhancing the vertical carbon fluxes in this region. They specifically observed in locations where there was no influence of the flowing rivers the ballast effect enhanced the vertical fluxes by 45%, whereas the same went up to 62% in the river-influenced regions of the northern Indian Ocean. These observations made Rixen et al. (2019a) infer that the future changes in the riverine and estuarine transport of carbonaceous matter might play a crucial role in governing the CO_2 sink potential of the oceanic region which receives such terrestrial inputs. Analyzing the data of 15 sediment trap sites in the northern Indian Ocean, Rixen et al. (2019b) inferred that the monsoon also plays a deciding role in enhancing the vertical carbon fluxes in this region. The monsoon-induced rainfall facilitates higher freshwater discharge from the adjoining rivers, and hence elevates the nutrient levels in the surface waters, which leads to higher productivity, and the lateral input of terrigenous matter accelerates the sinking rate of organic into the deep benthos. Based on a seventeen-year-long study, Rixen et al. (2019b) observed that the vertical fluxes of organic carbon towards the deep sea varied between 1.4 and 11.9 g C $m^{-2\,yr-1}$ (Table 2.5).

2.9 Indian Ocean Dipole and Carbon Biogeochemistry

An ocean–atmosphere coupled phenomenon like the ENSO was discovered by Saji et al. (1999) in the equatorial Indian Ocean basin, wherein they observed a difference in SST between the western Indian Ocean and the eastern Indian Ocean off Java coast and their variability in terms of both nature and magnitude regulates the SST of the overall Indian Ocean basin. The phenomenon based on the SST anomaly between the western and the eastern equatorial Indian Ocean was termed the Indian Ocean Dipole Mode. A positive dipole mode denotes warmer western waters than normal SST accompanied by cooler eastern waters than normal, and a negative dipole is just the reverse scenario. Follow-up studies like Saji and Yamagata (2003) and Ashok et al. (2003) claimed a substantial part of the fluctuations in dipole mode takes place independent of ENSO and it regulates the climate in many of the Indian Oceanic rim countries. Vinayachandran et al. (2002) emphasized that the IOD events govern the wave dynamics throughout the Indian Ocean basin and stimulate the phenomenon like upwelling and downwelling. Since several studies established that this oceanographic phenomenon has been taking place for a long time, (and that too independent of other global phenomena) and is capable of regulating the physical ocean mixing and wave dynamics, the question that was lurking that whether this phenomenon in some way regulate the carbon biogeochemistry of the basin or not. However, very few studies have been attempted to find a proper answer to this question. Sarma (2006) was perhaps the first to comment on this issue. He aimed at analyzing the correlation between the phytoplankton primary production and air-sea CO_2 flux with the changing dipoles in the Arabian Sea. He observed that positive SST anomalies during the IOD events led to a 30% reduction in the rate of primary productivity, which consequence an increase of 10 to 20 μatm in pCO_2(water). However, due to the weakening of the equatorial winds during these events, the air-sea fluxes reduced to a great extent also. These observations led Sarma (2006) to co-infer that primary production is the dominant process in the Arabain Sea that governs the pCO_2(water) magnitudes in the sea surface. In the Bay of Bengal, Sarma et al. (2015) observed that IOD events can significantly regulate the monsoon-induced rain and hence the river discharge from the perennial rivers that drain into this bay. They observed that a positive IOD event reduced the river discharge and thus, the coastal waters experienced a higher degree of upwelling which reduced the SST, and simultaneously decreased the pH by almost 0.01. This decrease in pH increased pCO_2(water) by about 60 μatm. Such observations indicate that IOD can play a crucial role in ocean acidification as well as in delineating the potential of coastal seawaters in absorbing or emitting CO_2. Recently, Valsala et al. (2020) modelled the impact of IOD in the carbon cycling in the surface waters of the Indian Ocean in the last six decades. They inferred that IOD can significantly alter the carbonate chemistry variables like TAlk (by ± 22 μmol kg^{-1}), DIC (by ± 35 μmol kg^{-1}), pCO_2(water) (by ± 20 μatm), and hence, the air-sea CO_2 fluxes (± 1 mC m^{-2} y^{-1}). Thus, overall, it can be inferred that the IOD events should be studied more to understand the future behaviour of this ocean basin to characterize its future CO_2 absorption potential.

2.10 Autotrophic Phytoplankton Community Dynamics

Phytoplankton, though kept outside the purview of the conventional blue carbon framework, in the strictest sense, plays a crucial role in the biological pump of CO_2 in any aquatic body, and the same is true for oceans also. Lovelock and Duarte (2019) advocated that at present there is a lack of understanding of how the phytoplankton stores carbon and mitigates greenhouse gas emission; however if we can overcome this lack in understanding these communities can very well be a part of the blue carbon ecosystems. Recently, Naik et al. (2020) observed the spatial variability in the phytoplankton community from the northern to the southern latitudes. They observed region-specific distinct dominance of the type of phytoplankton across the length of this ocean basin. The prokaryotes were found to dominate near the equator and the south of the equator, whereas the dinoflagellates were abundant in the north of the equator and the subtropical front. The diatoms on the contrary were thickly populated near the poles. They further inferred that an array of oceanographic parameters like SST, nutrient availability, salinity, and mixed-layer depth play a crucial role in regulating the species composition and the minute temporal variation in their abundance. Naik et al. (2015) previously studied the short-term temporal variability in the size fraction of the phytoplankton abundance in a vertical profile in the subtropical front and the Agulhas return front. The depth profiles (studied by pigment and chemotaxonomic analyses) showed an overall dominance of the picoplankton (i.e. the dinoflagellates); however, at subsurface depths, rapid changes among the three types of plankton, i.e. nano, pico, and micro were noticed with concomitant changes in the nutrient concentrations. The role of the grazer community was also found to be a significant role in governing the phytoplankton species composition in a place at a given time. Naik et al. (2015) further reported that a combination of two factors, top-down (regulated by zooplankton grazing) and bottom-up (upwelling or downwelling regulating nutrient availability) control the phytoplankton community structure in the Indian Ocean. However, over a long-term scale, Schubert et al. (1998) with the help of analyzing several biomarkers in the sediment columns retrieved from the Arabian Sea inferred that overall the phytoplankton species assemblage has not changed in the past 0.2 million years. They also concluded that the rate of organic carbon production by two of the most dominant groups, the dinoflagellates, and the diatoms, has also remained more or less the same in these years. Like Naik et al. (2020), Mengesha et al. (1998) also observed the dominance of diatoms in the seasonally ice-covered polar zones and the flagellates of various size categories in the permanently open sea zones. They further observed that phytoplankton biomasses were higher near the poles and gradually decreased towards the equator. Despite this variation in biomass, the community structure was more or less constant. The same was observed in the case of nitrogen uptake also. Though the nutrient regimes showed significant latitudinal differences and rate of nitrogen uptake, the same was not reflected in the species composition structure. Tarran et al. (1999), while working along a 1600 km transect in the Arabian Sea also observed that the overall species composition did not vary to a large extent. However, they observed some minute changes in the dominance between the monsoon and the inter-monsoon seasons. They also observed that the dominance of a particular type of phytoplankton (like diatoms, dinoflagellates, prokaryotes, or coccolithophores) depended on the trophic character of the seawater (Tarran et al. 1999). Thus, it can be inferred that the Indian Ocean basin has a stable phytoplankton community composition; however, some short-term variability in the composition is triggered due to an array of complex biogeochemical processes.

2.11 Biological Productivity

The autotrophic primary production is the fundamental process that absorbs CO_2 from the atmosphere. In the terrestrial compartments, the trees perform this process, and in the oceans, the billions and trillions of autotrophic primary producers are responsible for this phenomenon, which forms the basis of the biological pump of CO_2 dynamics in the oceans. The rate of primary production is usually defined as the unit carbon absorbed per unit area per unit time. Two terms are common in this domain; the gross primary production, and the net primary production. The former denotes the total carbon utilized in photosynthesis and the latter denotes the net carbon utilized after subtracting the carbon that is returned to the ocean through community respiration. Efforts to characterize the primary productivity of the Indian Ocean dates back to the 60s, wherein the International Indian Ocean Expedition collected the primary productivity data from various locations in the basin. Krey (1973) emphasized the importance of this parameter and advocated considering the physical forcings in this basin like the monsoonal rain, upwelling, and downwelling phenomenon to study the spatial variation in primary productivity. Later on, Qasim (1977) detailed a holistic measured data set for the entire Indian Ocean basin (25° N to 40° S) and observed that the average rate of primary production considering the water column up to the euphotic depth was 258 mg C m^{-2} d^{-1} (Fig. 2.6). They observed that due to the higher prevalence of cloud cover in the Bay of Bengal, the primary productivity in the sea surface of the Bay of Bengal was much higher than that observed in the Arabian Sea; however, the column productivity showed an opposite scenario. They also observed that the euphotic depth across the Indian Ocean basin varied between 40 and 120 m. However, in the coastal waters where turbidity remains substantially high the primary production is significantly compromised in comparison to what is observed in the open ocean waters. Varshney et al. (1983) reported an average water column primary productivity of 78.2 mg C m^{-2} d^{-1} off the Maharashtra coast (India) in the coastal Arabian Sea, where the euphotic depth never went beyond 2.5 m. During the southwest monsoon, the Bay of Bengal, on the contrary, despite having shallow euphotic depth and extremely high turbidity due to the river runoffs, exhibited substantially high primary productivity compared to the coastal waters of the Arabian Sea. Bhattathiri et al. (1980) observed a mean column primary productivity of 980 mg C m^{-2} d^{-1} in the August of 1978. Similarly, Devassy et al. (1983) observed the primary productivity in the Bay of Bengal to vary between

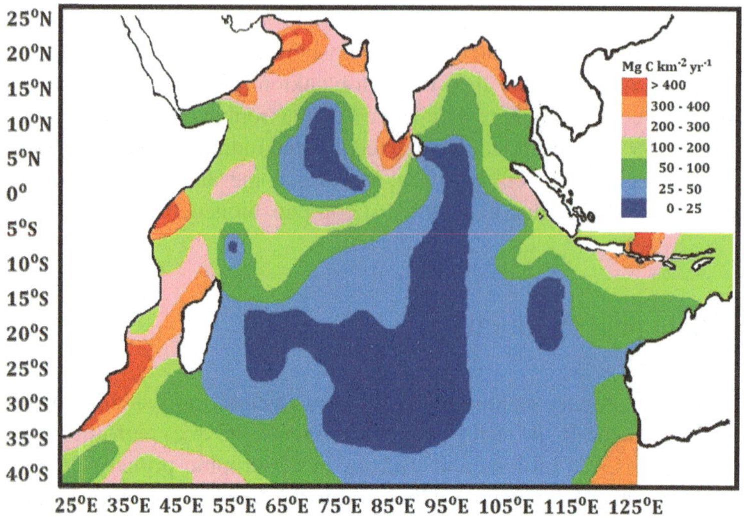

Fig. 2.6 The spatial distribution of primary productivity rates observed in the Indian Ocean (recreated from Qasim 1977)

140 and 5590 mg C m^{-2} d^{-1}. The substantial presence of nutrients in the Bay of Bengal has been attributed by most of the authors behind such high rates of primary productivity. Besides such spatial variability, the paleo-records have indicated that the average rate of primary production in the central Indian Ocean has varied between 70 and 390 g C m^{-2} yr^{-2} in the past 910,000 years (Beaufort et al. 1997). Sediment cores from the Arabian Sea also indicated such an oscillation in primary productivity in the past with comparatively higher magnitudes during the glacial phases and the lower during the interstadial phases (Schulte et al. 1999). Short-term seasonal variability in the primary production rates was also observed in the Indian Ocean. Veldhuis et al. (1997) observed that the average primary productivity rates in the Somali basin (the northwest Indian Ocean) varied between 0.8 and 1.25 g C m^{-2} d^{-1} during the southwest and northeast monsoon, respectively. Kumar et al. (2000) also observed that the physical-forcing-induced variability in the nutrient profile plays a crucial role in regulating the rates of primary productivity in the Indian Ocean. In this decade, several advanced model-based studies could accurately portray the holistic scenario of biological production throughout the ocean basin under upwelling conditions (Sreeush et al. 2018; 2020). Singh and Ramesh (2015) used tracers to characterize the new production rates in the Indian Ocean and they widened the number of parameters (as a part of the environmental control) that exhibits a significant relationship with the rate of primary productivity. They pointed out that SST, salinity levels, solar insolation, nitrate, phosphate, and silicate levels along with the concentration ratios between the nutrients can serve as indicators to characterize the f ratio, i.e. the ratio between new biomass production and the total biomass production. Thus, we can see that several pieces of research have revealed the characteristic traits of biological

production in the Indian Ocean basin; however, more endeavours are required to understand the role of phytoplankton in the long-term sequestration of carbon in this ocean reservoir if any.

2.12 Scope of Future Work

Compared to the coastal blue carbon habitats, the open oceans in the world altogether have received far less attention, as observed from the difference in the number of studies carried out in these two distinct sectors. It is undeniable that so far the scientific community has strongly established that the blue carbon habitats in the coastal periphery can store a massive amount of carbon and their rate of carbon removal from the atmosphere is much higher. However, the vast stretch of the oceans and the huge carbon reservoir can never be ignored. I believe that the global scientific community has realized or is about to realize very soon that drastic human intervention is required to enhance the natural carbon sink potential in whatever ecosystems lie around us. Under such circumstances, human beings are yet to mechanize fruitful avenues through which the open oceans can be utilized to enhance the carbon sink. The changes that we observed and continue to observe that take place in the open oceans are far beyond the reach of human beings to control or modulate. On the contrary, there are thousands of activities that can be done in ecosystems like mangroves, seagrass, tidal flats, and saltmarshes to protect, conserve, and enhance their carbon sequestration potential. This could be one of the prime reasons why open oceans are getting a cold shoulder compared to the coastal habitats. Moreover, research in the open ocean requires substantial funding and manpower. A collective effort including several nations is required to formulate joint expeditions, which in turn, can lead to successful data collections. Amidst the priorities of other research in the natural environment, the oceans might be lacking behind from the viewpoint of carbon biogeochemistry. The oceans are so vast that any amount of ground-measured data might seem insufficient to infer anything or about any phenomenon with confidence. Irrespective of all these factors, among all the other oceans of the world, the Indian Ocean is still considered one of the poorly sampled ocean basins of the world.

While collating the findings from the prominent studies that have been undertaken so far, I came to notice that some of the localized sectors of this basin have received substantial attention, whereas many of the regions are scarcely sampled. The prime point of interest in the present era of climate change is the role of the oceans in absorbing CO_2 from the atmosphere. More ground-measured data giving equal emphasis to the entire latitudinal and longitudinal stretch should be collected and there should be a provision for continuous monitoring in some selected sectors of this ocean basin (with the help of automated sensors and buoys) to properly understand the short to long-term variability in the carbon dynamics and other associated oceanographic parameters. In the present date, the initiative by National Oceanic and Atmospheric Administration (NOAA), USA has placed an array of moored buoys (known as RAMA buoys) in the Indian Ocean with an intent to study the monsoon pattern in Africa, Asia, and Australia. With every passing day, sensor-based ground measurements are witnessing newer technologies that are cost-effective and efficient in acquiring data. Parameters like salinity, pH, pCO_2(water), nutrients, chl-a, turbidity, and many more can be measured in one-second intervals. Such sensors should be deployed in networks of buoys to characterize the sea-surface dynamics. Automated depth profiling; however, remains a challenge, which we might overcome soon. Sediment traps should be placed in higher numbers to understand the capacity of this basin in sinking organic carbon to the deep sea, which serves as the long-term carbon sink in the oceans. Data gaps in the anthropogenic CO_2 inventory in the Indian Ocean should be minimized. Lastly, and most essentially, out-of-the-box thinking and approach are required to utilize this vast water reservoir in combating climate change.

2.13 Summary and Conclusion

Collating all the observations considered in this chapter, we can infer the following points. Though several studies are being conducted in the Indian Ocean basin with varying aims and objectives to characterize the carbon dynamics, this ocean basin is still undersampled in comparison to the other oceans of the world, hence more studies should be carried out in the future to minimize the data gaps remove the constraints in model-derived outputs. The observations from the last century till date indicate that this ocean basin, on the whole, acts as a sink for CO_2; however, the source/sink character varies significantly in both spatial and temporal scales. The northern latitudes are consistent sources of CO_2, whereas, as we move down south, the sink character strengthens. The Indian Ocean experiences a complicated seasonality in terms of water currents, wind flow, monsoon, and the inherent dipole mode. Physical forcing factors like upwelling, downwelling and tropical cyclones are also prominent in this basin. All these factors play a crucial role in regulating the partial pressure of CO_2 in the seawater and the water column as well, which in turn regulates the air-sea CO_2 fluxes. A combination of biological and physical pumps determines the carbon dynamics in the basin. A thorough understanding of all these physical, atmospheric, and hydrodynamic processes is required and should be taken into consideration while analyzing the carbon dynamics of this vast reservoir. This CO_2 concentration in this ocean is steadily increasing and at a rate higher than that observed in the increase in atmospheric concentration. This observation along with many others indicates that this ocean basin acts as an excellent inventory of anthropogenically-emitted CO_2 towards the atmosphere. Moreover, the sediment traps show that organic carbon both in dissolved and particulate carbon sinks in the deep seas, which serves as a long-term sink of carbon. Thus, altogether, several observations suggest that this basin is already playing and expected to play a substantial role in the future too, a crucial role in mitigating climate change. Hence, more endeavours are required from the scientific community to come up with novel inter-disciplinary ideas to best harness the potential of this ocean in combating climate change.

References

Alvarez M, Monaco CL, Tanhua T, Yool A, Oschlies A, Bullister JL et al (2009) Estimating the storage of anthropogenic carbon in the subtropical Indian Ocean: a comparison of five different approaches. Biogeosciences 6(4):681–703. https://doi.org/10.5194/bg-6-681-2009

Amante C, Eakins BW (2009) ETOPO1 1 Arc-minute global relief model: procedures, data sources, and analysis. NOAA technical memorandum NESDIS NGDC-24. National Geophysical Data Center, NOAA, Washington, DC. https://doi.org/10.7289/V5C8276M

Ashok K, Guan Z, Yamagata T (2003) A look at the relationship between the ENSO and the Indian Ocean dipole. J Met Soc Jpn Ser II 81(1):41–56. https://doi.org/10.2151/jmsj.81.41

Ashok K, Behera SK, Rao SA, Weng H, Yamagata T (2007) El Niño Modoki and its possible teleconnection. J Geophys Res 112:C11007. https://doi.org/10.1029/2006JC003798

Bates NR, Pequignet AC, Sabine CL (2006) Ocean carbon cycling in the Indian Ocean: 1. Spatiotemporal variability of inorganic carbon and air-sea CO_2 gas exchange. Glob Biogeochem Cycl 20(3). https://doi.org/10.1029/2005GB002491

Batrakov GF, Bezborodov AA, Eremeev V, Zemlyanoy AD (1981) CO_2 exchange between the Indian-ocean water and the atmosphere. Okeanologiya 21(1):63–69

Beaufort L, Lancelot Y, Camberlin P, Cayre O, Vincent E, Bassinot F, Labeyrie L (1997) Insolation cycles as a major control of equatorial Indian Ocean primary production. Science 278(5342):1451–1454. https://doi.org/10.1126/science.278.5342.1451

Bentaleb I, Fontugne M (1998) The role of the southern Indian Ocean in the glacial to interglacial atmospheric CO_2 change: organic carbon isotope evidences. Glob Planet Chang 16:25–36. https://doi.org/10.1016/S0921-8181(98)00004-6

Bercovici SK, McNichol AP, Xu L, Hansell DA (2018) Radiocarbon content of dissolved organic carbon in the South Indian Ocean. Geophys Res Lett 45(2):872–879. https://doi.org/10.1002/2017GL076295

Bhattathiri PMA, Devassy VP, Radhakrishna K (1980) Primary production in the Bay of Bengal during southwest monsoon of 1978. Mahasagar 13(4):315–323

Bradshaw AL, Brewer PG, Shafer DK, Williams RT (1981) Measurements of total carbon dioxide and alkalinity by potentiometric titration in the GEOSECS program. Earth Planet Sci Lett 55(1):99–115. https://doi.org/10.1016/0012-821X(81)90090-X

Broullón D, Pérez FF, Velo A, Hoppema M, Olsen A, Takahashi T, Key RM, Tanhua T, Santana-Casiano JM, Kozyr A (2020) A global monthly climatology of oceanic total dissolved inorganic carbon: a neural network approach. Earth Syst Sci Data 12:1725–1743. https://doi.org/10.5194/essd-12-1725-2020

Carroll D, Menemenlis D, Adkins JF, Bowman KW, Brix H, Dutkiewicz S et al (2020) The ECCO-Darwin data-assimilative global ocean biogeochemistry model: Estimates of seasonal to multidecadal surface ocean pCO_2 and air-sea CO_2 flux. J Adv Model Earth Syst 12(10):e2019MS001888. https://doi.org/10.1029/2019MS001888

Chen GT, Millero FJ (1979) Gradual increase of oceanic CO_2. Nature 277(5693):205–206. https://doi.org/10.1038/277205a0

Chen L, Xu S, Gao Z, Chen H, Zhang Y, Zhan J, Li W (2011) Estimation of monthly air-sea CO_2 flux in the southern Atlantic and Indian Ocean using in-situ and remotely sensed data. Remote Sens Environ 115(8):1935–1941. https://doi.org/10.1016/j.rse.2011.03.016

Chen S, Sutton AJ, Hu C, Chai F (2021) Quantifying the atmospheric CO_2 forcing effect on surface Ocean pCO_2 in the North Pacific subtropical gyre in the past two decades. Front Mar Sci 8:260. https://doi.org/10.3389/fmars.2021.636881

Ciais P, Sabine C, Govindasamy B, Bopp L, Brovkin V, Canadell J, Chhabra A, DeFries R, Galloway J, Heimann M, Jones C, Le Quéré C, Myneni R, Piao S, Thornton P (2013) Chapter 6: carbon and other biogeochemical cycles. In: Stocker T, Qin D, Platner G-K (eds) Climate change 2013 the physical science basis. Cambridge University Press, Cambridge

Denman KL, Brasseur G, Chidthaisong A, Ciais P, Cox PM, Dickinson RE, Hauglustaine D, Heinze C, Holland E, Jacob D, Lohmann U, Ramachandran S, Leite da Silva Dias P, Wofsy SC, Zhang

X (2007) Couplings between changes in the climate system and biogeochemistry. In: Solomon S, Qin D, Manning M, Marquis M, Averyt K, Tignor MMB, Miller HL, Chen ZL (eds) Climate change 2007: the physical science basis. Contribution of working Group I to the fourth assessment report of the intergovernmental panel on climate change. Cambridge University Press, Cambridge, UK and New York, USA, pp 499–587

Denvil-Sommer A, Gehlen M, Vrac M, Mejia C (2019) LSCE-FFNN-v1: a two-step neural network model for the reconstruction of surface ocean pCO_2 over the global ocean. Geosci Model Dev 12(5):2091–2105. https://doi.org/10.5194/gmd-12-2091-2019

Devassy VP, Bhattathiri PMA, Radhakrishna K (1983) Primary production in the Bay of Bengal during August 1977. Mahasagar 16(4):443–447

DeVries T, Le Quéré C, Andrews O, Berthet S, Hauck J, Ilyina T et al (2019) Decadal trends in the ocean carbon sink. Proc Natl Acad Sci 116(24):11646–11651. https://doi.org/10.1073/pnas.1900371116

Doi T, Osafune S, Sugiura N, Kouketsu S, Murata A, Masuda S, Toyoda T (2015) Multidecadal change in the dissolved inorganic carbon in a long-term ocean state estimation. J Adv Model Earth Syst 7(4):1885–1900. https://doi.org/10.1002/2015MS000462

Eakins BW, Sharman GF (2010) Volumes of the World's Oceans from ETOPO1. NOAA National Geophysical Data Center, Washington, DC. Available online at: http://www.ngdc.noaa.gov/mgg/global/etopo1_ocean_volumes.html (Accessed 17 May 2021)

Feely RA, Takahashi T, Wanninkhof R, McPhaden MJ, Cosca CE, Sutherland SC, Carr M-E (2006) Decadal variability of the air-sea CO_2 fluxes in the equatorial Pacific Ocean. J Geophys Res 111:C08S90. https://doi.org/10.1029/2005JC003129

Francis PA, Jithin AK, Effy JB, Chatterjee A, Chakraborty K, Paul A et al (2020) High-resolution operational ocean forecast and reanalysis system for the Indian ocean. Bull Am Met Soc 101(8):E1340–E1356. https://doi.org/10.1175/BAMS-D-19-0083.1

Friedlingstein P, Jones MW, O'sullivan M, Andrew RM, Hauck J, Peters GP et al (2019) Global carbon budget 2019. Earth Syst Sci Data 11(4):1783–1838. https://doi.org/10.5194/essd-11-1783-2019

Galy V, France-Lanord C, Beyssac O, Faure P, Kudrass H, Palhol F (2007) Efficient organic carbon burial in the Bengal fan sustained by the Himalayan erosional system. Nature 450(7168):407–410. https://doi.org/10.1038/nature06273

George MD, Kumar MD, Naqvi SWA, Banerjee S, Narvekar PV, De Sousa SN, Jayakumar DA (1994) A study of the carbon dioxide system in the northern Indian Ocean during premonsoon. Mar Chem 47(3–4):243–254. https://doi.org/10.1016/0304-4203(94)90023-X

Gloege L (2020) Global Ocean carbon dioxide flux mapping techniques: evaluation, development, and discrepancies (Doctoral dissertation, Columbia University)

Goyet C, Touratier F (2009) Challenges for present and future estimates of anthropogenic carbon in the Indian Ocean. Geophys Monogr Ser 185:231–237. https://doi.org/10.1029/2008GM000754

Goyet C, Beauverger C, Brunet C, Poisson A (1991) Distribution of carbon dioxide partial pressure in surface waters of the Southwest Indian Ocean. Tellus B 43(1):1–11. https://doi.org/10.1034/j.1600-0889.1991.00001.x

Gruber N, Sarmiento JL, Stocker TF (1996) An improved method for detecting anthropogenic CO_2 in the oceans. Glob Biogeochem Cycl 10(4):809–837. https://doi.org/10.1029/96GB01608

Gruber N, Clement D, Carter BR, Feely RA, Van Heuven S, Hoppema M et al (2019) The oceanic sink for anthropogenic CO_2 from 1994 to 2007. Science 363(6432):1193–1199. https://doi.org/10.1126/science.aau5153

Gruber N et al (2009) Oceanic sources, sinks, and transport of atmospheric CO_2. Glob Biogeochem Cycl 23:GB1005. https://doi.org/10.1029/2008GB003349

Gurney K, Baker D, Rayner P, Denning S (2008) Interannual variations in continental-scale net carbon exchange and sensitivity to observing networks estimated from atmospheric CO_2 inversions for the period 1980 to 2005. Glob Biogeochem Cycl 22:GB3025. https://doi.org/10.1029/2007GB003082

Haake B, Ittekkot V, Rixen T, Ramaswamy V, Nair RR, Curry WB (1993) Seasonality and inter-annual variability of particle fluxes to the deep Arabian Sea. Deep Sea Res Part i: Oceanogr Res Paper 40(7):1323–1344. https://doi.org/10.1016/0967-0637(93)90114-I

Hall TM, Primeau FW (2004) Separating the natural and anthropogenic air-sea flux of CO_2: The Indian Ocean. Geophys Res Lett 31(23). https://doi.org/10.1029/2004GL020589

Hall TM, Waugh DW, Haine TW, Robbins PE, Khatiwala S (2004) Estimates of anthropogenic carbon in the Indian Ocean with allowance for mixing and time-varying air-sea CO_2 disequilibrium. Glob Biogeochem Cycl 18(1). https://doi.org/10.1029/2003GB002120

Hansell DA (2009) Dissolved organic carbon in the carbon cycle of the Indian Ocean. Indian Ocean biogeochemical processes and ecological variability. Geophys Monogr Ser 185:217–230. https://doi.org/10.1029/2007GM000684

Hansell DA, Carlson CA (1998) Deep-ocean gradients in the concentration of dissolved organic carbon. Nature 395(6699):263–266. https://doi.org/10.1038/26200

Hansell DA, Carlson CA (2002) Biogeochemistry of marine dissolved organic matter. Academic, San Diego, California, pp 774

Hauck J, Zeising M, Le Quéré C, Gruber N, Bakker DC, Bopp L et al (2020) Consistency and challenges in the ocean carbon sink estimate for the Global Carbon Budget. Front Mar Sci 7:852. https://doi.org/10.3389/fmars.2020.571720

He YC, Tjiputra J, Langehaug HR, Jeansson E, Gao Y, Schwinger J, Olsen A (2018) A model-based evaluation of the inverse Gaussian transit-time distribution method for inferring anthropogenic carbon storage in the ocean. J Geophys Res Ocean 123(3):1777–1800. https://doi.org/10.1002/2017JC013504

Hedges JI (1992) Global biogeochemical cycles: progress and problems. Mar Chem 39:67–93. https://doi.org/10.1016/0304-4203(92)90096-S

Ho DT, Law CS, Smith MJ, Schlosser P, Harvey M, Hill P (2006) Measurements of air-sea gas exchange at high wind speeds in the Southern Ocean: Implications for global parameterizations. Geophys Res Lett 33(16). https://doi.org/10.1029/2006GL026817

Ho DT, Wanninkhof R, Schlosser P, Ullman DS, Hebert D, Sullivan KF (2011) Toward a universal relationship between wind speed and gas exchange: Gas transfer velocities measured with 3He/SF6 during the Southern Ocean Gas Exchange Experiment. J Geophys Res Ocean 116(C4). https://doi.org/10.1029/2010JC006854

Ho DT, Wanninkhof R (2016) Air–sea gas exchange in the North Atlantic: 3He/SF6 experiment during GasEx-98. Tellus B Chem Phys Met 68(1):30198. https://doi.org/10.3402/tellusb.v68.30198

Ivanenkov VN (1964) Distribution of elements of the carbonate system in the water of the northern part o£ the Indian Ocean. Trudy Instituta OkeanoloEii Akademiya Nauk SSSR 64:128–143 (in Russian)

Jabaud-Jan A, Metzl N, Brunet C, Poisson A, Schauer B (2004) Interannual variability of the carbon dioxide system in the southern Indian Ocean (20 S–60 S): The impact of a warm anomaly in austral summer 1998. Glob Biogeochem Cycl 18(1). https://doi.org/10.1029/2002GB002017

Jacobson AR, Mikaloff Fletcher SE, Gruber N, Sarmiento J, Gloor M (2007) A joint atmosphere-ocean inversion for surface fluxes of carbon dioxide: 2. Regional results. Glob Biogeochem Cycl 21:GB1020. https://doi.org/10.1029/2006GB002703

Johnson KM, Dickson AG, Eischeid G, Goyet C, Guenther P, Key RM et al (1998) Coulometric total carbon dioxide analysis for marine studies: assessment of the quality of total inorganic carbon measurements made during the US Indian Ocean CO_2 Survey 1994–1996. Mar Chem 63(1–2):21–37. https://doi.org/10.1016/S0304-4203(98)00048-6

Keeling CD, Waterman LS (1968) Carbon dioxide in surface ocean waters: 3. Measurements on Lusiad Expedition 1962–1963. J Geophys Res 73(14):4529–4541. https://doi.org/10.1029/JB073i014p04529

Khatiwala S, Tanhua T, Mikaloff Fletcher S, Gerber M, Doney SC, Graven HD et al (2013) Global ocean storage of anthropogenic carbon. Biogeosciences 10(4):2169–2191. https://doi.org/10.5194/bg-10-2169-2013

Kondo F, Tsukamoto O (2007) Air-sea CO_2 flux by eddy covariance technique in the equatorial Indian Ocean. J Oceanogr 63(3):449–456. https://doi.org/10.1007/s10872-007-0040-7
Krey J (1973) Primary Production in the Indian Ocean I. In: Zeitzschel B, Gerlach SA (eds) The Biology of the Indian Ocean. Ecological Studies (Analysis and Synthesis), vol 3. Springer, Berlin. https://doi.org/10.1007/978-3-642-65468-8_7
Krishna MS, Prasad VR, Sarma VVSS, Reddy NPC, Hemalatha KPJ, Rao YV (2015) Fluxes of dissolved organic carbon and nitrogen to the northern Indian Ocean from the Indian monsoonal rivers. J Geophys Res Biogeosci 120(10):2067–2080. https://doi.org/10.1002/2015JG002912
Krishna MS, Viswanadham R, Prasad MH, Kumari VR, Sarma VV (2019) Export fluxes of dissolved inorganic carbon to the northern Indian Ocean from the Indian monsoonal rivers. Biogeosciences 16(2):505–519. https://doi.org/10.5194/bg-16-505-2019
Krishna KV, Shanmugam P, Nagamani PV (2020) A multiparametric nonlinear regression approach for the estimation of global surface Ocean pCO_2 using satellite oceanographic data. IEEE J Sel Top Appl Earth Obs Remote Sens 13:6220–6235. https://doi.org/10.1109/JSTARS.2020.3026363
Kumar MD, Rajendran A, Somasundar K, Haake B, Jenisch A, Shuo Z et al (1990) Dynamics of dissolved organic carbon in the northwestern Indian Ocean. Mar Chem 31(1–3):299–316. https://doi.org/10.1016/0304-4203(90)90044-D
Kumar SP, Madhupratap M, Kumar MD, Gauns M, Muraleedharan PM, Sarma VVSS, De Souza SN (2000) Physical control of primary productivity on a seasonal scale in central and eastern Arabian Sea. J Earth Syst Sci 109(4):433–441. https://doi.org/10.1007/BF02708331
Kuwae T, Hori M (2018) Blue carbon in shallow coastal ecosystems. Springer, Singapore
Landschuetzer P, Gruber N, Bakker DC (2016) Decadal variations and trends of the global ocean carbon sink. Glob Biogeochem Cycl 30(10):1396–1417. https://doi.org/10.1002/2015GB005359
Landschützer P, Gruber N, Bakker DCE, Schuster U, Nakaoka S, Payne MR, Sasse T, Zeng J (2013) A neural network-based estimate of the seasonal to inter-annual variability of the Atlantic Ocean carbon sink. Biogeosciences 10:7793–7815. https://doi.org/10.5194/bg-10-7793-2013
Landschützer P, Gruber N, Bakker DC, Schuster U (2014) Recent variability of the global ocean carbon sink. Glob Biogeochem Cycl 28(9):927–949. https://doi.org/10.1002/2014GB004853
Landschützer P, Gruber N, Bakker DCE, Landschützer P (2017) An updated observation-based global monthly gridded sea surface pCO_2 and air-sea CO_2 flux product from 1982 through 2015 and its monthly climatology. National Centers for Environmental Information, Asheville, NC
Langlais CE, Lenton A, Matear R, Monselesan D, Legresy B, Cougnon E, Rintoul S (2017) Stationary Rossby waves dominate subduction of anthropogenic carbon in the Southern Ocean. Sci Rep 7(1):1–10. https://doi.org/10.1038/s41598-017-17292-3
Lauvset SK, Key RM, Olsen A, van Heuven S, Velo A, Lin X, Schirnick C, Kozyr A, Tanhua T, Hoppema M, Jutterström S, Steinfeldt R, Jeansson E, Ishii M, Perez FF, Suzuki T, Watelet S (2016) A new global interior ocean mapped climatology: the 1o×1o GLODAP version 2. Earth Syst Sci Data 8:325–340. https://doi.org/10.5194/essd-8-325-2016
Lenton A, Tilbrook B, Law RM, Bakker D, Doney SC, Gruber N et al (2013) Sea-air CO_2 fluxes in the Southern Ocean for the period 1990–2009. Biogeosciences 10:4037–4054. https://doi.org/10.5194/bg-10-4037-2013
Lerner P, Romanou A, Kelley M, Romanski J, Ruedy R, Russell G (2021) Drivers of air-sea CO_2 flux seasonality and its long-term changes in the NASA-GISS model CMIP6 submission. J Adv Model Earth Syst 13(2):e2019MS002028. https://doi.org/10.1029/2019MS002028
Le Quéré C, Moriarty R, Andrew RM, Peters GP, Ciais P, Friedlingstein P et al (2014) Global carbon budget 2013 Earth Syst Sci Data 6(1):235–263. https://doi.org/10.5194/essd-6-235-2014
Li H, Ilyina T (2018) Current and future decadal trends in the oceanic carbon uptake are dominated by internal variability. Geophys Res Lett 45(2):916–925. https://doi.org/10.1002/2017GL075370
Liss PS, Merlivat L (1986) Air-sea gas exchange rates: introduction and synthesis. The role of air-sea exchange in geochemical cycling. Springer, Dordrecht, pp 113–127
Louanchi F, Metzl N, Poisson A (1996) Modelling the monthly sea surface fCO_2 fields in the Indian Ocean. Mar Chem 55(3–4):265–279. https://doi.org/10.1016/S0304-4203(96)00066-7

Louanchi F (1995) Etude des variabilites spatio-temporelles de fCO_2, a la surface de l'ocean Indien: Processus et quantification. Ph.D. thesis, University of Paris 6, Paris, pp 234

Lovelock CE, Duarte CM (2019) Dimensions of blue carbon and emerging perspectives. Biol Lett 15(3):20180781. https://doi.org/10.1098/rsbl.2018.0781

Ma T, Cheng X, Qi Y, Chen J (2020) Interannual variability in the barrier layer and forcing mechanism in the eastern equatorial Indian Ocean and Bay of Bengal. Acta Oceanol Sin 39(7):19–31. https://doi.org/10.1007/s13131-020-1575-3

Matsumoto K (2007) Radiocarbon-based circulation age of the world oceans. J Geophys Res 112:C09004. https://doi.org/10.1029/2007JC004095

Mengesha S, Dehairs F, Fiala M, Elskens M, Goeyens L (1998) Seasonal variation of phytoplankton community structure and nitrogen uptake regime in the Indian Sector of the Southern Ocean. Polar Biol 20(4):259–272. https://doi.org/10.1007/s003000050302

Menzel DW (1964) The distribution of dissolved organic carbon in the Western Indian Ocean. Deep Sea Res Oceanogr Abstr 11(5):757–765. https://doi.org/10.1016/0011-7471(64)90948-9

Metzl N (2009) Decadal increase of oceanic carbon dioxide in Southern Indian Ocean surface waters (1991–2007). Deep Sea Res II Top Stud Oceanogr 56(8–10):607–619. https://doi.org/10.1016/j.dsr2.2008.12.007

Metzl N, Poisson A, Louanchi F, Brunet C, Schauer B, Bres B (1995) Spatiotemporal distributions of air-sea fluxes of CO_2 in the Indian and Antarctic Oceans: a first step. Tellus B 47(1–2):56–69. https://doi.org/10.1034/j.1600-0889.47.issue1.7.x

Mikaloff Fletcher SE, Gruber N, Jacobson AR, Doney SC, Dutkiewicz S, Gerber M et al (2006) Inverse estimates of anthropogenic CO_2 uptake, transport, and storage by the ocean. Glob Biogeochem Cycl 20:GB2002. https://doi.org/10.1029/2005GB002530

Miller SD, Marandino C, Saltzman ES (2010) Ship-based measurement of air-sea CO_2 exchange by eddy covariance. J Geophys Res Atmos 115(D2). https://doi.org/10.1029/2009JD012193

Mohapatra S, Gnanaseelan C, Deepa JS (2020) Multidecadal to decadal variability in the equatorial Indian Ocean subsurface temperature and the forcing mechanisms. Clim Dyn 54(7):3475–3487. https://doi.org/10.1007/s00382-020-05185-7

Naik RK, George JV, Soares MA, Devi A, Anilkumar N, Roy R et al (2015) Phytoplankton community structure at the juncture of the Agulhas return front and subtropical front in the Indian Ocean sector of Southern Ocean: bottom-up and top-down control. Deep Sea Res II Top Stud Oceanogr 118:233–239. https://doi.org/10.1016/j.dsr2.2015.01.002

Naik RK, George J, Soares MA, Anilkumar N, Mishra RK, Roy R et al (2020) Observations of surface water phytoplankton community in the Indian Ocean: a transect from tropics to polar latitudes. Deep Sea Res II Top Stud Oceanogr 178:104848. https://doi.org/10.1016/j.dsr2.2020.104848

Naqvi SWA, Reddy CVG (1975) Vertical distribution of the degree of calcium carbonate saturation in the Indian Ocean. India J Mar Sci 4:213–215

Naqvi SWA, Charles CD, Fairbanks RG (1994) Carbon and oxygen isotopic records of benthic foraminifera from the Northeast Indian Ocean: implications on glacial-interglacial atmospheric CO_2 changes. Earth Planet Sci Lett 121(1–2):99–110. https://doi.org/10.1016/0012-821X(94)90034-5

Nellemann C, Corcoran E, Duarte CM et al (2009) Blue carbon: a rapid response assessment. United Nations Environmental Programme, GRID-Arendal, Birkeland Trykkeri AS, Birkeland

Nightingale PD, Malin G, Law CS, Watson AJ, Liss PS, Liddicoat MI et al (2000) In situ evaluation of air-sea gas exchange parameterizations using novel conservative and volatile tracers. Glob Biogeochem Cycl 14(1):373–387. https://doi.org/10.1029/1999GB900091

Owens JD, Lyons TW, Lowery CM (2018) Quantifying the missing sink for global organic carbon burial during a Cretaceous oceanic anoxic event. Earth Planet Sci Lett 499:83–94. https://doi.org/10.1016/j.epsl.2018.07.021

Poisson A, Metzl N, Brunet C, Schauer B, Bres B, Ruiz-Pino D, Louanchi F (1993) Variability of sources and sinks of CO_2 in the Western Indian and Southern Oceans during the year 1991. J Geophys Res Ocean 98(C12):22759–22778. https://doi.org/10.1029/93JC02501

Qasim SZ (1977) Biological productivity of the Indian Ocean. Indian J Mar Sci 6:122–137
Regnier P, Friedlingstein P, Ciais P, Mackenzie FT, Gruber N, Janssens IA et al (2013) Anthropogenic perturbation of the carbon fluxes from land to ocean. Nat Geosci 6:597–607. https://doi.org/10.1038/ngeo1830
Resplandy L, Keeling RF, Rödenbeck C, Stephens BB, Khatiwala S, Rodgers KB et al (2018) Revision of global carbon fluxes based on a reassessment of oceanic and riverine carbon transport. Nat Geosci 11(7):504–509. https://doi.org/10.1038/s41561-018-0151-3
Rixen T, Haake B, Ittekkot V, Guptha MVS, Nair RR, Schlüssel P (1996) Coupling between SW monsoon-related surface and deep ocean processes as discerned from continuous particle flux measurements and correlated satellite data. J Geophys Res Oceans 101(C12):28569–28582. https://doi.org/10.1029/96JC02420
Rixen T, Gaye B, Emeis KC (2019a) The monsoon, carbon fluxes, and the organic carbon pump in the northern Indian Ocean. Prog Oceanogr 175:24–39. https://doi.org/10.1016/j.pocean.2019.03.001
Rixen T, Gaye B, Emeis KC, Ramaswamy V (2019b) The ballast effect of lithogenic matter and its influences on the carbon fluxes in the Indian Ocean. Biogeosciences 16(2):485–503. https://doi.org/10.5194/bg-16-485-2019
Sabine CL, Feely RA (2001) Comparison of recent Indian Ocean anthropogenic CO_2 estimates with a historical approach. Glob Biogeochem Cycl 15(1):31–42. https://doi.org/10.1029/2000GB001258
Sabine CL, Key RM, Johnson KM, Millero FJ, Poisson A, Sarmiento JL et al (1999) Anthropogenic CO_2 inventory of the Indian Ocean. Glob Biogeochem Cycl 13(1):179–198. https://doi.org/10.1029/1998GB900022
Sabine CL, Wanninkhof R, Key RM, Goyet C, Millero FJ (2000) Seasonal CO_2 fluxes in the tropical and subtropical Indian Ocean. Mar Chem 72(1):33–53. https://doi.org/10.1016/S0304-4203(00)00064-5
Sabine CL, Key RM, Feely RA, Greeley D (2002) Inorganic carbon in the Indian Ocean: distribution and dissolution processes. Glob Biogeochem Cycl 16(4):15–21. https://doi.org/10.1029/2002GB001869
Saji NH, Yamagata T (2003) Structure of SST and surface wind variability during Indian Ocean dipole mode events: COADS observations. J Clim 16(16):2735–2751. https://doi.org/10.1175/1520-0442(2003)016%3c2735:SOSASW%3e2.0.CO;2
Saji NH, Goswami BN, Vinayachandran PN, Yamagata T (1999) A dipole mode in the tropical Indian Ocean. Nature 401(6751):360–363. https://doi.org/10.1038/43854
Sardessai S, De Sousa SN (2001) Dissolved organic carbon in the INDEX area of the Central Indian Basin. Deep Sea Res II Top Stud Oceanogr 48(16):3353–3361. https://doi.org/10.1016/S0967-0645(01)00045-5
Sarkar A, Bhattacharya SK, Sarin MM (1993) Geochemical evidence for anoxic deep water in the Arabian Sea during the last glaciation. Geochim Cosmochim Acta 57(5):1009–1016. https://doi.org/10.1016/0016-7037(93)90036-V
Sarma VVSS (2006) The influence of Indian Ocean Dipole (IOD) on biogeochemistry of carbon in the Arabian Sea during 1997–1998. J Earth Syst Sci 115(4):433–450. https://doi.org/10.1007/BF02702872
Sarma VVSS, Dileep Kumar M, Saino T (2007) Impact of sinking carbon flux on accumulation of deep-ocean carbon in the Northern Indian Ocean. Biogeochemistry 82(1):89–100. https://doi.org/10.1007/s10533-006-9055-1
Sarma VVSS, Lenton A, Law RM, Metzl N, Patra PK, Doney S, Lima ID, Dlugokencky E, Ramonet M, Valsala V (2013) Sea–air CO_2 fluxes in the Indian Ocean between 1990 and 2009. Biogeosciences 10(11):7035–7052. https://doi.org/10.5194/bg-10-7035-2013
Sarma VVSS, Paul YS, Vani DG, Murty VSN (2015) Impact of river discharge on the coastal water pH and pCO_2 levels during the Indian Ocean Dipole (IOD) years in the western Bay of Bengal. Cont Shelf Res 107:132–140. https://doi.org/10.1016/j.csr.2015.07.015

Sarma VVSS, Kumari VR, Srinivas TNR, Krishna MS, Ganapathi P, Murty VSN (2018) East India Coastal Current controls the dissolved inorganic carbon in the coastal Bay of Bengal. Mar Chem 205:37–47. https://doi.org/10.1016/j.marchem.2018.07.010

Sarma VVSS (2003) Monthly variability in surface pCO_2 and net air-sea CO_2 flux in the Arabian Sea. J Geophys Res Ocean 108(C8). https://doi.org/10.1029/2001JC001062

Schubert CJ, Villanueva J, Calvert SE, Cowie GL, Von Rad U, Schulz H et al (1998) Stable phytoplankton community structure in the Arabian Sea over the past 200,000 years. Nature 394(6693):563–566. https://doi.org/10.1038/29047

Schulte S, Rostek F, Bard E, Rullkötter J, Marchal O (1999) Variations of oxygen-minimum and primary productivity recorded in sediments of the Arabian Sea. Earth Planet Sci Lett 173(3):205–221. https://doi.org/10.1016/S0012-821X(99)00232-0

Sen R, Pandey S, Dandapat S, Francis PA, Chakraborty A (2020) A numerical study on seasonal transport variability of the North Indian Ocean boundary currents using Regional Ocean Modeling System (ROMS). J Operat Oceanogr. https://doi.org/10.1080/1755876X.2020.1846266

Sengupta RS, Naqvi SWA (1984) Chemical oceanography of the Indian Ocean, north of the equator. Deep Sea Res Oceanogr Res Pap 31(6–8):671–706. https://doi.org/10.1016/0198-0149(84)90035-9

Shah C, Sudheer AK, Bhushan R (2018) Distribution of dissolved organic carbon in the Bay of Bengal: Influence of sediment discharge, fresh water flux, and productivity. Mar Chem 203:91–101. https://doi.org/10.1016/j.marchem.2018.04.004

Shetye SS, Mohan R, Patil S, Jena B, Chacko R, George JV et al (2015) Oceanic pCO_2 in the Indian sector of the Southern Ocean during the austral summer-winter transition phase. Deep Sea Res II Top Stud Oceanogr 118:250–260. https://doi.org/10.1016/j.dsr2.2015.05.017

Singh A, Ramesh R (2015) Environmental controls on new and primary production in the northern Indian Ocean. Prog Oceanogr 131:138–145. https://doi.org/10.1016/j.pocean.2014.12.006

Socci AD (1986) Potential productivity and surface circulation in the northern Indian Ocean during the last glacial-to-interglacial transition: application of a new paleoceanographic model. Palaeogeogr Palaeoclimatol Palaeoecol 53(1):63–95. https://doi.org/10.1016/0031-0182(86)90039-8

Sreeush MG, Valsala V, Pentakota S, Prasad KVSR, Murtugudde R (2018) Biological production in the Indian Ocean upwelling zones–Part 1: refined estimation via the use of a variable compensation depth in ocean carbon models. Biogeosciences 15(7):1895–1918. https://doi.org/10.5194/bg-15-1895-2018

Sreeush MG, Valsala V, Santanu H, Pentakota S, Prasad KVSR, Naidu CV, Murtugudde R (2020) Biological production in the Indian Ocean upwelling zones-Part 2: Data based estimates of variable compensation depth for ocean carbon models via cyclo-stationary Bayesian Inversion. Deep Sea Res II Top Stud Oceanogr 179:104619. https://doi.org/10.1016/j.dsr2.2019.07.007

Sweeney C, Gloor E, Jacobson AR, Key RM, McKinley G, Sarmiento JL, Wanninkhof R (2007) Constraining global air-sea gas exchange for CO_2 with recent bomb 14C measurements. Glob Biogeochem Cycl 21(2). https://doi.org/10.1029/2006GB002784

Takahashi TT, Feely RA, Weiss RF, Wanninkhof RH, Chipman DW, Sutherland SC, Takahashi TT (1997) Global air-sea flux of CO_2: an estimate based on measurements of sea-air pCO_2 difference. Proc Natl Acad Sci USA 94:8292–8299. https://doi.org/10.1073/pnas.94.16.8292

Takahashi T, Sutherland SC, Wanninkhof R, Sweeney C, Feely RA, Chipman DW et al (2009) Climatological mean and decadal change in surface ocean pCO_2, and net sea-air CO_2 flux over the global oceans. Deep-Sea Res II 56:554–577. https://doi.org/10.1016/j.dsr2.2008.12.009

Takahashi T, Sutherland SC, Chipman DW, Goddard JG, Ho C (2014) Climatological distributions of pH, pCO_2, total CO_2, alkalinity, and $CaCO_3$ saturation in the global surface ocean, and temporal changes at selected locations. Mar Chem 164:95–125. https://doi.org/10.1016/j.marchem.2014.06.004

Takahashi T, Wanninkhof RH, Feely RA, Weiss RF, Chipman DW, Bates N, Olafsson J, Sabine C, Sutherland SC (1999) Net air-sea CO2 flux over global oceans: an improved estimate based on sea-air pCO_2 difference. In: Nojiri Y (ed) Proceedings of 2nd international symposium on CO_2

in the oceans. Center for Global Environmental Research, National Institute for Environmental Studies, Tsukuba, Japan, pp 9–14
Tarran GA, Burkill PH, Edwards ES, Woodward EMS (1999) Phytoplankton community structure in the Arabian Sea during and after the SW monsoon, 1994. Deep Sea Res II Top Stud Oceanogr 46(3–4):655–676. https://doi.org/10.1016/S0967-0645(98)00122-2
Terhaar J, Frölicher TL, Joos F (2021) Southern Ocean anthropogenic carbon sink constrained by sea surface salinity. Sci Adv 7(18):eabd5964. https://doi.org/10.1126/sciadv.abd5964
Valsala V, Maksyutov S, Murtugudde R (2012) A window for carbon uptake in the southern subtropical Indian Ocean. Geophys Res Lett 39(17). https://doi.org/10.1029/2012GL052857
Valsala V, Sreeush MG, Chakraborty K (2020) The IOD impacts on the Indian Ocean Carbon cycle. J Geophys Res Ocean 125(11):e2020JC016485. https://doi.org/10.1029/2020JC016485
Valsala V, Sreeush MG, Anju M, Sreenivas P, Tiwari YK, Chakraborty K, Sijikumar S (2021) An observing system simulation experiment for Indian Ocean surface pCO_2 measurements. Prog Oceanogr 194:102570. https://doi.org/10.1016/j.pocean.2021.102570
Varshney PK, Nair VR, Abidi SAH (1983) Primary productivity in nearshore waters of Thal, Maharashtra coast. Indian J Mar Sci 12:27–30
Veldhuis MJ, Kraay GW, Van Bleijswijk JD, Baars MA (1997) Seasonal and spatial variability in phytoplankton biomass, productivity and growth in the northwestern Indian Ocean: the southwest and northeast monsoon, 1992–1993. Deep Sea Res i: Oceanogr Res Paper 44(3):425–449. https://doi.org/10.1016/S0967-0637(96)00116-1
Vianello P, Ternon JF, Demarcq H, Herbette S, Roberts MJ (2020) Ocean currents and gradients of surface layer properties in the vicinity of the Madagascar Ridge (including seamounts) in the South West Indian Ocean. Deep-Sea Res II Top Stud Oceanogr 176:104816. https://doi.org/10.1016/j.dsr2.2020.104816
Vinayachandran PN, Iizuka S, Yamagata T (2002) Indian Ocean dipole mode events in an ocean general circulation model. Deep-Sea Res II Top Stud Oceanogr 49(7–8):1573–1596. https://doi.org/10.1016/S0967-0645(01)00157-6
Wang H, Hu X, Sterba-Boatwright B (2016) A new statistical approach for interpreting oceanic fCO_2 data. Mar Chem 183:41–49. https://doi.org/10.1016/j.marchem.2016.05.007
Wanninkhof R (1992) Relationship between wind speed and gas exchange over the ocean. J Geophys Res Ocean 97(C5):7373–7382. https://doi.org/10.1029/92JC00188
Wanninkhof R (2014) Relationship between wind speed and gas exchange over the ocean revisited. Limnol Oceanogr Method 12(6):351–362. https://doi.org/10.4319/lom.2014.12.351
Wanninkhof R, McGillis WR (1999) A cubic relationship between air-sea CO_2 exchange and wind speed. Geophys Res Lett 26(13):1889–1892. https://doi.org/10.1029/1999GL900363
Wanninkhof R et al (2013) Global ocean carbon uptake: magnitude, variability, and trends. Biogeosciences 10:1983–2000. https://doi.org/10.5194/bg-10-1983-2013
Xue L, Yu W, Wang H, Jiang LQ, Feng L, Gao L et al (2014) Temporal changes in surface partial pressure of carbon dioxide and carbonate saturation state in the eastern equatorial Indian Ocean during the 1962–2012 period. Biogeosciences 11(22):6293–6305. https://doi.org/10.5194/bg-11-6293-2014
Xue L, Cai WJ (2020) Total alkalinity minus dissolved inorganic carbon as a proxy for deciphering ocean acidification mechanisms. Mar Chem 222:103791. https://doi.org/10.1016/j.marchem.2020.103791

Chapter 3
Carbon Dynamics in the Estuaries and Continental Shelves of Indian Ocean

Anirban Akhand and Abhra Chanda

Abstract Estuaries and coastal margins are one of the dynamic lotic ecosystems of this planet. The admixture of terrestrial inputs with the marine volume of seawater gives rise to a unique regime that leaves a signature on the global carbon dynamics. The blue carbon ecosystems grow and interact continuously with these shallow waters. The exchange of carbon and nutrients steadily goes on between the blue carbon ecosystems and these adjacent water bodies. These waters usually have a high residence time and thus, allow several biogeochemical reactions to occur within these systems. Developing a holistic understanding of the blue carbon strength in the coastal periphery is nigh impossible without these nearshore water bodies. The present chapter collated the crucial findings in this regard from the boundary of the Indian Ocean. The dissolved inorganic carbon, CO_2, and CH_4 dynamics come under the purview of this chapter. The chapter discussed the lateral fluxes of dissolved inorganic carbon and organic carbon from the river estuaries to the adjacent ocean basin. The data on air-water CO_2 and CH_4 fluxes in the estuaries that drain into the Indian Ocean have been collated in this chapter. It also discussed their regulating factors. The drivers of autochthonous carbon production and allochthonous organic substrates that lead to net heterotrophy are discussed in the light of Indian Ocean estuaries. Lastly, the chapter also indicated the lacunae that need to be addressed in days to come to help build an understanding of the carbon dynamics in this regime.

Keywords Dissolved inorganic carbon · Dissolved organic carbon · Particulate inorganic carbon · Particulate organic carbon · pCO_2(water) · Air-water CO_2 flux · Estuary · Nearshore waters · Runoff · Biogeochemistry

---

A. Akhand
Coastal and Estuarine Environment Research Group, Port and Airport Research Institute, 3-1-1, Nagase, Kanagawa, Yokosuka 239-0826, Japan

A. Chanda (✉)
School of Oceanographic Studies, Jadavpur University, 188 Raja S.C. Mullick Road, Kolkata, West Bengal 700084, India
e-mail: abhrachanda.sos.rs@jadavpuruniversity.in

A. Chanda et al. (eds.), *Blue Carbon Dynamics of the Indian Ocean*,
https://doi.org/10.1007/978-3-030-96558-7_3

3.1 Carbon Speciation in the Estuaries and Continental Shelves

Estuaries and extended estuarine reaches toward the continental shelves are some of the most biogeochemically active aquatic ecosystems of this planet (Dutta et al. 2019). The terrigenous input in the dissolved and particulate form reaches the rivers from the adjoining landmasses through precipitation and runoff (Howarth et al. 1991). The estuaries and continental shelves act as junction points between the terrestrial and marine spheres. The water mass in these regions usually has a high residence time (Wang et al. 2004; de Brauwere et al. 2011). Dissolved and particulate matters undergo several transformations and interchange takes place between different species (Bianchi 2006). The biological competition between the autotrophs and the consumers gives rise to a complex productivity scenario in these lotic ecosystems (Watanabe et al. 2018). Physical drivers like geomorphologic setting, water temperature, tidal pattern and amplitude, the shape of the estuarine mouth, and several other factors play a crucial role in characterizing carbon biogeochemistry (Borges et al. 2018; Laruelle et al. 2019). The estuaries, in general, act as strong sources of both carbon dioxide (CO_2) and methane (CH_4); however, the magnitude of fluxes exhibit significant spatiotemporal variability within an estuary and also among estuaries (Cai 2011; Laruelle et al. 2017; Li et al. 2017; Krauss et al. 2018). Several authors indicated that continental shelves, though adjacent to the estuaries, often act as net sinks for CO_2 (Laruelle et al. 2018). Pieces of research indicated that the upper reaches of the estuary acts as a larger source of CO_2, whereas near the estuarine mouth the flux magnitudes decrease (Crosswell et al. 2017; Oliveira et al. 2017; Jeffrey et al. 2018). Besides such spatial variability, the estuaries and nearshore coastal waters often exhibit significant seasonal and interannual variability (Friedman et al. 2020; Akhtar et al. 2021). Thus, characterizing the carbon dynamics of these lotic ecosystems has become a challenging endeavor altogether.

Dissolved inorganic carbon (DIC), dissolved organic carbon (DOC), particulate inorganic carbon (PIC), and particulate organic carbon (POC) are the four fundamental chemical species of carbon that exist in any aquatic ecosystem. The relative composition of these constituents depends on a plethora of autochthonous processes and allochthonous sources (Ray et al. 2018; Guardiani et al. 2021). Total alkalinity (TAlk) is another crucial parameter and it maintains an equilibrium with DIC depending on the pH of the water column. The ratio of TAlk and DIC often indicates the carbonate buffering capacity of an estuary (Akhand et al. 2021a). The dissolved carbon dioxide gas, that the researchers measure as the partial pressure of CO_2 in water [pCO_2(water)] or fugacity of CO_2 in water [fCO_2(water)] has always remained a point of concern. The magnitude of pCO_2(water) and its difference concerning the concentration of CO_2 in ambient air determines whether an estuary would act as a source or sink for CO_2. The same holds for any other gases like CH_4 or O_2. The partial pressure of CO_2 in the air [pCO_2(air)] does not exhibit large short-term variability in both space and time. However, the pCO_2(water) exhibits a wide spectrum throughout the global estuaries (Cai 2011). Besides the air-water CO_2 and CH_4 flux that takes place directly between the atmosphere and the underlying hydrosphere, lateral fluxes of carbon species also play a key role in governing the overall carbon biogeochemistry (Santos et al. 2019; Ohtsuka et al. 2020). Though the carbon in the aquatic medium is not stable in a single chemical form, several authors argue that these are key components of the blue carbon system (Maher et al. 2018; Akhand et al. 2021b). All the blue carbon ecosystems thrive in either estuarine or coastal setups where they come in contact with the water mass, and the exchange of several ingredients takes place through tidal and inundation mechanisms. Twilley et al. (2017) emphasized that the exchange of carbon between the aquatic compartments and the adjacent mangrove wetlands remains an active domain of research for coastal oceanographers. Similarly, Taillardat et al. (2018) suggested that porewater exchange between the open estuarine system and the mangrove sediments plays a crucial role in regulating the overall blue carbon biogeochemistry in mangrove habitats. The characteristic properties of the water column adjoining the seagrass beds directly impact the carbon sequestration potential of this unique blue carbon ecosystem (Ganguly et al. 2017). Especially, the transparency of the water column provides direct feedback to the autochthonous productivity of the seagrass meadows and regulates the blue carbon stock in their biomass and sediment bed (Belshe et al. 2017). Salt marshes have received much less attention in this regard; however, few studies indicate that water and marsh interaction in the coastal plains plays a fundamental role in their blue carbon sequestration capability (Sapkota and White 2019). Thus, it is evident that ignoring the estuarine and nearshores water column carbon biogeochemistry would not facilitate the holistic understanding of the blue carbon dynamics altogether. The scientific is undoubtedly more interested in quantifying the long-term carbon sink within the umbrella of blue carbon. However, to come up with a master plan to conserve, manage, and restore the blue carbon stock of these ecosystems under threat, we need to understand the short-term carbon dynamics as well. Chemical oceanographers and marine chemists have conducted several studies in this domain. However, compared to the number of estuaries that drain in the Indian Ocean and the extent of nearshore coastal waters in this ocean basin, the number of studies is not sufficient. This chapter tried to generate a snapshot of the present state-of-the-art related to this topic in the periphery of the Indian Ocean.

3.1.1 DIC Variability and Its Lateral Fluxes to the Oceans

The northern part of the Indian Ocean has two prominent flanks, namely the Bay of Bengal and the Arabian Sea. Indian east coast has several perennial estuaries that drain into the Bay of Bengal. This carbon biogeochemistry of this sector received substantial attention. Krishna et al. (2019) quantified the DIC dynamics in twenty-seven rivers in India and estimated the DIC input from these estuaries to the northern Indian Ocean. They observed a varying DIC of 56–1226 $\mu mol\ kg^{-1}$ in these river systems and an estimated lateral flux of 10.3 Tg DIC per year (Table 3.1). Monsoon

Table 3.1 Lateral DIC fluxes observed from rivers and estuaries in the Indian Ocean bordering countries

Country	Estuary	DIC Input (Tg C yr^{-1})	Authors
India	27 river estuaries	10.3	Krishna et al. (2019)
India	Hooghly	2.7	Ghosh et al. (2021)
India	Sundarban	3.69	Ray et al. (2018)
Malaysia	Rajang*	0.371	Müller-Dum et al. (2019)
Tanzania	Pangani	0.16	Selemani et al. (2018)
Kenya	Tana	0.9	Bouillon et al. (2007a)

*flows into the South China Sea

plays a critical role in the carbonate discharge to the adjacent Indian Ocean in countries like India, Bangladesh, and Sri Lanka. Indian rivers are accountable for only 1.3% of the total riverine discharge of this planet. However, they export 2.5% of the global DIC input to the oceans. Thus, the Indian river estuarine systems are some of the significant lotic ecosystems that regulate the carbon biogeochemistry of the Indian Ocean altogether. According to Ghosh et al. (2021), the Hooghly estuary, which is the tail end of the River Ganges, exports almost $4.59 \pm 1.70 \times 10^{11}$ mol DIC per year to the adjacent Bay of Bengal. They also observed that this magnitude could be 60 times the atmosphere-hydrosphere CO_2 efflux from the entire estuary. Akhand et al. (2016) observed a DIC range of 1700–5200 μmol kg^{-1} in the Hooghly estuary, whereas in the adjoining Matla estuary flowing through the Sundarban mangroves, the DIC varied from 1400 to 2000 μmol kg^{-1} (Fig. 3.1). The Sundarban mangroves include an intricate network of several estuaries that continues to the Bangladesh counterpart. In totality, this gives rise to the largest deltaic platform (the Ganga-Brahmaputra-Meghna Delta) and the largest mangrove forest in the world as well. Ray et al. (2018) estimated the DIC flux from this estuarine complex and estimated an annual lateral export rate of 3.69 Tg C yr^{-1}. They inferred that mineralization of mangrove-derived carbon led to substantial enhancement of DIC levels in the estuaries of Indian Sundarban. Pattanaik et al. (2020) researched two more estuaries flowing through the east coast of India, namely Mahanadi and Dhamra over an annual cycle. The DIC range in Mahanadi and Dhamra was 1201–1854 and 903–2070 $^{\mu mol\ kg-1}$, respectively. Similar measurements of DIC and their lateral flux toward the Indian Ocean were not available for Bangladesh and Myanmar. However, Miyajima et al. (2009) studied the DIC export from two river estuaries of Thailand that flow through mangrove stands. They implemented a stable isotopic technique and developed a model whereby the conservative mixing of DIC along the salinity gradient could indicate and differentiate between the autochthonous DIC production and the lateral DIC input from the upstream. They inferred that the mangrove forests

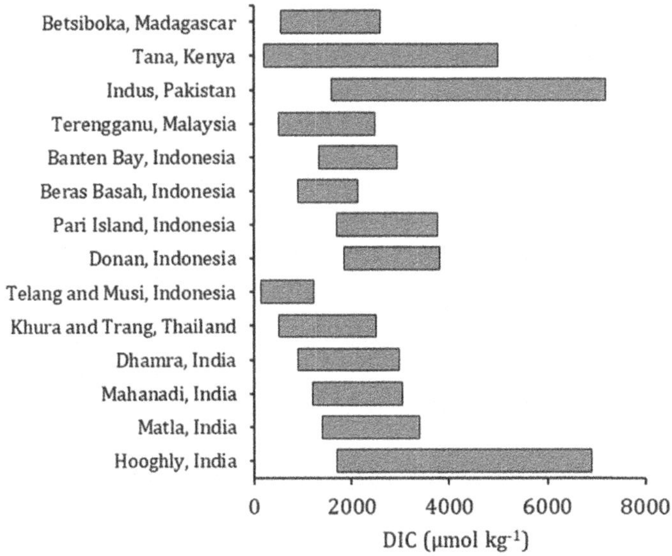

Fig. 3.1 Range of DIC concentration observed in the estuaries draining into the Indian Ocean (Data collated from Karim and Veizer 2000; Ralison et al. 2008; Miyajima et al. 2009; Afdal and Noerjito 2011; Ramawijaya et al. 2012; Ritonga 2012; Tamooh et al. 2013; Rustam et al. 2014; Akhand et al. 2016; Purwiyanto and Agustriani 2016; Pattanaik et al. 2020; Yan et al. 2020)

act as a CO_2 pump wherein they absorb atmospheric CO_2 molecules and release the same in the form of DIC into the adjacent estuaries. Reports of DIC concentration are hard to find in Bangladesh, Myanmar, and Singapore; however, Indonesian and Malaysian coastal waters have some studies in this regard. Purwiyanto and Agustriani (2016) observed a modest DIC concentration range of 145–1081 μmol kg^{-1} in the Telang and Musi River, whereas Afdal and Noerjito (2011) observed higher concentrations but a narrow range of 1842–1958 μmol kg^{-1} in the Donan Estuary of Indonesia. Rustam et al. (2014) observed a similar DIC range in the nearshore waters of Pari Island, Indonesia. However, the obervations of Ramawijaya et al. (2012) and Ritonga (2012) showed much lower DIC ranges in Beras Basah and Banten Bay, Indonesia. Thus, Indonesian estuaries exhibited substantial spatial variability in DIC. Lately, Yan et al. (2020) developed an *in-situ* underwater DIC measurement system and operated it in the Terengganu River Estuary. They observed a DIC range of 510 to 1980 μmol kg-1 from the upper estuarine reaches to the river mouth. Müller-Dum et al. (2019), while working in the Rajang River Estuary of Malaysia observed a DIC input of 0.371 Tg C per year. However, this river estuary though situated in Malaysia drains into the South China Sea, which is technically a part of the Pacific Ocean. Studies related to DIC measurement in river estuaries are scanty in the northwestern part of the Indian Ocean. However, Karim and Veizer (2000) observed a wide range of DIC (1600–5600 μmol kg^{-1}) in the Indus River Basin flowing through Pakistan. The carbonate-rich rocks adjoining the Indus Basin were primarily responsible for the high DIC values obtained at some selected locations in the basin. The high concentrations of DIC were also indicators of strong terrain weathering in the Indus River.

Among the east-African countries, Tamooh et al. (2013) observed a wide range of DIC (200–4800 μmol kg^{-1}) in the Tana River Estuary flowing through Kenya. Bouillon et al. (2007a), while working in the same Tana Estuary observed a DIC input of 0.9 Tg C per year to the Indian Ocean. Selemani et al. (2018), while working in the Pangani River Estuary of Tanzania observed a lateral DIC flux of 0.16 Tg C per year; however, they did not report the DIC range in the same region. Ralison et al. (2008), while working in the largest estuary of Madagascar, Betsiboka, observed a DIC range of 537–2050 μmol kg^{-1} that exhibited an almost conservative mixing throughout the estuary, except a few spikes in the mesohaline region.

3.1.2 DOC and POC in the Estuarine and Continental Shelf Waters

DOC and POC altogether received much less attention than DIC or pCO_2(water) estimation in the estuaries. One of the probable reasons could be that the organic counterpart does not readily participate in instantaneous processes like air-water CO_2/CH_4 exchange. However, the mineralization and remineralization of DOC and POC play a crucial role in regulating the DIC and hence, the atmospheric exchange of carbonaceous greenhouse gases from these bodies (Abril et al. 1999; Moran et al. 2000; Weston et al. 2006). These two parameters are reported from many estuaries that drain into the Indian Ocean. However, the number of studies is still scarce compared to the number of estuaries present in this region. Ray et al. (2018) estimated a POC and DOC lateral input of 0.58 Tg C yr^{-1} and 3.03 Tg C yr^{-1} from the estuaries of Indian Sundarban to the adjacent Bay of Bengal. Ray et al. (2018) observed that mangrove litter decomposition played an active role in generating DOC, whereas, soil erosion led to the enhancement of POC levels in the Sundarban estuaries. Bird et al. (2008) researched two major river systems of Myanmar, namely Thanlwin and Ayeyarwady, and observed substantially higher POC ranges than Ray et al. (2018). However, the DOC range was much lower than the magnitudes in Sundarban, India. Despite such high values, the estimated POC and DOC flux to the Indian Ocean was much less due to their smaller size compared to the estuarine network in Sundarbans (Table 3.2). However, Bird et al. (2008) inferred that the organic matter export from these two rivers of Myanmar is one of the highest reported so far from river estuaries of similar size and shape. They principally attributed the monsoon-driven runoff, the steep slopes of forest, their high level of productivity, and the small area of the coastal flood plain behind such a high rate of lateral fluxes of DOC and POC. Yoshioka et al. (2002), while working in the Bang Nara River of Thailand reported a mean DOC and POC of 516 $\pm$ 108 and 241 $\pm$ 166 μmol kg^{-1}; however, they did not compute the lateral fluxes of these parameters to the ocean. Indonesia

Table 3.2 Lateral DOC and POC fluxes observed from rivers and estuaries in the Indian Ocean bordering countries

Country	Estuary	DOC (Tg C yr^{-1})	POC (Tg C yr^{-1})	Authors
India (AS)	Sundarban	0.58	3.03	Ray et al. (2018)
Myanmar (AS)	Thanlwin and Ayeyarwady	0.000000032	0.0000077	Bird et al. (2008)
Indonesia (AS)	Dumai	0.03		Alkhatib et al. (2007)
	Sebangau	0.43	0.03	Moore et al. (2011)
Malaysia (AS)	Rajang*	0.269	0.246	Müller-Dum et al. (2019)
Tanzania (AF)	Pangani	0.00174	0.00073	Selemani et al. (2018)
South Africa (AF)	Orange	0.03	0.01	Martins and Probst (1991)
Kenya	Tana	0.01	0.03	Bouillon et al. (2007a)

*flows into the South China Sea

received a fair bit of attention in this regard. Alkhatib et al. (2007) reported a mean DOC concentration of 4983 μmol kg^{-1} from the Dumai River of Indonesia, flowing through Sumatra. Baum et al. (2007) reported a lower range of DOC (554–2594) from the Siak River of Sumatra. Moore et al. (2011) researched the Sebangau River Estuary in the Kalimantan region of Indonesia and observed a lateral DOC and POC flux of 0.43 Tg C and 0.03 Tg C per year. They further observed a DOC and POC range of 2333–4333 and 58–400 μmol kg^{-1}, respectively (Figs. 3.2 and 3.3). Lately, Triana et al. (2021) sampled at 43 locations in Indonesian coastal waters and reported a POC range of 1.6–45.3 μmol kg^{-1}. Müller-Dum et al. (2019), while working in the Rajang River Estuary of Malaysia observed a DOC and POC input of 0.269 and 0.246 Tg C per year, respectively. However, this river estuary though situated in Malaysia drains into the South China Sea, which is technically a part of the Pacific Ocean.

Studies in this regard are scarce in the northwestern sector of the Indian Ocean. Ray and Shahraki (2016) observed substantially high DOC and POC concentrations (315 ± 25 and 303 ± 37 μmol kg^{-1}) in water columns adjoining the Queshm Islands in Iran. They observed substantially high autotrophic productivity and long residence time of the water mass that led to such high concentrations of these two parameters. According to Ray and Shahraki (2016), the enhanced suspended load is also responsible for such high magnitudes. Ittekkot and Arain (1986) observed a varying POC range of 25–1333 μmol kg^{-1} in the Indus River Estuary flowing through Pakistan. They also measured DOC and stated that often POC/DOC maintained a ratio of 1; however, they did not report the DOC concentration ranges. Among the

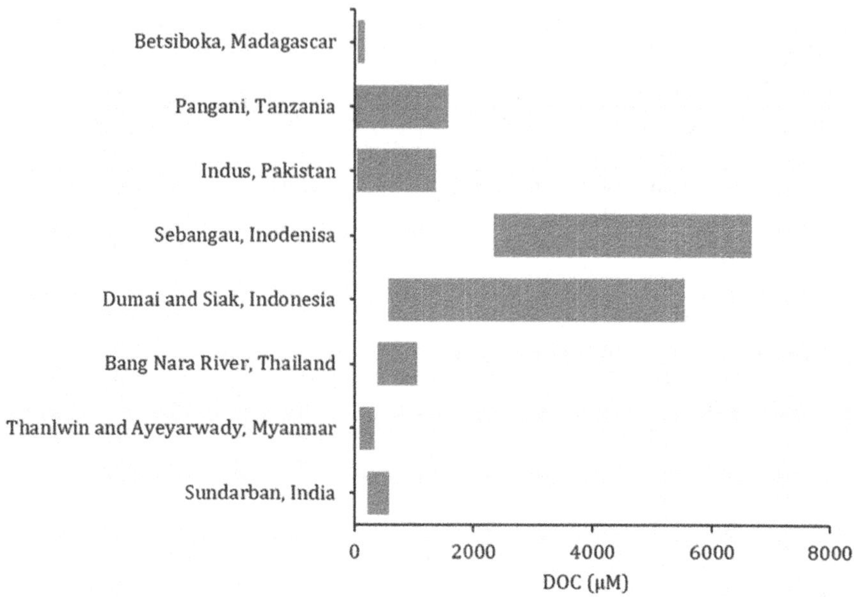

Fig. 3.2 Range of DOC concentration observed in the estuaries draining into the Indian Ocean (Data collated from Ittekkot and Arain 1986; Yoshioka et al. 2002; Alkhatib et al. 2007; Baum et al. 2007; Bird et al. 2008; Ralison et al. 2008; Moore et al. 2011; Ray et al. 2018; Selemani et al. 2018)

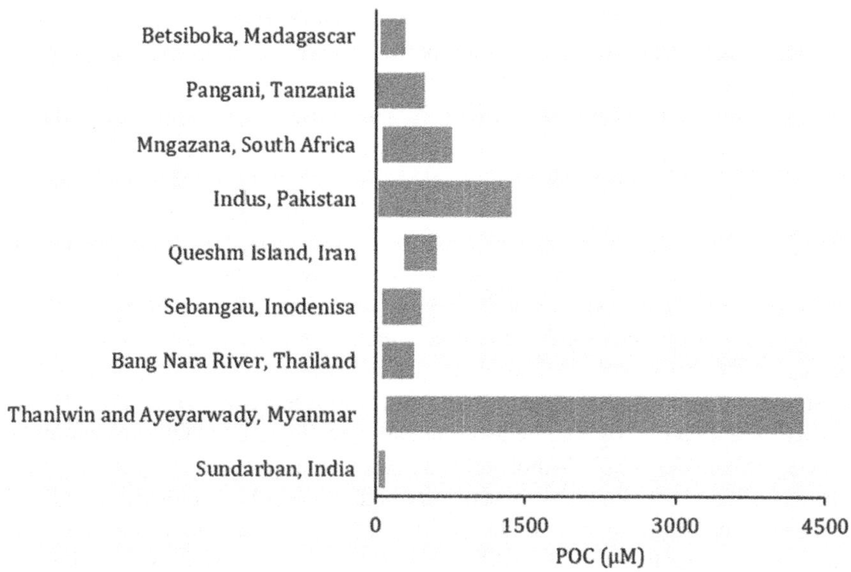

Fig. 3.3 Range of DOC concentration observed in the estuaries draining into the Indian Ocean (Data collated from Ittekkot and Arain 1986; Yoshioka et al. 2002; Rajkaran and Adams 2007; Bird et al. 2008; Ralison et al. 2008; Moore et al. 2011; Ray and Shahraki 2016; Ray et al. 2018; Selemani et al. 2018)

east-African countries, Rajkaran and Adams (2007) quantified DOC and POC in the Mngazana Estuary, South Africa. They observed a wide range of POC (60–700 μmol kg^{-1}), but the DOC levels were below the detection limit. They inferred that this estuary acted as a net sink for DOC. Rajkaran and Adams (2007) also indicated that mangrove harvesting is inversely proportional to the POC load in the estuarine water column, and it plays a crucial role in the sustenance of the marine food web in the nearshore coastal waters. Martins and Probst (1991) observed a DOC and POC flux of 0.03 and 0.01 Tg C per year in the Orange Estuary of South Africa. Bouillon et al. (2007a) observed an exactly opposite estimate of DOC and POC lateral flux in the Tana Estuary of Kenya. Selemani et al. (2018), while working in the Pangani River Estuary of Tanzania observed a DOC and POC flux of 0.00174 and 0.00073 Tg C per year, respectively. They also recorded a DOC and POC range of 9–1560 and 8–480 μmol kg^{-1}, respectively. Among the island countries in the Indian Ocean, Ralison et al. (2008) researched the largest estuary of Madagascar, Betsiboka. They observed a DOC and POC range of 40–120 and 42–250 μmol kg^{-1}, respectively.

3.1.3 Air-Water CO_2 Exchange and Degassing of CO_2 from the Estuaries

Characterizing inter-spheral dynamics of greenhouse gases in several ecosystem compartments has gained impetus in the last three to four decades. The estuarine ecosystems have received substantial attention as these have been mostly sources of CO_2 toward the atmosphere. pCO_2(water) estimation and trying to understand the regulators of this crucial parameter that governs the air-water CO_2 exchange have been prioritized by many scholars working in this domain. Several studies measured carbonate chemistry parameters like pH, TAlk, salinity, water temperature, nutrients, and DIC to estimate the pCO_2(water). However, in the last few decades, the direct measurement of in-situ pCO_2(water) with the help of a non-dispersive infrared gas analyzer and equilibrator, has replaced the indirect estimation techniques (DeGrandpre et al. 1995; Hunt et al. 2017). However, measuring the air-water CO_2 exchange continues to be a challenging endeavor. The bulk formula method has been quite popular among the scholarly communities but this approach has several limitations and scopes of uncertainties. Eddy covariance techniques have largely replaced the bulk formula methods; however, these approaches are also not free from uncertainties (Dong et al., 2021). Quantifying the gas transfer velocities has always been the most critical aspect of computing the air-water CO_2 fluxes. Several oceanographers came up with region-specific models that have been implemented in several estuaries depending on the geophysical and geomorphologic characteristics of the seascape (Wanninkhof 2014; Van Dam et al. 2019; Tokoro and Kuwae 2018). Chemical tracers have found an all-around utility in characterizing the region-specific gas transfer velocities (Ho et al. 2011, 2014; 2016). Several workers carried out

pCO_2(water) assessments followed by estimation of air-water CO_2 exchange in many estuaries that drain into the Indian Ocean. However, the number of such studies and the locations remain biased and skewed to certain places.

Indian peninsula drains several estuaries to the northern Indian Ocean. Sarma et al. (2012a) carried out exhaustive work in this regard on the Indian estuaries. They sampled twenty-seven estuaries of this country and observed a pCO_2(water) range of from ~ 300 to ~ 18,500 μatm (Fig. 3.4), which corresponded to a flux range of from −0.2 to 362 mmol m^{-2} d^{-1} (Fig. 3.5). Akhand et al. (2016) researched two adjacent estuaries on the east coast of India.

The Hooghly estuary, which maintains substantial freshwater throughout the year exhibited much higher pCO_2(water) and CO_2 effluxes than the Matla estuary, which suffers from freshwater scarcity, except the monsoon season. Akhand et al. (2016) observed much higher monsoon effluxes in the inner estuarine stations of Hooghly than that reported by Sarma et al. (2012a). Sarma et al. (2011) observed pCO_2(water)

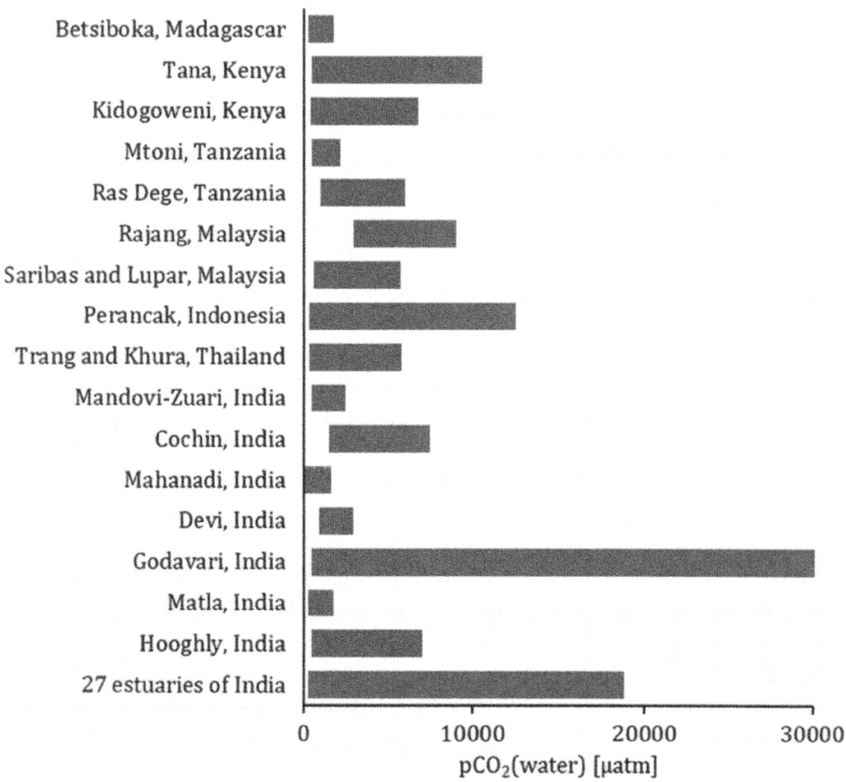

Fig. 3.4 Range of pCO_2(water) observed in the estuaries draining into the Indian Ocean (Data collated from Sarma et al. 2001; 2011; 2012; Borges et al. 2004; Bouillon et al. 2007; Kristensen et al. 2008; Ralison et al. 2008; Gupta et al. 2009; Miyajima et al. 2009; Akhand et al. 2016; Müller et al. 2016; Müller-Dum et al. 2019; Macklin et al. 2020; Pattanaik et al. 2020b; Akhtar et al. 2021)

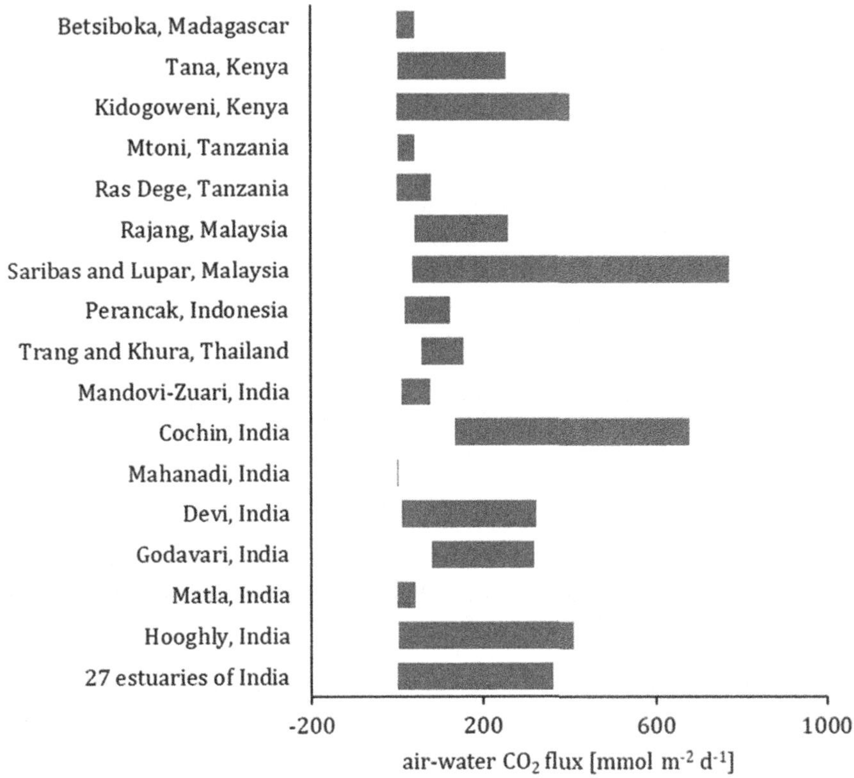

Fig. 3.5 Range of air-water CO_2 fluxes observed in the estuaries draining into the Indian Ocean (Data collated from Sarma et al. 2001; 2011; 2012; Borges et al. 2004; Bouillon et al. 2007; Kristensen et al. 2008; Ralison et al. 2008; Gupta et al. 2009; Miyajima et al. 2009; Akhand et al. 2016; Müller et al. 2016; Müller-Dum et al. 2019; Macklin et al. 2020; Pattanaik et al. 2020b; Akhtar et al. 2021)

as high as 30,000 μatm in the Godavari estuary of India during the monsoon season. Akhtar et al. (2021) observed a modest seasonal mean pCO_2(water) range of 950–~ 2000 μatm corresponding to a flux range of 12–312 mmol m^{-2} d^{-1}. Pattanaik et al. (2020b) observed a pCO_2(water) range of 100–1571 μatm. However, the CO_2 flux range varied between a small range of − 0.14 and 0.31 mmol m^{-2} d^{-1}. Gupta et al. (2009) observed a pCO_2(water) range of 1500–6000 μatm and a net negative ecosystem production of 136–541 mmol m^{-2} d^{-1} in the Cochin estuary on the west coast of India. In another west coast estuary of India, namely Mandovi-Zuari, Sarma et al. (2011) observed a pCO_2(water) range of 500–2000 μatm, followed by a CO_2 flux range of 11–67 mmol m^{-2} d^{-1}. Studies in this regard are hard to find in Bangladesh and Myanmar estuaries to date. Miyajima et al. (2009) observed a pCO_2(water) range of 365–5428 μatm in the Khura and Trang estuaries of Thailand. Chen et al. (2013) carried out a meta-analysis of pCO_2(water) and air_water

CO_2 fluxes. In this review, Chen et al. (2013) estimated the annual mean CO_2 flux from the Khura estuary from the pCO_2(water) magnitudes of Miyajima et al. (2009) to be 98 mmol m^{-2} d^{-1}. Müller et al. (2016) worked in the Saribas and Lupar River Estuaries of Malaysia and observed a pCO_2(water) range of 600–5000 μatm corresponding to an emission range of 38–734 mmol m^{-2} d^{-1}. They observed minimal seasonal variability. Müller-Dum et al. (2019) observed similar nominal variability in pCO_2(water) while working in the Rajang estuary of Malaysia. They observed a pCO_2(water) range of 3000–6000 μatm followed by a flux range of 42–217 mmol m^{-2} d^{-1}. Macklin et al. (2020) characterized the pCO_2(water) and CO_2 emission rates of the Perancak Estuary of Indonesia. pCO_2(water) varied from 330 to 12,126 μatm, whereas the estuarine CO_2 fluxes varied between 19.1 and 107.3 mmol m^{-2} d^{-1}. In a synoptic-scale study, Wit et al. (2015) observed that Malaysian and Indonesian river estuaries emit around 67 ± 16 Tg C yr^{-1}.

The northern half of the western Indian Ocean bordering Asian countries like Oman, Iran, UAE, Yemen, and Pakistan have almost no studies in this regard. However, the eastern coastline of the African continent received a tad bit of attention. Borges et al. (2004) reported the pCO_2 load and CO_2 emission from the Kidogoweni estuary of Kenya. They sampled only in one month (July 2003) and observed a varying pCO_2(water) of 430–6320 μatm, and a corresponding air-water CO_2 flux of 1–400 mmol m^{-2} d^{-1}. Bouillon et al. (2007a) observed pCO_2(water) values as high as 10,000 μatm in the Tana estuary of Kenya. The deltaic regions of this estuary exhibited an emission range of 3–252 mmol m^{-2} d^{-1}. Bouillon et al. (2007b) observed a pCO_2(water) range of 430–2450 μatm and a mean air-water CO_2 flux of 39 ± 116 mmol m^{-2} d^{-1}, respectively, in the Gazi Bay region of Kenya. Kristensen et al. (2008) worked on the mangrove waters of two forests in Tanzania, namely Mtoni and Ras Dege. They observed much higher pCO_2(water) and CO_2 emission ranges from the pristine mangrove site (Ras Dege) than that observed in the site having significant anthropogenic intervention. Ralison et al. (2008) observed a comparatively lower range of pCO_2(water) and air-water CO_2 exchange (mean: 9.1 ± 14.2 mmol m^{-2} d^{-1}) in the Betsiboka estuary of Madagascar than that observed in many tropical and subtropical estuaries of the world. Apart from Madagascar, similar studies are not available in any other island nations of the Indian Ocean.

3.1.4 Factors Regulating the Air-Water CO_2 Exchange

Several atmospheric, hydrological, and biogeochemical factors regulate the pCO_2(water) dynamics and the exchange rate of CO_2 between the water surface and the atmosphere. Often these factors and their regulating role exhibit a wide spatiotemporal variability. Sarma et al. (2012a) observed that monsoon-induced precipitation played a significant role in enhancing the CO_2 fluxes from the estuaries of India. They observed a significant positive correlation between the freshwater discharge and pCO_2(water). Incidentally, POC also increased during the monsoon season in almost all Indian estuaries. Moreover, they observed the highest CO_2 emissions toward the atmosphere from estuaries, which facilitated unhindered wind flow. Thus, estuarine stretches with high wind velocity acted as strong sources of CO_2. The dry seasons recorded much lower pCO_2(water) and lesser air-water CO_2 fluxes from all the Indian estuaries compared to that observed during the monsoon season. Akhand et al. (2016), while comparing the pCO2(water) and CO_2 fluxes between a freshwater abundant and a freshwater deprived estuary of India, observed that freshwater discharge promoted net heterotrophy in the water column and enhanced the CO_2 fluxes toward the atmosphere. They also inferred that salinity plays a crucial role in acting as a proxy of pCO_2(water). Highly saline waters were deficient in pCO_2(water), and in places having acute scarcity of freshwater from the upstream and a predominance of seawater (the outer estuarine reaches of Matla), the estuary often acted as a net sink for CO_2, especially in the post-monsoon months (Akhand et al. 2021a). Sarma et al. (2011) observed that unfavorable photosynthetic conditions and an abundance of organic substrates facilitated by high freshwater discharge led to severe net heterotrophy and enhanced CO_2 emissions from the Godavari estuary of India. Akhtar et al. (2021), while working in the principal estuary of the Mahanadi estuary (Devi estuary), observed a contradicting seasonal trend in CO_2 fluxes. They observed higher effluxes in the post-monsoon season, followed by pre-monsoon, and the least in the monsoon season. They observed that the Devi estuary remained supersaturated with oxygen throughout the year. However, in the dry seasons, they observed a decrease in pH coupled with an increase in pCO_2(water), and they attributed the degradation of organic matter as the prime factor behind the CO_2 emissions from Devi estuary. Pattanaik et al. (2020b) observed a near-neutral CO_2 flux in this estuary and they attributed the nutrient input and heterotrophic processes behind the CO_2 emission from this estuary. Gupta et al. (2009) mentioned that the Cochin estuary used to act as a net sink; however, in the past two decades, it has turned into a net heterotrophic ecosystem and started emitting CO_2 toward the atmosphere. They held indiscriminate and unplanned urbanization responsible for such a switchover of CO_2 flux character. Sarma et al. (2011), while working in the Mandovi-Zuari estuarine system on the west coast of India, emphasized the role of pH regulation due to the soil-water biogeochemical interaction. They observed that this interaction significantly regulates the pCO_2(water) in the surface water of the estuary. Miyajima et al. (2009), while researching the Khura and Trang estuaries of Thailand, recorded the highest pCO_2(water) in the low saline upper reaches, whereas the lowest values prevailed near the estuarine mouths. This type of spatial variability was the most common among the global estuaries indicating the significance of freshwater in enhancing the pCO_2 load of the water column. Miyajima et al. (2009) also underpinned the significance of the blue carbon ecosystems like mangroves, seagrass, and saltmarshes in regulating the carbonate chemistry of the estuarine and the nearshore coastal waters. Müller et al. (2016a), while researching the Saribas and Lupar River Estuaries of Malaysia, observed that wind velocity-based gas transfer velocities lead to underestimation of CO_2 fluxes, whereas turbulence-driven gas transfer estimation based on estuarine water current portrays the real scenario. Müller-Dum et al. (2019), while researching the Rajang estuary of Malaysia emphasized the significance of water draining through peatlands in regulating the pCO_2(water) load in the estuarine water column. Macklin et al. (2020), while working in the Perancak Estuary of Indonesia, observed that groundwater discharge and episodic precipitation events played a crucial role in governing the CO_2 emission dynamics. Wit et al. (2015) argued that the Southeast Asian countries like Malaysia and Indonesia comprise substantially high riverine and estuarine areas, yet they emit moderate quantities of CO_2. They attributed the short residence time of organic carbon in the waters drained through peatlands behind such nominal CO_2 emission rates. The short residence does not facilitate ample time for the required remineralization of DOC to DIC, and hence the pCO_2 load in the water column does not climb up to higher levels. Borges et al. (2004) during their research in the Kidogoweni estuary of Kenya observed a significant non-conservative variation of DIC and TAlk throughout the length of the estuary. They observed that the diagenetic processes may a larger role in regulating the DIC and TAlk of the water column of this estuary. Bouillon et al. (2007) also inferred that anaerobic diagenetic processes are key controlling factors of the CO_2 supersaturation in the east African estuaries and deltaic regimes. Kristensen et al. (2008), while working in the Tanzanian mangrove adjacent waters observed that tidal amplitude, wind velocity, and pore water seepage played a crucial role in governing the water column CO_2 dynamics.

3.1.5 Continental Shelves: Source or Sink of CO_2 and CH_4

The water mass above the continental shelves or the nearshore waters experiences unique biogeochemistry that is much different than the open ocean and the freshwater end members. These regions derive substantial terrestrial input and process it to several end products before getting diluted with the ocean water volume. Several studies indicated that continental shelf waters are the most biogeochemically active hydrospheres on earth (Painter 2020; Das and Ghosh 2021). However, these regions are difficult to study as well. Most of these waters remain rough and turbulent throughout the year. Accessibility and navigation with research vessels often face several hindrances. These reasons have perhaps delimited the number of research papers on these regions and it is scarce in the Indian Ocean's shelf waters too. Sudheesh et al. (2020), while working in the nearshore coastal waters of the Arabian Sea off the Indian coastline, observed an elevated concentration of CH_4(water) in the land to the sea transition zone. They observed that methanogenesis in the sediment compartment and supply of organic substrate through estuarine discharge plays a crucial role in regulating the fluxes. Signatures of methanotrophy and anaerobic oxidation were available in their findings; however, the entire nearshore water regime bordering the southwest coastline of India acted as a source of CH_4 toward the atmosphere. Rao and Sharma (2017) observed that the coastal waters of the Bay of Bengal adjoining the Indian coastline also acted as a source of CH_4. However, the northern part of the coastal waters emitted much less CH_4 than the southwestern part throughout the year. Sarma et al. (2012b) focused on the same region from the perspective of CO_2 dynamics. They inferred that contrary to the belief that the shelf waters act as CO_2 sink, the coastal Bay of Bengal waters acted as a net source of CO_2 toward the atmosphere. In an annual cycle, this region act as a net sink during the retreating monsoon phase. However, during the peak of the southwest monsoon, this region emits a substantial amount of CO_2 that supersedes the sink strength. Sarma et al. (2018) inferred that the East India Coastal Current plays a fundamental role in governing the DIC dynamics along the coastal waters of this ocean basin. Linto et al. (2014) observed substantial CO_2 emission from the mangrove adjacent nearshore waters of the Andaman Islands in the eastern Bay of Bengal; however, the CH_4 emission was much lower than that reported from the western Bay of Bengal. Laruelle et al. (2014) compartmentalized the air-water CO_2 fluxes in the nearshore shelf waters of the globe into basic subdivisions. According to their model-derived estimates, both eastern and western Arabian Sea continental margins acted as a net source of CO_2. However, they inferred the tropical eastern Indian Ocean region and the Bay of Bengal to be quasi-neutral CO_2 sinks. Laruelle et al. (2014) further indicated that the east-African shelf waters acted as a moderate sink for CO_2; however, data was not sufficient for the western boundary of the Indian Ocean to comment on the Persian Gulf and Gulf of Oman waters.

3.1.6 CH_4 Dynamics in the Estuarine and Shelf Waters

The wetlands are the principal natural sources of CH_4 toward the atmosphere (Zhang et al. 2017; Ganesan et al. 2018) and these ecosystems received due attention from the global scientific communities. However, characterizing CH_4 dynamics in estuaries and continental shelf waters remains understudied. The number of such studies is disproportionately less in the periphery of the Indian Ocean. Very little research exists in some of the countries bordering this ocean basin. However, in India, several researchers studied the CH_4 dynamics in the estuarine water column. Shalini et al. (2006) observed a CH_4(water) concentration's variability of 94 to 501 nM, corresponding to a flux range of 35–280 μmol m^{-2} d^{-1} in the shallow Pulicat estuary of India. They observed that diffusion was of secondary importance in this estuary and ebullition was the principal pathway of methane emission. Araujo et al. (2018) studied the spatial variability of the CH_4 dynamics along the salinity gradient in the Mandovi estuary. They observed that water column CH_4 concentration varied between 28 and 360 nM (Fig. 3.6) and the CH_4 flux ranged between 4.7 and 16.0 μmol m^{-2} d^{-1} (Fig. 3.7). Besides, they also observed a significant spatial and depth-wise variability of pore water CH_4 concentration that ranged from 2 to 9529 $_{\mu M}$. Overall, they inferred that freshwater dominance led to higher CH_4 emission as salinity plays the most crucial role in enhancing methanotrophy. The monsoon season emits substantial CH_4 throughout the entire stretch of the estuary due to the excessive freshwater dominance induced by runoffs and discharge. Dutta et al. (2017) studied the CH_4 biogeochemistry in the mangrove estuaries of Sundarban. They observed a CH_4(water) variability of 13–91 $_{nM}$ in the estuarine surface and subsurface waters.

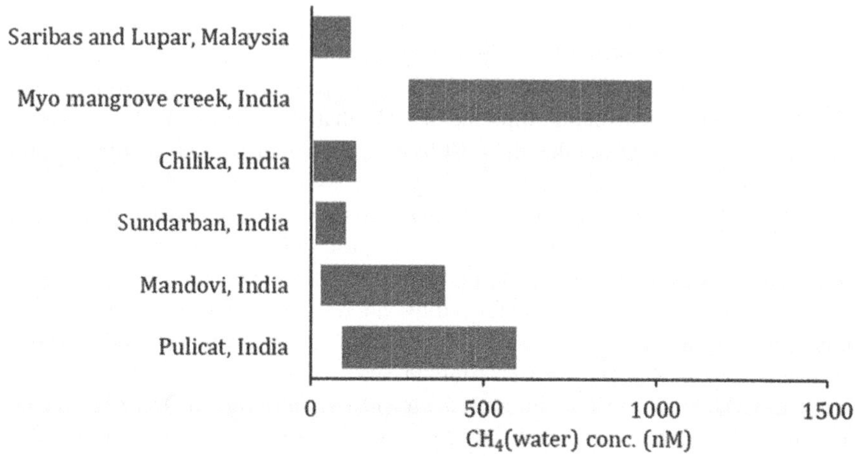

Fig. 3.6 Range of CH_4 concentration in water observed in the estuaries draining into the Indian Ocean (Data collated from Barnes et al. 2006; Shalini et al. 2006; Müller et al., 2016b; Dutta et al. 2017; Araujo et al. 2018; Banerjee et al. 2018)

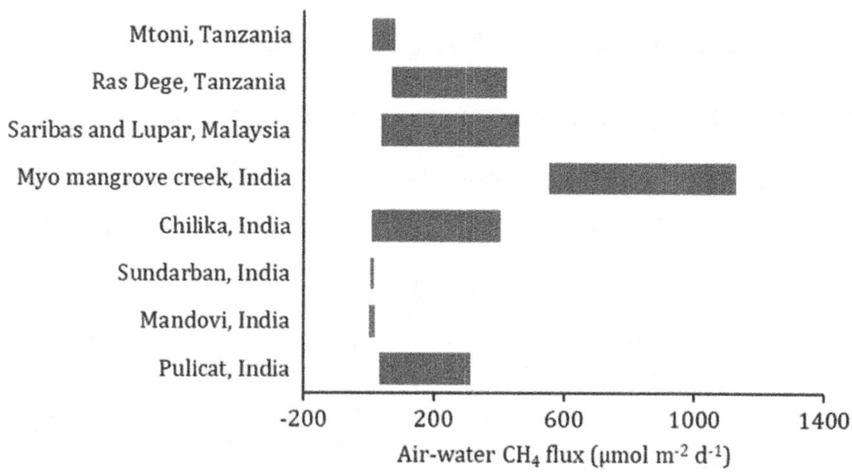

Fig. 3.7 Range of air-water CH_4 fluxes observed in the estuaries draining into the Indian Ocean (Data collated from Barnes et al. 2006; Shalini et al. 2006; Müller et al., 2016b; Dutta et al. 2017; Araujo et al. 2018; Banerjee et al. 2018)

They reported pore water CH_4 values more than 50 times of the surface and subsurface CH_4 concentration. The fluxes ranged between 6.3 and 9.7 μmol m^{-2} d^{-1}. They inferred that anaerobic CH_4 oxidation played a vital role in reducing the net CH_4 efflux from the Sundarban mangrove forest toward the atmosphere. Banerjee et al. (2018) measured the CH_4 concentration in the aquatic column adjacent to the seagrass and macrophyte-laden Chilika lagoon on the east coast of India. They observed a significant seasonality in the CH_4 fluxes and spatial variability within the lagoon. The CH_4(water) concentration ranged between 7 and 127 nM, whereas the fluxes varied from 11 to 391 μmol m^{-2} d^{-1}. They observed that the seagrass meadows emitted a modest amount of CH_4 toward the atmosphere. Barnes et al. (2006) measured the CH_4(water) concentration in the shallow tidal creeks of Myo mangrove forest in the Andaman Islands. CH_4(water) in this creek varied between 282 and 704 $_{nM}$ across two seasons. However, the CH_4 fluxes varied in a narrow range of 552–574 μmol m^{-2} d^{-1}. They observed that tidal fluctuation and rainfall significantly regulated the diffusive fluxes of CH_4 in these creeks.

Besides India, the number of studies characterizing CH_4 emission from estuarine stretches is scarce among the Indian Ocean peripheral countries. Müller et al. (2016b) studied the CH_4 dynamics in the Saribas and Lupar estuaries of the Borneo sector in Sarawak, Malaysia. CH_4(water) concentration varied between 3.7 and 113.9 nM and air-water CH_4 fluxes ranged from 39 to 418 μmol m^{-2} d^{-1}. They inferred that the monsoon-induced variability in the fluxes from these estuaries is crucial as they contribute to the annual budget of CH_4. Moreover, the disturbance posed to the peat soils adjacent to these estuaries might have a long-term effect on the emission potential of this region. Kristensen et al. (2008) estimated CH_4 concentration in water and computed the air-water CH_4 emission from the mangrove adjacent waters (creeks) of Ras Dege and Mtoni forest of Tanzania. They observed a mean CH_4(water) concentration of 150 nM and 1000 nM in Mtoni and Ras Dege mangrove creeks, respectively. The water to air CH_4 effluxes varied from 10 to 70 μmol m^{-2} d^{-1} and 70 to 350$^{\mu mol\ m^{-2}}$ d^{-1} in Mtoni and Ras Dege waters, respectively. They inferred that the mangrove-adjacent waters that are eutrophic produce substantially higher quantities of CH_4 than the meso- or oligotrophic regions.

3.1.7 *Scope of Future Work*

The mining of literature on the estuaries and nearshore waters of the Indian Ocean indicates a need for more studies to characterize the carbon dynamics and understand the holistic scenario in this sector. Indian coastline has many studies in this regard. Malaysian and Indonesian estuaries and nearshore coastal waters have also received a fair bit of attention. However, the western Indian Ocean periphery remains data deficient. Compared to the coast length of East Africa, the number of studies is not sufficient. Several small to large estuaries exist in the periphery of the Indian Ocean. The model-derived estimates indicate the pros and cons of the CO_2 and CH_4 budget from the estuaries and continental shelf waters of the globe. However, to eliminate the uncertainties of such models, ground-level data from all the estuaries are ideally essential. However, several estuaries have not yet been studied from this viewpoint. The marine scientist community of all these countries should strive to monitor the labile carbon dynamics in these fluvial sectors to comment on the recalcitrant counterpart that remains stored for the millennial scale. Buoys and automated sensors are gradually replacing the erstwhile cumbersome measurement protocols.

Today, long-term field data acquisition is possible from remote places without human presence. Such fixed station monitoring for a long period has become an utmost necessity to understand the natural carbon dynamics in such hydrologically active regimes. Bearing the cost for such monitoring programs often becomes difficult for researchers from developing and under-developed countries. Thus, government-level intervention is best to build a data repository that can undergo subsequent analysis by specialists that can add to the global data pool.

References

Abril G, Etcheber H, Le Hir P, Bassoullet P, Boutier B, Frankignoulle M (1999) Oxic/anoxic oscillations and organic carbon mineralization in an estuarine maximum turbidity zone (The Gironde, France). Limnol Oceanogr 44(5):1304–1315. https://doi.org/10.4319/lo.1999.44.5.1304

Afdal PLMG, Noerjito DR (2011) Fluks Karbon Dioksida, Hubungannya dengan Produktifitas Primer Fitoplankton di Perairan Estuari Donan. Oseanologi Dan Limnologi Di Indonesia 37:323–337

Akhand A, Chanda A, Manna S, Das S, Hazra S, Roy R et al (2016) A comparison of CO_2 dynamics and air-water fluxes in a river-dominated estuary and a mangrove-dominated marine estuary. Geophys Res Lett 43(22):11–726. https://doi.org/10.1002/2016GL070716

Akhand A, Chanda A, Watanabe K, Das S, Tokoro T, Hazra S, Kuwae T (2021a) Reduction in riverine freshwater supply changes inorganic and organic carbon dynamics and air-water CO_2 fluxes in a tropical mangrove dominated estuary. J Geophys Res Biogeosci 126(5): e2020JG006144. https://doi.org/10.1029/2020JG006144

Akhand A, Watanabe K, Chanda A, Tokoro T, Chakraborty K, Moki H et al (2021b) Lateral carbon fluxes and CO_2 evasion from a subtropical mangrove-seagrass-coral continuum. Sci Total Environ 752:142190. https://doi.org/10.1016/j.scitotenv.2020.142190

Akhtar S, Equeenuddin SM, Bastia F (2021) Distribution of pCO_2 and air-sea CO_2 flux in Devi estuary, eastern India. Appl Geochem 131:105003. https://doi.org/10.1016/j.apgeochem.2021.105003

Alkhatib M, Jennerjahn TC, Samiaji J (2007) Biogeochemistry of the Dumai River estuary, Sumatra, Indonesia, a tropical black-water river. Limnol Oceanogr 52(6):2410–2417. https://doi.org/10.4319/lo.2007.52.6.2410

Araujo J, Pratihary A, Naik R, Naik H, Naqvi SWA (2018) Benthic fluxes of methane along the salinity gradient of a tropical monsoonal estuary: Implications for CH_4 supersaturation and emission. Mar Chem 202:73–85. https://doi.org/10.1016/j.marchem.2018.03.008

Banerjee K, Paneerselvam A, Ramachandran P, Ganguly D, Singh G, Ramesh R (2018) Seagrass and macrophyte mediated CO_2 and CH_4 dynamics in shallow coastal waters. PLoS ONE 13(10):e0203922. https://doi.org/10.1371/journal.pone.0203922

Barnes J, Ramesh R, Purvaja R, Nirmal Rajkumar A, Senthil Kumar B, Krithika K et al (2006) Tidal dynamics and rainfall control N_2O and CH_4 emissions from a pristine mangrove creek. Geophys Res Lett 33(15). https://doi.org/10.1029/2006GL026829

Baum A, Rixen T, Samiaji J (2007) Relevance of peat-draining rivers in central Sumatra for the riverine input of dissolved organic carbon into the ocean. Estuar Coast Shelf Sci 73(3–4):563–570. https://doi.org/10.1016/j.ecss.2007.02.012

Belshe EF, Mateo MA, Gillis L, Zimmer M, Teichberg M (2017) Muddy waters: unintentional consequences of blue carbon research obscure our understanding of organic carbon dynamics in seagrass ecosystems. Front Mar Sci 4:125. https://doi.org/10.3389/fmars.2017.00125

Bianchi TS (2006) Biogeochemistry of estuaries. Oxford University Press, pp 720

Bird MI, Robinson RAJ, Oo NW, Aye MM, Lu XX, Higgitt DL et al (2008) A preliminary estimate of organic carbon transport by the Ayeyarwady (Irrawaddy) and Thanlwin (Salween) Rivers of Myanmar. Quat Int 186(1):113–122. https://doi.org/10.1016/j.quaint.2007.08.003

Borges AV, Abril G, Bouillon S (2018) Carbon dynamics and CO_2 and CH_4 outgassing in the Mekong delta. Biogeosciences 15(4):1093–1114. https://doi.org/10.5194/bg-15-1093-2018

Borges A, Frankignoulle M, Delille B, Bouillon S (2004) DIC dynamics in a tropical estuary (Kiddogoweni, Kenya). European Geosciences Union, 1st General Assembly, (Nice, France, 25–30 April 2004)

Bouillon S, Dehairs F, Schiettecatte L, Borges AV (2007a) Biogeochemistry of the Tana estuary and delta (northern Kenya). Limnol Oceanogr 52:46–59. https://doi.org/10.4319/LO.2007.52.1.0046

Bouillon S, Dehairs F, Velimirov B, Abril G, Borges AV (2007b) Dynamics of organic and inorganic carbon across contiguous mangrove and seagrass systems (Gazi Bay, Kenya). J Geophys Res Biogeosci 112(G2). https://doi.org/10.1029/2006JG000325

Cai WJ (2011) Estuarine and coastal ocean carbon paradox: CO_2 sinks or sites of terrestrial carbon incineration? Ann Rev Mar Sci 3:123–145. https://doi.org/10.1146/annurev-marine-120709-142723

Chen CT, Huang TH, Chen YC, Bai Y, He X, Kang Y (2013) Air-sea exchanges of CO_2 in the world's coastal seas. Biogeosciences 10(10):6509–6544. https://doi.org/10.5194/bg-10-6509-2013

Crosswell JR, Anderson IC, Stanhope JW, Van Dam B, Brush MJ, Ensign S et al (2017) Carbon budget of a shallow, lagoonal estuary: transformations and source-sink dynamics along the river-estuary-ocean continuum. Limnol Oceanogr 62(S1):S29–S45. https://doi.org/10.1002/lno.10631

Das S, Ghosh T (eds) (2021) Estuarine biogeochemical dynamics of the east coast of India. Springer International Publishing, pp 268. https://doi.org/10.1007/978-3-030-68980-3

de Brauwere A, De Brye B, Blaise S, Deleersnijder E (2011) Residence time, exposure time and connectivity in the Scheldt Estuary. J Mar Syst 84(3–4):85–95. https://doi.org/10.1016/j.jmarsys.2010.10.001

DeGrandpre MD, Hammar TR, Smith SP, Sayles FL (1995) In situ measurements of seawater pCO_2. Limnol Oceanogr 40(5):969–975. https://doi.org/10.4319/lo.1995.40.5.0969

Dong Y, Yang M, Bakker DC, Kitidis V, Bell TG (2021) Uncertainties in eddy covariance air-sea CO_2 flux measurements and implications for gas transfer velocity parameterizations. Atmospheric Chem Phys 21(10):8089–8110. https://doi.org/10.5194/acp-21-8089-2021

Dutta MK, Kumar S, Mukherjee R, Sharma N, Acharya A, Sanyal P et al (2019) Diurnal carbon dynamics in a mangrove-dominated tropical estuary (Sundarbans, India). Estuar Coast Shelf Sci 229:106426. https://doi.org/10.1016/j.ecss.2019.106426

Dutta MK, Bianchi TS, Mukhopadhyay SK (2017) Mangrove methane biogeochemistry in the Indian Sundarbans: a proposed budget. Front Mar Sci 4:187. https://doi.org/10.3389/fmars.2017.00187

Friedman JR, Shadwick EH, Friedrichs MA, Najjar RG, De Meo OA, Da F, Smith JL (2020) Seasonal variability of the CO_2 system in a large coastal plain estuary. J Geophys Res Oceans 125(1):e2019JC015609. https://doi.org/10.1029/2019JC015609

Ganesan AL, Stell AC, Gedney N, Comyn-Platt E, Hayman G, Rigby M et al (2018) Spatially resolved isotopic source signatures of wetland methane emissions. Geophys Res Lett 45(8):3737–3745. https://doi.org/10.1002/2018GL077536

Ganguly D, Singh G, Ramachandran P, Selvam AP, Banerjee K, Ramachandran R (2017) Seagrass metabolism and carbon dynamics in a tropical coastal embayment. Ambio 46(6):667–679. https://doi.org/10.1007/s13280-017-0916-8

Ghosh J, Chakraborty K, Chanda A, Akhand A, Bhattacharya T, Das S et al (2021) Outwelling of total alkalinity and dissolved inorganic carbon from the Hooghly River to the adjacent coastal Bay of Bengal. Environ Monit Assess 193(7):1–14. https://doi.org/10.1007/s10661-021-09191-y

Guardiani J, Tobias C, Smith R (2021) Seasonal carbon dynamics in a temperate Lagoonal Estuary: New River, NC, USA. Estuar Coast 1–21. https://doi.org/10.1007/s12237-021-00992-5

Gupta GVM, Thottathil SD, Balachandran KK, Madhu NV, Madeswaran P, Nair S (2009) CO_2 supersaturation and net heterotrophy in a tropical estuary (Cochin, India): influence of anthropogenic effect. Ecosystems 12(7):1145–1157. https://doi.org/10.1007/s10021-009-9280-2

Ho DT, Coffineau N, Hickman B, Chow N, Koffman T, Schlosser P (2016) Influence of current velocity and wind speed on air-water gas exchange in a mangrove estuary. Geophys Res Lett 43(8):3813–3821. https://doi.org/10.1002/2016GL068727

Ho DT, Ferrón S, Engel VC, Larsen LG, Barr JG (2014) Air-water gas exchange and CO_2 flux in a mangrove-dominated estuary. Geophys Res Lett 41(1):108–113. https://doi.org/10.1002/2013GL058785

Ho DT, Schlosser P, Orton PM (2011) On factors controlling air-water gas exchange in a large tidal river. Estuar Coast 34(6):1103–1116. https://doi.org/10.1007/s12237-011-9396-4

Howarth RW, Fruci JR, Sherman D (1991) Inputs of sediment and carbon to an estuarine ecosystem: influence of land use. Ecol Appl 1(1):27–39. https://doi.org/10.2307/1941845

Hunt CW, Snyder L, Salisbury JE, Vandemark D, McDowell WH (2017) SIPCO2: a simple, inexpensive surface water pCO_2 sensor. Limnol Oceanogr Method 15(3):291–301. https://doi.org/10.1002/lom3.10157

Ittekkot V, Arain R (1986) Nature of particulate organic matter in the river Indus, Pakistan. Geochim Cosmochim Acta 50(8):1643–1653. https://doi.org/10.1016/0016-7037(86)90127-4

Jeffrey LC, Maher DT, Santos IR, Call M, Reading MJ, Holloway C, Tait DR (2018) The spatial and temporal drivers of pCO_2, pCH_4, and gas transfer velocity within a subtropical estuary. Estuar Coast Shelf Sci 208:83–95. https://doi.org/10.1016/j.ecss.2018.04.022

Joesoef A, Huang W-J, Gao Y, Cai W-J (2015) Air-water fluxes and sources of carbon dioxide in the Delaware Estuary: spatial and seasonal variability. Biogeosciences 12:6085–6101. https://doi.org/10.5194/bg-12-6085-2015

Joesoef A, Kirchman DL, Sommerfield CK, Cai W-J (2017) Seasonal variability of the inorganic carbon system in a large coastal plain estuary. Biogeosciences 14:4949–4963. https://doi.org/10.5194/bg-14-4949-2017

Karim A, Veizer J (2000) Weathering processes in the Indus River Basin: implications from riverine carbon, sulfur, oxygen, and strontium isotopes. Chem Geol 170(1–4):153–177. https://doi.org/10.1016/S0009-2541(99)00246-6

Krauss KW, Noe GB, Duberstein JA, Conner WH, Stagg CL, Cormier N et al (2018) The role of the upper tidal estuary in wetland blue carbon storage and flux. Glob Biogeochem Cycl 32(5):817–839. https://doi.org/10.1029/2018GB005897

Krishna MS, Viswanadham R, Prasad MH, Kumari VR, Sarma VV (2019) Export fluxes of dissolved inorganic carbon to the northern Indian Ocean from the Indian monsoonal rivers. Biogeosciences 16(2):505–519. https://doi.org/10.5194/bg-16-505-2019

Kristensen E, Flindt MR, Ulomi S, Borges AV, Abril G, Bouillon S (2008) Emission of CO_2 and CH_4 to the atmosphere by sediments and open waters in two Tanzanian mangrove forests. Mar Ecol Prog Ser 370:53–67. https://doi.org/10.3354/meps07642

Laruelle GG, Cai WJ, Hu X, Gruber N, Mackenzie FT, Regnier P (2018) Continental shelves as a variable but increasing global sink for atmospheric carbon dioxide. Nature Commun 9(1):1–11. https://doi.org/10.1038/s41467-017-02738-z

Laruelle GG, Goossens N, Arndt S, Cai WJ, Regnier P (2017) Air-water CO_2 evasion from US East Coast estuaries. Biogeosciences 14(9):2441–2468. https://doi.org/10.5194/bg-14-2441-2017

Laruelle GG, Lauerwald R, Pfeil B, Regnier P (2014) Regionalized global budget of the CO_2 exchange at the air-water interface in continental shelf seas. Glob Biogeochem Cycl 28(11):1199–1214. https://doi.org/10.1002/2014GB004832

Laruelle GG, Marescaux A, Le Gendre R, Garnier J, Rabouille C, Thieu V (2019) Carbon dynamics along the Seine River network: insight from a coupled estuarine/river modeling approach. Front Mar Sci 6:216. https://doi.org/10.3389/fmars.2019.00216

Li M, Peng C, Wang M, Xue W, Zhang K, Wang K et al (2017) The carbon flux of global rivers: a re-evaluation of amount and spatial patterns. Ecol Indic 80:40–51. https://doi.org/10.1016/j.ecolind.2017.04.049

Linto N, Barnes J, Ramachandran R, Divia J, Ramachandran P, Upstill-Goddard RC (2014) Carbon dioxide and methane emissions from mangrove-associated waters of the Andaman Islands, Bay of Bengal. Estuar Coast 37(2):381–398. https://doi.org/10.1007/s12237-013-9674-4

Macklin PA, Suryaputra IGNA, Maher DT, Sidik F, Santos IR (2020) Carbon dioxide dynamics in a tropical estuary over seasonal and rain-event time scales. Cont Shelf Res 206:104196. https://doi.org/10.1016/j.csr.2020.104196

Maher DT, Call M, Santos IR, Sanders CJ (2018) Beyond burial: lateral exchange is a significant atmospheric carbon sink in mangrove forests. Biol Lett 14(7):20180200. https://doi.org/10.1098/rsbl.2018.0200

Martins O, Probst JL (1991) Biogeochemistry of major African rivers: carbon and mineral transport. In: Degens ET, Kempe S, Richey JE (eds) Biogeochemistry of major world rivers. Wiley, New York, pp 127–155

Miyajima T, Tsuboi Y, Tanaka Y, Koike I (2009) Export of inorganic carbon from two Southeast Asian mangrove forests to adjacent estuaries as estimated by the stable isotope composition of dissolved inorganic carbon. J Geophys Res Biogeosci 114(G1). https://doi.org/10.1029/2008JG000861

Moore S, Gauci V, Evans CD, Page SE (2011) Fluvial organic carbon losses from a Bornean blackwater river. Biogeosciences 8(4):901–909. https://doi.org/10.5194/bg-8-901-2011

Moran MA, Sheldon WM Jr, Zepp RG (2000) Carbon loss and optical property changes during long-term photochemical and biological degradation of estuarine dissolved organic matter. Limnol Oceanogr 45(6):1254–1264. https://doi.org/10.4319/lo.2000.45.6.1254

Müller D, Bange HW, Warneke T, Rixen T, Müller M, Mujahid A, Notholt J (2016a) Nitrous oxide and methane in two tropical estuaries in a peat-dominated region of northwestern Borneo. Biogeosciences 13(8):2415–2428. https://doi.org/10.5194/bg-13-2415-2016

Müller D, Warneke T, Rixen T, Müller M, Mujahid A, Bange HW, Notholt J (2016b) Fate of terrestrial organic carbon and associated CO_2 and CO emissions from two Southeast Asian estuaries. Biogeosciences 13(3):691–705. https://doi.org/10.5194/bg-13-691-2016

Müller-Dum D, Warneke T, Rixen T, Müller M, Baum A, Christodoulou A et al (2019) Impact of peatlands on carbon dioxide (CO_2) emissions from the Rajang River and Estuary. Malaysia. Biogeosciences 16(1):17–32. https://doi.org/10.5194/bg-16-17-2019

Ohtsuka T, Onishi T, Yoshitake S, Tomotsune M, Kida M, Iimura Y et al (2020) Lateral export of dissolved inorganic and organic carbon from a small mangrove estuary with tidal fluctuation. Forests 11(10):1041. https://doi.org/10.3390/f11101041

Oliveira AP, Cabeçadas G, Mateus MD (2017) Inorganic carbon distribution and CO_2 fluxes in a large European estuary (Tagus, Portugal). Sci Rep 7(1):1–14. https://doi.org/10.1038/s41598-017-06758-z

Painter SC (2020) The biogeochemistry and oceanography of the East African Coastal Current. Prog Oceanogr 186:102374. https://doi.org/10.1016/j.pocean.2020.102374

Pattanaik S, Chanda A, Sahoo RK, Swain S, Satapathy DR, Panda CR et al (2020a) Contrasting intra-annual inorganic carbon dynamics and air-water CO_2 exchange in Dhamra and Mahanadi Estuaries of northern Bay of Bengal, India. Limnology 21(1):129–138. https://doi.org/10.1007/s10201-019-00592-0

Pattanaik S, Roy R, Sahoo RK, Choudhury SB, Panda CR, Satapathy DR et al (2020b) Air-Sea CO_2 dynamics from tropical estuarine system Mahanadi. India. Reg Stud Mar Sci 36:101284. https://doi.org/10.1016/j.rsma.2020.101284

Purwiyanto AIS, Agustriani F (2016) Asessment of carbon status in marine protected area of Payung Island waters, South Sumatera Province, Indonesia. Ilmu Kelautan 22(1):1–6

Rajkaran A, Adams JB (2007) Mangrove litter production and organic carbon pools in the Mngazana Estuary. South Africa. Afr J Aquat Sci 32(1):17–25. https://doi.org/10.2989/AJAS.2007.32.1.3.140

Ralison OH, Borges AV, Dehairs F, Middelburg JJ, Bouillon S (2008) Carbon biogeochemistry of the Betsiboka estuary (north-western Madagascar). Organic Geochem 39(12):1649–1658. https://doi.org/10.1016/j.orggeochem.2008.01.010

Ramawijaya R, Awaludin MY, Pranowo WS (2012) Variabilitas Parameter Oseanografi dan Karbon Laut di Teluk Banten. J Perikanan Dan Kelautan 3(3):339–346
Rao GD, Sarma VVSS (2017) Influence of river discharge on the distribution and flux of methane in the coastal Bay of Bengal. Mar Chem 197:1–10. https://doi.org/10.1016/j.marchem.2017.11.002
Ray R, Shahraki M (2016) Multiple sources driving the organic matter dynamics in two contrasting tropical mangroves. Sci Total Environ 571:218–227. https://doi.org/10.1016/j.scitotenv.2016.07.157
Ray R, Baum A, Rixen T, Gleixner G, Jana TK (2018a) Exportation of dissolved (inorganic and organic) and particulate carbon from mangroves and its implication to the carbon budget in the Indian Sundarbans. Sci Total Environ 621:535–547. https://doi.org/10.1016/j.scitotenv.2017.11.225
Ray R, Michaud E, Aller RC, Vantrepotte V, Gleixner G, Walcker R et al (2018b) The sources and distribution of carbon (DOC, POC, DIC) in a mangrove-dominated estuary (French Guiana, South America). Biogeochemistry 138(3):297–321. https://doi.org/10.1007/s10533-018-0447-9
Ritonga IR (2012) Distribusi Karbon Anorganik Dan Fluks CO_2 Di Perairan Beras Basah. Kota Bontang. Lingkungan Tropis 6(2):149–158. https://doi.org/10.14499/jml.v23i1.609
Rustam A, Bengen DG, Arifin Z, Gaol JL (2014) Dinamika Dissolved Inorganic Carbon (DIC) Di Ekosistem Lamun Pulau Pari. J Segara 10(1):31–41. https://doi.org/10.15578/segara.v10i1.13
Santos IR, Maher DT, Larkin R, Webb JR, Sanders CJ (2019) Carbon outwelling and outgassing versus burial in an estuarine tidal creek surrounded by mangrove and saltmarsh wetlands. Limnol Oceanogr 64(3):996–1013. https://doi.org/10.1002/lno.11090
Sapkota Y, White JR (2019) Marsh edge erosion and associated carbon dynamics in coastal Louisiana: a proxy for future wetland-dominated coastlines worldwide. Estuar Coast Shelf Sci 226:106289. https://doi.org/10.1016/j.ecss.2019.106289
Sarma VVSS, Krishna MS, Rao VD, Viswanadham R, Kumar NA, Kumari TR et al (2012a) Sources and sinks of CO_2 in the west coast of Bay of Bengal. Tellus B Chem Phys Meteorol 64(1):10961. https://doi.org/10.3402/tellusb.v64i0.10961
Sarma VVSS, Viswanadham R, Rao GD, Prasad VR, Kumar BSK, Naidu SA et al (2012b) Carbon dioxide emissions from Indian monsoonal estuaries. Geophys Res Lett 39(3). https://doi.org/10.1029/2011GL050709
Sarma VVSS, Kumar NA, Prasad VR, Venkataramana V, Appalanaidu S, Sridevi B et al (2011) High CO_2 emissions from the tropical Godavari estuary (India) associated with monsoon river discharges. Geophys Res Lett 38(8). https://doi.org/10.1029/2011GL046928
Sarma VVSS, Kumari VR, Srinivas TNR, Krishna MS, Ganapathi P, Murty VSN (2018) East India Coastal Current controls the dissolved inorganic carbon in the coastal Bay of Bengal. Mar Chem 205:37–47. https://doi.org/10.1016/j.marchem.2018.07.010
Sarma VV, Kumar MD, Manerikar M (2001) Emission of carbon dioxide from a tropical estuarine system, Goa, India. Geophys Res Lett 28(7):1239–1242. https://doi.org/10.1029/2000GL006114
Selemani JR, Zhang J, Wu Y, Muzuka AN, Njau KN, Zhang G et al (2018) Distribution of organic carbon: possible causes and impacts in the Pangani River Basin ecosystem, Tanzania. Environ Chem 15(3):137–149. https://doi.org/10.1071/EN17185
Shalini A, Ramesh R, Purvaja R, Barnes J (2006) Spatial and temporal distribution of methane in an extensive shallow estuary, south India. J Earth Syst Sci 115(4):451–460. https://doi.org/10.1007/BF02702873
Sudheesh V, Gupta GVM, Naqvi SWA (2020) Massive methane loss during seasonal hypoxia/anoxia in the nearshore waters of southeastern Arabian Sea. Front Mar Sci 7:324. https://doi.org/10.3389/fmars.2020.00324
Taillardat P, Ziegler AD, Friess DA, Widory D, Van VT, David F et al (2018) Carbon dynamics and inconstant porewater input in a mangrove tidal creek over contrasting seasons and tidal amplitudes. Geochim Cosmochim Acta 237:32–48. https://doi.org/10.1016/j.gca.2018.06.012
Tamooh F, Borges AV, Meysman FJR, Meersche K, Dehairs F, Merckx R, Bouillon S (2013) Dynamics of dissolved inorganic carbon and aquatic metabolism in the Tana River basin, Kenya. Biogeosciences 10(11):6911–6928. https://doi.org/10.5194/bg-10-6911-2013

Tokoro T, Kuwae T (2018) Improved post-processing of eddy-covariance data to quantify atmosphere–aquatic ecosystem CO_2 exchanges. Front Mar Sci 5:286. https://doi.org/10.3389/fmars.2018.00286
Triana K, Murakami-Sugihara N, Ogawa H (2021) Spatial and temporal variations in particulate organic carbon in Indonesian waters over two decades. Mar Freshw Res. https://doi.org/10.1071/MF20264
Twilley RR, Castañeda-Moya E, Rivera-Monroy VH, Rovai A (2017) Productivity and carbon dynamics in mangrove wetlands. In: Rivera-Monroy V, Lee S, Kristensen E, Twilley R (eds) Mangrove ecosystems: a global biogeographic perspective. Springer, Cham, pp 113–162. https://doi.org/10.1007/978-3-319-62206-4_5
Van Dam BR, Edson JB, Tobias C (2019) Parameterizing air-water gas exchange in the shallow, microtidal New River estuary. J Geophys Res Biogeosci 124(7):2351–2363. https://doi.org/10.1029/2018JG004908
Wang CF, Hsu MH, Kuo AY (2004) Residence time of the Danshuei River estuary, Taiwan. Estuar Coast Shelf Sci 60(3):381–393. https://doi.org/10.1016/j.ecss.2004.01.013
Wanninkhof R (2014) Relationship between wind speed and gas exchange over the ocean revisited. Limnol Oceanogr Method 12(6):351–362. https://doi.org/10.4319/lom.2014.12.351
Watanabe Y, Kawamura T, Yamashita Y (2018) Introduction: the coastal ecosystem complex as a unit of structure and function of biological productivity in coastal areas. Fish Sci 84(2):149–152. https://doi.org/10.1007/s12562-018-1176-7
Watson SA, Fabricius KE, Munday PL (2017) Quantifying pCO_2 in biological ocean acidification experiments: a comparison of four methods. PLoS ONE 12(9):e0185469. https://doi.org/10.1371/journal.pone.0185469
Weston NB, Porubsky WP, Samarkin VA, Erickson M, Macavoy SE, Joye SB (2006) Porewater stoichiometry of terminal metabolic products, sulfate, and dissolved organic carbon and nitrogen in estuarine intertidal creek-bank sediments. Biogeochemistry 77(3):375–408. https://doi.org/10.1007/s10533-005-1640-1
Wit F, Müller D, Baum A, Warneke T, Pranowo WS, Müller M, Rixen T (2015) The impact of disturbed peatlands on river outgassing in Southeast Asia. Nat Commun 6(1):1–9. https://doi.org/10.1038/ncomms10155
Yan J, Lin Q, Poh SC, Li Y, Zhan L (2020) Underway measurement of dissolved inorganic carbon (DIC) in estuarine waters. J Mar Sci Eng 8(10):765. https://doi.org/10.3390/jmse8100765
Yoshioka T, Ueda S, Miyajima T, Wada E, Yoshida N, Sugimoto A et al (2002) Biogeochemical properties of a tropical swamp forest ecosystem in southern Thailand. Limnology 3(1):51–59. https://doi.org/10.1007/s102010200007
Zhang Z, Zimmermann NE, Stenke A, Li X, Hodson EL, Zhu G et al (2017) Emerging role of wetland methane emissions in driving 21st-century climate change. Proc Natl Acad Sci 114(36):9647–9652. https://doi.org/10.1073/pnas.1618765114

Chapter 4
Blue Carbon Dynamics in the Indian Ocean Mangrove Ecosystems

Anirban Akhand, Abhra Chanda, and Sourav Das

Abstract The global scientific community unequivocally accepts the significance of mangroves in offsetting anthropogenic carbon emissions. This unique vegetation sequesters disproportionately high carbon than many other natural ecosystems, and the countries bordering the Indian Ocean host varying extents of mangrove cover in their coastlines. Small countries like Comoros encompass as little as 1 km^2 mangrove, whereas Indonesia shelters the largest share (~30,000 km^2). Many of these mangrove forests have witnessed a substantial reduction in the last century and the recent past decades. However, the scientific community has not paid equal attention to studying several aspects of these forests throughout the periphery of the Indian Ocean. The eastern Indian Ocean mangroves have received much more attention than their western counterparts. The authors have assembled all the data on these mangroves and explicitly discussed the present state-of-the-art of this crucial habitat and the spatial extent of these mangroves in the respective countries and their aboveground and belowground carbon stock. The chapter also focused on the soil organic carbon pool of these mangrove forests. Finally, the authors collated the findings on the factors that regulate the carbon dynamics within the mangrove forests of the Indian Ocean. This chapter would be beneficial for future researchers who wish to research this sector in the coming days. Besides illustrating the present scenario of knowledge base, the chapter also identified the lacunae that need to be taken into account soon.

Keyword Aboveground biomass · Belowground biomass · Soil organic carbon pool · Spatial distribution · Species assemblage · Mangrove health · Remote Sensing · GIS · *Avicennia marina* · Dominant species

---

A. Akhand
Coastal and Estuarine Environment Research Group, Port and Airport Research Institute, 3-1-1, Nagase, Yokosuka 239-0826, Kanagawa, Japan

A. Chanda (✉) · S. Das
School of Oceanographic Studies, Jadavpur University, 188 Raja S.C. Mullick Road, Kolkata, West Bengal 700084, India
e-mail: abhrachanda.sos.rs@jadavpuruniversity.in

A. Chanda et al. (eds.), *Blue Carbon Dynamics of the Indian Ocean*,
https://doi.org/10.1007/978-3-030-96558-7_4

4.1 Mapping of Mangroves

Mangroves are one of the most crucial blue carbon ecosystems of the world, capable of storing substantial quantities of carbon. According to the latest estimates of Hamilton and Casey (2016), mangroves encompass an area of 131,931 km^2 (as of the year 2014). Mangroves approximately store 739 Mg C ha^{-1}, which equals a global mangrove carbon stock of 6.2 Pg (Alongi 2020). The mean carbon density in mangrove ecosystems is one of the highest among the natural forest ecosystems. Thus, this vegetation acts as an efficient tool to combat climate change. However, despite realizing the significance of this habitat, the mangrove area has continued to reduce throughout the last century. Due to the ongoing coastal development, agriculture, and aquaculture activities, mangroves have become vulnerable in almost all corners of the world (Friess and Webb 2014). The mangroves underwent substantial fragmentation that led to the deterioration of their ecosystem services (Bryan-Brown et al. 2020). Several anthropogenic, natural, and climatic factors are accelerating the net reduction in global mangrove cover. Thus, the scientific community unequivocally presses that these unique vegetative habitats require the utmost attention in the current date. The conservation, restoration, and forestation of mangroves have become a top priority, given their potential in offsetting carbon emissions.

Protection of any natural vegetation requires a thorough understanding of its spatial extent. Hence, accurate mapping is an essential prerequisite for framing a proper conservation strategy. Mangroves grow in the land–water interfaces of the estuaries, intertidal seashores, lagoons, and several other coastal habitats. The transient effect of tides makes mapping mangroves a challenging endeavor. However, it is easier than mapping submerged blue carbon ecosystems like seagrasses and saltmarshes. The advent of tools like remote sensing has made it possible for synoptic-scale mapping of mangroves and acquires holistic information on large patches (Wang et al. 2019). The latest technological advancements in remote sensing and geographical information system enable the researchers to map the mangrove cover and develop an idea about their species composition and biomass (Pham et al. 2019). Remote sensing also provides suitable tools to study the phenology (Pastor-Guzman et al. 2018) and structural traits of mangrove forests (Lucas et al. 2020). Accurate delineation of the spatial extent often fails to provide a holistic scenario. It is essential to monitor and characterize the health of mangroves (Razali et al. 2019; Maurya et al. 2021). Distinguishing mangroves from non-mangroves remain a challenge in this field (Wang and Sousa 2009). However, several remote-sensing-based indices effectively discriminate the mangroves from the adjacent non-mangroves (Gupta et al. 2018).

Several studies implemented remote sensing tools on the Indian mangroves (Nayak and Bahuguna 2001). Chellamani et al. (2014) assessed the overall health of the Indian mangroves. Giri et al. (2014) studied the abundance and distribution of mangrove species in the Indian Sundarbans (the world's largest mangrove forest). Characterization of mangrove stands in the Bhitarkanika sanctuary (Reddy et al. 2007), adjoining the Mahanadi Estuary (Pattanaik and Prasad 2011), Pichavaram forest (Selvam et al. 2003), and Gulf of Kutch (Srivastava et al. 2015) involved the application of remotely sensed images and tools. Lately, Bindu et al. (2020) estimated carbon stock in the mangroves of Kerala, and Thakur et al. (2020) reviewed the utility of multispectral remote sensing in mapping mangroves. Ghosh et al. (2020) estimated the mangrove canopy height using sentinel images and machine learning techniques, whereas Vaghela et al. (2021) estimated the aboveground biomass using the dual-pol SAR data in the Indian mangroves. There are numerous studies in this regard from the Indian mangroves. Several researchers have come up from India, and they are incessantly striving hard to develop newer protocols to monitor mangrove dynamics throughout the country and the world. The Indian Space Research Organization (ISRO) and many of their sister concerns have stressed research related to the blue carbon ecosystems.

Bangladesh Sundarban has received substantial attention in this regard (Azad et al. 2020). Lately, Hussain and Islam (2020) assessed the vulnerability of the Sundarban mangrove forest using the hotspot model, and Islam et al. (2019) comprehensively monitored the forest dynamics for a span of 40 years from 1976 to 2015. Small and Sousa (2019) studied the mangrove phenology and response to natural disturbances in Bangladesh Sundarban. Islam and Bhuiyan (2018) characterized the causes of mangrove degradation and cultivated sustainable management options with the help of remote sensing tools. The rapidly diminishing mangroves of Myanmar have become a point of concern. Veettil et al. (2018) conducted a review in this regard. Maung and Sasaki (2021) investigated the natural recovery potential of the Myanmar mangroves after facing anthropogenic disturbances using sentinel data and neural network classification techniques. Gaw et al. (2018) and Estoque et al. (2018) studied the deforestation dynamics and the role of political and economic scenarios in Myanmar. Studies based on the application of remote sensing tools in the scientific literature are available from Thailand. Songsom et al. (2019) assessed the phenology and environmental drivers in the mangroves of Thailand, and Pimple et al. (2018) used Google Earth Engine to analyze the temporal changes in the mangrove forests of the Trat Province in Thailand. In Singapore, Bird et al. (2004) studied the mangrove forest dynamics of the Sungei Buloh–Kranji. Lately, Friess et al. (2016) assessed the carbon stock in the mangroves of Singapore with the help of remote sensing. The Indonesian mangroves encompass a substantial portion of the Southeast Asian mangroves and received attention in this regard (Kristensen et al. 2008, 2011; Chen et al. 2014; Kristiningrum 2019; Analuddin et al. 2020). Aslan et al. (2016) studied the forest structure of West Papua by coupling the active and passive remote sensing protocols. Darmawan et al. (2020) studied the after-effects of forest fire and predicted the spatiotemporal dynamics of the South Sumatra forest. They implemented the cellular automata and the Markov chain approach. The mangroves of Malaysia have also attracted the attention of the scientific community (Putz and Chan 1986; Ashton et al. 1999; Zhila et al. 2014; Pradisty et al. 2021). Otero et al. (2019) studied the regeneration of Matang mangrove forest in Malaysia, whereas Satyanarayana et al. (2018) assessed the status of the undisturbed mangrove stands

Brunei Bay in East Malaysia. Overall, the number of studies on the mangroves situated in the periphery of the eastern Indian Ocean is substantially higher than that in the western counterpart.

The northwestern periphery of the Indian Ocean has Yemen, Oman, Iran, UAE, and Pakistan. Remote-sensing-based studies on mangroves of these regions are fewer than those carried out in the eastern part. Kumar et al. (2011) delineated the spatial distribution of Yemeni mangroves. Lately, Veettil et al. (2020) studied the spatiotemporal dynamics of the mangroves in the Socotra Archipelago, Yemen. Milani (2018) assembled the spatial distribution scenario of the mangroves in the Persian Gulf and the Gulf of Oman. Habshi et al. (2007), Howari et al. (2009), and Alsumaiti et al. (2017) studied the mangroves along the coastline of the UAE. Lately, Elmahdy et al. (2020) characterized the temporal change in the mangrove forest cover of UAE between 1990 and 2019. Iranian mangroves have received substantial attention in this regard (Ghasemi et al. 2013). Etemadi et al. (2018a) assessed the land-use changes in the mangrove forests of Iran, whereas Mafi-Gholami et al. (2020) predicted the plausible mangrove scenario in Iran by the end of the twenty-first century. Lately, Toosi et al. (2019) have examined several classification algorithms in the mangrove forests of Iran along with Vafaei et al. (2018). Pakistani mangroves have also received substantial attention in this regard. Masood et al. (2015) compared a suite of remote sensing techniques to monitor the mangroves of the Indus Delta in Pakistan, whereas Abbas et al. (2013) assessed the entire mangrove cover of Pakistan with the help of remote sensing tools. Qasim et al. (2016) analyzed the land-use dynamics in the mangroves of Thatta District, Pakistan. Remote-sensing-based studies are also scarce in the East African periphery. Rajkaran et al. (2004) used remotely sensed data to monitor the harvesting of mangroves in South Africa. Kairo et al. (2002) implemented both remote sensing and GIS to study the management of mangroves within the protected areas of Kenya. Wang et al. (2003) comprehensively studied the mangroves along the Tanzanian coastline, whereas Shapiro et al. (2015) characterized the temporal dynamics of mangrove forests in Mozambique. The literature from Djibouti and Somalia is scarcely available in this domain. Among the Island nations in the open Indian Ocean, Sri Lankan mangroves received the due attention of the scientific community (Jayakody et al. 2008). Satyanarayana et al. (2011) validated the remotely sensed data on mangrove development in the Sri Lankan coastline. Dahdouh-Guebas et al. (2002) studied the mangrove and aquaculture pond dynamics in the Pambala–Chilaw lagoon of Sri Lanka. The Maldives has no such studies to date. However, Constance et al. (2021) lately studied the wave exposure on the atoll mangroves of Seychelles. Clarisse et al. (2016) distinguished the spatial distribution of *Bruguiera gymnorrhiza* and *Rhizophora mucronata* in the Mauritian mangroves. Rasolofoharinoro et al. (1998) studied the mangrove forest dynamics, whereas Giri and Muhlhausen (2008) focused on the temporal change in Madagascar. Like the Maldives, Comoros also has no record of such studies.

4.2 Spatial Distribution of Mangrove Patches

The Indian Ocean has a vast expanse, and it shelters substantial mangrove cover in the peripheral coastlines. Mangroves occupy around 85,000 km^2 area bordering the Indian Ocean. This estimate is equivalent to about 47% of the world's mangrove area (18×10^6 ha; Spalding et al. 2010). About 30 countries of the Indian Ocean region have mangrove patches. Some countries like Comoros shelter small mangrove patches of about 1 km^2. However, a country like Indonesia covers 32,500 km^2. Myanmar, Malaysia, and India host substantially large areas of mangroves within their country's periphery (Kathiresan and Rajendran 2005). The mangroves in India cover an area of 4,975 km^2 (Table 4.1). The east coast and west coast harbor

Table 4.1 Latest estimates of mangrove forest area in the Indian Ocean countries

Country	Area (km^2)	Authors
India (AS)	4975	Forest Survey of India (2019)
Bangladesh (AS)	4950	Islam et al. (2019)
Myanmar (AS)	6287	De Alban et al. (2020)
Thailand (AS)	2455	Wanthongchai and Pongruktham (2019)
Singapore (AS)	6.5	Lai et al. (2015)
Indonesia (AS)	32,500	Bakosurtanal (2009)
Malaysia (AS)	5800	Kanniah et al. (2015)
Yemen (AS)	22 (approx)	PERSGA/GEF (2004), Khalil (2015)
Oman (AS)	5.3	Milani (2018)
UAE (AS)	40	Habshi et al. (2007)
Iran (AS)	94	Milani (2018)
Pakistan (AS)	1464	Gilani et al. (2021)
South Africa (AF)	20	Hoppe-Speer et al. (2015)
Mozambique (AF)	2909	Fatoyinbo et al. (2008)
Tanzania (AF)	1581	MNRT (2015)
Kenya (AF)	456	Kirui et al. (2013)
Somalia (AF)	49	Hutchison et al. (2014)
Djibouti (AF)	10	Spalding et al. (2010)
Sri Lanka (AS)	91	Hutchison et al. (2014)
Maldives (AS)	No data	
Seychelles (AF)	32	Hutchison et al. (2014)
Madagascar (AF)	2797	Giri and Muhlhausen (2008)
Comoros (AF)	1.08	Ahamada (1997)
Mauritius (AF)	1.45	Appadoo (2003), Foolmaun et al. (2011)

60% and 14% of these mangroves, respectively. The Andaman and Nicobar Islands shelter the remaining 26%. India (40%) and Bangladesh (60%) share the largest at a stretch mangrove patch of the world, Sundarbans. Sundarban (2112 km^2) followed by Bhitarkanika (130 km^2) and Pichavaram (10 km^2) are some of the prominent mangrove forests of India. According to Kumar (2000), more than 40% of the Indian mangroves perished in the twentieth century. However, recent assessments indicated that the mangrove forest area has increased between the years 1987 and 2013. Estimates show that the mean increase in these twenty-six years was 24.25 $\pm$ 82.57 km^2. Almost all the coastal states of India reported an area enhancement, except for Andaman and Nicobar Islands (-3.41 ± 52.32 km^2) and Andhra Pradesh (-5.95 ± 15.70 km^2), which recorded a decrease in the area (Sahu et al. 2015).

Bangladesh shelters approximately 60% of the world's largest mangrove forest, Sundarban. According to Hoque and Datta (2005), Bangladesh Sundarban and the adjoining mangroves encompassed around 5700 km^2. However, mangroves suffered considerable loss during the mid of the last century (Mijanuddin et al. 2014). Recent analyses indicated that Bangladesh mangroves increased in the area between the years 1976 and 2015. According to Islam et al. (2019), Bangladesh mangroves comprised approximately 4950 km^2 in 2015. Between 1976 and 2015, the nation witnessed an enhancement of 3.1% in the mangrove cover. Though Bangladesh recorded a marginal increase in mangrove forest cover, some regions like Chakaria Sundarbans exhibited complete disappearance during 1976–89. Myanmar used to have a dense mangrove even in the 1980s. This country has experienced one of the most catastrophic deforestation in the last few decades. De Alban et al. (2020) reported a total mangrove area of 6287 km^2 in Myanmar in 2016. They observed that the mangrove area in Myanmar used to cover almost 13,000 km^2 at the end of the twentieth century. The nation witnessed a loss of around 4000 km^2 within the next ten years, followed by another 2000 km^2 in the following decade. In total, Myanmar lost more than half of its pre-existing mangrove cover within the last two decades. Though the deforestation rate has decreased lately, it continues to be at around 3.60–3.87% per year.

Thailand has a coastline of about 2800 km. Almost half of the entire coastal stretch shelters mangrove patches in the current date. The first mangrove survey in this country dates back to 1961. Since then, all the surveys indicated a rampant loss of mangroves along the Thai coastal belt. However, the estuaries, lagoons, and intertidal sediments still encompass substantial mangrove cover (Alongi and Dixon 2000; Poungparn et al. 2009, 2020). According to Wanthongchai and Pongruktham (2019), Thailand shelters at present a mangrove area of almost 2450 km^2. According to them, 43% of the entire country used to be forests in 1961. Mangrove area in 1961 used to be around 3,700 km^2. Rampant forest clearing took place all over Thailand in the last century (Charruppat and Charruppat 1997). However, surveys during the past three decades indicate a marginal increase in the mangrove forest cover (Aksomkoae 1993; Wanthongchai and Pongruktham 2019). Like Thailand, Singapore, a small nation, encompassed almost 75 km^2 mangrove areas during the colonial era. Widespread deforestation for timber and fuel drastically reduced the forest area to 60 km^2 in the 1950s and less than 6 km^2 by the end of the twentieth century (Hilton and Manning 1995; Ng and Low 1994). According to the latest estimates, Singapore encompasses almost 6.5 km^2 of mangroves (Friess et al. 2012; Lai et al. 2015).

Indonesia, as a country, holds the credit of sheltering the world's highest mangrove cover, which encompasses around 32,500 km^2 (Bakosurtanal 2009). In the mid of the twentieth century, Indonesia used to have almost 42,500 km^2 mangrove area. Like all other countries in the world, Indonesia also witnessed widespread forest clearing in the last century. However, since 2000, Indonesia showed improvements by enhancing mangrove cover from 29,300 km^2 to 32,500 km^2 (Giesen et al. 2006; Ilman et al. 2011). The country has enormous potential to accommodate more mangroves and enhance the area in the future to almost 78,000 km^2. The five major islands of this country, namely Java, Sumatra, Sulawesi, Kalimantan, and Papua, have substantial mangrove cover (Kusmana 2014). Malaysia hosts the third largest mangrove cover in the Asia–Pacific realm (Wong et al. 2020). Johor, Kedah, and Perak states are the significant mangrove-dominated regions of the country. Kelantan, Pahang, and Terengganu also shelter smaller patches. The Sarawak and Trusan Rivers and the Rajang Delta are the principal storehouses of mangroves in Malaysia. According to Kanniah et al. (2015), the total mangrove cover in Malaysia comes around 5800 km^2, out of which the Sabah region shelters close to 60% of the national mangrove cover. Sarawak region and the Peninsular Malaysia shelter the rest of the 40%.

The coastline of Yemen and its exclusive economic zone (EEZ) has a high productivity rate and shelters a wide range of biodiversity (Hariri 2010). Fringe mangroves cover a substantial part of the coastal periphery of this country. Four islands together form the Socotra Archipelago, which happens to be within the jurisdiction of Yemen, though situated 350 km away from the Yemen coastline in the Arabian Sea. The scientific community often refers to these islands as the "Galapagos of the Indian Ocean" due to their high biodiversity, exotic and endemic floras, and faunas. UNESCO inscribed Socotra Island as a protected area into their list of "Man and the Biosphere" reserves (UNEP 2008). Mangroves occur on five different locations of the Socotra Island, mainly on the landward side of the beach consisting of almost 4 m high *Avicennia marina* and other halophytic plants that cover an area of merely 3 km^2 (Klaus and Turner 2004). Ghubbah di-Nit, in the southwestern part of the Island, shelters one of the dense *A. marina* patches (Klaus and Turner 2004). A comprehensive report by PERSGA/GEF (2004) mentioned in Khalil (2015) indicated that a coastal stretch of 128 km in Yemen hosted thin mangrove patches varying between 50 and 300 m. If we consider the mean width, Yemen approximately shelters a mangrove area of 22 km^2. Oman has a long coastline facing the Arabian Sea, which includes the Gulf of Oman. However, the mangrove vegetation in this country occurs in small patches and is of narrow fringe types. *A. marina* is the only species available in this region. This species mainly thrive adjoining the tidal creeks (Fouda and Al-Muharrami 1995). Muscat shelters some such small patches located in the northeastern sector of the Gulf of Oman. Mahout Island shelters the most luxurious mangrove patch (<1.7 km^2) (Fouda and Ali-Muharrami 1996). Overall, the Oman coastline is deficient in mangroves, principally because of the rocky substratum. Mangroves always prefer soft-sediment columns, which is virtually absent in Oman.

Oman, in totality, used to have a small mangrove cover of only 2 km^2 in 1977; however, in 2017, the mangrove cover in Oman increased to 5.3 km^2 (Milani 2018).

Like Oman, UAE also offers an arid to the semi-arid coastal environment for the mangroves to flourish. Though the climatic features in UAE are almost similar to that observed in Oman, mangrove cover in the former is substantially higher than the latter. UAE shelters around 40 km^2 of mangroves in their coastal periphery (Habshi et al. 2007). However, the distribution is principally patchy. A substantial portion of the UAE's mangrove vegetation thrives near Abu Dhabi. Researchers carried out extensive surveys on the mangroves of Khor Umm Al-Quwain, Khor Kalba, and Al Dabbiya. In the last two to three decades, Elmahdy et al. (2020) observed a substantial increase in mangrove cover in several pockets of the UAE. However, the species variety remains restricted to mainly *A. marina*, according to the latest reports. Iran has a long coastline that faces the Persian Gulf and the Oman Sea and experiences a typical semi-arid climate. Unlike the other Gulf countries discussed above, mangroves in Iran cover substantial areas. Mangrove forest types available in Iran range from discrete patchy covers to well-connected stretches. The region between Khamir Port and Qeshm Islands shelters the prominent mangrove patches of Iran (Milani 2018; Mafi-Gholami et al. 2020). Mangroves of Iran have lately received substantial attention from the scientific community in mapping and predicting future scenarios (Mafi-Gholami et al. 2019, 2020; Toosi et al. 2019). According to Milani (2018), Iran used to shelter only 47 km^2 of mangroves in 1977, and now, it increased to 94 km^2 in 2017.

Pakistan shelters luxuriant mangrove patches in the coastal periphery. Deltaic environments with soft muddy substratum and typical tropical climate are suitable for mangrove growth (Siddiqui et al. 2009; Farooqui et al. 2012, 2014). The Indus Delta in Pakistan has a similar geomorphologic setting to the Ganges–Brahmaputra-Meghna Delta that shelters the Sundarban mangroves in the India-Bangladesh border. Abbas et al. (2013) reported that the Indus Delta encompasses more than 95% of the country's mangrove cover. The rest remains scattered in the Makran coast (Balochistan Province), Kalmat Khor, Sonmiani Bay, and Jiwani. Abbas et al. (2013) indicated that Pakistan used to shelter approximately 1580 km^2 in 1997–2001, which reduced to 981 km^2 in 2009. However, recent estimates by Gilani et al. (2021) suggested that mangroves in Pakistan cover around 1464 km^2 in 2020. According to Gilani et al. (2021), plantation and conservation endeavors led to the mangrove cover enhancement.

South African coastal periphery has several small to large estuaries (total 183 estuaries), mostly permanently open, and some are semi-closed (Ward and Steinke 1982; Steinke et al. 1983; Rajkaran and Adams 2007; Naidoo 2016). Adams et al. (2004) investigated 76 estuaries and confirmed the presence of mangroves in 17 of them. The mangroves in South Africa mainly concentrate in the eastern coastal periphery. Sparse tree cover and occasional dense mangroves are characteristic traits of this coastline. Hoppe-Speer et al. (2015) indicated a total mangrove area of 20 km^2 in the South African coastlines. Mthatha, Mtentu, Mngazana, and Mbashe estuaries host the most prominent mangrove forests of South Africa. However, these forests have suffered substantial mangrove losses in the recent past due to enhanced anthropogenic pressure. Hoppe-Speer et al. (2015) further mentioned that only 19% of the country's coastline comes under the protected category, which encompasses only 7% of the total mangroves of South Africa. Among all the east African countries, Mozambique has the longest coastline. The coastal periphery of this country has many swamps and muddy patches, which are ideal for the mangroves to grow. Saket and Matusse (1994) estimated back in the 1970s to 90s that Mozambique shelters 3960 km^2 of mangroves. Fatoyinbo et al. (2008) later fine-tuned the estimate to 2909 km^2. The Save-Zambezi River complex shelters around 65% of the country's total mangrove cover (Trettin et al. 2016). The central portion of the Mozambique coastline has many deltaic regimes and adjoining estuaries, where mangroves can grow. The mudflats in the intertidal zones and the typical tropical environment favor the luxuriant growth of mangroves in Mozambique (Fernando and Bandeira 2009). Mozambique has rich mangrove diversity, and recent studies indicate that they are under severe threat (Barbosa et al. 2001; Charrua et al. 2020).

Tanzania has an 800 km long coastline having a gentle slope that provides a suitable habitat for mangroves (Shunula and Whittick 1999; Minu et al. 2020). The coastal periphery has several small to large river estuaries and bays having intricate tidal creek networks. Tanzania hosts the Rufiji River Delta, which is home to approximately 500 km^2 of mangroves. The Rufiji mangrove forest is the largest in entire Eastern Africa. Earlier studies estimated that mainland Tanzania used to have 1096 km^2 mangroves during 1988–1990, which decreased to 1081 km^2 in 2000 (as mentioned in Wang et al. (2003)). Wang et al. (2003) fine-tuned the estimate to a total of 1120 km^2, and FAO (2007) revised the figure to 1250 km^2 in 2005. According to the latest survey (MNRT 2015), Tanzania shelters around 1581 km^2 of mangroves. Even the latest studies like Njana (2020) mentioned this figure. After Rufiji, the Kilwa region shelters the second largest mangrove patches in Tanzania.

On the African east coast, Kenya has a long coastline facing the western Indian Ocean. Kokwaro (1985) observed that Kenya used to have approximately 530 km^2 of mangroves in their coastal periphery. However, since then, the country has lost substantial mangrove cover due to anthropogenic pressure and natural drivers. Kirui et al. (2013) carried out a pan-Kenya estimate of mangrove spatial extent and observed that the mangrove stands in 2010 encompassed around 456 km^2. They further added that the rates of losses varied spatiotemporally throughout the country. In the 2000–2010 decade, the overall rate of loss has reduced substantially. Compared to the southern African countries along the east coast, knowledge of mangroves in Djibouti and Somalia is poor, even at the current date. Spalding et al. (2010) mentioned that Djibouti shelters around 10 km^2 mangroves and the area did not change much in the last few decades. However, given the shallow shape of the country, almost the entire population of the country relies on coastal ecosystem services. Such overdependence can pose a significant threat to the mangrove cover of this country (Witsen 2012). Somalia has a long coastline of almost 3025 km and shelters mangroves in their coastal periphery. However, the pieces of research are very few to develop a holistic scenario (Carbone and Accordi 2000). According to the latest modeled estimates of Hutchison et al. (2014), Somalia encompasses approximately 49 km^2 of mangroves.

Sri Lanka is one of the significant island countries in the open Indian Ocean. The coastal periphery of this country has several tidal inlets and lagoons that are suitable habitats for mangroves. However, surveys on mangroves are few compared to the possibility of their existence in this country. Due to political tension, the eastern and northern fronts received far less attention in the last two to three decades. However, situations at present are much better. The Puttlam–Kalpitiya lagoon hosts the most prominent forest patch within Sri Lanka. Jayewardene (1986) claimed a mangrove cover of only 30 km^2, whereas CCD (1986) reported approximately 120 km^2. However, according to the present modeled estimates of Hutchison et al. (2014), 91 km^2 mangroves remain in Sri Lanka. The coastal periphery of Sri Lanka experiences overall micro-tidal amplitude, which confines the vegetation to shallow intertidal zones (Jayatissa et al. 2002; Karunathilake 2003).

The Maldives, an archipelago of 1190 islands, has emerged as a popular tourist destination for its exotic beaches and natural beauty. This country has several atolls and is known for corals. Fe pieces of research show that mangrove occupies many of these islands in the Maldives. The atolls of Maldives shelter the rich biodiversity of mangroves (Sivakumar et al. 2018). Shazra et al. (2008) also reported such unique biodiversity. However, the forest types are predominantly patchy in this country. Exact estimates of mangrove areas do not exist for the Maldives to date. However, Saleem and Nileysha (2011) reported that mangroves thrive in almost 12% of islands of the entire archipelago. Like the Maldives, Seychelles is a small island country in the Indian Ocean and shelters mangroves in patches. Sivakumar et al. (2018) reported that 115 islands in Seychelles shelter fringe mangroves. However, the best possible estimate of mangrove cover in Seychelles is the model-derived figure by Hutchison et al. (2014).

Madagascar is a large island adjacent to the continental landmass of Africa. According to the earliest estimates, this country used to shelter approximately 3000–4000 km^2 of mangroves (Roger and Andrianasaolo 2003). Modeled estimates of Hutchison et al. (2014) indicate a mangrove cover of around 3273 km^2 (Fig. 4.1). However, ground-truthed estimates suggest a modest mangrove area of 2797 km^2 (Giri and Muhlhausen 2008). The western coastline of Madagascar hosts more than 98% of the country's total mangrove cover. Compared to the area, the species diversity is less in Madagascar. In the last few decades, the country witnessed a substantial reduction in mangrove cover (Scales and Friess 2019); however, the rate of forest loss is much less than the global average (Rakotomavo and Fromard 2010; Arias-Ortiz et al. 2021). Mauritius, another exotic tourist destination, encompasses an area of 1865 km^2. Jagtap (1993) reported a mangrove cover of 20 km^2 in Mauritius. The model-derived estimates of Hutchison et al. (2014) indicate 8 km^2 mangrove areas. However, the ground-validated figures suggest that the mangroves in this island occupy 1.45 km^2 (Appadoo 2003; Foolmaun et al. 2011). Ferney and Mahebourg have prominent mangrove covers in Mauritius (Sauer 1962; Fagoonee, 1990; Seeruttun et al. 2021). Comoros is another small island in the western Indian Ocean. However, very little information is available at the present date related to the mangrove cover. Ahamada (1997) reported a 1.1 km^2 mangrove cover in this country. Grande Comose,

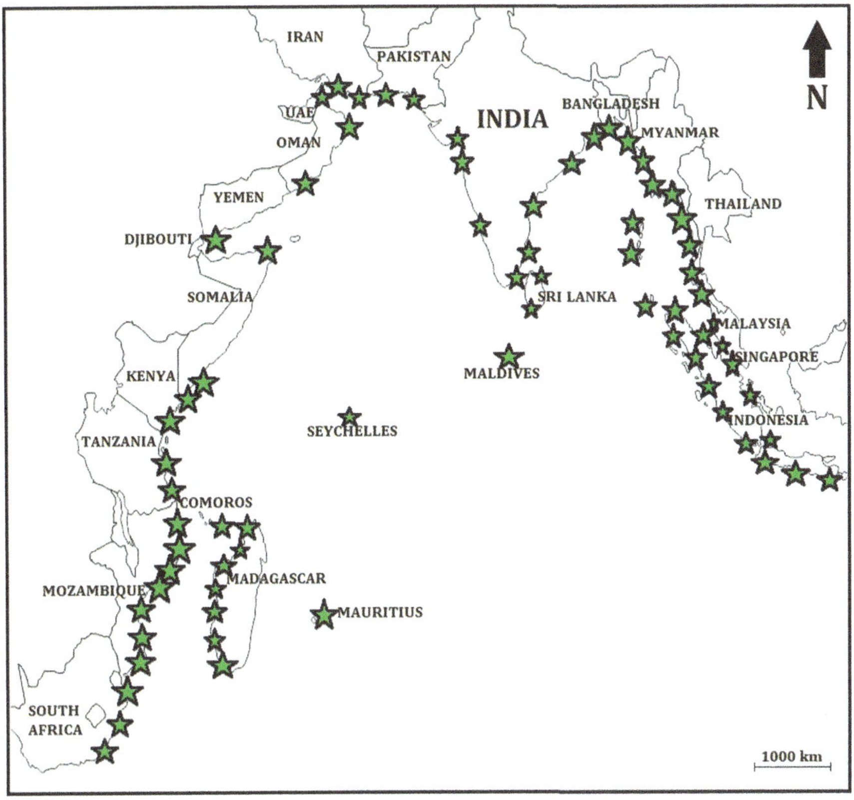

Fig. 4.1 The study area map showing the prominent location of mangrove forests in the periphery of the Indian Ocean

Moheli Island, and Anjouna have some of the prominent patches of mangroves in Comoros.

4.3 Geomorphologic Settings

The mangrove patches in the periphery of the Indian Ocean occupy a wide variety of coastal habitats. The mangroves in the western Indian Ocean thrive in arid areas. However, in the eastern part, they mainly grow along the estuaries. Several lagoons and deltaic regimes also encompass substantial mangrove cover. Mangroves are also common in many of the coastal fringes. The forest structure types are over-wash mangrove forests, fringing mangrove forests, basin mangrove forests, hammock mangrove forests, and scrub mangrove forests (Kathiresan and Rajendran 2005). The East coast of India and the entire Bangladesh coastline have several river mouths that lead to the formation of deltas. The Ganga-Meghna-Brahmaputra Delta occupies parts of India and a substantial portion of Bangladesh (Rogers and Goodbred 2014; Rahman et al. 2020). The deltaic sediments provide an ideal substratum for mangroves (Ellison 1999). The entire east coast of India has many such deltas like the Mahanadi, Godavari, and Krishna that shelter substantial mangrove cover. The Indus Delta provides similar habitats in Pakistan. Most of the Indian and Pakistan shoreline experience substantially high tidal amplitudes, which enabled the mangroves to expand their horizon (Pitchaikani 2020). For instance, the mangrove vegetation in the Sundarban stretches almost 70 km inshore from the Bay of Bengal. The flat and extensive cover of flood plains is the fundamental feature of Myanmar's coastal geomorphology (Brakenridge et al. 2017). The submerged coastal sectors of Myanmar provide ideal habitat conditions for mangroves to flourish (Giardino et al. 2016; Cameron et al. 2021). The other eastern Indian Ocean nations like Singapore, Malaysia, Indonesia, and Thailand provide a similar setup. The geomorphic settings exhibit a spectrum in Indonesia (Cameron et al. 2019), which facilitates the rich biodiversity of the mangrove forests. The species diversity of mangroves depends on the sediment substratum (Sarker et al. 2021), salinity (Basyuni et al. 2017), elevation (Oh et al. 2017), and physical conditions of the intertidal habitats. The more the spatial variability of these geomorphic features, the richer the biodiversity, allowing several species to accommodate within a single niche. Due to all these potential factors, the eastern Indian Ocean exhibit comparatively high species richness than that observed in the western counterpart. The Asian countries like Yemen, Oman, UAE, and Iran, and few Eastern African countries like Somalia and Djibouti experience arid to semi-arid environments. Besides, regions like the Persian Gulf and the Gulf of Oman experience micro-tidal amplitudes (Hereher et al. 2020). Only a few mangrove species can adapt to the typical arid environments, which led to lower biodiversity in the western counterpart of the Indian Ocean. Shallow tidal amplitudes restricted the spatial extent of mangroves in these coastlines. Most studies observed narrow fringing mangroves along the coastal periphery in these countries (Milani 2018). The southern African countries like Mozambique, Tanzania, and Madagascar have extensive swampy areas with a typical subtropical climate. The large-scale flood plains and tidal inlets, including creeks and bays, enabled dense mangrove patches in few pockets of these countries.

4.4 Species Diversity of Mangroves

India exhibits rich mangrove biodiversity. The floral stands in India comprise 46 true mangrove species (Fig. 4.2) under 22 genera and 14 families (Naskar and Mandal 1999). The species richness in the east coast of India is much higher than that observed in the western counterpart. The Andaman and Nicobar Islands exhibit the highest mangrove species diversity in the entire country (Ragavan et al. 2016). Some of the Rhizophora, Lumnitzera, and Sonneratia species are available only in the Andaman and Nicobar and do not exist in mainland India (Ragavan et al. 2016). Such a variety

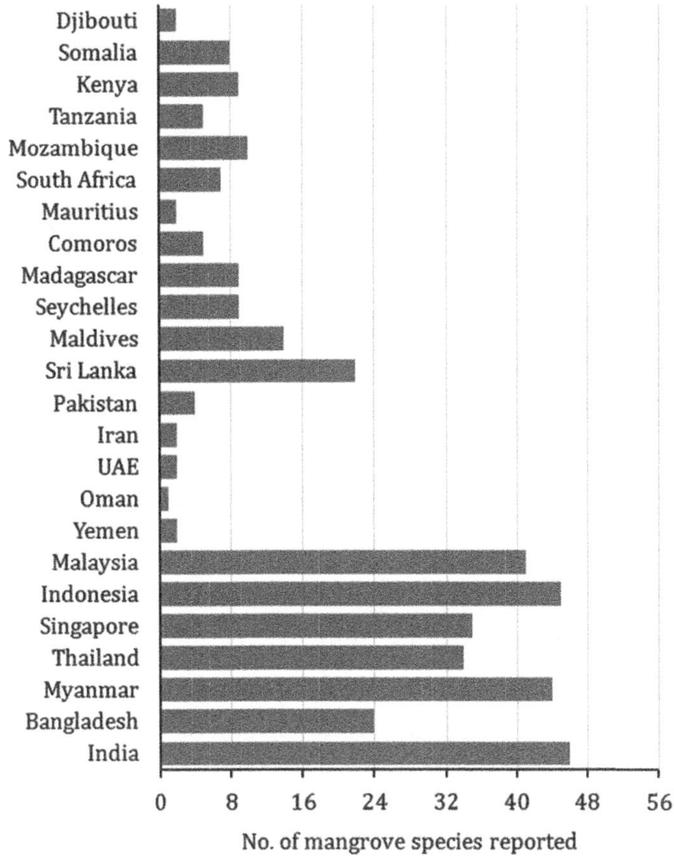

Fig. 4.2 The number of true mangrove species reported from the countries bordering the Indian Ocean

of mangroves makes India the third richest after Indonesia and Australia. Table 4.2 provides a list of reported dominant species in the respective countries bordering the Indian Ocean. The Table contains the true mangrove species listed by Quadros and Zimmer (2017) and some others.

Bangladesh shelters the Sundarbans, which is famous for its heterogeneous forest stand and biodiversity. Rahman et al. (2015) reported 24 true mangroves from Bangladesh Sundarbans. However, in both India and Bangladesh, *Avicennia*, *Rhizophora*, and *Bruguiera* are the dominant genera. Myanmar also has rich floral biodiversity comprising 44 true mangrove species (Zöckler and Aung 2019). *Rhizophora*, *Heritiera*, and *Xylocarpus* sp are the dominant species in Myanmar. The Ayeyarwady Delta has a predominance of *Avicennia*, *Sonneratia*, and *Ceriops* spp. (Veettil et al. 2018). In the eastern part of the Indian Ocean, Thailand hosts substantial mangrove cover (Srisunont et al. 2017).

Table 4.2 Mangrove species composition in the Indian Ocean countries

Sr No	Species name	IN	BD	MM	TH	SG	ID	MY	IR	YM	OM	UAE	SL
1	*Acanthus ebracteatus*			+	+	+	+	+					
2	*Acanthus ilicifolius*	+	+	+	+	+	+	+					+
3	*Acanthus volubilis*	+		+		+	+						
4	*Acrostichum speciosum*			+	+	+	+	+					
5	*Acrostichum aureum*	+	+	+	+	+	+	+					+
6	*Aegialitis annulata*												
7	*Aegialitis rotundifolia*	+	+	+	+								
8	*Aegiceras corniculatum*	+	+	+	+	+	+	+					+
9	*Aegiceras floridum*						+	+					
10	*Aglaia cucullata*	+	+	+									
11	*Avicennia alba*	+		+	+	+	+	+					
12	*Avicennia bicolor*												
13	*Avicennia eucalyptifolia*												
14	*Avicennia germinans*	+											
15	*Avicennia integra*												
16	*Avicennia lanata*						+						
17	*Avicennia marina*	+	+	+	+	+	+	+	+	+	+	+	+
18	*Avicennia officinalis*	+	+	+	+	+	+	+					+
19	*Avicennia rumphiana*					+		+					
20	*Avicennia schaueriana*												
21	*Brownlowia lanceolata*												
22	*Brownlowia tersa*		+	+		+							
23	*Bruguiera cylindrica*	+		+	+	+	+	+					+
24	*Bruguiera exaristata*						+						
25	*Bruguiera gymnorrhiza*	+	+	+	+	+	+	+					+
26	*Bruguiera hainesii*			+	+	+	+	+					

(continued)

Table 4.2 (continued)

Sr No	Species name	I N	BD	MM	TH	SG	I D	MY	I R	YM	OM	UAE	SL
27	*Bruguiera parviflora*	+		+	+	+	+	+					
28	*Bruguiera rhynchopetala*												
29	*Bruguiera sexangula*	+	+	+	+	+		+					+
30	*Camptostemon schultzii*												
31	*Ceriops australis*												
32	*Ceriops decandra*	+	+	+	+			+					+
33	*Ceriops roxburghiana*	+											
34	*Ceriops tagal*	+		+	+			+					+
35	*Ceriops somalensis*												
36	*Ceriops zippeliana*					+							
37	*Conocarpus erectus*	+											
38	*Cynometra iripa*			+	+			+					
39	*Cynometra ramiflora*	+	+		+			+					
40	*Dolichandrone spathacea*			+		+	+						
41	*Excoecaria agallocha*	+	+	+	+	+	+	+					+
42	*Excoecaria indica*							+					
43	*Heritiera fomes*	+	+	+	+		+						
44	*Heritiera globosa*							+					
45	*Heritiera littoralis*			+	+	+	+	+					+
46	*Kandelia candel*	+	+	+		+	+	+					
47	*Kandelia obovata*	+					+						
48	*Laguncularia racemosa*	+					+						
49	*Lumnitzera littorea*	+		+		+	+	+					
50	*Lumnitzera racemosa*	+	+	+	+	+	+	+					+
51	*Nypa fruticans*	+	+	+	+	+	+	+					+
52	*Osbornia octodonta*						+						
53	*Pelliciera rhizophorae*												

(continued)

Table 4.2 (continued)

Sr No	Species name	IN	BD	MM	TH	SG	ID	MY	IR	YM	OM	UAE	SL
54	*Pemphis acidula*			+		+	+	+					+
55	*Phoenix paludosa*	+	+	+	+		+						
56	*Rhizophora annamalaya*			+				+					
57	*Rhizophora apiculata*	+	+	+	+	+	+	+					+
58	*Rhizophora harrisonii*						+						
59	*Rhizophora lamarckii*	+		+			+	+					
60	*Rhizophora mangle*	+					+						
61	*Rhizophora mucronata*	+	+	+	+	+	+	+	+	+		+	+
62	*Rhizophora racemosa*	+					+						
63	*Rhizophora samoensis*												
64	*Rhizophora stylosa*	+		+		+		+					
65	*Scyphiphora hydrophylacea*			+	+	+		+					+
66	*Sonneratia alba*	+		+		+	+	+					+
67	*Sonneratia apetala*	+	+	+			+						+
68	*Sonneratia caseolaris*	+	+	+	+	+	+	+					+
69	*Sonneratia griffithii*	+		+	+		+	+					
70	*Sonneratia gulngai*	+											
71	*Sonneratia hainanensis*	+											
72	*Sonneratia lanceolata*	+											
73	*Sonneratia ovata*	+		+	+	+	+	+					
74	*Xylocarpus granatum*	+	+		+	+	+	+					+
75	*Xylocarpus mekongensis*	+		+				+					
76	*Xylocarpus moluccensis*		+	+	+	+	+						
77	*Xylocarpus rumphii*						+	+					

(continued)

Table 4.2 (continued)

Sr No	Species name	MD	SC	MG	CM	KN	MA	MZ	SM	SA	TZ	DJ	PK
1	*Acanthus ebracteatus*												
2	*Acanthus ilicifolius*												
3	*Acanthus volubilis*												
4	*Acrostichum speciosum*												
5	*Acrostichum aureum*	+	+	+				+		+	+		
6	*Aegialitis annulata*												
7	*Aegialitis rotundifolia*												
8	*Aegiceras corniculatum*												+
9	*Aegiceras floridum*												
10	*Aglaia cucullata*												
11	*Avicennia alba*												
12	*Avicennia bicolor*												
13	*Avicennia eucalyptifolia*												
14	*Avicennia germinans*												
15	*Avicennia integra*												
16	*Avicennia lanata*												
17	*Avicennia marina*	+	+	+	+	+		+	+	+	+	+	+
18	*Avicennia officinalis*												
19	*Avicennia rumphiana*												
20	*Avicennia schaueriana*												
21	*Brownlowia lanceolata*												
22	*Brownlowia tersa*												
23	*Bruguiera cylindrica*	+						+					
24	*Bruguiera exaristata*												

(continued)

Table 4.2 (continued)

Sr No	Species name	MD	SC	MG	CM	KN	MA	MZ	SM	SA	TZ	DJ	PK
25	*Bruguiera gymnorrhiza*	+	+	+	+	+	+	+	+	+			
26	*Bruguiera hainesii*												
27	*Bruguiera parviflora*												
28	*Bruguiera rhynchopetala*												
29	*Bruguiera sexangula*	+											
30	*Camptostemon schultzii*												
31	*Ceriops australis*												
32	*Ceriops decandra*												
33	*Ceriops roxburghiana*												
34	*Ceriops tagal*	+	+	+		+		+	+	+	+	+	+
35	*Ceriops somalensis*								+				
36	*Ceriops zippeliana*												
37	*Conocarpus erectus*												
38	*Cynometra iripa*												
39	*Cynometra ramiflora*												
40	*Dolichandrone spathacea*												
41	*Excoecaria agallocha*	+											
42	*Excoecaria indica*												
43	*Heritiera fomes*	+											
44	*Heritiera globosa*												
45	*Heritiera littoralis*	+		+		+		+					
46	*Kandelia candel*												
47	*Kandelia obovata*												
48	*Laguncularia racemosa*												
49	*Lumnitzera littorea*												
50	*Lumnitzera racemosa*	+	+	+	+	+		+	+	+	+		
51	*Nypa fruticans*												

(continued)

Table 4.2 (continued)

Sr No	Species name	MD	SC	MG	CM	KN	MA	MZ	SM	SA	TZ	DJ	PK
52	*Osbornia octodonta*												
53	*Pelliciera rhizophorae*												
54	*Pemphis acidula*	+	+										
55	*Phoenix paludosa*												
56	*Rhizophora annamalaya*												
57	*Rhizophora apiculata*	+											
58	*Rhizophora harrisonii*												
59	*Rhizophora lamarckii*												
60	*Rhizophora mangle*												
61	*Rhizophora mucronata*	+	+	+	+	+	+	+	+	+	+		+
62	*Rhizophora racemosa*												
63	*Rhizophora samoensis*												
64	*Rhizophora stylosa*												
65	*Scyphiphora hydrophylacea*												
66	*Sonneratia alba*		+	+	+	+		+	+				
67	*Sonneratia apetala*												
68	*Sonneratia caseolaris*	+											
69	*Sonneratia griffithii*												
70	*Sonneratia gulngai*												
71	*Sonneratia hainanensis*												
72	*Sonneratia lanceolata*												
73	*Sonneratia ovata*												
74	*Xylocarpus granatum*		+	+		+		+	+	+			
75	*Xylocarpus mekongensis*					+							
76	*Xylocarpus moluccensis*												

(continued)

Table 4.2 (continued)

Sr No	Species name	MD	SC	MG	CM	KN	MA	MZ	SM	SA	TZ	DJ	PK
77	*Xylocarpus rumphii*	+											

IN—India; BD—Bangladesh; MM—Myanmar; TH—Thailand; SG—Singapore; ID—Indonesia; MY—Malaysia; IR—Iran; YM—Yemen; OM—Oman; UAE—United Arab Emirates; SL—Sri Lanka; MD—Maldives; SC—Seychelles; MG—Madagascar; CM—Comoros; KN—Kenya; MA—Mauritius; MZ—Mozambique; SM—Somalia; SA—South Africa; TZ—Tanzania; DJ—Djibouti; PK—Pakistan

According to the survey of Santisuk (1983), 34 true mangrove species thrive in Thailand. Recent observations by Wanthongchai and Pongruktham (2019) conform to these estimates. Singapore shelters one of the tiniest mangrove covers in this part of the world. Yet this country reported as many as 35 true mangrove species (Yang et al. 2011). They observed that the mangrove area diminished to a great extent during the last century. However, only one species, namely Brownlowia argentata, became nationally extinct in Singapore. Ilman et al. (2011) indicate that 45 true mangrove species thrive in Indonesia. However, Kusmana (2014) reported 43 mangrove species from Indonesia. Malaysia also has substantially high species diversity. FAO (2005) indicated the presence of 41 true mangrove species in Malaysia.

The species diversity observed in the western Indian Ocean countries is much less than that observed in the eastern counterpart. Nagi and Abubakr (2013) reported only two mangrove species in the Yemeni coastline, namely *A. marina* and *Rhizophora mucronata*. Out of these two species, the *A. marina* was the most dominant one. Small patches of *Rhizophora mucronata* are visible in some pockets of this country, like the north Hodeidah and Kamaran regions. The patchy mangroves of Oman are even more species deficient, with only one recorded species so far, namely *A. marina* (Fouda and Al-Muharrami 1995; Milani 2018). Like Yemen, Iran also has the same two mangrove species all along their coastal stretch. *Rhizophora mucronata* is only available in the Sirik region, and the rest of the coastline has homogeneous patches of the *A. marina* (Danehkar 1996, 1998, 2001). The UAE also has the same scenario (Schile et al. 2017). Milani (2018) mentioned that *Rhizophora mucronata* was almost extinct in UAE; however, a series of fruitful forestation programs reinstated them. *A. marina* is also the dominant species in Pakistan. This species accounts for more than 95% of the country's mangrove cover. However, in total, Pakistan shelters four true mangroves. The other three less prominent mangroves are *Rhizophora mucronata*, *Ceriops tagal*, and *Aegiceras corniculatum* (Khan and Aziz 2001; Mukhtar and Hannan 2012).

Among the East African countries, Djibouti has a similar scenario to Iran, UAE, and Yemen. *A. marina* is the only dominant species with a sporadic presence of *Rhizophora mucronata* in few pockets. These two species comprise the narrow fringe mangroves of Djibouti (Witsen 2012; Khalil 2015). On the whole, these two species dominate the entire east African coastline. However, as we move southwards, the presence of other mangroves in patches became prominent. FAO (2007) indicated that the east coast of the African continent has a much mangrove floristic diversity

(14 species) than that observed in the western African coastline (7 species). Somalia, Kenya, Tanzania, Mozambique, and South Africa recorded eight, nine, five, ten, and seven true mangrove species, respectively (FAO 2007). *Xylocarpus granatum*, *Xylocarpus mekongensis*, *Bruguiera gymnorrhiza*, *Lumnitzera racemosa*, and *Sonneratia alba* are some of the other species available in these African countries. Similarly, among the African Island nations, Madagascar and Seychelles recorded nine species, followed by Comoros (five species) and Mauritius (two species) (FAO 2007). On the contrary, the two Asian Island nations, namely Sri Lanka and the Maldives, have twenty-two and fourteen species, respectively, in their coastlines.

4.5 Spatial Variability of Aboveground Biomass Carbon Stock

Among the conventional blue carbon ecosystems, mangroves have received substantial attention concerning the characterization of biomass carbon stock (Richards et al. 2020). Mangrove ecosystems store bulk carbon in the aboveground tree structures like stems and shoots (Mitra et al. 2011). Besides, the belowground compartments that mainly include the luxuriant root structures are also storehouses of carbon. Aboveground carbon (AGC) estimation is much easier than quantifying the belowground carbon stock. Ground-based allometric approaches and remote sensing enabled the present-day scientific community to carry out comprehensive research on the mangroves in this regard. The peninsular part of India hosts several prominent mangrove forests, and many of these sites gained attention from researchers. The Sundarbans, being the world's largest mangrove forest, deserved significant research, and it has been the prime focus for many researchers in the last two to three decades. Mitra et al. (2011) observed that *Sonneratia apetala* had higher aboveground biomass than *Avicennia alba* and *Excoecaria agallocha* in the Indian Sundarban. They further mentioned that the aboveground carbon stock exhibited significant spatial variability within the forest. Ray et al. (2011) carried out comprehensive research in the Indian Sundarbans and measured the aboveground biomass carbon stock in various locations having different species compositions. They recorded a mean AGC of 39.93 $\pm$ 14.05 Mg C ha^{-1} (Table 4.3). Harishma et al. (2020) observed a similar AGC while working in the Kerala mangroves down south in the peninsular regions (40.11 $\pm$ 0.40 Mg C ha^{-1}). Contrary to Mitra et al. (2011), Harishma et al. (2020) observed the highest AGC for *A. marina* and lowest for *Sonneratia alba*. Sahu et al. (2016) reported one of the highest net ecosystem carbon stocks from the mangrove forests adjoining the Mahanadi Estuary in Odisha, India. They observed a mean AGC of 62.45 $\pm$ 6.4 Mg C ha^{-1}. *A. marina*, *Excoecaria agallocha*, *Lumnitzera racemosa*, *Rhizophora apiculata*, and *Heritiera fomes* were the principal contributors. The Bhitarkanika mangrove forest in India is the second largest in India. Rasquinha and Mishra (2021) reported the highest AGC of the country from this region (179.7 $\pm$ 67.0–196.1 $\pm$ 182.1 Mg C ha^{-1}). Several studies did not measure the AGC, BGC,

Table 4.3 Carbon stocks in the aboveground and belowground biomass and soil organic carbon pool observed in the mangrove patches of the Indian Ocean countries (as mean ± standard deviation or minimum to maximum)

Country	Study site	AGC (Mg C ha^{-1})	BGC (Mg C ha^{-1})	SOC (Mg C ha^{-1})	Authors
India (AS)	Sundarban	39.93 ± 14.05	9.61 ± 3.37	11.25	Ray et al. (2011)
	Kerala	40.11 ± 0.40	18.45 ± 0.12	81.26	Harishma et al. (2020)
	Mahanadi	62.45 ± 6.4	26.69 ± 2.6	54.3 ± 7.4	Sahu et al. (2016)
	Bhitarkanika	179.7 ± 67.0 to 196.1 ± 182.1	64.0 ± 55.0 to 68.3 ± 17.1	15.63 ± 7.22 to 16.49 ± 6.59	Rasquinha and Mishra (2021)
Bangladesh (AS)	Sundarbans	76.80	41.10	48.3 ± 10.3 to 74.8 ± 14.9	Kamruzzaman et al. (2018); Rahman et al. (2021)
Myanmar (AS)	Ayeyarwady Delta	21 – 70	26 – 104	167 ± 58	Thant et al. (2012)
Thailand (AS)	Andaman coast	125	47.5		Jachowski et al. (2013)
	Ranong	140.6	58.8		Tamai et al. (1986)
	Trang District			193.62 ± 15.31 to 317.75 ± 16.18	Cadiz et al. (2020)
Singapore (AS)	Entire country	25.36 – 172.09	11.21 – 54.81	307.4	Friess et al. (2016)
Indonesia (AS)	Entire Country	159.1	16.7	774.7	
	South Sulawesi	125.48 ± 93.48	60.23 ± 44.87		Malik et al. (2020)
	West Papua	96 ± 65	17 ± 16	750	Sasmito et al. (2020)
Malaysia (AS)	Kuala Selangor and Sungai Haji Dorani	48.17 – 121.82	3.26 – 13.12	90.11 – 119.69	Hong et al. (2017)
	Matang	49.70		385.2 – 545.0	Tanouchi et al. (2000); Hamdan et al. (2014); Adame et al. (2018)

(continued)

Table 4.3 (continued)

Country	Study site	AGC (Mg C ha^{-1})	BGC (Mg C ha^{-1})	SOC (Mg C ha^{-1})	Authors
	Kelantan Delta and Johor Park	35.43 ± 2.54 to 73.05 ± 10.19	34.90 - 62.64	384.57 - 413.33	Rozainah et al. (2018)
	Strait of Malacca	71.00 ± 2.30	36.07 ± 0.52		Azman et al. (2021)
Yemen (AS)	ND	ND	ND	ND	
Oman (AS)	Al-Qurm Nature Reserve, Muscat	70	36		Al-Nadabi and Sulaiman (2018)
UAE (AS)	Entire coastline	37.3	24.6	156.3	Schile et al. (2017)
Iran (AS)	Gowatr Bay	14.15 ± 4.51	32.81 ± 6.40	227.2 ± 11.79	Savari et al. (2020)
	Azini creek, Sirik	111.1	41.5	188.3	Askari et al. (2021)
	Nayband National Park	0.13	0.06		Amiri (2021)
Pakistan (AS)	Sandspit forest	93.33 ± 2.60	39 ± 4.50	51.66 ± 3.92	Nazim et al. (2018)
	Sindh, Karachi	24.4	10.2	233.73	Khan et al. (2021)
South Africa (AF)	Nxaxo Estuary	2.3 ± 1.0 to 9.6 ± 3.6		176.9 ± 4.5 to 262.5 ± 18.9	Johnson et al. (2020)
	Nahoon Estuary			110.14	Raw et al. (2019)
Mozambique (AF)	Sofala Bay	28.02 ± 9.2	25.22 ± 5.3	160	Sitoe et al. (2014)
	Zambezi River Delta	75.4 ± 12.6 to 268.5 ± 36.6	23.8 ± 3.1 to 72.8 ± 9.4	274.6 ± 25.0 to 314.1 ± 14.8	Stringer et al. (2015)
	Maputo City	4.59 ± 0.53		43.08 ± 2.49	Magalhães (2019)
Tanzania (AF)	Mtimbwani and Geza	54.9 ± 19.2 to 167.8 ± 57.1	5.5 to 16.8	311–462	Alavaisha and Mangora (2016)
	Entire coastline	33.5 ± 5.8	30.0 ± 4.5		Njana et al. (2018)
	Rufiji River Delta	40.5	21.1	98.6	Lupembe (2014)

(continued)

Table 4.3 (continued)

Country	Study site	AGC (Mg C ha^{-1})	BGC (Mg C ha^{-1})	SOC (Mg C ha^{-1})	Authors
Kenya (AF)	Gazi Bay	53.33	5.5		Kairo et al. (2008)
	Gazi Bay		3.6–38		Tamooh et al. (2008)
	Gazi Bay			36–297	Andreetta et al. (2014)
Somalia (AF)	Entire coastline	72			Fatoyinbo and Simard (2013)
Djibouti (AF)	Entire coastline	45			Fatoyinbo and Simard (2013)
Sri Lanka (AS)	Micro-tidal estuaries and lagoons			316–580	Perera and Amarasinghe (2019)
	Dry, wet, and intermediate zones	75.5–189.1	7.9–14.3	643.6–1253.6	Cooray et al. (2021)
Maldives (AS)		ND	ND	ND	
Seychelles (AF)		ND	ND	ND	
Madagascar (AF)	Tsimipaika Bay	66–122	24–39	220 ± 10	Arias-Ortiz et al. (2021)
	Ambanja and Ambaro bays	147 (including BGC)		446	Jones et al. (2014)
	Helodrano Fagnemotse	46.2–73.9 (including BGC)		294.6–381.0	Benson et al. (2017)
Comoros (AF)		ND	ND	ND	
Mauritius (AF)	Mahebourg and Le Morne	19.4 ± 2.7 to 461.9 ± 53.4		331.3 ± 0.9 to 470.1 ± 0.9	Ramdhun and Appadoo (2020)

and SOC but reported the net ecosystem carbon stock. Pandey and Pandey (2013) observed that the Gujarat mangroves on the west coast of India store around 8.2 million tons of carbon. Bangladesh Sundarbans record the most pieces of research concerning the characterization of blue carbon stock (Hossain et al. 2014, 2015, 2016; Kamruzzaman et al. 2017, 2018, 2019; Mahmood et al. 2019; Ahmed et al. 2021). The recent estimates of Kamruzzaman et al. (2018) indicate that the mean AGC is around 76.80 Mg C ha^{-1} in Sundarbans (Fig. 4.3). It is worth observing that

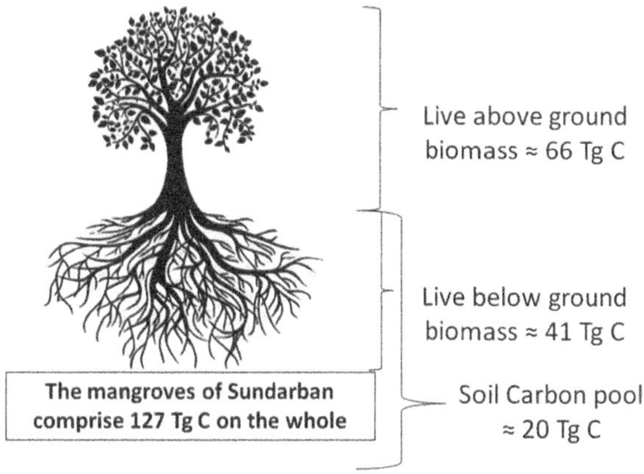

Fig. 4.3 The total carbon content in the world's largest contiguous mangrove forest of Sundarbans

the Bangladesh part of Sundarbans stores substantially high AGC compared to the Indian counterpart.

Myanmar shelters a luxuriant mangrove cover. However, the number of studies on characterizing the carbon content is scarce. Thant et al. (2012) measured the AGC across different species-specific stands in the Ayeyarwady Delta, and their estimates ranged between 21 and 70 Mg C ha^{-1}. The mixed mangrove stands had more carbon content than the mono-species floral stands. Very few studies on computing AGC exist in Thailand. Jachowski et al. (2013) implemented machine learning techniques to model the AGC in the mangroves of the Andaman coast, Thailand. They estimated a mean AGC of 125 Mg C ha^{-1}. Besides this study, Kridiborworn et al. (2012) observed that *Rhizophora apiculata* stands in the Samut Songkram region of central Thailand stored 140 Mg C ha^{-1} within twelve years from planting the saplings. Tamai et al. (1986) reported similar observations from the Ranong region of Thailand. Friess et al. (2016) carried out a comprehensive study across the urban landscape of Singapore and observed an AGC range of 25.36–172.09 Mg C ha^{-1}. The mangroves of Singapore border around the developed urban setup. Yet, these mangroves hold substantial quantities of carbon. Friess et al. (2016) estimated a total of 450,572 Mg C for the entire 9.6 km^2 mangrove area of Singapore. Indonesia encompasses the world's largest share of mangroves and has received substantial attention in this regard. Alongi et al. (2016) observed a median AGC of 159.1 Mg C ha^{-1} for the entire country. Extrapolation of the net ecosystem carbon stock for the Indonesian mangroves led to a figure of 3.0 Pg C. Lately, Malik et al. (2020) recorded a mean AGC of 125.48 ± 93.48 Mg C ha^{-1} in the rehabilitated mangroves of South Sulawesi, Indonesia. Sasmito et al. (2020) studied the undisturbed mangroves of West Papua in Indonesia. They observed a mean AGC of 96 ± 65 Mg C ha^{-1}. This mangrove forest holds a substantial share of entire Indonesia's mangrove cover. Besides the studies discussed so far, several other studies characterized the ecosystem carbon stocks in different pockets of Indonesia (Wicaksono et al. 2016; Martuti et al. 2017; Nehren and Wicaksono 2018; Widyastuti et al. 2018; Indrayani et al. 2021 and many more). Malaysia adjacent to Indonesia hosts substantial mangrove cover (Hossain et al. 2008). Hong et al. (2017), while working in the Kuala Selangor Nature Park and Sungai Haji Dorani regions of Malaysia, observed an AGC range of 48.17–121.82 Mg C ha^{-1}. Hamdan et al. (2014) estimated a mean AGC of 49.70 Mg C ha^{-1} in the Matang mangrove forests of Malaysia. They approximated a total of 2.1 million Mg C for the entire Matang mangrove reserve. Rozainah et al. (2018) observed substantial variability in AGC between the Avicennia and the Rhizophora stands of the Johor Park and Kelantan Delta. Their estimated AGC ranged between 35.43 $\pm$ 2.54 and 73.05 $\pm$ 10.19 Mg C ha^{-1}. Azman et al. (2021) reported a mean AGC of 71.00 $\pm$ 2.30 Mg C ha^{-1} from the regenerated mangroves of the Strait of Malacca.

Yemen has a record of few studies that surveyed the spatial extent of mangroves. However, endeavors of characterizing AGC are not available in the existing literature. Some assessments exist in Oman. Al-Nadabi and Sulaiman (2018) observed a mean AGC of 70 Mg C ha^{-1} in the Al-Qurm Nature Reserve of Muscat (Oman). Some studies focused on the Iranian mangroves. Savari et al. (2020) reported a mean AGC of 14.15 $\pm$ 4.51 Mg C ha^{-1} from the Gowatr Bay, Iran. They did not observe any significant difference in AGC between the seasons. However, they recorded a significant spatial variability. Askari et al. (2021) observed a substantially higher mean AGC than Savari et al. (2020) in the Azini creek of Sirik mangrove forest, Iran (111.1 Mg C ha^{-1}). However, in the sparse mangrove setups of Iran, AGC was substantially low. Amiri (2021) reported a mean AGC of 0.13 Mg C ha^{-1} from the Nayband National Park of Bushehr province in Iran. Schile et al. (2017) carried out a comprehensive study on the blue carbon estimation throughout the coastline of the UAE. They observed a mean AGC of 37.3 Mg C ha^{-1} in the UAE coasts. Nazim et al. (2018) reported a substantially higher mean AGC of 93.33 $\pm$ 2.60 Mg C ha^{-1} from the Sandspit mangrove forest of Pakistan. Pakistan Forest Institute extensively studied the mangroves of the Sindh region near Karachi (Khan et al. 2021). They recorded a mean AGC of 24.4 Mg C ha^{-1}.

The number of studies focusing on the characterization of AGC from the east African border is scarce. Lately, Johnson et al. (2020) reported the mean AGC from the temperate mangroves of South Africa for the first time by sampling at five locations of the Nxaxo Estuary. According to their estimate, live tree biomass contributed the most, and leaf litter and pneumatophores had a negligible contribution. The mean AGC varied between 2.25 $\pm$ 1.04 and 9.56 $\pm$ 3.57 Mg C ha^{-1}. Sitoe et al. (2014) recorded substantially higher mean AGC (28.02 $\pm$ 9.2 Mg C ha^{-1}) in the mangroves of Mozambique while working in the Sofala Bay. Stringer et al. (2015) reported a wide range of AGC (75.4 $\pm$ 12.6 to 268.5 $\pm$ 36.6 Mg C ha^{-1}) from the Zambezi River Delta of Mozambique. In contrary to these observations, Magalhães (2019) recorded an unusually low mean AGC (4.59 $\pm$ 0.53 Mg C ha^{-1}) from the mangroves of Maputo City, Mozambique. Few studies focused on AGC computation in the mangroves of Tanzania. Alavaisha and Mangora (2016) observed an AGC range of 54.9 $\pm$ 19.2 to 167.8 $\pm$ 57.1 Mg C ha^{-1} in the Mtimbwani and Geza mangroves of Tanzania. Njana et al. (2018) sampled nine stations throughout the coastline of Tanzania. They observed a modest mean AGC of 33.5 $\pm$ 5.8 Mg C ha^{-1}. Lupembe (2014) reported a similar mean AGC of 40.5 Mg C ha^{-1} from the mangroves of the Rufiji River

Delta. A few studies in this regard concentrated on the Kenyan mangroves too. Kairo et al. (2008) observed a mean AGC of 53.33 Mg C ha^{-1} from the Gazi Bay, Kenya, and Cohen et al. (2013) extrapolated the landscape level AGB of Kenyan mangroves. They reported an AGB range of 5.4–7.2 Tg from the entire coastline of Kenya, which is equivalent to 2.6–3.6 Tg C. Djibouti and Somalia are the two countries that received the least attention in this regard. Fatoyinbo and Simard (2013) carried out exhaustive research on the African mangroves. They reported an AGC of 45 and 72 Mg C ha^{-1} in Djibouti and Somalia, respectively.

Cooray et al. (2021) carried out an exhaustive study on the AGC of mangroves in the different climatic zones of Sri Lanka. They reported an AGC range of 75.5–189.1 Mg C ha^{-1}. Maldives is yet to document any estimate of AGC from their atoll mangroves. Seychelles also has no literature in this regard. Lately, Ramdhun and Appadoo (2020) carried out a study in this regard in two sites of Mauritius (Mahebourg and Le Morne). They reported a wide range of 19.4 $\pm$ 2.7–461.9 $\pm$ 53.4 Mg C ha^{-1} in the Mauritian mangroves. Similar studies are not available from Comoros as well. However, the mangroves of Madagascar received substantial attention in this regard. Lately, Arias-Ortiz et al. (2021) characterized the effect of mangrove losses on soil organic carbon pool. They reported an AGC range of 66–122 Mg C ha^{-1} in the intact mangrove stands. Jones et al. (2014) recorded a mean biomass carbon content of 147 Mg C ha^{-1}, including the BGC from the Ambanja and Ambaro bays of Madagascar. Benson et al. (2017) reported a much low biomass carbon content of 46.2–73.9 Mg C ha^{-1} (including the BGC) from the Helodrano Fagnemotse mangroves in Madagascar.

4.6 Spatial Variability of Belowground Biomass Carbon Stock

Belowground biomass carbon (BGC) constitutes the intricate root system. Several researchers observed that the mangrove root systems store substantially high quantities of carbon (Lovelock 2008). However, quantifying belowground carbon is a challenging endeavor. It involves uncertainties in estimation (Adame et al. 2017). Ray et al. (2011) implemented the allometric approach and observed that the mean BGC in Indian Sundarban was around 9.61 $\pm$ 3.37 Mg C ha^{-1}, substantially less than the measured AGC. Harishma et al. (2020), while working in Kerala, observed a double BGC (18.45 $\pm$ 0.12 Mg C ha^{-1}). Sahu et al. (2016) reported an even higher mean BGC (26.69 $\pm$ 2.6 Mg C ha^{-1}) from the Mahanadi mangroves of India. They observed that the 66 km^2 mangrove cover adjoining the Mahanadi Estuary holds almost one metric ton of carbon. Rasquinha and Mishra (2021) reported the highest BGC in Indian mangroves from the Bhitarkanika region (64.0 $\pm$ 55.0 to 68.3 $\pm$ 17.1 Mg C ha^{-1}). The Bangladesh Sundarbans, which spreads along almost the entire coastline of Bangladesh, received substantial attention concerning the carbon sequestration potential. Kamruzzaman et al. (2018) reported a mean BGC of 41.1 Mg C ha^{-1}, which is substantially higher than that observed in the Indian part of Sundarbans.

Thant et al. (2012) observed substantial BGC in the Ayeyarwady Delta of Mynamar, ranging between 26 and 104 Mg C ha^{-1}. The BGC in mixed mangrove stands was higher than those reported in the mono-species patches. Jachowski et al. (2013) applied machine learning techniques on the mangroves of the Andaman coast to evaluate the mean BGC (47.5 Mg C ha^{-1}). Tamai et al. (1986) observed slightly higher mean BGC from the *Rhizophora* stands of the Ranong region in Thailand. Friess et al. (2016) observed a BGC range of 11.21–54.81 Mg C ha^{-1} in the urban landscape of Singapore mangroves. However, Alongi et al. (2016) observed a median BGC of only 16.7 Mg C ha^{-1} for entire Indonesian mangroves. On the contrary, Malik et al. (2020) recorded a mean BGC of 60.23 ± 44.87 Mg C ha^{-1} in the rehabilitated mangroves of South Sulawesi, Indonesia. The observations of Sasmito et al. (2020) in the West Papua region (17 ± 16 Mg C ha^{-1}) conformed to Alongi et al. (2016). Hong et al. (2017), while working in the Kuala Selangor Nature Park and Sungai Haji Dorani regions of Malaysia, observed a BGC range of 3.26–13.12 Mg C ha^{-1}. However, Rozainah et al. (2018) observed a much range of SOC (34.90–62.64 Mg C ha^{-1}) in the Kelantan Delta and Johor Park of Malaysia. Azman et al. (2021) reported a mean BGC of 36.07 ± 0.52 Mg C ha^{-1} from the regenerated mangroves of the Strait of Malacca.

Yemen is yet to report any estimates of belowground carbon stock in the mangrove patches. Al-Nadabi and Sulaiman (2018) observed a mean BGC of 36 Mg C ha^{-1} in the Al-Qurm Nature Reserve of Muscat (Oman). Few studies concentrated on the Iranian mangroves in this regard. Savari et al. (2020) reported a mean BGC of 32.81 ± 6.40 Mg C ha^{-1} from the Gowatr Bay, Iran. They observed more than double BGC compared to AGC in these mangrove stands. Askari et al. (2021) observed an even higher mean AGC than Savari et al. (2020) in the Azini creek of Sirik mangrove forest, Iran (41.5 Mg C ha^{-1}). However, Amiri (2021) observed a substantially low mean BGC of 0.06 Mg C ha^{-1} in the Nayband National Park of Iran. Schile et al. (2017) carried out an exhaustive study to characterize the blue carbon content throughout the coastline of the UAE. They observed a mean BGC of 24.6 Mg C ha^{-1} in the UAE coasts. Nazim et al. (2018) reported a substantially higher mean BGC of 39 ± 4.50 Mg C ha^{-1} from the Sandspit mangrove forest of Pakistan. However, Khan et al. (2021) mentioned a much lower mean BGC of 10.2 Mg C ha^{-1} from the Sindh region near Karachi.

The studies on mangrove biomass and carbon stock assessment are scarce from the African continental boundary. Recently, Johnson et al. (2020) estimated the AGC and SOC; however, they did not mention anything about the belowground biomass from the temperate mangroves of South Africa. However, Sitoe et al. (2014) reported a mean BGC of 25.22 ± 5.3 Mg C ha^{-1} from the mangroves of Mozambique while working in the Sofala Bay. Stringer et al. (2015) recorded a wide range of BGC (23.8 ± 3.1 to 72.8 ± 9.4 Mg C ha^{-1}) from the Zambezi River Delta mangroves of Mozambique. Alavaisha and Mangora (2016) observed a BGC range of 5.5–16.8 Mg C ha^{-1} in the Mtimbwani and Geza mangroves of Tanzania. They observed that the BGC magnitudes were close to or less than 10% of the corresponding AGC.

Njana et al. (2018) sampled nine stations throughout the coastline of Tanzania. They observed a modest mean BGC of 30.0 ± 4.5 Mg C ha^{-1}, which is comparable to the mean AGC observed in this study. However, Lupembe (2014) reported a low mean AGC of 21.1 Mg C ha^{-1} from the mangroves of the Rufiji River Delta. Kenyan mangroves received substantial attention in this regard (Rao et al. 1994; Lang'at et al. 2014). Several studies focused on the Gazi Bay region of this country. Kairo et al. (2008) observed a mean BGC of 5.5 Mg C ha^{-1} from the Gazi Bay, Kenya. However, Tamooh et al. (2008) reported a wide range of BGC from 3.6 to 38 Mg C ha^{-1}. They observed that the species-specific BGC varied across the three dominant species, namely *R. mucronata*, *S. alba*, and *A. marina*.

Cooray et al. (2021) carried out an exhaustive study on the BGC of mangroves in the different climatic zones of Sri Lanka. They reported a BGC range of 7.9–14.3 Mg C ha^{-1}. Maldives and Seychelles are yet to document any estimate of BGC from their mangrove habitats. Ramdhun and Appadoo (2020) reported some findings from the Mauritian mangroves; however, they did not consider BGC in their study. Similar studies are not available from Comoros as well. However, the mangroves of Madagascar received substantial attention in this regard. Lately, Arias-Ortiz et al. (2021) characterized the effect of mangrove losses on soil organic carbon pool. They reported an AGC range of 24–39 Mg C ha^{-1} in the intact mangrove stands.

4.7 Soil Organic Carbon Pool Assessment

Leaf litter and root decay add substantial quantities of carbon to the mangrove soils (Liu et al. 2017). The labile part of this litter-derived carbon washes away with the tides; however, the recalcitrant portion accumulates and remains there for a long time (Kida and Fujitake 2020). Ray et al. (2011) measured the soil organic carbon stock (SOC) in the Sundarbans by sampling at different locations. They observed a moderate mean SOC of 11.25 Mg C ha^{-1}. Harishma et al. (2020), while working in the Kerala mangroves, observed an eight-fold high SOC (81.26 Mg C ha^{-1}) compared to that observed in the Sundarbans. However, the estimates of Sahu et al. (2016) in the Mahanadi mangroves (54.3 ± 7.4 Mg C ha^{-1}) were lower than those observed in Kerala. The Bhitarkanika mangrove forest adjacent to the Mahanadi Estuary recorded a much lower mean SOC of 15.63 ± 7.22 to 16.49 ± 6.59 Mg C ha^{-1} (Rasquinha and Mishra 2021). Lately, Rahman et al. (2021) measured the SOC throughout the Bangladesh Sundarbans. Their estimates ranged between 48.3 ± 10.3 and 74.8 ± 14.9 Mg C ha^{-1}. Thus, it is evident that AGC, BGC, and SOC in the Bangladesh Sundarbans are substantially higher than the Indian counterpart.

Thant et al. (2012) observed a substantially high mean SOC in the Ayeyarwady Delta of Mynamar (167 ± 58 Mg C ha^{-1}). The swampy soils of Myanmar store substantial quantities of carbon than that observed in the deltaic soils of India and Bangladesh. Studies on Thai mangroves are scarce in this regard. Lately, Cadiz et al. (2020) reported a high range of SOC from the mangroves situated in the Tran District of Thailand (193.62 ± 15.31 to 317.75 ± 16.18 Mg C ha^{-1}). Friess et al. (2016) observed a substantially high mean SOC in the fringe mangroves of Singapore (307.4 Mg C ha^{-1}). However, the observations of Alongi et al. (2016) for the overall Indonesian mangroves were even higher than this estimate. They observed a median SOC of 774.7 Mg C ha^{-1} in Indonesian mangroves. Sasmito et al. (2020), while working in the undisturbed mangrove patches of West Papua, observed that 89% of the total ecosystem carbon stock exists in the form of SOC (750 Mg C ha^{-1}). The remaining 11% remains in the live biomass and deadwood. The SOC in Malaysia was comparatively less than that observed in Indonesia, though the two countries experience almost similar climates. Hong et al. (2017) observed a SOC range of 90.11–119.69 Mg C ha^{-1} in the Kuala Selangor Nature Park and Sungai Haji Dorani regions of Malaysia. Adame et al. (2018) reported a higher range of SOC (385.2–545.0 Mg C ha^{-1}) from the Matang mangrove forest.

Yemen is yet to report any estimates of soil organic carbon from its mangrove sediments. A few studies reported the SOC from Iranian mangroves. Savari et al. (2020), while working on the mangroves of Gowatr Bay, observed a mean SOC of 227.2 ± 11.79 Mg C ha^{-1}. The observed mean SOC was sixteen and seven times higher than the mean AGC and BGC, respectively. Askari et al. (2021) observed similar mean SOC of 188.3 Mg C ha^{-1} in the Azini creek of Sirik mangrove forest, Iran. Schile et al. (2017) sampled eighteen sites throughout the coastline of the UAE. They reported a mean SOC of 156.3 Mg C ha^{-1} from this region, almost double the biomass carbon stock observed in the natural mangrove stands. Nazim et al. (2018) reported a much lower mean SOC of 51.66 ± 3.92 Mg C ha^{-1} from the Sandspit mangrove forest of Pakistan. However, Khan et al. (2021) mentioned the results of the measurements carried out by the Pakistan Forest Institute, which indicates a mean SOC of 233.7 Mg C ha^{-1} from the mangroves of Sindh, Karachi.

The African coastline has very few studies on mangrove biomass and carbon stock estimation. Johnson et al. (2020) reported a SOC range of 176.9 ± 4.5 to 262.5 ± 18.9 Mg C ha^{-1} from the temperate mangroves of the Nxaxo Estuary, South Africa. Raw et al. (2019) observed a much less mean SOC of 110.14 Mg C ha^{-1} in the Nahoon Estuary of South Africa. Sitoe et al. (2014) examined a similar mean SOC of 160 Mg C ha^{-1} from the Sofala Bay mangroves of Mozambique. However, Stringer et al. (2015) recorded a much higher SOC range (274.6 ± 25.0 to 314.1 ± 14.8 Mg C ha^{-1}) from the mangrove soils of the Zambezi River Delta in Mozambique. Contrary to these observations, Magalhães (2019) recorded an unusually low mean SOC (43.08 ± 2.49 Mg C ha^{-1}) from the mangroves of Maputo City, Mozambique. Few SOC assessments focused on the mangroves of Tanzania. Alavaisha and Mangora (2016) reported a SOC range of 311–462 Mg C ha^{-1} in the Mtimbwani and Geza mangroves of Tanzania. Lupembe (2014) reported a much less mean SOC of 98.6 Mg C ha^{-1} from the mangroves of the Rufiji River Delta of Tanzania. The Gazi Bay in Kenya received substantial attention in this regard. Andreetta et al. (2014) recorded a SOC range of 36–297 Mg C ha^{-1} in the mangrove forests of this region.

Among the island nations of the open Indian Ocean, Sri Lanka received substantial attention in this regard (Amarasinghe and Balasubramaniam 1992). Lately, Perera and Amarasinghe (2019) exhaustively studied the mangrove soil carbon stock in the micro-tidal estuaries and lagoons in the Sri Lankan periphery. They reported a SOC range of 316–580 Mg C ha^{-1}. Cooray et al. (2021) reported a spectrum of SOC from the Sri Lankan mangroves (643.6–1253.6 Mg C ha^{-1}). Jonsson and Hedman (2019) analyzed an excessively high total ecosystem carbon stock of 7864 Mg C ha^{-1} from the Batticaloa Lagoon of Sri Lanka. SOC in the top 30 cm significantly contributed to the ecosystem carbon stock of this region. Maldives is yet to report any estimate of SOC from their atoll mangroves. A similar scenario exists in Seychelles. Ramdhun and Appadoo (2020) reported a substantially high SOC range (331.3 $\pm$ 0.9 to 470.1 $\pm$ 0.9 Mg C ha^{-1}) from the only study on the Mauritian mangroves to date. Similar studies are not available from Comoros as well. Arias-Ortiz et al. (2021) recently studied the effect of mangrove losses on the soil organic carbon pool in the Tsimipaika Bay of Madagascar. They reported a mean SOC of 220 $\pm$ 10 Mg C ha^{-1}. Jones et al. (2014) reported a much higher mean SOC of 446 Mg C ha^{-1} from Ambanja and Ambaro bays of Madagascar. Benson et al. (2017) recorded a marginally lower SOC range of 294.6–381.0 Mg C ha^{-1} in the Helodrano Fagnemotse mangroves within the same country.

4.8 Factors Regulating the Carbon Stock and Primary Productivity

The carbon stock in the mangrove forest and soil depths vary spatiotemporally throughout the world. Several physicochemical, geomorphologic, and atmospheric factors regulate the mangrove ecosystem's carbon sink potential. The scientific community strives hard to understand these governing factors to formulate an ideal management strategy and enhance the carbon sink capacity of these forests. Mitra et al. (2011), while working in the Indian Sundarbans, observed that salinity played a crucial role in varying the biomass carbon stock. The highly saline central sector of the Indian Sundarbans displayed low biomass carbon stock, whereas the freshwater-dominated soils had high biomass carbon. Ray et al. (2011) observed substantially low soil organic carbon in the Indian Sundarbans, usually not observed in mangrove ecosystems. They argued that the slow process of humification of organic carbon behind such low SOC. Sahu et al. (2016) observed a significant relationship between the soil organic carbon stock and the biomass carbon, which showed that higher carbon in the aboveground structures leads to enhanced soil organic carbon pools. Rahman et al. (2021) observed that low saline zones had the highest SOC in the Bangladesh Sundarbans. The moderately saline and the highly saline zones had much lower SOC. They reported that SOC exhibited a significant relationship with the soil electrical conductivity, C/N ratio, and the tree diameter (a proxy of biomass). Chanda et al. (2016) indicated that the oligohaline (freshwater loving) species had more carbon stock than the polyhaline species. Both the Indian and Bangladesh Sundarbans suffer from freshwater scarcity (Bhadra et al. 2017; Dasgupta et al. 2017; Murshed and Kaluarachchi 2018). However, the scenario is worse in the Indian part. Except for the Hooghly River flowing through the western boundary of the Indian Sundarbans, none of the estuaries receive freshwater from upstream (Chatterjee et al., 2013). Several pieces of research indicate that aquatic salinization significantly deteriorated the carbon stock of this crucial forest.

Cadiz et al. (2020) concluded that the age of the mangrove trees governs the biomass carbon content (while studying the mangroves of Thailand). They observed that young planted mangrove trees had much less biomass than old natural trees. They also inferred that higher tree density leads to enhanced accumulation of SOC. However, Malik et al. (2020) concluded that the low planting distance negatively affected the carbon stock of the rehabilitated mangroves of South Sulawesi, Indonesia. Lately, Murdiyarso et al. (2021) stated that selective and sustainable logging of mangroves effectively retained the biomass carbon in the West Papua's Bintuni Bay mangrove forests of Indonesia. They showed that scientifically logged mangrove forests exhibit equivalent carbon stock to that of undisturbed forests. Sasmito et al. (2020) observed that geomorphologic settings play a crucial role in governing the SOC. They inferred that organic carbon in the estuarine sediments goes deep down compared to the inward forest soils. Rozainah et al. (2018), while working in the Malaysian mangroves, observed that seasons play a crucial role in regulating the SOC in the top layer. Azman et al. (2021) lately observed that natural regeneration of mangroves leads to higher ecosystem carbon stock than artificially restored mangroves in the Strait of Malacca.

Savari et al. (2020), while working on the Iranian mangroves that thrive in an arid environment, appreciated the role of biomass carbon stock in governing the SOC. Though the AGC and BGC were substantially low, the SOC was significantly high in these mangrove stands. This study showed that even the arid and semi-arid mangroves store substantial quantities of carbon, and most of it remains locked in the soils. Al-Nadabi and Sulaiman (2018) observed that the seaward mangroves of Al-Qurm Nature Reserve, Muscat, had lower ecosystem carbon stock than their landward counterparts. Amiri (2021), while studying the arid mangroves of Iran, observed that soil electrical conductivity, soil texture, and total neutralizing value governed the concentrations of SOC. Etemadi et al. (2018b) inferred that regional sea-level rise in the Jask and Gabrik mangrove sites of Iran facilitated landward migration of mangroves. This observation indicates that mangroves can enhance their habitat margin stimulated by the ongoing rise in sea level, given the presence of adequate landmass for this vegetation to colonize. Schile et al. (2017), while working throughout the coastline of the UAE, observed that natural mangrove stands had substantially higher biomass carbon stock than the planted ones. However, the difference in SOC between the natural mangroves and planted ones was not significant.

The average height of the mangrove trees and their density (number of trees per unit area) are two crucial parameters that govern the AGC, BGC, and SOC in the mangrove habitats. Johnson et al. (2020), while working in the Nxaxo Estuary mangroves of South Africa, observed that tree density played the most significant role in regulating the biomass carbon stock of the *A. marina* stands. Sitoe et al. (2014) inferred similar findings from the mangroves of Sofala Bay, Mozambique. Raw et al. (2019), while working in the Nahoon Estuary of South Africa, observed that SOC values vary substantially across the top one-meter depth. Sitoe et al. (2014)

also conformed to similar findings. In contrary to Johnson et al. (2020), who emphasized tree density, Stringer et al. (2015) observed that mangroves of specific tree height have significantly different ecosystem carbon stock in the Zambezi River Delta (Mozambique). The net ecosystem carbon stock in the lowest height class was 373.8 Mg C ha^{-1}, whereas the highest height class had 620.8 ^{Mg} C ha^{-1}. Magalhães (2019) observed that secondary mangroves of less height store substantially lower AGC and SOC in the mangroves of Maputo City, Mozambique. Tamooh et al. (2008) examined the root structure of the three dominant species in the Gazi Bay, Kenya. They inferred that the BGC significantly varied with the species and their stand age. Andreetta et al. (2014) also researched in the Gazi Bay. They concluded that the macrobenthos played a crucial role in regulating the SOC in mangrove soils along with Woitchik et al. (1997).

Perera and Amarasinghe (2019), while working in the Sri Lankan mangroves, observed that the climatic regimes played a crucial role in governing the SOC. They observed that intermediate climate conditions favored the accumulation of SOC, whereas the dry belts have very low SOC. They also reported a significant positive relationship between the annual rainfall and the mean SOC. Perera and Amarasinghe (2019) stated that mangroves adjoining micro-tidal estuaries store substantially high SOC than the macro-tidal ones. Cooray et al. (2021) seconded the observations of Perera and Amarasinghe (2019). They also commented that climatic variability plays a crucial role in regulating AGC and BGC. Ramdhun and Appadoo (2020), while working in the Mauritian mangroves, observed that inundation zones played a significant role in governing the SOC. Arias-Ortiz et al. (2021), while working in the mangroves of Madagascar, observed that clearing of mangroves substantially reduces the SOC in the topsoil layers.

4.9 Conclusion

The present chapter indicates that the mangroves of the Indian Ocean received substantial attention; however, several countries are yet to carry out adequate research. Few countries reported total carbon stock in their nationwide mangrove forests. Such holistic carbon estimation has become a dire need to plan for effective conservation measures. Characterizing carbon stock has become a part of mainstream research in many countries bordering the Indian Ocean. However, studies finding avenues to enhance the carbon density by increasing mangrove cover are scarce. The mangroves of African countries and the arid western regions have promising potential to act as carbon sinks. However, the number of studies in these belts needs to increase in the future. This chapter collated the present findings on this issue and also pointed out the lacunae as of now. Future researchers would get a better picture of the areas that require focus and priority.

References

Abbas S, Qamer FM, Ali G, Tripathi NK, Shehzad K, Saleem R, Gilani H (2013) An assessment of status and distribution of mangrove forest cover in Pakistan. J Biodiver Environ Sci 3(6):64–78

Adame MF, Cherian S, Reef R, Stewart-Koster B (2017) Mangrove root biomass and the uncertainty of belowground carbon estimations. For Ecol Manag 403:52–60. https://doi.org/10.1016/j.foreco.2017.08.016

Adame MF, Zakaria RM, Fry B, Chong VC, Then YHA, Brown CJ, Lee SY (2018) Loss and recovery of carbon and nitrogen after mangrove clearing. Ocean Coast Manage 161:117–126. https://doi.org/10.1016/j.ocecoaman.2018.04.019

Adams JB, Colloty BM, Bate GC (2004) The distribution and state of mangroves along the coast of Transkei, eastern Cape Province, South Africa. Wetl Ecol Manag 12:531–541. https://doi.org/10.1007/s11273-005-5165-0

Ahamada S (1997) Situation de la mangroves des Comores. SAREC-SIDA Regional Workshop on mangrove ecology, physiology and management, Zanzibar

Ahmed S, Kamruzzaman M, Azad MS, Khan MNI (2021) Fine root biomass and its contribution to the mangrove communities in three saline zones of Sundarbans, Bangladesh. Rhizosphere 17:100294. https://doi.org/10.1016/j.rhisph.2020.100294

Aksomkoae S (1993) Ecology and management of mangroves. IUCN, Bangkok, Thailand

Alavaisha E, Mangora MM (2016) Carbon stocks in the small estuarine mangroves of Geza and Mtimbwani, Tanga, Tanzania. Int J Res https://doi.org/10.1155/2016/2068283

Al-Nadabi A, Sulaiman H (2018). Carbon sink potential of *Avicennia marina* in the Al-Qurm Nature Reserve, Muscat, Oman. In: IOP Conf Ser Earth Environ Sci 151(1):012003. https://doi.org/10.1088/1755-1315/151/1/012003

Alongi DM (2020) Global significance of mangrove blue carbon in climate change mitigation. Sci 2(3):67. https://doi.org/10.3390/sci2030067

Alongi DM, Dixon P (2000) Mangrove primary production and above-and below-ground biomass in Sawi Bay, southern Thailand. Phuket Mar Biol C Spec Publ 22(3)

Alongi DM, Murdiyarso D, Fourqurean JW, Kauffman JB, Hutahaean A, Crooks S et al (2016) Indonesia's blue carbon: a globally significant and vulnerable sink for seagrass and mangrove carbon. Wetl Ecol Manag 24(1):3–13. https://doi.org/10.1007/s11273-015-9446-y

Alsumaiti TS, Hussein K, Al-Sumaiti AS (2017) Mangroves of Abu Dhabi Emirate, UAE, in a global context: a review. Int J Environ Sci 6(4):110–121

Amarasinghe MD, Balasubramaniam S (1992) Net primary productivity of two mangrove forest stands on the northwestern coast of Sri Lanka. In: Jaccarini V, Martens E (eds) the ecology of mangrove and related ecosystems. Developments in hydrobiology, 80, Springer, Dordrecht. https://doi.org/10.1007/978-94-017-3288-8_5

Amiri F (2021) Carbon storage potential of *Avicennia marina* as influenced by soil factors in National Park Nayband, South Coast of Iran. Acta Ecol Sin. https://doi.org/10.1016/j.chnaes.2021.06.005

Analuddin K, La Ode K, La Ode MYH, Andi S, Idin S, La S et al (2020) Aboveground biomass, productivity and carbon sequestration in Rhizophora stylosa mangrove forest of Southeast Sulawesi, Indonesia. Biodivers J Biol Divers 21(4). https://doi.org/10.13057/biodiv/d210407

Andreetta A, Fusi M, Cameldi I, Cimò F, Carnicelli S, Cannicci S (2014) Mangrove carbon sink. Do burrowing crabs contribute to sediment carbon storage? Evidence from a Kenyan mangrove system. J Sea Res 85:524–533. https://doi.org/10.1016/j.seares.2013.08.010

Appadoo C (2003) Status of mangroves in Mauritius. J Coast Dev 7(1):1–4

Arias-Ortiz A, Masqué P, Glass L, Benson L, Kennedy H, Duarte CM et al (2021) Losses of soil organic carbon with deforestation in mangroves of Madagascar. Ecosystems 24(1):1–19. https://doi.org/10.1007/s10021-020-00500-z

Ashton EC, Hogarth PJ, Ormond R (1999) Breakdown of mangrove leaf litter in a managed mangrove forest in Peninsular Malaysia. In: Dodd RS (eds) Diversity and function in mangrove ecosystems. Developments in hydrobiology, vol 145. Springer, Dordrecht. https://doi.org/10.1007/978-94-011-4078-2_8

Askari M, Homaei A, Kamrani E, Zeinali F, Andreetta A (2021) Estimation of carbon pools in the biomass and soil of mangrove forests in Sirik Azini Creek, Hormozgan Province (Iran). Res Sq. https://doi.org/10.21203/rs.3.rs-518020/v1

Aslan A, Rahman AF, Warren MW, Robeson SM (2016) Mapping spatial distribution and biomass of coastal wetland vegetation in Indonesian Papua by combining active and passive remotely sensed data. Remote Sens Environ 183:65–81. https://doi.org/10.1016/j.rse.2016.04.026

Azad MS, Kamruzzaman M, Osawa A (2020) Quantification and understanding of above and belowground biomass in medium saline zone of the Sundarbans, Bangladesh: the relationships with forest attributes. J Sustain for 39(4):331–345. https://doi.org/10.1080/10549811.2019.1664307

Azman MS, Sharma S, Shaharudin MAM, Hamzah ML, Adibah SN, Zakaria RM, MacKenzie RA (2021) Stand structure, biomass and dynamics of naturally regenerated and restored mangroves in Malaysia. For Ecol Manag 482:118852. https://doi.org/10.1016/j.foreco.2020.118852

Bakosurtanal (2009) Peta mangroves Indonesia. Pusat Survey Sumberdaya Alam Laut. Badan Koordinasi Survey dan Pemetaan Nasional.

Barbosa FM, Cuambe CC, Bandeira SO (2001) Status and distribution of mangroves in Mozambique. S Afr J Bot 67(3):393–398. https://doi.org/10.1016/S0254-6299(15)31155-8

Basyuni M, Sagami H, Baba S, Putri LA, Wati R, Oku H (2017) Salinity alters the polyisoprenoid alcohol content and composition of both salt-secreting and non–salt-secreting mangrove seedlings. Hayati J Biosci 24(4):206–214. https://doi.org/10.1016/j.hjb.2017.11.006

Benson L, Glass L, Jones TG, Ravaoarinorotsihoarana L, Rakotomahazo C (2017) Mangrove carbon stocks and ecosystem cover dynamics in southwest Madagascar and the implications for local management. Forests 8(6):190. https://doi.org/10.3390/f8060190

Bhadra T, Mukhopadhyay A, Hazra S (2017) Identification of river discontinuity using geo-informatics to improve freshwater flow and ecosystem services in Indian Sundarban Delta. In: Hazra S, Mukhopadhyay A, Ghosh A, Mitra D, Dadhwal V (eds) Environment and earth observation. Springer Remote Sensing/Photogrammetry. Springer, Cham. https://doi.org/10.1007/978-3-319-46010-9_10

Bindu G, Rajan P, Jishnu ES, Joseph KA (2020) Carbon stock assessment of mangroves using remote sensing and geographic information system. Egypt J Remote Sens Sp Sci 23(1):1–9. https://doi.org/10.1016/j.ejrs.2018.04.006

Bird M, Chua S, Fifield LK, Teh TS, Lai J (2004) Evolution of the Sungei Buloh-Kranji mangrove coast, Singapore. Appl Geogr 24(3):181–198. https://doi.org/10.1016/j.apgeog.2004.04.002

Brakenridge GR, Syvitski JPM, Niebuhr E, Overeem I, Higgins SA, Kettner AJ, Prades L (2017) Design with nature: Causation and avoidance of catastrophic flooding, Myanmar. Earth-Sci Rev 165:81–109. https://doi.org/10.1016/j.earscirev.2016.12.009

Bryan-Brown DN, Connolly RM, Richards DR, Adame F, Friess DA, Brown CJ (2020) Global trends in mangrove forest fragmentation. Sci Rep 10(1):1–8. https://doi.org/10.1038/s41598-020-63880-1

Cadiz PL, Calumpong HP, Sinutok S, Chotikarn P (2020) Carbon storage potential of natural and planted Manglas in Trang. Thailand. Appl Ecol Environ Res 18(3):4383–4403

Cameron C, Hutley LB, Friess DA, Brown B (2019) Community structure dynamics and carbon stock change of rehabilitated mangrove forests in Sulawesi, Indonesia. Ecol Appl 29(1):e01810. https://doi.org/10.1002/eap.1810

Cameron C, Hutley LB, Munksgaard NC, Phan S, Aung T, Thinn T et al (2021) Impact of an extreme monsoon on CO_2 and CH_4 fluxes from mangrove soils of the Ayeyarwady Delta, Myanmar. Sci Total Environ 760:143422. https://doi.org/10.1016/j.scitotenv.2020.143422

Carbone F, Accordi G (2000) The Indian Ocean coast of Somalia. Mar Pollut Bull 41(1–6):141–159. https://doi.org/10.1016/S0025-326X(00)00107-7

CCD (1986) Master plan on Coastal erosion management, Coastal conversion department and Danish Hydraulic Institute. Colombo 1:142

Chanda A, Mukhopadhyay A, Ghosh T, Akhand A, Mondal P, Ghosh S et al (2016) Blue carbon stock of the Bangladesh Sundarban mangroves: what could be the scenario after a century? Wetlands 36(6):1033–1045. https://doi.org/10.1007/s13157-016-0819-7

Charrua AB, Bandeira SO, Catarino S, Cabral P, Romeiras MM (2020) Assessment of the vulnerability of coastal mangrove ecosystems in Mozambique. Ocean Coast Manag 189:105145. https://doi.org/10.1016/j.ocecoaman.2020.105145

Charuppat T, Charuppat J (1997) Application of Lansat-5 (TM) for changing mangrove forest areas of Thailand. Royal Forest Development Press, Bangkok, Thailand, p 78

Chatterjee M, Shankar D, Sen GK, Sanyal P, Sundar D, Michael GS et al (2013) Tidal variations in the Sundarbans Estuarine System, India. J Earth Syst Sci 122(4):899–933

Chellamani P, Singh CP, Panigrahy S (2014) Assessment of the health status of Indian mangrove ecosystems using multi temporal remote sensing data. Trop Ecol 55(2):245–253

Chen GC, Ulumuddin YI, Pramudji S, Chen SY, Chen B, Ye Y et al (2014) Rich soil carbon and nitrogen but low atmospheric greenhouse gas fluxes from North Sulawesi mangrove swamps in Indonesia. Sci Total Environ 487:91–96. https://doi.org/10.1016/j.scitotenv.2014.03.140

Clarisse JO, Appadoo C, Boojhawon R (2016) Spatial distribution of *Rhizophora mucronata* and *Bruguiera gymnorrhiza* mangroves using kriging technique: a contribution to understanding forest structure in south-east coast of Mauritius (Indian Ocean). Univ Maurit Res J 22:458–483

Cohen R, Kaino J, Okello JA, Bosire JO, Kairo JG, Huxham M, Mencuccini M (2013) Propagating uncertainty to estimates of above-ground biomass for Kenyan mangroves: a scaling procedure from tree to landscape level. For Ecol Manag 310:968–982. https://doi.org/10.1016/j.foreco.2013.09.047

Constance A, Haverkamp PJ, Bunbury N, Schaepman-Strub G (2021) Extent change of protected mangrove forest and its relation to wave power exposure on Aldabra Atoll. Glob Ecol Conserv 27:e01564. https://doi.org/10.1016/j.gecco.2021.e01564

Cooray PLIGM, Kodikara KAS, Kumara MP, Jayasinghe UI, Madarasinghe SK, Dahdouh-Guebas F et al (2021) Climate and intertidal zonation drive variability in the carbon stocks of Sri Lankan mangrove forests. Geoderma 389:114929. https://doi.org/10.1016/j.geoderma.2021.114929

Dahdouh-Guebas F, Zetterström T, Rönnbäck P, Troell M, Wickramasinghe A, Koedam N (2002) Recent changes in land-use in the Pambala–Chilaw lagoon complex (Sri Lanka) investigated using remote sensing and GIS: conservation of mangroves vs. development of shrimp farming. Environ Dev Sustain 4(2):185–200. https://doi.org/10.1023/A:1020854413866

Danehkar A (1996) Iranian mangroves forests. Environ Sci Q J 20:8–22

Danehkar A (1998) Marine sensitive areas of Iran. Environ Sci Q J 24:28–38

Danehkar A (2001) Mangroves forests zonation in Gaz and Harra international wetlands. Environ Sci Q J 34:43–49

Darmawan S, Sari DK, Wikantika K, Tridawati A, Hernawati R, Sedu MK (2020) Identification before-after forest fire and prediction of mangrove forest based on Markov-cellular automata in part of Sembilang national park, Banyuasin, South Sumatra, Indonesia. Remote Sens 12(22):3700. https://doi.org/10.3390/rs12223700

Dasgupta S, Sobhan I, Wheeler D (2017) The impact of climate change and aquatic salinization on mangrove species in the Bangladesh Sundarbans. Ambio 46(6):680–694. https://doi.org/10.1007/s13280-017-0911-0

De Alban JDT, Jamaludin J, de Wen DW, Than MM, Webb EL (2020) Improved estimates of mangrove cover and change reveal catastrophic deforestation in Myanmar. Environ Res Lett 15(3):034034. https://doi.org/10.1088/1748-9326/ab666d

Ellison JC (1999) Impacts of sediment burial on mangroves. Mar Pollut Bull 37(8–12):420–426. https://doi.org/10.1016/S0025-326X(98)00122-2

Elmahdy SI, Ali TA, Mohamed MM, Howari FM, Abouleish M, Simonet D (2020) Spatiotemporal mapping and monitoring of mangrove forests changes From 1990 to 2019 in the Northern Emirates, UAE using random forest, kernel logistic regression and Naive Bayes tree models. Front Environ Sci 8:102. https://doi.org/10.3389/fenvs.2020.00102

Estoque RC, Myint SW, Wang C, Ishtiaque A, Aung TT, Emerton L et al (2018) Assessing environmental impacts and change in Myanmar's mangrove ecosystem service value due to deforestation (2000–2014). Glob Chang Biol 24(11):5391–5410. https://doi.org/10.1111/gcb.14409
Etemadi H, Smoak JM, Karami J (2018a) Land use change assessment in coastal mangrove forests of Iran utilizing satellite imagery and CA–Markov algorithms to monitor and predict future change. Environ Earth Sci 77(5):1–13. https://doi.org/10.1007/s12665-018-7392-8
Etemadi H, Smoak JM, Sanders CJ (2018b) Forest migration and carbon sources to Iranian mangrove soils. J Arid Environ 157:57–65. https://doi.org/10.1016/j.jaridenv.2018.06.005
Fagoonee I (1990) Coastal marine ecosystems of Mauritius. Hydrobiologia 203:55–62. https://doi.org/10.1007/BF00008443
FAO (2005) Global forest resources assessment: thematic study on mangroves of Malaysia, Country Profile. Forestry Department Food and Agriculture Organization of the United Nations, FAO, Rome, Italy
FAO (2007) Food and Agriculture Organization of the United Nations. The world's mangroves 1980–2005. FAO Forestry Paper 153, Rome, Italy
Farooqui Z, Shafique S, Khan KL, Ali A, Iqbal P, Siddiqui PJ (2012) Assessment of litter production in semi arid mangroves forests near active Indus river mouth (Hajambro creek) and Karachi backwaters, Pakistan. Pak J Bot 44(5):1763–1768
Farooqui Z, Siddiqui PJ, Rasheed M (2014) Changes in organic, inorganic contents, carbon nitrogen ratio in decomposing *Avicennia marina* and *Rhizophora mucronata* leaves on tidal mudflats in Hajambro creek, Indus delta, Pakistan. J Trop Life Sci 4(1):37–45
Fatoyinbo TE, Simard M (2013) Height and biomass of mangroves in Africa from ICESat/GLAS and SRTM. Int J Remote Sens 34(2):668–681. https://doi.org/10.1080/01431161.2012.712224
Fatoyinbo TE, Simard M, Washington-Allen RA, Shugart HH (2008) Landscape-scale extent, height, biomass, and carbon estimation of Mozambique's mangrove forests with Landsat ETM+ and Shuttle Radar Topography Mission elevation data. J Geophys Res Biogeosci 113(G2). https://doi.org/10.1029/2007JG000551
Fernando SM, Bandeira SO (2009) Litter fall and decomposition of mangrove species *Avicennia marina* and *Rhizophora mucronata* in Maputo Bay, Mozambique. West India Ocean J Mar Sci 8(2). https://doi.org/10.4314/wiojms.v8i2.56975
Foolmaun RK, Chamilall DS, Munhurrun G (2011) Overview of non-hazardous solid waste in the small island state of Mauritius. Resour Conserv Recycl 55(11):966–972. https://doi.org/10.1016/j.resconrec.2011.05.004
Forest Survey of India (2019) India State of Forest Report, Forest Survey of India, Dehradun, India
Fouda MM, Al-Muharrami M (1995) An initial assessment of mangrove resources and human activities at Mahout Island, Arabian Sea, Oman. In: Wong YS, Tam NFY (eds) Asia-Pacific symposium on mangrove ecosystems. Developments in hydrobiology, 106, Springer, Dordrecht. https://doi.org/10.1007/978-94-011-0289-6_39
Fouda MM, Ali-Muharrami MM (1996) Significance of mangroves in the arid environments of the Sultanate of Oman. Agric Res 1:41–49. https://journals.squ.edu.om/index.php/jams/article/view/484
Friess DA, Webb EL (2014) Variability in mangrove change estimates and implications for the assessment of ecosystem service provision. Global Ecol Biogeogr 23(7):715–725. https://doi.org/10.1111/geb.12140
Friess DA, Phelps J, Leong RC, Lee WK, Wee AKS, Sivasothi N et al (2012) Mandai mangrove, Singapore: lessons for the conservation of Southeast Asia's mangroves. Raffles Bull Zool 25:55–65
Friess DA, Richards DR, Phang VX (2016) Mangrove forests store high densities of carbon across the tropical urban landscape of Singapore. Urban Ecosyst 19(2):795–810. https://doi.org/10.1007/s11252-015-0511-3
Gaw LY, Linkie M, Friess DA (2018) Mangrove forest dynamics in Tanintharyi, Myanmar from 1989–2014, and the role of future economic and political developments. Singap J Trop Geogr 39(2):224–243. https://doi.org/10.1111/sjtg.12228

Ghasemi S, Mola N, Zakaria M (2013) Aboveground biomass, litterfall, and forest structure in the mangroves of Hormozgan Province. Iran. Nat Area J 33(3):339–343. https://doi.org/10.3375/043.033.0312

Ghosh SM, Behera MD, Paramanik S (2020) Canopy height estimation using sentinel series images through machine learning models in a mangrove forest. Remote Sens 12(9):1519. https://doi.org/10.3390/rs12091519

Giardino C, Bresciani M, Fava F, Matta E, Brando VE, Colombo R (2016) Mapping submerged habitats and mangroves of Lampi Island Marine National Park (Myanmar) from in situ and satellite observations. Remote Sens 8(1):2. https://doi.org/10.3390/rs8010002

Giesen W, Wulffraat S, Zieren M, Scholten L (2006) Mangrove guidebook for Southeast Asia (Part II). FAO and Wetlands International, The Netherlands

Gilani H, Naz HI, Arshad M, Nazim K, Akram U, Abrar A, Asif M (2021) Evaluating mangrove conservation and sustainability through spatiotemporal (1990–2020) mangrove cover change analysis in Pakistan. Estuar Coast Shelf Sci 249:107128. https://doi.org/10.1016/j.ecss.2020.107128

Giri C, Muhlhausen J (2008) Mangrove forest distributions and dynamics in Madagascar (1975–2005). Sensors 8(4):2104–2117. https://doi.org/10.3390/s8042104

Giri S, Mukhopadhyay A, Hazra S, Mukherjee S, Roy D, Ghosh S et al (2014) A study on abundance and distribution of mangrove species in Indian Sundarban using remote sensing technique. J Coast Conserv 18(4):359–367. https://doi.org/10.1007/s11852-014-0322-3

Gupta K, Mukhopadhyay A, Giri S, Chanda A, Majumdar SD, Samanta S et al (2018) An index for discrimination of mangroves from non-mangroves using LANDSAT 8 OLI imagery. MethodsX 5:1129–1139. https://doi.org/10.1016/j.mex.2018.09.011

Habshi AA, Youssef T, Aizpuru M, Blasco F (2007) New mangrove ecosystem data along the UAE coast using remote sensing. Aquat Ecosyst Health Manag 10(3):309–319. https://doi.org/10.1080/14634980701512525

Hamdan O, Aziz HK, Hasmadi IM (2014) L-band ALOS PALSAR for biomass estimation of Matang Mangroves, Malaysia. Remote Sens Environ 155:69–78. https://doi.org/10.1016/j.rse.2014.04.029

Hamilton SE, Casey D (2016) Creation of a high spatio-temporal resolution global database of continuous mangrove forest cover for the 21st century (CGMFC-21). Glob Ecol Biogeogr 25(6):729–738. https://doi.org/10.1111/geb.12449

Hariri K (2010) Environmental impact assessment of fisheries investment project. Unpublished Report Submitted to International Fund for Agricultural Development (IFAD)

Harishma KM, Sandeep S, Sreekumar VB (2020) Biomass and carbon stocks in mangrove ecosystems of Kerala, southwest coast of India. Ecol Process 9(1):1–9. https://doi.org/10.1186/s13717-020-00227-8

Hereher M, Al-Awadhi T, Al-Hatrushi S, Charabi Y, Mansour S, Al-Nasiri N et al (2020) Assessment of the coastal vulnerability to sea level rise: sultanate of Oman. Environ Earth Sci 79(15):1–12. https://doi.org/10.1007/s12665-020-09113-0

Hilton MJ, Manning SS (1995) Conversion of coastal habitats in Singapore: indications of unsustainable development. Environ Conserv 22(4):307–322. https://doi.org/10.1017/S0376892900034883

Hong LC, Hemati Z, Zakaria R (2017) Carbon stock evaluation of selected mangrove forests in peninsular Malaysia and its potential market value. J Environ Sci Manag 20(2):77–87

Hoppe-Speer SC, Adams JB, Bailey D (2015) Present state of mangrove forests along the Eastern Cape coast, South Africa. Wetl Ecol Manag 23(3):371–383. https://doi.org/10.1007/s11273-014-9387-x

Hoque AF, Datta DK (2005) The mangroves of Bangladesh. Int J Ecol Environ Sci 31(3):245–253

Hossain M, Othman S, Bujang JS, Kusnan M (2008) Net primary productivity of *Bruguiera parviflora* (Wight &Arn.) dominated mangrove forest at Kuala Selangor, Malaysia. For Ecol Manag 255(1):179–182. https://doi.org/10.1016/j.foreco.2007.09.011

Hossain M, Siddique MRH, Abdullah SR, Saha S, Ghosh DC, Rahman MS, Limon SH (2014) Nutrient dynamics associated with leaching and microbial decomposition of four abundant mangrove species leaf litter of the Sundarbans, Bangladesh. Wetlands 34(3):439–448. https://doi.org/10.1007/s13157-013-0510-1

Hossain M, Siddique MRH, Saha S, Abdullah SR (2015) Allometric models for biomass, nutrients and carbon stock in *Excoecaria agallocha* of the Sundarbans, Bangladesh. Wetl Ecol Manage 23(4):765–774. https://doi.org/10.1007/s11273-015-9419-1

Hossain M, Saha C, Abdullah SR, Saha S, Siddique MRH (2016) Allometric biomass, nutrient and carbon stock models for Kandelia candel of the Sundarbans, Bangladesh. Trees 30(3):709–717. https://doi.org/10.1007/s00468-015-1314-0

Howari FM, Jordan BR, Bouhouche N, Wyllie-Echeverria S (2009) Field and remote-sensing assessment of mangrove forests and seagrass beds in the northwestern part of the United Arab Emirates. J Coast Res 25(1):48–56. https://doi.org/10.2112/07-0867.1

Hussain N, Islam MN (2020) Hot spot (G_i^*) model for forest vulnerability assessment: a remote sensing-based geo-statistical investigation of the Sundarbans mangrove forest, Bangladesh. Model Earth Syst Environ 6:2141–2151

Hutchison J, Manica A, Swetna R, Balmford A, Spalding M (2014) Predicting global patterns in mangrove forest biomass. Conserv Lett 7:233–240. https://doi.org/10.1111/conl.12060

Ilman M, Wibisono ITC, Suryadiputra INN (2011) State of the art information on mangrove ecosystems in Indonesia. Wetlands International-Indonesia Programme, Bogor

Indrayani E, Kalor JD, Warpur M, Hamuna B (2021) Using allometric equations to estimate mangrove biomass and carbon stock in Demta Bay, Papua Province, Indonesia. J Ecol Eng 22(5):263–271. https://doi.org/10.12911/22998993/135945

Islam SDU, Bhuiyan MAH (2018) Sundarbans mangrove forest of Bangladesh: causes of degradation and sustainable management options. Environ Sustain 1(2):113–131. https://doi.org/10.1007/s42398-018-0018-y

Islam MM, Borgqvist H, Kumar L (2019) Monitoring Mangrove forest landcover changes in the coastline of Bangladesh from 1976 to 2015. Geocarto Int 34(13):1458–1476. https://doi.org/10.1080/10106049.2018.1489423

Jachowski NR, Quak MS, Friess DA, Duangnamon D, Webb EL, Ziegler AD (2013) Mangrove biomass estimation in Southwest Thailand using machine learning. Appl Geogr 45:311–321. https://doi.org/10.1016/j.apgeog.2013.09.024

Jagtap TG (1993) Studies on littoral and sublittoral macrophytes around the Mauritius coast. Atoll Res Bull 382:1–23

Jayakody JMAL, Amarasinghe MD, PahalawatttaArachchi V, De Silva KHWL (2008) Vegetation structure and potential gross primary productivity of mangroves at Kadolkele in Meegamuwa (Negombo) estuary, Sri Lanka. Sri Lanka J Aquat Sci 13:95–108

Jayatissa LP, Dahdouh-Guebas F, Koedam N (2002) A review of the floral composition and distribution of mangroves in Sri Lanka. Bot J Linnean Soc 138(1):29–43. https://doi.org/10.1046/j.1095-8339.2002.00002.x

Jayewardene RP (1986) Sri Lanka. In: Umali RM, Zamora PM, Gotera RR, Jara RS, Camacho AS, Vannucci M (eds) Mangroves of Asia and the Pacific: status and management. Philippines, UNESCO/UNDP, Manila, pp 219–230

Johnson JL, Raw JL, Adams JB (2020) First report on carbon storage in a warm-temperate mangrove forest in South Africa. Estuar Coast Shelf Sci 235:106566. https://doi.org/10.1016/j.ecss.2019.106566

Jones TG, Ratsimba HR, Ravaoarinorotsihoarana L, Cripps G, Bey A (2014) Ecological variability and carbon stock estimates of mangrove ecosystems in northwestern Madagascar. Forests 5(1):177–205. https://doi.org/10.3390/f5010177

Jonsson MN, Hedman AM (2019) Carbon stock assessment of mangrove ecosystems in Batticaloa Lagoon, Sri Lanka, with different degrees of human disturbances. Singap J Trop Geogr 40(2):199–218. https://doi.org/10.1111/sjtg.12267

Kairo JG, Kivyatu B, Koedam N (2002) Application of remote sensing and GIS in the management of mangrove forests within and adjacent to Kiunga Marine Protected Area, Lamu, Kenya. Environ Dev Sustain 4(2):153–166. https://doi.org/10.1023/A:1020890711588

Kairo JG, Lang'at JK, Dahdouh-Guebas F, Bosire J, Karachi M (2008) Structural development and productivity of replanted mangrove plantations in Kenya. For Ecol Manage 255(7):2670–2677. https://doi.org/10.1016/j.foreco.2008.01.031

Kamruzzaman M, Ahmed S, Osawa A (2017) Biomass and net primary productivity of mangrove communities along the Oligohaline zone of Sundarbans, Bangladesh. For Ecosyst 4(1):1–9. https://doi.org/10.1186/s40663-017-0104-0

Kamruzzaman M, Ahmed S, Paul S, Rahman MM, Osawa A (2018) Stand structure and carbon storage in the oligohaline zone of the Sundarbans mangrove forest, Bangladesh. For Sci Technol 14(1):23–28. https://doi.org/10.1080/21580103.2017.1417920

Kamruzzaman MD, Basak K, Paul SK, Ahmed S, Osawa A (2019) Litterfall production, decomposition and nutrient accumulation in Sundarbans mangrove forests, Bangladesh. For Sci Technol 15(1):24–32. https://doi.org/10.1080/21580103.2018.1557566

Kanniah KD, Sheikhi A, Cracknell AP, Goh HC, Tan KP, Ho CS, Rasli FN (2015) Satellite images for monitoring mangrove cover changes in a fast growing economic region in southern Peninsular Malaysia. Remote Sens 7(11):14360–14385. https://doi.org/10.3390/rs71114360

Karunathilake KMB (2003) Status of mangroves in Sri Lanka. J Coast Dev 7(1):5–9

Kathiresan K, Rajendran N (2005) Mangrove ecosystems of the Indian Ocean region. India J Mar Sci 34(1):104–113

Khalil ASM (2015) Mangroves of the Red Sea. In: Rasul N, Stewart I (eds) The Red Sea. Springer Earth System Sciences, Springer, Berlin, Heidelberg. https://doi.org/10.1007/978-3-662-45201-1_33

Khan MA, Aziz I (2001) Salinity tolerance in some mangrove species from Pakistan. Wetl Ecol Manag 9(3):229–233. https://doi.org/10.1023/A:1011112908069

Khan IA, Khan WR, Ali A, Nazre M (2021) Assessment of above-ground biomass in Pakistan Forest Ecosystem's carbon pool: a review. Forests 12(5):586. https://doi.org/10.3390/f12050586

Kida M, Fujitake N (2020) Organic carbon stabilization mechanisms in mangrove soils: a review. Forests 11(9):981. https://doi.org/10.3390/f11090981

Kirui KB, Kairo JG, Bosire J, Viergever KM, Rudra S, Huxham M, Briers RA (2013) Mapping of mangrove forest land cover change along the Kenya coastline using Landsat imagery. Ocean Coastal Manag 83:19–24. https://doi.org/10.1016/j.ocecoaman.2011.12.004

Klaus R, Turner JR (2004) The marine biotopes of the Socotra Archipelago. In: Friedhelm K (ed.) Fauna of Arabia, Socotra Archipelago, SenckenbergischeNaturforschende gesellschaft and KACST, Riyadh, KSA, 20, pp 45–115

Kokwaro JO (1985) The distribution and economic importance of the mangrove forests in Kenya. J East Afr Nat Hist Soc Natl Mus 75(188):1–12. http://hdl.handle.net/1834/8465

Kridiborworn P, Chidthaisong A, Yuttitham M, Tripetchkul S (2012) Carbon sequestration by mangrove forest planted specifically for charcoal production in Yeesarn, Samut Songkram. J Sustain Energy Environ 3(2):87–92

Kristensen E, Flindt MR, Ulomi S, Borges AV, Abril G, Bouillon S (2008) Emission of CO_2 and CH_4 to the atmosphere by sediments and open waters in two Tanzanian mangrove forests. Mar Ecol Prog Ser 370:53–67. https://doi.org/10.3354/meps07642

Kristensen E, Mangion P, Tang M, Flindt MR, Holmer M, Ulomi S (2011) Microbial carbon oxidation rates and pathways in sediments of two Tanzanian mangrove forests. Biogeochemistry 103(1):143–158. https://doi.org/10.1007/s10533-010-9453-2

Kristiningrum R (2019) Species diversity, stand productivity, aboveground biomass, and economic value of mangrove ecosystem in Mentawir Village, East Kalimantan, Indonesia. Biodiversitas 20(10):2848–2857. http://repository.unmul.ac.id/handle/123456789/6136

Kumar R (2000) Distribution of mangroves in Goa. Indian J Forestry 23:360–365

Kumar A, Khan MA, Muqtadir A (2011) Distribution of mangroves along the Red Sea Coast of the Arabian Peninsula: part 3: Coast of Yemen. Earth Sci India 4(II):29–38

Kusmana C (2014) Distribution and current status of mangrove forests in Indonesia. In: Faridah-Hanum I, Latiff A, Hakeem K, Ozturk M (eds) Mangrove ecosystems of Asia, Springer, New York. https://doi.org/10.1007/978-1-4614-8582-7_3

Lai S, Loke LH, Hilton MJ, Bouma TJ, Todd PA (2015) The effects of urbanisation on coastal habitats and the potential for ecological engineering: a Singapore case study. Ocean Coast Manag 103:78–85. https://doi.org/10.1016/j.ocecoaman.2014.11.006

Lang'at JKS, Kairo JG, Mencuccini M, Bouillon S, Skov MW, Waldron S, Huxham M (2014) Rapid losses of surface elevation following tree girdling and cutting in tropical mangroves. Plos One 9(9):e107868. https://doi.org/10.1371/journal.pone.0107868

Liu X, Xiong Y, Liao B (2017) Relative contributions of leaf litter and fine roots to soil organic matter accumulation in mangrove forests. Plant Soil 421(1):493–503. https://doi.org/10.1007/s11104-017-3477-5

Lovelock CE (2008) Soil respiration and belowground carbon allocation in mangrove forests. Ecosystems 11(2):342–354. https://doi.org/10.1007/s10021-008-9125-4

Lucas R, Van De Kerchove R, Otero V, Lagomasino D, Fatoyinbo L, Omar H et al (2020). Structural characterisation of mangrove forests achieved through combining multiple sources of remote sensing data. Remote Sens Environ 237:111543. https://doi.org/10.1016/j.rse.2019.111543

Lupembe IB (2014) Carbon stocks in the Mangrove ecosystem of Rufiji river delta, Rufiji district, Tanzania (Doctoral dissertation, Sokoine University of Agriculture).

Mafi-Gholami D, Zenner EK, Jaafari A, Ward RD (2019) Modeling multi-decadal mangrove leaf area index in response to drought along the semi-arid southern coasts of Iran. Sci Total Environ 656:1326–1336. https://doi.org/10.1016/j.scitotenv.2018.11.462

Mafi-Gholami D, Zenner EK, Jaafari A, Bui DT (2020) Spatially explicit predictions of changes in the extent of mangroves of Iran at the end of the 21st century. Estuar Coast Shelf Sci 237:106644. https://doi.org/10.1016/j.ecss.2020.106644

Magalhães TM (2019) Carbon storage in secondary mangroves along the West Coastline of Maputo City, Mozambique. Wetlands 39(2):239–249. https://doi.org/10.1007/s13157-018-1104-8

Mahmood H, Siddique MR, Costello L, Birigazzi L, Abdullah SR, Henry M et al (2019) Allometric models for estimating biomass, carbon and nutrient stock in the Sal zone of Bangladesh. Iforest-Biogeosci for 12(1):69–75. https://doi.org/10.3832/ifor2758-011

Malik A, Jalil AR, Arifuddin A, Syahmuddin A (2020) Biomass carbon stocks in the mangrove rehabilitated area of Sinjai District, South Sulawesi, Indonesia. Geogr Environ Sustain 13(3):32–38. https://doi.org/10.24057/2071-9388-2019-131

Martuti NKT, Setyowati DL, Nugraha SB, Mutiatari DP (2017) Carbon stock potency of mangrove ecosystem at Tapak Sub-village, Semarang, Indonesia. Aquac Aquar Conserv Legis 10(6):1524–1533

Masood H, Afsar S, Zamir UB, Kazmi JH (2015) Application of comparative remote sensing techniques for monitoring mangroves in Indus Delta, Sindh, Pakistan. Biol Forum Int J 7(1):783–792

Maung WS, Sasaki J (2021) Assessing the natural recovery of mangroves after human disturbance using neural network classification and Sentinel-2 Imagery in Wunbaik Mangrove Forest, Myanmar. Remote Sens 13(1):52. https://doi.org/10.3390/rs13010052

Maurya K, Mahajan S, Chaube N (2021) Remote sensing techniques: mapping and monitoring of mangrove ecosystem—a review. Complex Intell Syst. https://doi.org/10.1007/s40747-021-00457-z

Mijanuddin SM, Rafiqul Hoque ATM, Abdullah SA (2014) The changing landscape of mangroves in Bangladesh compared to four other countries in tropical regions. J for Res 25:605–611. https://doi.org/10.1007/s11676-014-0448-z

Milani AS (2018) Mangrove forests of the Persian Gulf and the Gulf of Oman. In: Makowski C, Finkl C (eds) Threats to mangrove forests. Coastal Research Library, 25, Springer, Cham. https://doi.org/10.1007/978-3-319-73016-5_3

Minu A, Routh J, Machiwa JF, Pamba S (2020) Spatial variation of nutrients and primary productivity in the Rufiji Delta mangroves, Tanzania. Afr J Mar Sci 42(2):221–232. https://doi.org/10.2989/1814232X.2020.1776391

Mitra A, Sengupta K, Banerjee K (2011) Standing biomass and carbon storage of above-ground structures in dominant mangrove trees in the Sundarbans. For Ecol Manage 261(7):1325–1335. https://doi.org/10.1016/j.foreco.2011.01.012

MNRT (2015) National Forest Resources Monitoring and Assessment (NAFORMA) main results. Dar es Salaam: Tanzania Forest Services, Ministry of Natural Resources and Tourism

Mukhtar I, Hannan A (2012) Constrains on mangrove forests and conservation projects in Pakistan. J Coast Conserv 16(1):51–62. https://doi.org/10.1007/s11852-011-0168-x

Murdiyarso D, Sasmito SD, Sillanpää M, MacKenzie R, Gaveau D (2021) Mangrove selective logging sustains biomass carbon recovery, soil carbon, and sediment. Sci Rep 11(1):1–10. https://doi.org/10.1038/s41598-021-91502-x

Murshed SB, Kaluarachchi JJ (2018) Scarcity of fresh water resources in the Ganges Delta of Bangladesh. Water Secur 4:8–18. https://doi.org/10.1016/j.wasec.2018.11.002

Nagi HM, Abubakr MM (2013) Threats status to the mangrove ecosystem along the coastal zone of Yemen. J King Abdulaziz Univ 24(1):101–117

Naidoo G (2016) The mangroves of South Africa: an ecophysiological review. S Afr J Bot 107:101–113. https://doi.org/10.1016/j.sajb.2016.04.014

Naskar K, Mandal R (1999) Ecology and biodiversity of Indian mangroves. Daya Books, India

Nayak S, Bahuguna A (2001) Application of remote sensing data to monitor mangroves and other coastal vegetation of India. Indian J Mar Sci 30(4):195–2013

Nazim K, Ahmed M, Khan MU (2018) Carbon squestration potential of sandspit backwater mangrove and sediment, Pakistab. FUUAST J Biol 8(2):209–215

Nehren U, Wicaksono P (2018) Mapping soil carbon stocks in an oceanic mangrove ecosystem in Karimunjawa Islands, Indonesia. Estuar Coast Shelf Sci 214:185–193. https://doi.org/10.1016/j.ecss.2018.09.022

Ng PKL, Low JKY (1994) Status of mangroves in Singapore: conservation beyond the year 2000. In: Proceedings of the Third ASEAN–Australia symposium on living coastal resources. Chulalongkorn University, Bangkok, Thailand, pp 229–232

Njana MA (2020) Structure, growth, and sustainability of mangrove forests of mainland Tanzania. Glob Ecol Conserv 24:e01394. https://doi.org/10.1016/j.gecco.2020.e01394

Njana MA, Zahabu E, Malimbwi RE (2018) Carbon stocks and productivity of mangrove forests in Tanzania. South for J for Sci 80(3):217–232. https://doi.org/10.2989/20702620.2017.1334314

Oh RRY, Friess DA, Brown BM (2017) The role of surface elevation in the rehabilitation of abandoned aquaculture ponds to mangrove forests, Sulawesi, Indonesia. Ecol Eng 100:325–334. https://doi.org/10.1016/j.ecoleng.2016.12.021

Otero V, Van De Kerchove R, Satyanarayana B, Mohd-Lokman H, Lucas R, Dahdouh-Guebas F (2019) An analysis of the early regeneration of mangrove forests using Landsat time series in the Matang Mangrove Forest Reserve, Peninsular Malaysia. Remote Sens 11(7):774. https://doi.org/10.3390/rs11070774

Pandey CN, Pandey R (2013) Carbon sequestration in mangroves of Gujarat, India. Int J Bot Res 3(2):57–70

Pastor-Guzman J, Dash J, Atkinson PM (2018) Remote sensing of mangrove forest phenology and its environmental drivers. Remote Sens Environ 205:71–84. https://doi.org/10.1016/j.rse.2017.11.009

Pattanaik C, Prasad SN (2011) Assessment of aquaculture impact on mangroves of Mahanadi delta (Orissa), East coast of India using remote sensing and GIS. Ocean Coast Manag 54(11):789–795. https://doi.org/10.1016/j.ocecoaman.2011.07.013

Perera KARS, Amarasinghe MD (2019) Carbon sequestration capacity of mangrove soils in micro tidal estuaries and lagoons: a case study from Sri Lanka. Geoderma 347:80–89. https://doi.org/10.1016/j.geoderma.2019.03.041

PERSGA/GEF (2004) Status of mangroves in the Red Sea and Gulf of Aden. PERSGA Technical Series No, 11, PERSGA, Jeddah

Pham TD, Yokoya N, Bui DT, Yoshino K, Friess DA (2019) Remote sensing approaches for monitoring mangrove species, structure, and biomass: opportunities and challenges. Remote Sens 11(3):230. https://doi.org/10.3390/rs11030230

Pimple U, Simonetti D, Sitthi A, Pungkul S, Leadprathom K, Skupek H et al (2018) Google earth engine based three decadal landsat imagery analysis for mapping of mangrove forests and its surroundings in the trat province of Thailand. J Comput Commun 6:247–264. https://doi.org/10.4236/jcc.2018.61025

Pitchaikani JS (2020) Vertical current structure in a macro-tidal, well mixed Sundarban ecosystem, India. J Coast Conserv 24(5):1–11. https://doi.org/10.1007/s11852-020-00782-4

Poungparn S, Komiyama A, Tanaka A, Sangtiean T, Maknual C, Kato S et al (2009) Carbon dioxide emission through soil respiration in a secondary mangrove forest of eastern Thailand. J Trop Ecol 25(4):393–400. https://doi.org/10.1017/S0266467409006154

Poungparn S, Komiyama A, Umnouysin S, Rodtassana C, Sangtiean T, Maknual C et al (2020) Ten-year estimation of net primary productivity in a mangrove forest under a tropical monsoon climate in Eastern Thailand: significance of the temperature environment in the dry season. Forests 11(9):987. https://doi.org/10.3390/f11090987

Pradisty NA, Amir AA, Zimmer M (2021) Plant species-and stage-specific differences in microbial decay of mangrove leaf litter: the older the better? Oecologia 195(4):843–858. https://doi.org/10.1007/s00442-021-04865-3

Putz FE, Chan HT (1986) Tree growth, dynamics, and productivity in a mature mangrove forest in Malaysia. Forest Ecol Manag 17(2–3):211–230. https://doi.org/10.1016/0378-1127(86)90113-1

Qasim H, Luqman M, Khan S (2016) A study of forest land cover changes using satellite remote sensing in Thatta District Pakistan. Sci Int 28(4):4069–4075

Quadros AF, Zimmer M (2017) Dataset of true mangroves plant species traits. Biodivers Data J 5:e22089. https://doi.org/10.3897/BDJ.5.e22089

Ragavan P, Saxena A, Jayaraj RSC, Mohan PM, Ravichandran K, Saravanan S, Vijayaraghavan A (2016) A review of the mangrove floristics of India. Taiwania 61(3):224–242

Rahman MS, Hossain GM, Khan SA, Uddin SN (2015) An annotated checklist of the vascular plants of Sundarban Mangrove Forest of Bangladesh, Bangladesh. J Plant Taxon 22(1):17–41. https://doi.org/10.3329/bjpt.v22i1.23862

Rahman MM, Ghosh T, Salehin M, Ghosh A, Haque A, Hossain MA et al (2020) Ganges-Brahmaputra-Meghna delta, Bangladesh and India: a transnational mega-delta. In: Nicholls RJ, Adger WN, Hutton CW, Hanson SE (eds) Deltas in the anthropocene. Palgrave Macmillan, Cham, pp 23–51

Rahman MS, Donoghue DN, Bracken LJ (2021) Is soil organic carbon underestimated in the largest mangrove forest ecosystems? Evidence from the Bangladesh Sundarbans. Catena 200:105159. https://doi.org/10.1016/j.catena.2021.105159

Rajkaran A, Adams JB (2007) Mangrove litter production and organic carbon pools in the Mngazana Estuary, South Africa. Afr J Aquat Sci 32(1):17–25. https://doi.org/10.2989/AJAS.2007.32.1.3.140

Rajkaran A, Adams JB, du Preez DR (2004) A method for monitoring mangrove harvesting at the Mngazana estuary, South Africa. Afr J Aquat Sci 29(1):57–65. https://doi.org/10.2989/16085910409503792

Rakotomavo A, Fromard F (2010) Dynamics of mangrove forests in the Mangoky River delta, Madagascar, under the influence of natural and human factors. For Ecol Manag 259(6):1161–1169. https://doi.org/10.1016/j.foreco.2010.01.002

Ramdhun D, Appadoo C (2020) A contribution to understanding blue carbon sequestration and forest structure in mangroves of different ages in a small island (Mauritius). Indo Pac J Ocean Life 4(2). https://doi.org/10.13057/oceanlife/o040203

Rao RG, Woitchik AF, Goeyens L, Van Riet A, Kazungu J, Dehairs F (1994) Carbon, nitrogen contents and stable carbon isotope abundance in mangrove leaves from an east African coastal lagoon (Kenya). Aquat Bot 47(2):175–183. https://doi.org/10.1016/0304-3770(94)90012-4

Rasolofoharinoro M, Blasco F, Bellan MF, Aizpuru M, Gauquelin T, Denis J (1998) A remote sensing based methodology for mangrove studies in Madagascar. Int J Remote Sens 19(10):1873–1886. https://doi.org/10.1080/014311698215036

Rasquinha DN, Mishra DR (2021) Impact of wood harvesting on mangrove forest structure, composition and biomass dynamics in India. Estuar Coast Shelf Sci 248:106974. https://doi.org/10.1016/j.ecss.2020.106974

Raw JL, Julie CL, Adams JB (2019) A comparison of soil carbon pools across a mangrove-salt marsh ecotone at the southern African warm-temperate range limit. S Afr J Bot 127:301–307. https://doi.org/10.1016/j.sajb.2019.11.005

Ray R, Ganguly D, Chowdhury C, Dey M, Das S, Dutta MK et al (2011) Carbon sequestration and annual increase of carbon stock in a mangrove forest. Atmos Environ 45(28):5016–5024. https://doi.org/10.1016/j.atmosenv.2011.04.074

Razali SM, Nuruddin AA, Lion M (2019) Mangrove vegetation health assessment based on remote sensing indices for Tanjung Piai, Malay Peninsular. J Landsc Ecol 12(2):26–40. https://doi.org/10.1016/j.atmosenv.2011.04.074

Reddy CS, Pattanaik C, Murthy MSR (2007) Assessment and monitoring of mangroves of Bhitarkanika Wildlife Sanctuary, Orissa, India using remote sensing and GIS. Current Sci 92(10):1409–1415. https://www.jstor.org/stable/24097770

Richards DR, Thompson BS, Wijedasa L (2020) Quantifying net loss of global mangrove carbon stocks from 20 years of land cover change. Nat Commun 11(1):1–7. https://doi.org/10.1038/s41467-020-18118-z

Roger E, Andrianasaolo M (2003) Mangroves and Salt Marshes. In: Goodman S, Benstead J (eds) Natural history of Madagascar. Chicago University Press, Chicago, pp 209–210

Rogers KG, Goodbred SL (2014) The Sundarbans and Bengal Delta: the world's largest tidal mangrove and delta system. In: Kale V (eds) Landscapes and landforms of India. World Geomorphological Landscapes. Springer, Dordrecht, pp 181–187. https://doi.org/10.1007/978-94-017-8029-2_18

Rozainah MZ, Nazri MN, Sofawi AB, Hemati Z, Juliana WA (2018) Estimation of carbon pool in soil, above and below ground vegetation at different types of mangrove forests in Peninsular Malaysia. Mar Pollut Bull 137:237–245. https://doi.org/10.1016/j.marpolbul.2018.10.023

Sahu SC, Suresh HS, Murthy IK, Ravindranath NH (2015) Mangrove area assessment in India: implications of loss of mangroves. J Earth Sci Clim Chang 6(5). https://doi.org/10.4172/2157-7617.1000280

Sahu SC, Kumar M, Ravindranath NH (2016) Carbon stocks in natural and planted mangrove forests of Mahanadi Mangrove Wetland, East Coast of India. Curr Sci 110(12):2253–2260

Saket M, Matusse R (1994) Study for the determination of the rate of deforestation of the mangrove vegetation in Mozambique. FAO/PNUD/MOZ/92/013: 9, DNFFB

Saleem A, Nileysha A (2011) Characteristics, status and need for conservation of mangrove ecosystems in the Republic of Maldives, Indian Ocean. J Natl Sci Found Sri Lanka 31(1–2):201–213

Santisuk T (1983) Taxonomy and distribution of terrestrial trees and shrubs in the mangrove formations in Thailand. Nat Hist Bull Siam Soc 31:63–91

Sarker S, Masud-Ul-Alam M, Hossain MS, Rahman Chowdhury S, Sharifuzzaman SM (2021) A review of bioturbation and sediment organic geochemistry in mangroves. Geol J 56(5):2439–2450. https://doi.org/10.1002/gj.3808

Sasmito SD, Sillanpää M, Hayes MA, Bachri S, Saragi-Sasmito MF, Sidik F et al (2020) Mangrove blue carbon stocks and dynamics are controlled by hydrogeomorphic settings and land-use change. Glob Chang Biol 26(5):3028–3039. https://doi.org/10.1111/gcb.15056

Satyanarayana B, Koedam N, De Smet K, Di Nitto D, Bauwens M, Jayatissa LP et al (2011) Long-term mangrove forest development in Sri Lanka: early predictions evaluated against outcomes using VHR remote sensing and VHR ground-truth data. Mar Ecol Prog Ser 443:51–63. https://doi.org/10.3354/meps09397

Satyanarayana B, Muslim AM, Horsali NAI, Zauki NAM, Otero V, Nadzri MI et al (2018) Status of the undisturbed mangroves at Brunei Bay, East Malaysia: a preliminary assessment based on remote sensing and ground-truth observations. Peer J 6:e4397. https://doi.org/10.7717/peerj.4397

Sauer JD (1962) Effects of recent tropical cyclones on the coastal vegetation of Mauritius. J Ecol 50:275–290. https://doi.org/10.2307/2257445

Savari A, Khaleghi M, Safahieh AR, Hamidian Pour M, Ghaemmaghami S (2020) Estimation of biomass, carbon stocks and soil sequestration of Gowatr mangrove forests, Gulf of Oman. Iran J Fish Sci 19(4):1657–1680. https://doi.org/10.22092/IJFS.2020.121484

Scales IR, Friess DA (2019) Patterns of mangrove forest disturbance and biomass removal due to small-scale harvesting in southwestern Madagascar. Wetl Ecol Manag 27(5):609–625. https://doi.org/10.1007/s11273-019-09680-5

Schile LM, Kauffman JB, Crooks S, Fourqurean JW, Glavan J, Megonigal JP (2017) Limits on carbon sequestration in arid blue carbon ecosystems. Ecol Appl 27(3):859–874. https://doi.org/10.1002/eap.1489

Seeruttun LD, Raghbor P, Appadoo C (2021) First assessment of anthropogenic marine debris in mangrove forests of Mauritius, a small oceanic island. Mar Pollut Bull 164:112019. https://doi.org/10.1016/j.marpolbul.2021.112019

Selvam V, Ravichandran KK, Gnanappazham L, Navamuniyammal M (2003) Assessment of community-based restoration of Pichavaram mangrove wetland using remote sensing data. Curr Sci 85(6):794–798. https://www.jstor.org/stable/24109889

Shapiro AC, Trettin CC, Küchly H, Alavinapanah S, Bandeira S (2015) The mangroves of the Zambezi Delta: increase in extent observed via satellite from 1994 to 2013. Remote Sens 7(12):16504–16518. https://doi.org/10.3390/rs71215838

Shazra A, Rasheed S, Ansari AA (2008) Study on the mangrove ecosystem in Maldives. Glob J Environ Res 2(2):84–86

Shunula JP, Whittick A (1999) Aspects of litter production in mangroves from Unguja Island, Zanzibar, Tanzania. Estuar Coast Shelf Sci 49:51–54. https://doi.org/10.1016/S0272-7714(99)80008 0

Siddiqui PJA, Farooqui Z, Valeem EE, Rasheed M, Shafique S (2009) Studies on decomposition rates of Avicennia and Rhizophora leaves on tidal mudflats in active Indus Deltaic area of Pakistan. Int J Phycol Phycocem 5(1):93–98

Sitoe AA, Mandlate LJC, Guedes BS (2014) Biomass and carbon stocks of Sofala bay mangrove forests. Forests 5(8):1967–1981. https://doi.org/10.3390/f5081967

Sivakumar K, Rilwan A, Priyanka K, Salah M, Kathiresan K (2018) Mangroves of the atolls of the Maldives, rich among the atoll groups of the Indian Ocean. ISME/GLOMIS Electron J 16(3):11–18

Small C, Sousa D (2019) Spatiotemporal characterization of mangrove phenology and disturbance response: the Bangladesh Sundarban. Remote Sens 11(17):2063. https://doi.org/10.3390/rs11172063

Songsom V, Koedsin W, Ritchie RJ, Huete A (2019) Mangrove phenology and environmental drivers derived from remote sensing in southern Thailand. Remote Sens 11(8):955. https://doi.org/10.3390/rs11080955

Spalding M, Kainuma M, Collins L (2010) World atlas of Mangroves. Earthscan, p 319

Srisunont C, Jaiyen T, Tenrung M, Likitchaikul M, Srisunont T (2017) Nutrient accumulation by litterfall in mangrove forest at Klong Khone, Thailand. Sci Technol Asia 22(1):9–18. https://doi.org/10.14456/tijsat.2017.2

Srivastava PK, Mehta A, Gupta M, Singh SK, Islam T (2015) Assessing impact of climate change on Mundra mangrove forest ecosystem, Gulf of Kutch, western coast of India: a synergistic evaluation using remote sensing. Theor Appl Climatol 120(3):685–700. https://doi.org/10.1007/s00704-014-1206-z

Steinke TD, Naidoo G, Charles LM (1983) Degradation of mangrove leaf and stem tissues in situ in Mgeni Estuary, South Africa. In: Teas HJ (eds) Biology and ecology of mangroves. Tasks for vegetation science, vol 8, Springer, Dordrecht. https://doi.org/10.1007/978-94-017-0914-9_16

Stringer CE, Trettin CC, Zarnoch SJ, Tang W (2015) Carbon stocks of mangroves within the Zambezi River Delta, Mozambique. For Ecol Manag 354:139–148. https://doi.org/10.1016/j.foreco.2015.06.027

Tamai S, Nakasuga T, Tabauchi R, Ogino K (1986) Standing biomass of mangrove forests in southern Thailand. J Jpn For Soc 68(9):384–388. https://doi.org/10.11519/jjfs1953.68.9_384

Tamooh F, Huxham M, Karachi M, Mencuccini M, Kairo JG, Kirui B (2008) Below-ground root yield and distribution in natural and replanted mangrove forests at Gazi Bay, Kenya. For Ecol Manag 256(6):1290–1297. https://doi.org/10.1016/j.foreco.2008.06.026

Tanouchi H, Nakamura S, Ochiai Y, Azman H (2000) Profiles of species composition and aboveground biomass in a mangrove forest, Matang. Peninsular Malaysia. Jpn Agric Res Q 34(2):147–151

Thakur S, Mondal I, Ghosh PB, Das P, De TK (2020) A review of the application of multispectral remote sensing in the study of mangrove ecosystems with special emphasis on image processing techniques. Spat Inf Res 28(1):39–51. https://doi.org/10.1007/s41324-019-00268-y

Thant YM, Kanzaki M, Ohta S, Than MM (2012) Carbon sequestration by mangrove plantations and a natural regeneration stand in the Ayeyarwady Delta, Myanmar. Tropics 21(1):1–10. https://doi.org/10.3759/tropics.21.1

Toosi NB, Soffianian AR, Fakheran S, Pourmanafi S, Ginzler C, Waser LT (2019) Comparing different classification algorithms for monitoring mangrove cover changes in southern Iran. Glob Ecol Conserv 19:e00662. https://doi.org/10.1016/j.gecco.2019.e00662

Trettin CC, Stringer CE, Zarnoch SJ (2016) Composition, biomass and structure of mangroves within the Zambezi River Delta. Wetl Ecol Manag 24(2):173–186. https://doi.org/10.1007/s11273-015-9465-8

Vafaei S, Soosani J, Adeli K, Fadaei H, Naghavi H, Pham TD, Tien Bui D (2018) Improving accuracy estimation of Forest aboveground biomass based on incorporation of ALOS-2 PALSAR-2 and Sentinel-2A imagery and machine learning: a case study of the Hyrcanian forest area (Iran). Remote Sens 10(2):172. https://doi.org/10.3390/rs10020172

Vaghela B, Chirakkal S, Putrevu D, Solanki H (2021) Modelling above ground biomass of Indian mangrove forest using dual-pol SAR data. Remote Sens Appl Soc Environ 21:100457. https://doi.org/10.1016/j.rsase.2020.100457

Veettil BK, Pereira SFR, Quang NX (2018) Rapidly diminishing mangrove forests in Myanmar (Burma): a review. Hydrobiologia 822:19–35. https://doi.org/10.1007/s10750-018-3673-1

Veettil BK, Van DD, Quang NX (2020) Biodiversity and nature conservation in island ecosystems: spatiotemporal changes in Socotra archipelago (Yemen). J Coast Conserv 24(5):1–9. https://doi.org/10.1007/s11852-020-00774-4

Wang LE, Sousa WP (2009) Distinguishing mangrove species with laboratory measurements of hyperspectral leaf reflectance. Int J Remote Sens 30(5):1267–1281. https://doi.org/10.1080/01431160802474014

Wang Y, Bonynge G, Nugranad J, Traber M, Ngusaru A, Tobey J et al (2003) Remote sensing of mangrove change along the Tanzania coast. Mar Geod 26(1–2):35–48. https://doi.org/10.1080/01490410306708

Wang L, Jia M, Yin D, Tian J (2019) A review of remote sensing for mangrove forests: 1956–2018. Remote Sens Environ 231:111223. https://doi.org/10.1016/j.rse.2019.111223

Wanthongchai P, Pongruktham O (2019) Mangrove cover, biodiversity, and carbon storage of mangrove forests in Thailand. In: Gul B, Böer B, Khan M, Clüsener-Godt M, Hameed A (eds) sabkha ecosystems. tasks for vegetation science, vol 49, Springer, Cham. https://doi.org/10.1007/978-3-030-04417-6_28

Ward CJ, Steinke TD (1982) A note on the distribution and approximate areas of mangroves in South Africa. S Afr J Bot 1(3):51–53

Wicaksono P, Danoedoro P, Hartono Nehren U (2016) Mangrove biomass carbon stock mapping of the Karimunjawa Islands using multispectral remote sensing. Int J Remote Sens 37(1):26–52. https://doi.org/10.1080/01431161.2015.1117679

Widyastuti A, Yani E, Nasution EK, Rochmatino R (2018) Diversity of mangrove vegetation and carbon sink estimation of Segara Anakan Mangrove Forest, Cilacap, Central Java, Indonesia. Biodivers J Biol Divers 19(1):246–252. https://doi.org/10.13057/biodiv/d190133

Witsen DB (2012) Mangrove restoration and management in djibouti: criteria and conditions for success. Masters Dissertation, Virginia Polytechnic Institute and State University

Woitchik AF, Ohowa B, Kazungu JM, Rao RG, Goeyens L, Dehairs F (1997) Nitrogen enrichment during decomposition of mangrove leaf litter in an east African coastal lagoon (Kenya): relative importance of biological nitrogen fixation. Biogeochemistry 39(1):15–35. https://doi.org/10.1023/A:1005850032254

Wong CJ, James D, Besar NA, Kamlun KU, Tangah J, Tsuyuki S, Phua MH (2020) Estimating mangrove above-ground biomass loss due to deforestation in Malaysian Northern Borneo between 2000 and 2015 Using SRTM and Landsat Images. Forests 11(9):1018. https://doi.org/10.3390/f11091018

Yang S, Lim RL, Sheue CR, Yong JW (2011). The current status of mangrove forests in Singapore. In: Proceedings of Nature Society, Singapore's conference on 'nature conservation for a sustainable, Singapore 16 Oct, pp 99–120

Zhila H, Mahmood H, Rozainah MZ (2014) Biodiversity and biomass of a natural and degraded mangrove forest of Peninsular Malaysia. Environ Earth Sci 71(11):4629–4635. https://doi.org/10.1007/s12665-013-2853-6

Zöckler C, Aung C (2019) The Mangroves of Myanmar. In: Gul B, Böer B, Khan M, Clüsener-Godt M, Hameed A (eds) Sabkha ecosystems. Tasks for vegetation science, 49, Springer, Cham. https://doi.org/10.1007/978-3-030-04417-6_16

Chapter 5
Blue Carbon Dynamics in the Indian Ocean Seagrass Ecosystems

Abhra Chanda

Abstract Seagrasses are flowering plants that thrive in marine environments. The high productivity of the seagrass meadows enables these vegetative covers to act as a net repository of blue carbon. In this context, this ecosystem has received substantial attention in the last decade. However, information is mostly available from the seagrasses of the Mediterranean Sea, the Atlantic Ocean, and the Pacific Ocean. This chapter has detailed the present state of the art about seagrass and associated blue carbon dynamics in the Indian Ocean. The chapter has focused on the observations made so far on the aspects like spatial extent, species diversity, and carbon stock in the live biomass and the seagrass beds, along with the ecosystem productivity of the Indian Ocean's seagrass meadows. This chapter also considered the inorganic and organic carbon dynamics in the water column lying above the seagrass meadows. The chapter has collated all the observations related to the effect of climatic variables and anthropogenic stressors on the carbon sequestration potential of the seagrasses. Lastly, based on scrutiny of the present findings, the chapter tried to detail out the major lacunas and the primary research agendas in the Indian Ocean region that needs emphasis shortly. The review of the literature indicates that the Indian Ocean countries can act as leaders to battle out climate change by conserving and restoring this crucial marine ecosystem; however, lack of research and baseline investigation in many of the countries is barring such initiatives.

Keywords Seagrass · Asia · Africa · Carbon stock · Living biomass · Seagrass sediments · Net ecosystem productivity · Anthropogenic stressors · Research agenda

5.1 Spatial Extent of Seagrass Ecosystems

Unlike many vegetated ecosystems, which mostly remain exposed, seagrass ecosystems are quite difficult to map, as they remain submerged under the water and that

A. Chanda (✉)
School of Oceanographic Studies, Jadavpur University, 188 Raja S.C. Mullick Road, Kolkata, West Bengal 700084, India
e-mail: abhrachanda.sos.rs@jadavpuruniversity.in

A. Chanda et al. (eds.), *Blue Carbon Dynamics of the Indian Ocean*,
https://doi.org/10.1007/978-3-030-96558-7_5

too in terrains, which are difficult to access (Torres-Pulliza et al. 2013). Three to four decades back, ground-based surveys used to be the only option to map the spatial extent of seagrass. There used to be several drawbacks in such assessment as physically accessing a large area was always beyond the logistic feasibility. Moreover, baseline surveys by ground-based methods mostly relied on the aboveground evidence of seagrass only, which attributed to significant scope of uncertainty, as below ground seagrass beyond a certain depth was often overlooked (Roelofs et al. 2001). With the advent of tools like remote sensing, mapping of seagrass beds has become much easy (Roelfsema et al. 2009; Traganos et al. 2018). High-resolution satellite images and aerial photography, coupled with the advancement in geospatial software and geographical information system has enabled us to accomplish a very tedious and tiresome work like seagrass mapping with much ease (Pasqualini et al. 2001). There are several challenges in mapping seagrass with remotely sensed images that arise due to varying turbidity and variability in the bathymetry of the seagrass meadows. Variability of the biophysical characteristics of the seagrass bed substrate and the water column lying above in both spatial and temporal scales also lead to considerable uncertainty. However, a wide range of remote sensing protocols came into operation, overruling these challenges to delineate the extent of submerged vegetation (Dekker et al. 2007; Andréfouët 2008; Roelfsema et al. 2013). However, despite the availability of such remote technologies for the measurement of seagrass available at hand, stringent ground-truthing is always required for accurate delineation of seagrass extents (McKenzie et al. 2001). In this regard, the integration of remote sensing technology and ground-based surveys to formulate newer, easy, replicable, and comparable protocols for mapping seagrass vegetation remains an active wing of future research in this domain.

As far as the seagrass ecosystems bordering the Indian Ocean is concerned, 22 countries out of the 24 countries considered in this book have a varying extent of seagrass meadows along the respective country's shoreline. There is no information regarding any presence of seagrass meadows from Pakistan and Djibouti. However, ten countries among these 22 countries (which have seagrass meadows in their coastal regions) have no proper estimate of seagrass extent (see Table 5.1).

When it comes to seagrass mapping, the endeavor of the United Nations Environment World Conservation Monitoring Centre led by Frederick Short and other collaborating organizations and projects deserves a special mention. According to their latest model-derived estimate (UNEP-WCMC, Short 2018), 17 countries out of the 24 countries considered in this book encompasses a seagrass cover of 51,680 km^2. Their country-wise estimated seagrass covers reflect that Madagascar, followed by India, Indonesia, and UAE is one of the leading countries in terms of seagrass area (>5000 km^2) (Vanderklift et al. 2019).

However, the ground-truthed conservative estimates by various researchers show much less area coverage of seagrass in the countries bordering the Indian Ocean, and many countries do not have any measured data as well. In India, there are several endeavors of seagrass mapping by integrating remote sensing tools and field measurements carried out in different locations like the Gulf of Mannar (Thangaradjou et al. 2008; Umamaheswari et al. 2009), the Lakshadweep Islands (Jagtap and Inamdar

Table 5.1 Areal coverage seagrass in the Indian Ocean countries

Country	Area (km^2) [a]	Authors	Estimated area (km^2) [b]
India (AS)	516.6	Geevarghese et al. (2018), Thangaradjou and Bhatt (2018)	7361
Bangladesh (AS)	No data	–	No data
Myanmar (AS)	4.3	BOBLME (2015)	3284
Thailand (AS)	148.5	UNEP (2008), Rattanachot and Prathep (2011)	1873 (IO 1538) [c]
Singapore (AS)	3	Fortes et al. (2018)	128
Indonesia (AS)	8813	Fortes et al. (2018)	18868 (IO 6420)[c]
Malaysia (AS)	16.3	Fortes et al. (2018)	545 (IO 440)[c]
Yemen (AS)	No data	–	5862
Oman (AS)	No data	–	No data
UAE (AS)	5500	Campbell et al. (2015)	5516
Iran (AS)	160	Maghsoodloo and Eghtesadi (2008)	3515
Pakistan (AS)	No data	Patro et al. (2017)	No data
South Africa (AF)	9.8	Adams (2016)	515
Mozambique (AF)	439	Bandeira and Gell (2003)	857
Tanzania (AF)	No data	–	47
Kenya (AF)	34	Dahdouh-Guebas et al. (1999), ASCLME Kenya (2012)	114
Somalia (AF)	No data	–	No data
Djibouti (AF)	No data	–	No data
Sri Lanka (AS)	No data	–	2934
Maldives (AS)	No data	–	2431
Seychelles (AF)	No data	–	No data
Madagascar (AF)	No data	–	9385
Comoros (AF)	No data	–	1333
Mauritius (AF)	7	ASCLME Mauritius (2012)	No data

[a] Conservative estimates validated by ground-truthing
[b] Model derived estimates by UNEP-WCMC, Short (2018) [mentioned in Vanderklift et al. (2019)]
[c] Thailand, Indonesia, and Malaysia are the three countries considered in this book that encompasses coastlines beyond the Indian Ocean. The magnitudes in parentheses indicate the estimated seagrass cover only in the periphery of the Indian Ocean

1991), and the Andaman and Nicobar Islands (Paulose et al. 2013). Paulose et al. (2013) estimated a loss of about 16.19 km^2 of seagrass beds in and around the Andaman and Nicobar Islands due to the Indian Ocean tsunami that took place in the year 2004. However, their estimates also indicated that there are about 29.43 km^2 of seagrass beds that surround the Andaman and Nicobar Islands. Several endeavors of seagrass mapping reported from Thailand (Koedsin et al. 2016; Chayhard et al. 2018; Yamakita et al. 2019) utilized remote sensing techniques. Koedsin et al. (2016) also successfully quantified the live biomass of seagrass in southern Thailand. Similar literature, though fewer in number, are available for Asian countries like Myanmar (Giardino et al. 2016), Singapore (Yaakub et al. 2013), Malaysia (Douven et al. 2003; Fauzan et al. 2017), Malaysia (Hossain et al. 2015; 2019), and UAE (Howari et al. 2009; Erftemeijer and Shuail 2012). However, such types of assessments are yet to make in Iran, Yemen, Oman, and Bangladesh.

Among the African countries, Somalia and Djibouti have no record of remotely sensed seagrass maps, available on the internet. Only two studies are available for the Tanzania coast (Gullström et al. 2006; Knudby and Nordlund 2011) that mapped the seagrass vegetation cover along with the other submerged vegetation with the help of remote sensing, in and around the Zanzibar Island. However, information related to the total area of seagrass cover for Tanzania is not available. Adams (2016) reported the area distribution of only one seagrass species, namely *Zostera capensis*, in the South African estuaries. However, detailed information on South African seagrass beds is not available so far. Quite a few numbers of studies dealt with seagrass mapping in Mozambique and Kenyan coastlines, and few of these assessments took the help of the remotely sensed data (Dahdouh-Guebas et al. 1999; Amone-Mabuto et al. 2017; Harcourt et al. 2018). Overall, the African coastlines need more exhaustive monitoring to update the database of seagrass coverage. Bandeira and Björk (2001) mentioned that the entire east African coastline is a potential habitat for seagrass beds; however, until the year 2001, only 60 research articles were available on the seagrass dynamics of this region. Even in the present date, there is no noticeable improvement in the scenario, and seagrass research in this area remains to be in its infancy. The estimate of seagrass coverage remains poorly constrained in the island countries of the Indian Ocean. Among the six island countries, only Mauritius has a record of seagrass cover of seven km^2 (ASCLME Mauritius 2012), whereas, no conclusive report is available for the other countries—Sri Lanka, Maldives, Seychelles, Madagascar, and Comoros.

5.2 Physicochemical and Hydrogeological Settings

Seagrass mostly prefers to thrive in the shallow region of the continental shelves and outer estuaries and lagoons. Like terrestrial plants, seagrasses also rely on photosynthesis. Excessive turbidity bars the penetration of light to deeper layers, and this restricts the seagrass meadows to a maximum depth of 10 m in most of the estuarine areas in the Indian Ocean (Patro et al. 2017). In India, there are dense seagrass vegetation in Chilika lagoon and Palk Bay (two sites in India where substantial seagrass vegetation thrives all-round the year) in depths ranging from 0.9 to 3.7 m (Banerjee et al. 2018; Ganguly et al. 2017, 2018). Ganguly et al. (2017) reported that the presence of seagrass is up to depths of seven meters in India. Ochieng and Erftemeijer (2003) also observed similar depth ranges (1–6 m) in the case of African seagrasses. Only along the Arabian coastlines, seagrasses exist in much deeper zones. Erftemeijer and Shuail (2012) reported seagrasses at depths of about 22 m in some places along the Arabian coasts. Campbell et al. (2015) also observed seagrass meadows in depths ranging between 2.7 and 14 m in Abu Dhabi, UAE. The existence of seagrass meadows at such depths is only possible when the above lying water column is transparent enough to allow the light to penetrate to such depths.

Seagrasses in the periphery of the Indian Ocean thrive at a varying range of salinities. Banerjee et al. (2018) observed dense seagrass meadows in Chilika lagoon, India having a salinity range of 7.0–15.0, whereas, Campbell et al. (2015) reported very high salinity varying between 42.5 and 46.6, in the seagrass water column of Abu Dhabi, UAE. Hena et al. (2007) observed the flowering of the seagrass species, *H. beccarii*, in a salinity range of 16.0–35.0, in the coastal waters of Bangladesh. Seagrass species distribution varies with varying salinity. Adams (2016) observed that above a salinity of 15, the seagrass species, *Z. capensis*, dominates the South African estuaries; however when the salinity falls below 15, other submerged macrophytes replace this species. Soe-Htun et al. (2009) also observed that the salinity ranges of 5.0–15.0 reduced the luxuriant growth of seagrass in Myanmar.

Unlike salinity, most of the seagrass communities adjoining the Indian Ocean prefer to thrive in narrow ranges of water temperature. Ganguly et al. (2017) observed the seagrass in India thriving between 26 and 34 °C. Soe-Htun et al. (2018) reported that the seagrass along the Myanmar coast flourished well between 30 and 32 °C. Similarly, Uku and Björk (2005) recorded a mean water temperature range of 29.0–33.6 adjacent to the seagrass meadows of Kenya. Campbell et al. (2015) reported a water temperature range of 25.0–28.6 °C along the seagrass meadows of Abu Dhabi, UAE. However, Martins and Bandeira (2001) observed a comparatively broader range of water temperatures (20–39 °C) in the case of the seagrasses of Mozambique. These observations indicate that most of the seagrass species found in the Indian Ocean region do not prefer wide variability in temperature.

5.3 Species Diversity

Short et al. (2011) carried out an exhaustive study involving several eminent seagrass experts from all over the world and prepared a list of 72 seagrass species found in mother Earth. Out of these 72 reported species of seagrass, only 22 thrives in the Indian Ocean countries (as observed by mining the existing literature related to seagrass diversity reported from the Indian Ocean countries). Figure 5.1 shows the country-wise availability of seagrass species.

Sr. No.	Species name	IN	BD	MM	TH	SG	ID	MY	IR	YM	OM	UAE	SL	MD	SC	MG	CM	KN	MA	MZ	SM	SA	TZ	DJ	PK
1	*E. acoroides*	+		+	+	+	+	+					+		+			+		+			+		
2	*T. hemprichii*	+		+	+	+	+	+		+			+	+	+	+	+	+		+	+	+	+		
3	*H. beccarii*	+	+	+	+	+	+	+																	
4	*H. decipiens*	+		+	+	+	+	+					+												
5	*H. major*			+	+		+	+																	
6	*H. minor*	+		+	+	+	+	+																	
7	*H. ovalis*	+		+	+	+	+	+	+	+	+	+	+		+	+	+	+	+	+	+	+	+		
8	*H. ovalis ramamurthiana*	+																							
9	*H. ovata*	+																							
10	*H. spinulosa*			+		+	+	+																	
11	*H. sulawesii*						+																		
12	*H. stipulacea*	+								+		+				+	+	+	+	+	+		+		
13	*H. tricostata*							+																	
14	*C. rotundata*	+		+	+	+	+	+		+			+		+	+	+	+		+	+	+	+		
15	*C. serrulata*	+		+	+	+	+	+		+			+		+	+	+	+	+	+	+	+	+		
16	*H. pinifolia*	+		+	+	+	+	+																	
17	*H. uninervis*	+		+	+	+	+	+	+	+	+	+	+		+	+	+	+	+	+	+		+		
18	*H. wrightii*	+																							
19	*S. isoetifolium*	+		+	+	+	+	+					+	+	+	+	+	+	+	+	+		+		
20	*T. ciliatum*						+	+			+				+	+	+	+	+	+			+		
21	*R. maritima*	+			+		+	+																	
22	*Z. capensis*																	+		+		+	+		

Fig. 5.1 The seagrass species diversity observed in IN—India [Thangaradjou and Bhatt (2018)]; BD—Bangladesh [Abu Hena et al. (2007)]; MM—Myanmar [Soe-Htun et al. (2018), Thu-Rein Aung et al. (2019), BOBLME (2015)]; TH—Thailand [Nguyen et al. (2014), Tuntiprapas et al. (2015)]; SG—Singapore [Yaakub et al. (2013)]; ID—Indonesia [Kuo (2007), Wawan (2011), Tuntiprapas et al. (2015)]; MY—Malaysia [Japar Sidik and Muta Harah (2003)]; IR—Iran [Maghsoodloo and Eghtesadi (2008)]; YM—Yemen [Bawazir (2009)]; OM—Oman [Jupp et al. (1996)]; UAE—United Arab Emirates [Campbell et al. (2015)]; SL—Sri Lanka [Ranahewa et al. (2018)]; MD—Maldives [Payri et al. (2012)]; SC—Seychelles [Ingram and Dawson (2001), ASCLME Seychelles (2012)]; MG—Madagascar [Hantanirina and Benbow (2013); ASCLME Madagascar (2012)]; CM—Comoros [ASCLME Comoros (2012)]; KN—Kenya [ASCLME Kenya (2012)]; MA—Mauritius [ASCLME Mauritius (2012)]; MZ—Mozambique [Bandeira and Gell (2003), ASCLME Mozambique (2012)]; SM—Somalia [ASCLME Somalia (2012)]; SA—South Africa [ASCLME South Africa (2012)]; TZ—Tanzania [ASCLME Tanzania (2012)]; DJ—Djibout [No data]; PK—Pakistan [No data]

Overall, the seagrass species diversity is much higher in a few of the Asian countries, compared to the African countries and open ocean island countries. Among the 24 countries considered in this book, Indonesia, Malaysia, and India, each of these countries, have the highest number of seagrass species (16) in their coastal periphery (Fig. 5.2). Bangladesh, Maldives, UAE, Iran, and Oman have the lowest species diversity. African countries like Kenya, Tanzania, and Mozambique and open ocean island countries like Comoros, Madagascar, Seychelles, and Sri Lanka have moderate species diversity (6–8).

Among the list of 22 seagrass species found in the Indian Ocean countries, 19 species belong to the least concern category, according to the IUCN Red List. *Halophila beccarii*, which is a vulnerable species according to the IUCN Red List, thrives in seven Asian countries, namely India, Bangladesh, Myanmar, Thailand, Singapore, and Myanmar. Whereas, *Zostera capensis*, another vulnerable species in the IUCN Red List, is only reported from four African countries, namely, Kenya, Mozambique, South Africa, and Tanzania. Out of the 22 seagrass species reported

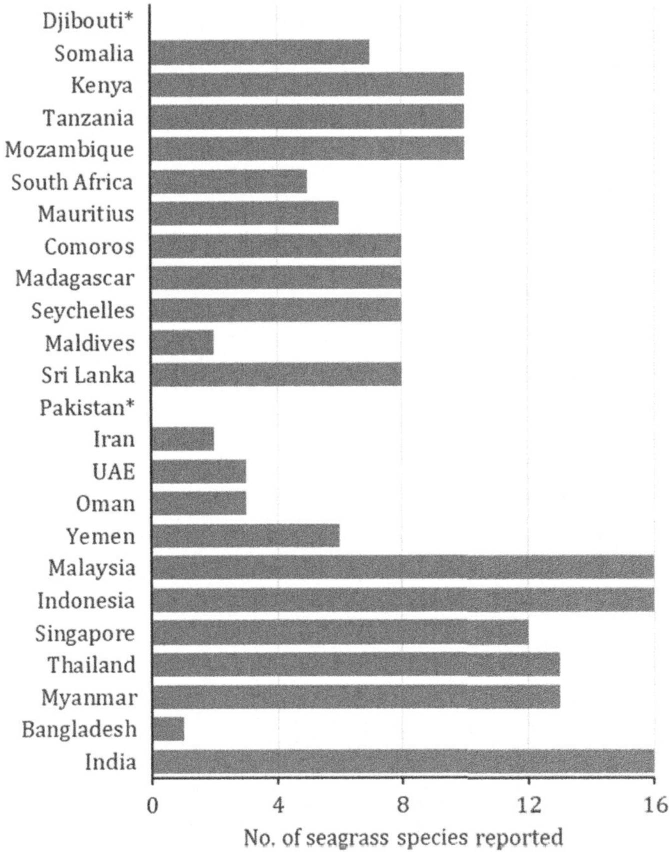

Fig. 5.2 Seagrass species count in the different Indian Ocean countries

in the Indian Ocean, *Halophila sulawesii* is the only species that belongs to the 'data deficient' category according to the IUCN Red List, and it thrives only in Indonesia. *Thalassia hemprichii*, *Halophila ovalis*, *Cymodocea rotundata*, *Cymodocea serrulata*, *Halodule uninervis*, and *Syringodium isoetifolium* are some of the most common seagrasses found in the Indian Ocean, with occurrences reported in 16 to 18 countries. Halophila ovata, *Halodule wrightii*, and *Halophila tricostata* that belong to the least concern category are found only in India and Malaysia. Figure 5.3 shows the most significant locations of the seagrass distribution along the periphery of the Indian Ocean.

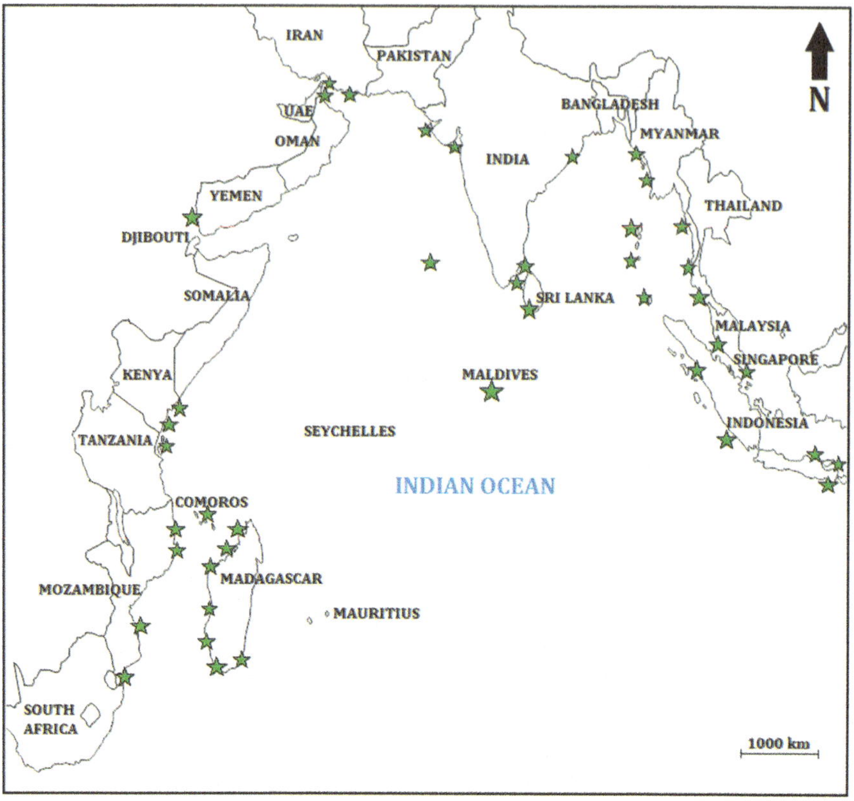

Fig. 5.3 Map showing the key locations [collated from WCMC, Short (2018) and Vanderklift et al. (2019)] of seagrass meadows in the Indian Ocean

5.4 Carbon Stock in Living Biomass

Conservation of organic carbon in both terrestrial and aquatic vegetated systems is one of the key routes for the mitigation of climate change. The scientific community recognized that the green carbon locked in the terrestrial ecosystems acts as one of the principal repositories of organic carbon on Earth. However, at the end of the past century, several studies indicated that the 'blue carbon' in the coastal vegetated ecosystems are equally significant sinks for carbon. Blue carbon ecosystems like that of seagrass meadows encompass <0.1% of the global ocean floor (in terms of surface area). Even then, this ecosystem contributes to almost 10–18% of the total burial of organic carbon in the oceans, which is equivalent to 48–112 Tg C yr^{-1} (Duarte et al. 2005; Kennedy et al. 2010; McLeod et al. 2011). Based on data assimilated from 946 seagrass meadows site, Fourqurean et al. (2012) estimated that globally seagrass meadows could be a repository of 19.9 Pg C; or at least 4.2–8.4 Pg, based on a more conservative approach.

Greiner et al. (2013) pointed out that the carbon assimilation rate in seagrass meadows exhibits marked spatial variability throughout the globe, and it is principally dependent on various factors like the species composition, the seagrass density, and the depth of the seagrass meadows. Moreover, the biomass of seagrass meadows also varies temporally over seasonal to annual scales. To conserve a particular blue carbon ecosystem and formulate strategies for its conservation, knowing its carbon strength is one of the essential prerequisites. That is why; monitoring the carbon dynamics of any and every seagrass meadows thriving in the world has become an utmost necessity.

Carbon stock in the live biomass of seagrass meadows in many of the Indian Ocean countries is measured and reported by several researchers (Table 5.2). Except, Yemen, Iran, and Pakistan, all the other Asian countries have assessed the carbon stock in their respective seagrass meadows. Indonesia, which represents almost 5% of the world's total seagrass cover (Kuriandewa et al. 2003), has a very high carbon stock in its seagrass meadows, ranging from 62 to 2450 Mg C ha^{-1}. Such a higher range of carbon stock in seagrass biomass is also reported from Sri Lanka, though there is no proper estimate about the area coverage of seagrass in Sri Lanka. Malaysia (mean: 251 Mg C ha^{-1}), Thailand (mean: 155 Mg C ha^{-1}), Myanmar (mean: 96.79 Mg C ha^{-1}), and Singapore (mean: 96.79 Mg C ha^{-1}) have substantially high carbon stock in their seagrass meadows (Thorhaug et al. 2020 and the references, therein). However, in countries like India, Bangladesh, Oman, and UAE, the live biomass carbon stock ranged from 0.01 to 9.59 Mg C ha^{-1} (Table 5.2). The carbon stock in seagrass biomass in the African countries is much lower than that observed in the Asian countries. The highest carbon stock per unit area, in the live seagrass biomass, is reported from Kenya followed by Tanzania, Mozambique, and South Africa. Among the open ocean island countries, Maldives, Madagascar, and Comoros have no measured data in this regard.

5.5 Seagrass Beds: A Strong Carbon Sink

Unlike the terrestrial forest ecosystems, where the biomass of the floras locks most of the carbon in organic form, the submerged aquatic vegetation tends to accumulate a larger proportion of the carbon in their sediments (Serrano et al. 2018). Seagrass meadows, being a type of vegetation that thrives under the water, are capable of reducing the water current and the flow velocity (Hansen and Reidenbach 2012). This lowering of flow velocity facilitates the seagrass meadows to trap the particulate materials from the ambient water column and promote a high rate of sedimentation (Gacia and Duarte 2001; Gacia et al. 2002). The particulate matters that are trapped and deposited by the seagrass meadows have high organic carbon content, and this leads to a high rate of organic carbon burial in the seagrass beds (Kennedy et al. 2010). However, the rate of deposition of the organic carbon in the seagrass beds depends on several factors like the terrain where the seagrass meadows are situated, the species composition, varying productivity, below-ground biomass, and the biodegradability

Table 5.2 Carbon stock of the seagrass meadows in the live biomass and seagrass bed sediments along with the ecosystem productivity observed in the Indian Ocean countries

Country	Live biomass carbon (Mg C ha^{-1})	Soil organic carbon (Tg C)	Ecosystem Productivity (Kg C ha^{-1} d^{-1})	Authors
India (AS)	0.20–2.90	6.7	9.11–33.3	Ganguly et al. (2018), Kaladharan and David Raj (1989)
Bangladesh (AS)	1.45–9.59	NA	NA	Abu Hena et al. (2007)
Myanmar (AS)	96.79	0.042	NA	Thorhaug et al. (2020)
Thailand (AS)	155	2.29	14.1 [a]	Thorhaug et al. (2020), Vichkovitten (1998)
Singapore (AS)	96.79	0.029	NA	Thorhaug et al. (2020)
Indonesia (AS)	62–2450	753	3.0–14.8 [a]	Thorhaug et al. (2020), Erftemeijer et al. (1993)
Malaysia (AS)	251	4.1	NA	Thorhaug et al. (2020)
Yemen (AS)	NA	NA	NA	
Oman (AS)	0.01–0.48 [a]	NA	NA	Jupp et al. (1996)
UAE (AS)	0.03–1.13	49.1	NA	Campbell et al. (2015)
Iran (AS)	NA	NA	NA	
Pakistan (AS)	NA	NA	NA	
South Africa (AF)	0.24–1.44 [a]	NA	3.12–6.65 [a]	Adams (2016), Hanekom and Baird (1988)
Mozambique (AF)	1.27–2.23 [a]	NA	0.60–3.62 [a]	Martins and Bandeira (2001), Githaiga et al. (2016)
Tanzania (AF)	0.28–3.20 [a]	NA	4.9–9.5 [a]	Githaiga et al. (2016), Ochieng and Erftemeijer (2003)
Kenya (AF)	0.13–7.10 [a]	160.7–233.8 [b]	6.05–27.5 [a]	Githaiga et al. (2016), Githaiga et al. (2017)
Somalia (AF)	NA	NA	NA	
Djibouti (AF)	NA	NA	NA	

(continued)

Table 5.2 (continued)

Country	Live biomass carbon (Mg C ha^{-1})	Soil organic carbon (Tg C)	Ecosystem Productivity (Kg C ha^{-1} d^{-1})	Authors
Sri Lanka (AS)	1.35–3006	NA	7.0	Faazil et al. (2019), Johnson and Johnstone (1995)
Maldives (AS)	NA	NA	NA	
Seychelles (AF)	0.01–1.38 [a]	NA	NA	Aleem (1984)
Madagascar (AF)	NA	NA	NA	
Comoros (AF)	NA	NA	NA	
Mauritius (AF)	0.34–0.43 [a]	NA	NA	Daby (2003)

[a] Dry biomass was converted to carbon content by multiplying with 0.336, assuming 33.6% of dry biomass is made up of carbon (Pérez et al. 1994)
[b] In Mg C ha^{-1}

of the organic carbon stored in the beds (Mateo et al. 2006; Carruthers et al. 2007; Lavery et al. 2013). However, the present understanding about the carbon burial potential of the seagrass beds is poorly constrained, as there are only a handful number of studies carried out mainly in the North Atlantic, Mediterranean, and Australian seagrasses (Nellemann et al. 2009; Fourqurean et al. 2012).

Exhaustive measurements to characterize the soil organic carbon stock in the seagrass beds are scarce from the Indian Ocean region. Among the African countries, only Kenya (Githaiga et al. 2017) has an estimate of sediment organic carbon stock. Githaiga et al. (2017), while working in the Gazi Bay (Kenya), reported the soil organic carbon stock from African seagrasses for the first time. They observed that the sediment carbon stock varied significantly between the four species that thrived in that region. They also reported that the seagrass vegetated part of the coastline has four to six times higher carbon stock than the un-vegetated parts. In an experimental study carried out in the same Gazi Bay (Kenya), Githaiga et al. (2019) observed that within 18 months, the top 5 cm seagrass beds suffered a loss of carbon stock at the rate of 2.21 Mg C ha^{-1} due to the removal of seagrass. However, apart from Kenya, none of the other African countries or open ocean island countries have any reported carbon stock from the seagrass sediment beds.

Indonesia is one of the strongest contenders of soil organic carbon stock in their seagrass meadows. Thorhaug et al. (2020) collated data from various literature and reported the total sedimentary carbon stock in Indonesia's seagrass meadows to be 753 Tg C. However, a significant portion of these carbon stocks is not part of the Indian Ocean, as Indonesia shares a fair proportion of coastline with the Pacific Ocean. However, no specific report of Indonesian seagrass's carbon stock only for the Indian Ocean part is not available. Ganguly et al. (2017), while characterizing the carbon stock in seagrass meadows of India, recorded that the soil organic carbon pool was 40 times higher than that observed in the seagrass biomass. At the same time, they also observed a decrease in organic carbon concentration with depth (0.012% cm^{-1}). In two of the most prominent seagrass sites in India, namely Palk Bay and Chilika Lagoon, Ganguly et al. (2018) observed soil carbon stock of 129 Mg C ha^{-1}, in the top 1 m soil profile. Multiplying this magnitude with the total seagrass coverage estimated for India, it comes out that Indian seagrass cover comprises about 6.7 Tg C in the seagrass beds. Malaysia and Thailand have an estimated total sediment carbon stock of 4.1 Tg C and 2.29 Tg C, respectively. Myanmar (0.042 Tg C) and Singapore (0.029 Tg C) have an order of magnitude less carbon stock in the sediments of their seagrass meadows. In the western part of the Indian Ocean, Campbell et al. (2015) inferred that sediment carbon stock in seagrass meadows of UAE per unit area (49.1 ± 7.1 Mg C ha^{-1}) is much less than that observed in other parts of the world. However, as the seagrass coverage is substantially high in this region, the total carbon stock in the seagrass sediments was estimated to be around 49.1 Tg C. Among the Asian countries, no report on the carbon stock of seagrass sediment is available from Bangladesh, Yemen, Oman, Pakistan, and Iran.

5.6 Seagrass Ecosystem Productivity

There are different protocols available for the measurement of the community ecosystem productivity in the seagrass meadows. The open water mass balance model by monitoring the diel changes in oxygen (Peeters et al. 2016), the benthic chamber model (Olivé et al. 2016), and the dissolved inorganic carbon (DIC) gradient model (Tokoro et al. 2014) are some of the well-accepted protocols. Many researchers have also adopted the approach of directly measuring the primary production of the leaves following the hole-punching method proposed by Kirkman and Reid (1979). In this way, one can measure the growth in leaf biomass within a stipulated time. The magnitudes of net ecosystem productivity in the seagrass meadows enable us to understand the carbon sequestration efficiency in a particular seagrass meadow or a particular seagrass species. The existing literature indicates that measurement of ecosystem productivity is carried out only in eight countries bordering the Indian Ocean (Table 5.2).

The seagrass meadows in India exhibited a wide range of seagrass ecosystem productivity. Ganguly et al. (2018) measured the seagrass ecosystem productivity in the Palk Bay and Chilika lagoon, India, by implementing the oxygen mass balance method. They observed mean net community productivity of 9.11 kg C ha^{-1} d^{-1}, whereas, Kaladharan and David Raj (1989) reported a mean seagrass ecosystem productivity of 33.3 kg C ha^{-1} d^{-1} from the Lakshadweep, India. In the Khung Kraben Bay of Thailand, Vichkovitten (1998) measured the seagrass ecosystem productivity in the mono-species stands of *Enhalus acoroides* by the hole-punching method. Vichkovitten (1998) recorded mean productivity of 14.1 kg C ha^{-1} d^{-1}. Erftemeijer et al. (1993) observed varying magnitudes of productivity in different species' stands in Indonesia. Besides these observations from Asian countries, Johnson and Johnstone (1995) reported mean productivity of 7 kg C ha^{-1} d^{-1} from the seagrasses of Sri Lanka. However, as of now, no other observations on the ecosystem productivity of seagrass meadows are available from Asia.

On the contrary, many of the African countries have measured this crucial parameter, except Somalia and Djibouti. Githaiga et al. (2016), in their review, have collated all the observations made so far concerning seagrass ecosystem productivity from the countries—Mozambique, Tanzania, and Kenya. As observed in the case of India, Kenyan seagrasses also exhibited a wide range of ecosystem productivity (6.05–27.5 kg C ha^{-1} d^{-1}). However, productivity observed in South Africa, Tanzania, and Mozambique varied over much narrower ranges (Table 5.2).

5.7 Carbon Dynamics in the Water Column of Seagrass Meadows

Ever since the coining of the term 'blue carbon' assessment of the carbon sequestration potential in the seagrass meadows and seagrass beds has gained impetus. However, the water mass lying above the seagrass meadows is also an integral part of the seagrass ecosystems. Characterizing the dynamics of carbon compounds and gases in this water column is also extremely crucial to have a holistic understanding of seagrass metabolism (Berg et al. 2019). Apart from the autochthonous carbon generated within the ecosystem, seagrass ecosystems often receive substantial allochthonous input from rivers (Ruesink et al. 2015), adjacent mangroves (Yates et al. 2007), other benthic ecosystems (Guo et al. 2009), and even coastal upwelling (Saderne et al. 2013). In this regard, Saderne et al. (2019) emphasized that inorganic carbon plays a very crucial role in determining the net carbon sink capacity of seagrass ecosystems, as many seagrass ecosystems have witnessed calcium carbonate ($CaCO_3$) precipitation, which can potentially offset 30% of the net carbon sequestered by seagrass meadows. $CaCO_3$ precipitation leads to the production of CO_2 that eventually escapes toward the atmosphere whereas, the dissolution of $CaCO_3$ leads to enhancement of total alkalinity, which in turn increases the buffering capacity and hence the CO_2 sink character. There are pieces of evidence that show both $CaCO_3$ precipitation and dissolution occur in the seagrass meadows (Mazarrasa et al. 2015; Howard et al. 2018), hence characterizing these processes in different seagrass ecosystems has become an utmost necessity (Macreadie et al. 2017).

Though we recognized that this aspect plays a crucial role in delineating the net sink character of the seagrass meadows, very few studies are available from the Indian Ocean seagrasses in this regard. Banerjee et al. (2018) measured the atmosphere-hydrosphere CO_2 and CH_4 exchange in the Chilika lagoon of India. They observed that the seagrass dominated sites in this region acted as a weak source of both CO_2 (9.8 mM m^{-2} d^{-1}) and CH_4 (0.12 mM m^{-2} d^{-1}); however, the net emissions were 14 times lower than that observed in the macrophyte-dominated sites, within the same study area. Banerjee et al. (2018) observed the highest dissolved inorganic carbon in the seagrass adjacent water column; however, the excess DIC (the DIC that is not consumed by the autotrophs during photosynthesis) was lowest in the same seagrass sites, when compared to other sites devoid of seagrass. Such an observation indicated that the water column adjoining the seagrass beds was efficient in the photosynthetic utilization of mineralized carbon forms. Despite this, the weak CO_2 source character of the seagrass waters originated due to the land-based discharges (allochthonous carbon input). These observations prompted Banerjee et al. (2018) to conclude that if we can regulate land-based pollution, we might enhance the CO_2 sink potential of these seagrass beds. Bouillon et al. (2007), while working in the seagrass-mangrove continuum stretch in the Gazi Bay, Kenya, observed that though the mangrove surrounding waters acted as a net source of CO_2, the water column above the seagrass beds often exhibited CO_2 sink character. Similarly, Macklin et al. (2019) observed a weak CO_2 source character of the seagrass water column (2.5 ± 3.4 mM m^{-2} d^{-1}) while working in an Indonesian seagrass-mangrove continuum situated in Bali Island.

5.8 Factors Regulating the Carbon Stock in Seagrass Ecosystems

Conditions favorable for photosynthesis enable any vegetation community to enhance their carbon stock; however, in the case of submerged vegetation, the multifarious processes, in turn, governs such conditions. Ganguly et al. (2017), while working in the Palk Bay, India, observed that partial pressure of CO_2 in water [pCO_2(water)] and photosynthetically active radiation (PAR) were the dominant factors that regulated the net seagrass community productivity. Since a substantial part of the Indian Ocean experiences the southwest monsoon, the functioning of seagrass ecosystems is significantly impacted by the monsoonal rain in different ways. Ganguly et al. (2018) reported that the productivity of the seagrass ecosystems in India reduced during the monsoon season due to a significant reduction in light availability. While researching the influence of seagrass in the sediment quality of eastern Malaysia, Rattanachot and Prathep (2015) pointed out that above-ground seagrass biomass was higher during the monsoon season; however, the monsoon induced strong wave action re-suspended the seagrass bed sediments that in turn, reduced the soil organic carbon density. Monsoonal discharge in the coastal waters brings with it a substantial quantity of particulate matter that enhances the turbidity of the water column adjoining the seagrass and hampers the photosynthetic rate. Light happens to be one of the most crucial factors that are necessary for the seagrasses to thrive and flourish (Longstaff and Dennison 1999). Several studies have indicated that light deprivation severely affects the productivity of the seagrass, and sometimes, prolonged absence of light can lead to the death of seagrasses (Longstaff et al. 1999; Longstaff and Dennison 1999). However, most of these studies concentrated on the seagrasses of the Mediterranean and Australia (Ruiz and Romero 2001; Ralph et al. 2007; Collier et al. 2009), with almost no observations from the Asian and African continental margins considered in this book.

Species diversity also plays a crucial role in the sediment carbon accumulation, as Gullström et al. (2018) observed a varying range of organic carbon content in the seagrass beds along the Western Indian Ocean (0.20–1.44%), and they attributed the species-specific variability as the main reason behind such variation of soil organic carbon stock. The species diversity depends on the aquatic physicochemical parameters of a specific site and the sediment texture, as Kanal and Short (2009) documented that *H. beccarii* preferred a sandy clay with occasional mixtures of silt for growing in the coastal waters of Bangladesh. Given the present scenario of ongoing global warming and ocean acidification, several experiments followed to see the effects of such individual parameters and their synergistic effect on seagrass productivity. However, the existing literature shows some mixed results. Egea et al. (2018) observed that warming enhanced the productivity of seagrass *Cymodocea nodosa*; however, it reduced the total carbon reserve in the ecosystem. Egea et al. (2018) also reported that acidification had no significant ill effect on the productivity of *Cymodocea nodosa*. They observed that ocean acidification increased the carbon stock marginally. Koch et al. (2007), while experimenting in Florida Bay, USA, and Collier and Waycott (2014), while working in the Great Barrier Reef, Australia, observed that increased temperature beyond 33 °C and 43 °C, respectively, proved to be detrimental for the seagrasses of the respective regions and lead to their death. Strydom et al. (2020) reported that occasional marine heatwaves destroyed as much as 1300 km^2 of seagrass vegetation in the western Australian shoreline. Though there are several studies on the effect of such stressors on the seagrass meadows, such studies are scarce from the Indian Ocean. There is only one study from Africa, wherein, Mvungi and Pillay (2019) observed that both warming and nutrient enrichment could significantly deteriorate the biomass and carbon stock of *Zostera capensis* (a vulnerable seagrass species) that thrives in South Africa. Among the natural factors that can affect seagrass productivity, submarine groundwater discharge is identified as one of the causes of concern, and it plays a significant role in regulating seagrass productivity in various parts of the world. Kamermans et al. (2002), while studying the seagrasses of Kenya and Zanzibar Island, observed that submarine groundwater discharge could significantly alter the salinity regime at places that deteriorated the productivity of several salt-tolerant seagrass species and could lead to change in species composition.

5.9 Biogeochemical Interactions with Other Blue Carbon Ecosystems

Several marine ecosystems like mangroves, corals, and macrophytes often lie adjacent to each other in a continuum with seagrass beds. Lately, it has been pointed out by many researchers that seagrass meadows exchange a significant amount of carbon in both of the particulates and dissolved form with other blue carbon ecosystems when found in a continuum (Walton et al. 2014; Chen et al. 2017; Guerra-Vargas et al. 2020; Akhand et al. 2021). However, such interaction-related studies between the blue carbon ecosystems concerning the water column carbon sequestration are scarce from the global seagrass meadows, not to mention the Indian Ocean also. So far, only three such studies based on this aspect are available from the periphery of the Indian Ocean, one from the Kenyan coastline, and the other two from the Indonesian coastal waters.

In Kenya, Bouillon et al. (2007) observed that there was a substantial lateral flux of particulate and dissolved carbon from the mangrove waters toward the offshore. However, in places where dense seagrass beds were in continuum with the mangroves, the seagrass meadows could trap and mineralize a significant portion of this laterally exported particulate organic carbon. Thus, the seagrass meadows could enhance the CO_2 uptake. In the absence of seagrass, this particulate carbon would have gone to the coastal offshore and eventually escaped toward the atmosphere as CO_2. Thus Bouillon et al. (2007) inferred that seagrass situated in continuity with mangroves acts as a barrier that absorbs substantial amounts of carbon exported from the mangroves and does not allow this carbon to flee offshore. In line with this observation, Chen et al. (2017) portrayed that the adjacent seagrass meadows efficiently trapped the mangrove-derived organic carbon export from North Sulawesi, Indonesia. The soil organic carbon in the mangrove adjacent seagrass meadows was substantially higher than those seagrass meadows that had no mangroves in their vicinity. Chen et al. (2017) observed a mangrove contribution of 34–83% to the soil carbon stock in the seagrass meadows adjacent to mangroves. This observation indicated that mangroves when present adjacent to seagrass can give positive feedback in mitigation of climate change by enhancing the soil carbon stock in the adjacent seagrass beds. Macklin et al. (2019) observed in an Indonesia mangrove-seagrass-coral continuum that the mangrove surrounding waters acted as a significant source of CO_2, and the outer bay mouth was also a weak source of CO_2 due to $CaCO_3$ precipitation by the adjoining corals. It was only the seagrass patches in between that were a net sink for CO_2, and it significantly reduced the emission of the overall continuum stretch. Thus, all the observations made so far in the periphery of the Indian Ocean indicated that seagrass ecosystems are capable of mitigating climate change, and the presence of other blue carbon ecosystems, especially mangroves, can significantly enhance their carbon sequestration potential.

5.10 Scope of Future Research

A scrutiny of the findings collated in this chapter shows that there are several aspects related to the blue carbon dynamics of the seagrasses of the Indian Ocean that remain understudied. Various studies have indicated promising blue carbon stock in the seagrass meadows that thrives in the periphery of the Indian Ocean; however, accurate delineation of the seagrass extent seems to be a lacuna for many of the countries. Conservation and restoration of the carbon stock in any blue carbon ecosystem require holistic knowledge about the bench strength of that ecosystem. A necessary research agenda in this regard should be to develop protocols that can not only lead to accurate mapping of seagrass but also facilitate monitoring of the seasonal seagrass biomass dynamics. In an era of climate change, marine waters are also undergoing dramatic changes in terms of salinity and temperature. Such changes in physicochemical parameters often consequence in a paradigm shift in the species composition. Many of the countries might have some species in their country's coastline, which remains unexplored due to a lack of proper monitoring and research. Hence regular monitoring of the species diversity needs emphasis. Given the importance of blue carbon in battling the ongoing climate change, every country situated on the rim of the Indian Ocean should strive hard to measure the carbon stock in the live biomass and the sediments of their respective country's seagrass meadows. The database generated so far is still not promising enough to carry out further advanced research. Characterizing the seagrass ecosystem productivity remains even less studied. Once we finish acquiring all this baseline information, we should focus on understanding the impact of anthropogenic and climate stressors on the future of the seagrasses from this region. To efficiently tackle the troublesome task of restoration of any natural submerged vegetated system, one has to have a proper understanding of the problems that this vegetation is facing at present, or likely to face in the future. Effects of natural phenomena like an increase in water temperature, rainfall, submarine groundwater discharge, and anthropogenic activities like nutrient enrichment and pollution on the functioning of seagrass need rigorous assessment. The scarcity of such studies leads to a lack of understanding. The Indian Ocean comprises a substantial proportion of the global seagrass meadows, and the absence of such studies from this region weakens the global database. On the whole, the knowledge about the seagrass blue carbon dynamics in the Indian Ocean is still in its infancy and needs more endeavors to develop a comprehensive understanding.

5.11 Summary and Conclusion

Twenty-four countries facing the Indian Ocean in their coastlines have substantial seagrass meadows that are capable of storing a large quantity of blue carbon. According to the existing literature, 22 seagrass species thrive in this region, under varying salinity, water temperature, and depth (Fig. 5.4). The carbon stock in the live biomass and seagrass bed sediments varies over a wide range. The net ecosystem productivity also shows wide spatial and temporal variability. The water column adjoining the seagrass acts as a weak source of both CO_2 and CH_4. Though several studies have focussed on characterizing the blue carbon dynamics in the seagrass meadows of the Indian Ocean, there is still a substantial gap in the quality and quantity of information required to draw a holistic scenario about this vast region.

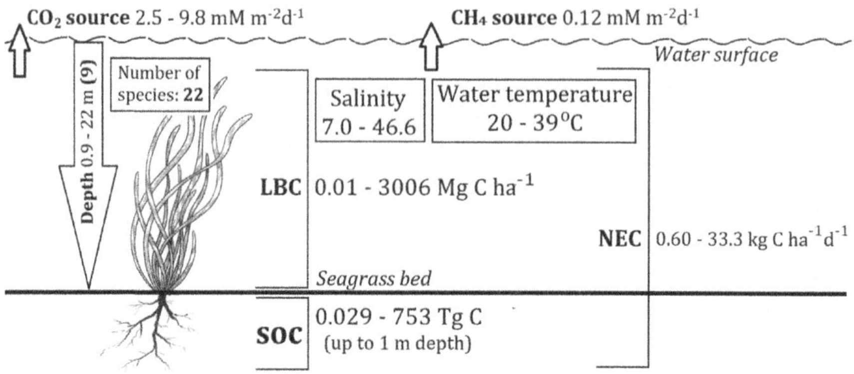

Fig. 5.4 A diagram showing the number of species, and ranges of depth, salinity, water temperature, live biomass carbon (LBC) stock, soil organic carbon (SOC) stock, net ecosystem productivity (NEC), air-water CO_2, and CH_4 exchange rates observed in the seagrass ecosystems in the Indian Ocean

References

Adams JB (2016) Distribution and status of *Zostera capensis* in South African estuaries—a review. S Afr J Bot 107:63–73. https://doi.org/10.1016/j.sajb.2016.07.007

Akhand A, Watanabe K, Chanda A, Tokoro T, Chakraborty K, Moki H, Tanaya T, Ghosh J, Kuwae T (2021) Lateral carbon fluxes and CO_2 evasion from a subtropical mangrove-seagrass-coral continuum. Sci Total Environ 752:142190. https://doi.org/10.1016/j.scitotenv.2020.142190

Aleem AA (1984) Distribution and ecology of seagrass communities. Deep Sea Res 31:919–933. https://doi.org/10.1016/0198-0149(84)90048-7

Amone-Mabuto M, Bandeira S, da Silva A (2017) Long-term changes in seagrass coverage and potential links to climate-related factors: the case of Inhambane Bay, southern Mozambique. West India Ocean J Mar Sci 16(2):13–25

Andréfouët S (2008) Coral reef habitat mapping using remote sensing: a user vs producer perspective. Implications for research, management and capacity building. J Spat Sci 53(1):113–129. https://doi.org/10.1080/14498596.2008.9635140

ASCLME Comoros (2012) National marine ecosystem diagnostic analysis. Comoros. Contribution to the Agulhas and Somali current large marine ecosystems project (supported by UNDP with GEF grant financing)

ASCLME Kenya (2012) National marine ecosystem diagnostic analysis. Kenya. Contribution to the Agulhas and Somali current large marine ecosystems project (supported by UNDP with GEF grant financing)

ASCLME Madagascar (2012) National marine ecosystem diagnostic analysis. Madagascar. Contribution to the Agulhas and Somali current large marine ecosystems project (supported by UNDP with GEF grant financing)

ASCLME Mauritius (2012) National marine ecosystem diagnostic analysis. Mauritius. Contribution to the Agulhas and Somali current large marine ecosystems project (supported by UNDP with GEF grant financing)

ASCLME Mozambique (2012) National marine ecosystem diagnostic analysis. Mozambique. Contribution to the Agulhas and Somali current large marine ecosystems project (supported by UNDP with GEF grant financing)

ASCLME Seychelles (2012) National marine ecosystem diagnostic analysis. Seychelles. Contribution to the Agulhas and Somali current large marine ecosystems project (supported by UNDP with GEF grant financing)

ASCLME Somalia (2012) National marine ecosystem diagnostic analysis. Somalia. Contribution to the Agulhas and Somali current large marine ecosystems project (supported by UNDP with GEF grant financing)

ASCLME South Africa (2012) National marine ecosystem diagnostic analysis. South Africa. Contribution to the Agulhas and Somali current large marine ecosystems project (supported by UNDP with GEF grant financing)

ASCLME Tanzania (2012) National marine ecosystem diagnostic analysis. Tanzania. Contribution to the Agulhas and Somali current large marine ecosystems project (supported by UNDP with GEF grant financing)

Bandeira SO, Björk M (2001) Seagrass research in the eastern Africa region: emphasis on diversity, ecology and ecophysiology. S Afr J Bot 67(3):420–425. https://doi.org/10.1016/S0254-6299(15)31158-3

Bandeira S, Gell F (2003) The seagrasses of Mozambique and southeastern Africa. World Atlas of Seagrasses. UNEO/WCMC Publications

Banerjee K, Paneerselvam A, Ramachandran P, Ganguly D, Singh G, Ramesh R (2018) Seagrass and macrophyte mediated CO_2 and CH_4 dynamics in shallow coastal waters. PLoS ONE 13(10):e0203922. https://doi.org/10.1371/journal.pone.0203922

Bawazir GM (2009) Marine Biodiversity of Aden Wetlands Protected Areas. Republic of Yemen, Yemen Society for the protection of wildlife, Environmental Protection Authority

Berg P, Delgard ML, Polsenaere P, McGlathery KJ, Doney SC, Berger AC (2019) Dynamics of benthic metabolism, O_2, and pCO_2 in a temperate seagrass meadow. Limnol Oceanogr 64(6):2586–2604. https://doi.org/10.1002/lno.11236

BOBLME (2015) Assessment of transboundary governance architecture for the Bay of Bengal LME. In: Mahon R, Fanning L (eds) BOBLME-2015-Governance-07

Bouillon S, Dehairs F, Velimirov B, Abril G, Borges AV (2007) Dynamics of organic and inorganic carbon across contiguous mangrove and seagrass systems (Gazi Bay, Kenya). J Geophys Res Biogeosci 112(G2). https://doi.org/10.1029/2006JG000325

Campbell JE, Lacey EA, Decker RA, Crooks S, Fourqurean JW (2015) Carbon storage in seagrass beds of Abu Dhabi, United Arab Emirates. Estuar Coast 38(1):242–251. https://doi.org/10.1007/s12237-014-9802-9

Carruthers TJB, Dennison WC, Kendrick GA, Waycott M, Walker DI, Cambridge ML (2007) Seagrasses of south-west Australia: a conceptual synthesis of the world's most diverse and extensive seagrass meadows. J Exp Mar Biol Ecol 350:21–45. https://doi.org/10.1016/j.jembe.2007.05.036

Chayhard S, Manthachitra V, Nualchawee K, Buranapratheprat A (2018) Multi-temporal mapping of seagrass distribution by using integrated remote sensing data in Kung Kraben Bay (KKB), Chanthaburi province, Thailand. Int J Agric Technol 14(2):161–170

Chen G, Azkab MH, Chmura GL, Chen S, Sastrosuwondo P, Ma Z et al (2017) Mangroves as a major source of soil carbon storage in adjacent seagrass meadows. Sci Rep 7:42406. https://doi.org/10.1038/srep42406

Collier CJ, Waycott M (2014) Temperature extremes reduce seagrass growth and induce mortality. Mar Poll Bull 83(2):483–490. https://doi.org/10.1016/j.marpolbul.2014.03.050

Collier CJ, Lavery PS, Ralph PJ, Masini RJ (2009) Shade-induced response and recovery of the seagrass Posidonia sinuosa. J Exp Mar Biol Ecol 370(1–2):89–103. https://doi.org/10.1016/j.jembe.2008.12.003

Daby D (2003) Effects of seagrass bed removal for tourism purposes in a Mauritian Bay. Environ Pollut 125:313–324. https://doi.org/10.1016/S0269-7491(03)00125-8

Dahdouh-Guebas F, Coppejans E, van Speybroeck D (1999) Remote sensing and zonation of seagrasses and algae along the Kenyan coast. Hydrobiologia 400:63–73. https://doi.org/10.1023/A:1003755012331

Dekker A, Brando V, Anstee J, Fyfe S, Malthus T, Karpouzli E (2007) Remote sensing of seagrass ecosystems: use of spaceborne and airborne sensors. In: Larkum AWD, Orth RJ, Duarte CM (eds) Seagrasses: biology, ecology and conservation. Springer, Dordrecht, pp 347–359. https://doi.org/10.1007/978-1-4020-2983-7_15

Douven WJAM, Buurman JJG, Kiswara W (2003) Spatial information for coastal zone management: the example of the Banten Bay seagrass ecosystem, Indonesia. Ocean Coast Manag 46(6–7):615–634. https://doi.org/10.1016/S0964-5691(03)00038-3

Duarte CM, Middelburg JJ, Caraco N (2005) Major role of marine vegetation on the oceanic carbon cycle. Biogeosciences 2:1–8. https://doi.org/10.5194/bgd-2-1-2005

Egea LG, Jiménez-Ramos R, Vergara JJ, Hernández I, Brun FG (2018) Interactive effect of temperature, acidification and ammonium enrichment on the seagrass *Cymodocea nodosa*. Mar Pollut Bull 134:14–26. https://doi.org/10.1016/j.marpolbul.2018.02.029

Erftemeijer PL, Shuail DA (2012) Seagrass habitats in the Arabian Gulf: distribution, tolerance thresholds and threats. Aquat Ecosyst Health Manag 15(sup1):73–83. https://doi.org/10.1080/14634988.2012.668479

Erftemeijer PL, Osinga R, Mars AE (1993) Primary production of seagrass beds in South Sulawesi (Indonesia): a comparison of habitats, methods and species. Aquat Bot 46(1):67–90. https://doi.org/10.1016/0304-3770(93)90065-5

Faazil MDM, Iroshanie RGA, Bandara WMDK (2019) Carbon Storage in Selected Seagrass Beds in Southern Coast of Sri Lanka. In: Proceedings of international forestry and environment symposium, vol 24. https://doi.org/10.31357/fesympo.v24i0.4226

Fauzan MA, Kumara IS, Yogyantoro R, Suwardana S, Fadhilah N, Nurmalasari I et al (2017) Assessing the capability of Sentinel-2A data for mapping seagrass percent cover in Jerowaru, East Lombok. Indones J Geogr 49(2):195–203. https://doi.org/10.22146/ijg.28407

Fortes MS, Ooi JLS, Tan YM, Prathep A, Bujan JS, Yaakub SM (2018) Seagrass in Southeast Asia: a review of status and knowledge gaps, and a road map for conservation. Bot Mar 61(3):269–288. https://doi.org/10.1515/bot-2018-0008

Fourqurean JW, Duarte CM, Kennedy H, Marbà N, Holmer M, Mateo MA et al (2012) Seagrass ecosystems as a globally significant carbon stock. Nat Geosci 5(7):505–509. https://doi.org/10.1038/ngeo1477

Gacia E, Duarte CM (2001) Sediment retention by a Mediterranean *Posidonia oceanica* meadow: the balance between deposition and resuspension. Estuar Coast Shelf Sci 52:505–514. https://doi.org/10.1006/ecss.2000.0753

Gacia E, Duarte CM, Middelburg JJ (2002) Carbon and nutrients deposition in a Mediterranean seagrass (*Posidonia oceanica*) meadow. Limnol Oceanogr 47:23–32. https://doi.org/10.4319/lo.2002.47.1.0023

Ganguly D, Singh G, Ramachandran P, Selvam AP, Banerjee K, Ramachandran R (2017) Seagrass metabolism and carbon dynamics in a tropical coastal embayment. Ambio 46(6):667–679. https://doi.org/10.1007/s13280-017-0916-8

Ganguly D, Singh G, Purvaja R, Bhatta R, Selvam AP, Banerjee K, Ramesh R (2018) Valuing the carbon sequestration regulation service by seagrass ecosystems of Palk Bay and Chilika, India. Ocean Coast Manag 159:26–33. https://doi.org/10.1016/j.ocecoaman.2017.11.009

Geevarghese GA, Akhil B, Magesh G, Krishnan P, Purvaja R, Ramesh R (2018) A comprehensive geospatial assessment of seagrass distribution in India. Ocean Coast Manag 159:16–25. https://doi.org/10.1016/j.ocecoaman.2017.10.032

Giardino C, Bresciani M, Fava F, Matta E, Brando VE, Colombo R (2016) Mapping submerged habitats and mangroves of Lampi Island Marine National Park (Myanmar) from in situ and satellite observations. Remote Sens 8(1):2. https://doi.org/10.3390/rs8010002

Githaiga MN, Kairo JG, Gilpin L, Huxham M (2017) Carbon storage in the seagrass meadows of Gazi Bay, Kenya. PLoS ONE 12(5):e0177001. https://doi.org/10.1371/journal.pone.0177001

Githaiga MN, Frouws AM, Kairo JG, Huxham M (2019) Seagrass removal leads to rapid changes in fauna and loss of carbon. Front Ecol Evol 7:62. https://doi.org/10.3389/fevo.2019.00062

Githaiga MN, Gilpin L, Kairo JG, Huxham M (2016) Biomass and productivity of seagrasses in Africa. Bot Mar 59(2–3):173–186. https://doi.org/10.1515/bot-2015-0075
Greiner JT, McGlathery KJ, Gunnell J, McKee BA (2013) Seagrass restoration enhances "blue carbon" sequestration in coastal waters. PLoS ONE 8(8):e72469. https://doi.org/10.1371/journal.pone.0072469
Guerra-Vargas LA, Gillis LG, Mancera-Pineda JE (2020) Stronger together: do coral reefs enhance seagrass meadows "blue carbon" potential? Front Mar Sci 7:628. https://doi.org/10.3389/fmars.2020.00628
Gullström M, Lundén B, Bodin M, Kangwe J, Öhman MC, Mtolera MS, Björk M (2006) Assessment of changes in the seagrass-dominated submerged vegetation of tropical Chwaka Bay (Zanzibar) using satellite remote sensing. Estuar Coast Shelf Sci 67(3):399–408. https://doi.org/10.1016/j.ecss.2005.11.020
Gullström M, Lyimo LD, Dahl M, Samuelsson GS, Eggertsen M, Anderberg E et al (2018) Blue carbon storage in tropical seagrass meadows relates to carbonate stock dynamics, plant–sediment processes, and landscape context: insights from the western Indian Ocean. Ecosystems 21(3):551–566. https://doi.org/10.1007/s10021-017-0170-8
Guo H, Noormets A, Zhao B, Chen J, Sun G, Gu Y, Li B, Chen J (2009) Tidal effects on net ecosystem exchange of carbon in an estuarine wetland. Agric for Meteorol 149:1820–1828. https://doi.org/10.1016/j.agrformet.2009.06.010
Hanekom N, Baird D (1988) Distribution and variations in seasonal biomass of eelgrass. *Zostera capensis* in the Kromme estuary, St Francis Bay South Africa. S Afr J Mar Sci 7:51–59. https://doi.org/10.2989/025776188784379099
Hansen JCR, Reidenbach MA (2012) Wave and tidally driven flows within *Zostera marina* seagrass beds and their impact on sediment suspension. Mar Ecol Prog Ser 448:271–287. https://doi.org/10.3354/meps09225
Hantanirina JMO, Benbow S (2013) Diversity and coverage of seagrass ecosystems in south-west Madagascar. Afr J Mar Sci 35(2):291–297. https://doi.org/10.2989/1814232X.2013.800580
Harcourt WD, Briers RA, Huxham M (2018) The thin (ning) green line? Investigating changes in Kenya's seagrass coverage. Biol Lett 14(11):20180227. https://doi.org/10.1098/rsbl.2018.0227
Hena MA, Short FT, Sharifuzzaman SM, Hasan M, Rezowan M, Ali M (2007) Salt marsh and seagrass communities of Bakkhali Estuary, Cox's Bazar, Bangladesh. Estuar Coast Shelf Sci 75(1–2):72–78. https://doi.org/10.1016/j.ecss.2007.01.022
Hossain MS, Bujang JS, Zakaria MH, Hashim M (2015) Application of Landsat images to seagrass areal cover change analysis for Lawas, Terengganu and Kelantan of Malaysia. Cont Shelf Res 110:124–148. https://doi.org/10.1016/j.csr.2015.10.009
Hossain MS, Hashim M, Bujang JS, Zakaria MH, Muslim AM (2019) Assessment of the impact of coastal reclamation activities on seagrass meadows in Sungai Pulai estuary, Malaysia, using Landsat data (1994–2017). Int J Remote Sens 40(9):3571–3605. https://doi.org/10.1080/01431161.2018.1547931
Howard JL, Creed JC, Aguiar MV, Fouqurean JW (2018) CO_2 released by carbonate sediment production in some coastal areas may offset the benefits of seagrass "blue carbon" storage. Limnol Oceanogr 63:160–172. https://doi.org/10.1002/lno.10621
Howari FM, Jordan BR, Bouhouche N, Wyllie-Echeverria S (2009) Field and remote-sensing assessment of mangrove forests and seagrass beds in the northwestern part of the United Arab Emirates. J Coast Res 25:48–56. https://doi.org/10.2112/07-0867.1
Ingram JC, Dawson TP (2001) The impacts of a river effluent on the coastal seagrass habitats of Mahé, Seychelles. S Afr J Bot 67(3):483–487. https://doi.org/10.1016/S0254-6299(15)31167-4
Jagtap TG, Inamdar SN (1991) Mapping of seagrass meadows from the Lakshadweep Islands (India), using aerial photographs. J India Soc Remote Sens 19(2):77–82. https://doi.org/10.1007/BF03008123
Japar Sidik B, Muta Harah Z (2003) The seagrasses of Malaysia. World atlas of seagrasses. UNEP World Conservation Monitoring Centre. University of California Press, Berkeley, pp 152–160

Johnson P, Johnstone R (1995) Productivity and nutrient dynamics of tropical sea-grass communities in Puttalam Lagoon, Sri Lanka. Ambio 24:7–8

Jupp BP, Durako MJ, Kenworthy WJ, Thayer GW, Schillak L (1996) Distribution, abundance, and species composition of seagrasses at several sites in Oman. Aquat Bot 53(3–4):199–213. https://doi.org/10.1016/0304-3770(96)01023-6

Kaladharan P, David Raj I (1989) Primary production of seagrass, *Cymodocea serrulata* and its contribution to productivity of Amini atoll, Lakshadweep Islands. India J Mar Sci 18:215–216

Kamermans P, Hemminga MA, Tack JF, Mateo MA, Marbà N, Mtolera M et al (2002) Groundwater effects on diversity and abundance of lagoonal seagrasses in Kenya and on Zanzibar Island (East Africa). Mar Ecol Prog Ser 231:75–83. https://doi.org/10.3354/meps231075

Kanal AHM, Short F (2009) A new record of seagrass *Halophila beccarii* Ascherson in Bangladesh. CMU J Nat Sci 8(2):201–206

Kennedy H, Beggins J, Duarte CM, Fourqurean JW, Holmer M et al (2010) Seagrass sediments as a global carbon sink: isotopic constraints. Glob Biogeochem Cycles 24:1–8. https://doi.org/10.1029/2010GB003848

Kirkman H, Reid DD (1979) A study of the role of the seagrass *Posidonia australis* in the carbon budget of an estuary. Aquat Bot 7:173–183. https://doi.org/10.1016/0304-3770(79)90020-2

Knudby A, Nordlund L (2011) Remote sensing of seagrasses in a patchy multi-species environment. Int J Remote Sens 32(8):2227–2244. https://doi.org/10.1080/01431161003692057

Koch MS, Schopmeyer S, Kyhn-Hansen C, Madden CJ (2007) Synergistic effects of high temperature and sulfide on tropical seagrass. J Exp Mar Biol Ecol 341(1):91–101. https://doi.org/10.1016/j.jembe.2006.10.004

Koedsin W, Intararuang W, Ritchie RJ, Huete A (2016) An integrated field and remote sensing method for mapping seagrass species, cover, and biomass in southern Thailand. Remote Sens 8(4):292. https://doi.org/10.3390/rs8040292

Kuo J (2007) New monoecious seagrass of *Halophila sulawesii* (Hydrocharitaceae) from Indonesia. Aquat Bot 87:171–175. https://doi.org/10.1016/j.aquabot.2007.04.006

Kuriandewa TE, Kiswara W, Hutomo M, Soemodihardjo S (2003) The seagrasses of Indonesia World Atlas of Seagrasses. University of California Press, Barkeley, pp 172–182

Lavery PS, Mateo MA, Serrano O, Rozaimi M (2013) Variability in the carbon storage of seagrass habitats and its implications for global estimates of blue carbon ecosystem service. PLoS ONE 8:e73748. https://doi.org/10.1371/journal.pone.0073748

Longstaff BJ, Dennison WC (1999) Seagrass survival during pulsed turbidity events: the effects of light deprivation on the seagrasses *Halodule pinifolia* and *Halophila ovalis*. Aquat Bot 65(1–4):105–121. https://doi.org/10.1016/S0304-3770(99)00035-2

Longstaff BJ, Loneragan NR, O'donohue MJ, Dennison WC (1999) Effects of light deprivation on the survival and recovery of the seagrass *Halophila ovalis* (R. Br.) Hook. J Exp Mar Biol Ecol 234(1):1–27. https://doi.org/10.1016/S0022-0981(98)00137-3

Macklin PA, Suryaputra IGNA, Maher DT, Murdiyarso D, Santos IR (2019) Drivers of CO_2 along a mangrove-seagrass transect in a tropical bay: delayed groundwater seepage and seagrass uptake. Cont Shelf Res 172:57–67. https://doi.org/10.1016/j.csr.2018.10.008

Macreadie PI, Serrano O, Maher DT, Duarte C, Beardall J (2017) Addressing calcium carbonate cycling in blue carbon accounting. Limnol Oceanogr Lett 2:195–201. https://doi.org/10.1002/lol2.10052

Maghsoodloo A, Eghtesadi P (2008) Coral reefs and related ecosystems in Iran: monitoring and management. Country report of the Islamic Republic of Iran. Powerpoint presentation, Marine Living Sciences Department, Iranian National Center for Oceanography (INCO)

Martins ARO, Bandeira SO (2001) Biomass distribution and leaf nutrient concentrations and resorption of Thalassia hemprichii at Inhaca Island, Mozambique. S Afr J Bot 67(3):439–442. https://doi.org/10.1016/S0254-6299(15)31161-3

Mateo MA, Cebrian J, Dunton K, Mutchler T (2006) Carbon flux in seagrass ecosystems. In: Larkum A, Orth R, Duarte C (eds) Seagrasses: biology, ecology and conservation. Springer, Netherlands, pp 159–192. https://doi.org/10.1007/978-1-4020-2983-7

Mazarrasa I, Marbà N, Lovelock CE, Serrano O, Lavery PS, Fourqurean JW et al (2015) Seagrass meadows as a globally significant carbonate reservoir. Biogeosciences 12:4993–5003. https://doi.org/10.5194/bg-12-4993-2015

McKenzie LJ, Finkbeiner MA, Kirkman H (2001) Methods for mapping seagrass distribution. In: Short FT, Coles RG (eds) Global seagrass research methods, pp 101–121

McLeod E, Chmura GL, Bouillon S, Salm R, Bjork M et al (2011) A blueprint for blue carbon: toward an improvement understanding of the role of vegetated coastal habitats in sequestering CO_2. Front Ecol Environ 9:552–560. https://doi.org/10.1890/110004

Mvungi EF, Pillay D (2019) Eutrophication overrides warming as a stressor for a temperate African seagrass (*Zostera capensis*). PLoS ONE 14(4):e0215129. https://doi.org/10.1371/journal.pone.0215129

Nellemann C et al (2009) Blue carbon. A rapid response assessment. United Nations Environment Programme, GRID-Arendal

Nguyen VX, Detcharoen M, Tuntiprapas P, Soe-Htun U, Sidik JB, Harah MZ et al (2014) Genetic species identification and population structure of *Halophila* (Hydrocharitaceae) from the Western Pacific to the Eastern Indian Ocean. BMC Evol Biol 14(1):92. https://doi.org/10.1186/1471-2148-14-92

Ochieng CA, Erftemeijer PLA (2003) Kenya and Tanzania. In: Green EP, Short FT (eds) World atlas of seagrasses, UNEP World Conservation Monitoring Centre, University of California Press, Berkeley, USA

Olivé I, Silva J, Costa MM, Santos R (2016) Estimating seagrass community metabolism using benthic chambers: the effect of incubation time. Estuar Coast 39(1):138–144. https://doi.org/10.1007/s12237-015-9973-z

Pasqualini V, Pergent-Martini C, Clabaut P, Marteel H, Pergent G (2001) Integration of aerial remote sensing, photogrammetry, and GIS technologies in seagrass mapping. Photogr Eng Remote Sens 67(1):99–105

Patro S, Krishnan P, Samuel VD, Purvaja R, Ramesh R (2017) Seagrass and salt marsh ecosystems in South Asia: an overview of diversity, distribution, threats and conservation status. In: Prusty B, Chandra R, Azeez P (eds) Wetland science. Springer, New Delhi, pp 87–104. https://doi.org/10.1007/978-81-322-3715-0_5

Paulose NE, Dilipan E, Thangaradjou T (2013) Integrating Indian remote sensing multi-spectral satellite and field data to estimate seagrass cover change in the Andaman and Nicobar Islands, India. Ocean Sci J 48(2):173–181. https://doi.org/10.1007/s12601-013-0014-1

Payri C, N'Yeurt ADR, Mattio L (2012) Benthic algal and seagrass communities in Baa atoll, Maldives. In: Andréfouët S (ed) Biodiversity, resources, and conservation of Baa atoll (Republic of Maldives): a UNESCO Man and Biosphere Reserve. Atoll Research Bulletin, pp 31–66

Peeters F, Atamanchuk D, Tengberg A, Encinas-Fernández J, Hofmann H (2016) Lake metabolism: comparison of lake metabolic rates estimated from a diel CO_2-and the common diel O_2-technique. PLoS ONE 11(12):0168393. https://doi.org/10.1371/journal

Pérez M, Duarte CM, Romero J, Sand-Jensen K, Alcoverro T (1994) Growth plasticity in *Cymodocea nodosa* stands: the importance of nutrient supply. Aquat Bot 47(3–4):249–264. https://doi.org/10.1016/0304-3770(94)90056-6

Ralph PJ, Durako MJ, Enriquez S, Collier CJ, Doblin MA (2007) Impact of light limitation on seagrasses. J Exp Mar Biol Ecol 350(1–2):176–193. https://doi.org/10.1016/j.jembe.2007.06.017

Ranahewa TH, Gunasekara AJM, Premarathna AD, Karunarathna SC, Jayamanne SC (2018) A comparative study on the diversity of seagrass species in selected areas of puttalam lagoon in Sri Lanka. J Oceanogr Mar Res 6(185):2

Rattanachot E, Prathep A (2011) Temporal variation in growth and reproduction of *Enhalus acoroides* (L.f.) Royle in a monospecific meadow in Haad Chao Mai National Park, Trang Province, Thailand. Bot Mar 54(2):201–207. https://doi.org/10.1515/bot.2011.018

Rattanachot E, Prathep A (2015) Species specific effects of three morphologically different below-ground seagrasses on sediment properties. Estuar Coast Shelf Sci 167:427–435. https://doi.org/10.1016/j.ecss.2015.10.019

Roelfsema CM, Phinn SR, Udy N, Maxwell P (2009) An integrated field and remote sensing approach for mapping seagrass cover, Moreton Bay, Australia. J Spat Sci 54(1):45–62. https://doi.org/10.1080/14498596.2009.9635166

Roelfsema C, Kovacs EM, Saunders MI, Phinn S, Lyons M, Maxwell P (2013) Challenges of remote sensing for quantifying changes in large complex seagrass environments. Estuar Coast Shelf Sci 133:161–171. https://doi.org/10.1016/j.ecss.2013.08.026

Roelofs A, Rasheed M, Thomas R (2001) Port of Weipa Seagrass monitoring: baseline surveys, April and September 2000. Ports Corporation of Queensland

Ruesink JL, Yang S, Trimble AC (2015) Variability in carbon availability and eelgrass (*Zostera marina*) biometrics along an estuarine gradient in Willapa Bay, WA, USA. Estuar Coast 38:1908–1917. https://doi.org/10.1007/s12237-014-9933-z

Ruiz JM, Romero J (2001) Effects of in situ experimental shading on the Mediterranean seagrass *Posidonia oceanica*. Mar Ecol Prog Ser 215:107–120. https://doi.org/10.3354/meps215107

Saderne V, Fietzek P, Herman PMJ (2013) Extreme variations of pCO_2 and pH in a macrophyte meadow of the Baltic Sea in summer: evidence of the effect of photosynthesis and local upwelling. PLoS ONE 8:e62689. https://doi.org/10.1371/journal.pone.0062689

Saderne V, Geraldi NR, Macreadie PI, Maher DT, Middelburg JJ, Serrano O et al (2019) Role of carbonate burial in Blue Carbon budgets. Nat Commun 10(1):1–9. https://doi.org/10.1038/s41467-019-08842-6

Serrano O, Almahasheer H, Duarte CM, Irigoien X (2018) Carbon stocks and accumulation rates in Red Sea seagrass meadows. Sci Rep 8(1):1–13. https://doi.org/10.1038/s41598-018-33182-8

Short FT, Polidoro B, Livingstone SR, Carpenter KE, Bandeira S, Bujang JS et al (2011) Extinction risk assessment of the world's seagrass species. Biol Conserv 144(7):1961–1971. https://doi.org/10.1016/j.biocon.2011.04.010

Soe-Htun U, Maung A, Mon S, Tun TZ, Hteik AA, Lwin ML, Tun Z, Lunn UZ, Murray-Jones S (2018) Seagrass surveys in the southern Rakhine coastal region, Myanmar: biodiversity, distribution and coverage. J Aquac Mar Biol 7(2):103–110

Soe-Htun U, Wai MK, Nyunt T, Kyaw SPP, Aye MM (2009) Seagrasses of Myanmar with special reference to the phytogeographic distribution of the species of ASEAN nations. J Myan Acad Art Sci 7(5)

Strydom S, Murray K, Wilson S, Huntley B, Rule M, Heithaus M et al (2020) Too hot to handle: unprecedented seagrass death driven by marine heatwave in a World Heritage Area. Glob Chang Biol 26(6):3525–3538. https://doi.org/10.1111/gcb.15065

Thangaradjou T, Bhatt JR (2018) Status of seagrass ecosystems in India. Ocean Coast Manag 159:7–15. https://doi.org/10.1016/j.ocecoaman.2017.11.025

Thangaradjou T, Sridhar R, Senthilkumar S, Kannan S (2008) Seagrass resource assessment in the Mandapam coast of the Gulf of Mannar Biosphere Reserve, India. Appl Ecol Environ Res 6(1):139–146

Thorhaug A, Gallagher JB, Kiswara W, Prathep A, Huang X, Yap TK et al (2020) Coastal and estuarine blue carbon stocks in the greater Southeast Asia region: seagrasses and mangroves per nation and sum of total. Mar Pollut Bull 160:111168. https://doi.org/10.1016/j.marpolbul.2020.111168

Thu-Rein Aung C, Murray-Jones S, Soe-Htun U (2019) Seagrass surveys in Shwe Thaung Yan coastal areas, the southern part of Rakhine Coastal Region, Myanmar: biodiversity, coverage and biomass. J Aquac Mar Biol 8(3):105–112

Tokoro T, Hosokawa S, Miyoshi E, Tada K, Watanabe K, Montani S, Kayanne H, Kuwae T (2014) Net uptake of atmospheric CO_2 by coastal submerged aquatic vegetation. Glob Chang Biol 20(6):1873–1884. https://doi.org/10.1111/gcb.12543

Torres-Pulliza D, Wilson JR, Darmawan A, Campbell SJ, Andréfouët S (2013) Ecoregional scale seagrass mapping: a tool to support resilient MPA network design in the Coral Triangle. Ocean Coast Manag 80:55–64. https://doi.org/10.1016/j.ocecoaman.2013.04.005

Traganos D, Aggarwal B, Poursanidis D, Topouzelis K, Chrysoulakis N, Reinartz P (2018) Towards global-scale seagrass mapping and monitoring using Sentinel-2 on Google Earth Engine: the case

study of the Aegean and Ionian seas. Remote Sens 10(8):1227. https://doi.org/10.3390/rs1008 1227
Tuntiprapas P, Shimada S, Pongparadon S, Prathep A, Saensouk P, Theerakulpisut P et al (2015) Is *Halophila major* (Zoll.) Miquel a big *H. ovalis* (R. Brown) JD Hooker? An evaluation based on age, morphology, and ITS sequence. Sci Asia 41(2):79–86
Uku J, Björk M (2005) Productivity aspects of three tropical seagrass species in areas of different nutrient levels in Kenya. Estuar Coast Shelf Sci 63(3):407–420. https://doi.org/10.1016/j.ecss.2004.12.007
Umamaheswari R, Ramach S, Nobi EP (2009) Mapping the extent of seagrass meadows of Gulf of Mannar Biosphere Reserve, India using IRS ID satellite imagery. Int J Biodivers Conserv 1(5):187–193
UNEP (2008) National Reports on Seagrass in the South China Sea. UNEP/GEF/SCS Technical Publication No. 12
UNEP-WCMC Short FT (2018) Global distribution of seagrasses (version 6). Sixth update to the data layer used in Green and Short (2003), superseding version 5. UN Environment World Conservation Monitoring Centre, Cambridge. http://data.unep-wcmc.org/datasets/7
Vanderklift MA, Gorman D, Steven AD (2019) Blue carbon in the Indian Ocean: a review and research agenda. J India Ocean Reg 15(2):129–138. https://doi.org/10.1080/19480881.2019.162 5209
Vichkovitten T (1998) Biomass, growth and productivity of seagrass; *Enhalus acoroides* (Linn. f) in Khung Kraben Bay, Chanthaburi, Thailand. Agric Nat Resour 32(1):109–115
Walton MEM, Al-Maslamani I, Skov MW, Al-Shaikh I, Al-Ansari IS, Kennedy HA, Le Vay L (2014) Outwelling from arid mangrove systems is sustained by inwelling of seagrass productivity. Mar Ecol Prog Ser 507:125–137. https://doi.org/10.3354/meps10827
Wawan K (2011) Seagrasses in Indonesia. In: Ogawa H, Japar Sidik B, Muta Harah Z (eds) Seagrasses: resource status and trends in Indonesia, Japan, Malaysia, Thailand and Vietnam. Seizando-Shoten Publishing Co., Ltd., Tokyo, pp 5–16
Yaakub SM, Lim RLF, Lim WL, Todd PA (2013) The diversity and distribution of seagrass in Singapore. Nat Singap 6:105–111
Yamakita T, Sodeyama F, Whanpetch N, Watanabe K, Nakaoka M (2019) Application of deep learning techniques for determining the spatial extent and classification of seagrass beds, Trang, Thailand. Bot Mar 62(4):291–307. https://doi.org/10.1515/bot-2018-0017
Yates KK, Dufore C, Smiley N, Jackson C, Halley RB (2007) Diurnal variation of oxygen and carbonate system parameters in Tampa Bay and Florida Bay. Mar Chem 104:110–124. https://doi.org/10.1016/j.marchem.2006.12.008

Chapter 6
Blue Carbon Dynamics in the Indian Ocean Salt Marshes and Tidal Flats

Abhra Chanda

Abstract Anecdotal information indicates that salt marshes and tidal flats are prevalent in almost all the countries on the Indian Ocean rim. However, this region has received very little attention compared to the stretch and potential that coastal habitats can offer to mitigate climate change. The present chapter collated the existing information on the country-wise spatial extent, species diversity (or most abundant species), carbon stock in the live biomass, and sediment carbon stock in these two crucial ecosystems. Most of the Indian Ocean nations have no proper estimates of the spatial extent of these two ecosystems. The present review indicates that India, Bangladesh, Indonesia, UAE, South Africa, Mozambique, and Tanzania are a few of the countries where carbon stock assessments in the salt marshes and tidal flats have been performed. Studies on the salt marsh species diversity exist for many of the countries; however, more emphasis on a nationwide scale can enhance the species count reported so far. The existing literature indicates that salt marshes and tidal flats can occupy varying morphological terrains with variable physicochemical and hydrodynamic conditions. The carbon stock in the live biomass of salt marshes and sediment column of both marshes and flats varies from 0.009 to 33.4 Mg C ha^{-1} and 2.61 to 163.7 Mg C ha^{-1}, respectively. On the whole, the present chapter indicates the data gaps in this regard from the Indian Ocean periphery and advocates the future focus and orientation in research that this region deserves.

Keywords Salt marshes · Tidal flats · Mudflats · Unvegetated coastal terrains · Species diversity · Spatial extent · Aboveground carbon · Belowground carbon · Soil organic carbon pool

---

A. Chanda (✉)
School of Oceanographic Studies, Jadavpur University, 188 Raja S.C. Mullick Road, Kolkata, West Bengal 700084, India
e-mail: abhrachanda.sos.rs@jadavpuruniversity.in

A. Chanda et al. (eds.), *Blue Carbon Dynamics of the Indian Ocean*,
https://doi.org/10.1007/978-3-030-96558-7_6

6.1 Brief Overview on Tidal Flat and Salt Marsh Ecosystems

Tidal flats are those regions that border the continents and islands where regular tidal activity exposes the terrain during the low tide and submerges during the high tide. These terrains, which almost lie parallel to the sea level, receive sediments from both the seaward end as well as from terrigenous sources and gives rise to a variety of landscapes (Fan et al. 2018; Lakshminarayana et al. 2020). These terrains based on the type of sedimentation can be rocky, muddy, or sandy (Healy et al. 2002). Like the other prominent coast-peripheral marine ecosystems (mangroves, seagrass, and coral), these tidal flats also play a crucial role in the global biogeochemical cycles (Schutte et al. 2019) and furnish a multitude of ecosystem services (Miththapala 2013). These natural terrains protect us from climatic disasters like storms and flooding (Reed et al. 2018), act as a natural purifier of organic matter and nutrients in the coastal water (Kim et al. 2020), support rich biodiversity (Liu et al. 2020a), and most essentially provide a variety of goods to human beings (Reise et al. 2010). Though the tidal flats are not considered a part of the blue carbon family in the truest sense (Macreadie et al. 2019; Serrano et al. 2019), this ecosystem offers tremendous potential to sequester and store substantial quantities of carbon (Endo and Otani 2019; Chen et al. 2020; Dong et al. 2020). Among the various types of tidal flats in terms of their geomorphic character, the mudflats are the most significant carbon sinks (Toniolo et al. 2021). These mudflats experience a flooded condition for a substantial time in the diurnal cycle (Bernal and Mitsch 2012), allowing the dissolved and particulate carbon load in the adjacent water masses to come in contact with the flats. Moreover, these flats usually witness a very high rate of sedimentation, which further increases the organic carbon load in these sediments owing to their intrinsic low permeable nature (Perillo 2019). In addition, the physiological functions of the microbial and the crustacean community, which usually exist in abundance in the mudflats, play a crucial role in enhancing the carbon sequestration potential (Botto and Iribarne 2000; Mok 2019; Sethi et al. 2020). Thus, altogether this ecosystem deserves special attention from the perspective of possibilities of mitigating climate change.

Tidal flats usually denote those intertidal areas, which are devoid of any vegetation; however, many of these areas are covered by vascular vegetation, known as salt marshes (Hopkinson et al. 2018). A substantial part of these tidal wetlands is covered by tidal halophytic plants known as mangroves; however, all these regions are not suitable for the growth of mangroves. Such typical areas, characterized by low energy and partially sheltered environment having an elevation where seawater reaches fortnightly and the high rate of evaporation leaves a salt encrustation over the soil surface, witnesses the flourishing of the salt marshes (Ruiz-Fernández et al. 2018). These marshes typically thrive in hypersaline soils and are in most cases found behind the mangrove lines (Flores-Verdugo et al. 2007; Costa et al. 2009). Like the other blue carbon habitats, this ecosystem also provides an array of ecosystem services (Chmura 2009, 2013). This ecosystem, like many other coastal vegetated ecosystems, provides refuge to fishes and acts as nursery grounds for the juveniles (Whitfield 2017; Friese et al. 2021). It acts as an abode to a multitude of avifaunas (Correll et al. 2017) and small to large mammalian communities (Pascual et al. 2019). Many species of these marshes are consumed by human beings and cattle (Davidson et al. 2017; Loconsole et al. 2019). Besides, this ecosystem is also valued for providing supporting services like protection from storms and surges (Rupprecht et al. 2017; Leonardi et al. 2018). The marshes also play a vital role in alleviating eutrophication in the coastal waters (Reef et al. 2017), by absorbing substantial quantities of nutrients while carrying out autotrophic activities. In addition, this ecosystem has been recognized as a potential sink for carbon (Macreadie et al. 2013; Bai et al. 2016). Not only the high productivity but the carbon trapping efficiency of these ecosystems laterally derived from the adjacent ecosystems makes this ecosystem worthy of conservation and protection (Tang et al. 2018; Dontis et al. 2020). Despite such significance, both the tidal flats and saltmarshes have witnessed a substantial reduction in areal cover due to both natural (Valiela et al., 2018; Burns et al. 2021) as well as anthropogenic disturbances (Buzzi et al. 2021; Zhang et al. 2021). The global scientific community unanimously realizes that it is time we looked into the conservation and management measures that these ecosystems direly need at the present date.

6.2 Spatial Extent of Tidal Flat and Salt Marsh Ecosystems

6.2.1 Mapping Tidal Flats: Challenges and Uncertainties

Like any other ecosystem of the world, conservation and management of tidal flats require an extensive mapping exercise to draw a clear picture of the spatial extent of this landscape throughout the world. However, mapping tidal flats is a difficult endeavor as this terrain remains submerged in water and gets exposed within a short duration. These transient states of tidal flats make it difficult for remotely sensed images to capture the total spatial extent accurately (Ryu et al. 2014). However, in recent times, several endeavors exist which reported successful characterization of tidal flats. Murray et al. (2019) quantified the area covered by the tidal flats throughout the world. They observed an estimated 130,000 km^2 area covered by this terrain as of the year 2016. Asia, North America, and South America together comprise more than two-thirds of the global tidal flats. They also observed that almost half of the total area covered by this ecosystem is concentrated in only eight countries, out of which four countries are in the periphery of the Indian Ocean (namely, India, Myanmar, Indonesia, and Australia). Murray et al. (2019) also mentioned that anthropogenic construction work (Liu et al. 2020b), drastic reduction in sediment flow through the rivers (Akhand et al. 2021) due to the increasing number of dams and barrages (Guo et al. 2018), regional sea-level rise, and shrinking of deltaic areas (Angamuthu et al. 2018) have led to an estimated loss of 16% of the global tidal flats between the years 1984 and 2016. Tseng et al. (2017) came up with some novel strategies temporally varying images were tested against the ground-measured tidal amplitude data to map a tidal flat in China. The regular tidal inundation of the flats, which makes this terrain a water body for some time and a land surface for the rest, has been the main challenge in mapping; however, this dependency has been eliminated by a novel approach proposed by Zhao et al. (2020). Hill et al. (2021), by overlaying the data on global protected areas on the present areal cover of the tidal flats, observed that 68% of the tidal flats experience moderate to severe human pressure, whereas the rest are protected. However, the rate of areal reduction in the last two decades has been the same in both the protected and unprotected areas.

6.2.2 *Spatial Extent of Tidal Flats in the Indian Ocean Countries*

Though China and many other countries have taken adequate steps to accurately delineate the spatial extent of the tidal flats within the respective countries, endeavors of this sort are lacking in the rim of the Indian Ocean. India encompassing a coastline of about 7500 km has a distal continental shelf and tidal flats on both its east and west coast; however, information on the total area is yet to be generated. The sedimentary structures of the tidal flats of India have received fair attention. Chakrabarti (2005) observed a wide variety of sedimentary structures in the tidal flats of India like shore-parallel tidal channels, exposed point bars, antidunes, and ripples of various kinds including rhomboidal, interference, sinuous bifurcating with eye-shaped patches, threadlike, pyramidal, crescent-shaped, and many more. These structures can play a crucial role in trapping and sequestering carbon (Ghadeer and Macquaker 2012) in suitable places with fewer disturbances. The geological features, such as grain size analysis (Saha and Sinha 2021) and geomorphological characteristics (Saha et al. 2011) of the tidal flats of India are well studied. A substantial part of the entire Bangladesh coastline encompasses the world's largest mangrove forest of Sundarban. However, Payo et al. (2016), while predicting the future changes in mangrove area observed that Bangladesh Sundarbans occupy around 238 km^2 of tidal flats, which are most susceptible to erosion and loss by regional sea-level rise. Beyond the Sundarbans, Kabir et al. (2020) observed a temporally dynamic tidal flat while studying the shoreline dynamics of the Hatiya Island, situated on the western border of Bangladesh. Islam et al. (2020), while studying the resilience of the Kalapara region of Bangladesh observed a tidal flat cover of ~5 km^2, which they found to be extremely susceptible to losses arising due to tropical cyclones. They noticed that most of the tidal flats of this area were unvegetated and muddy. Followed by Bangladesh, Myanmar has a long coastline of almost 2300 km. A large part of their economy relies on fisheries, and this country is known to have large-scale tidal flats. The Gulf of Mottama and the adjoining areas cover one of the largest stretches of mudflats in the world. Zöckler et al. (2014) reported that this part of the world approximately covers 4000 km^2 of tidal mudflats. This region is an abode for a multitude of migratory birds, and especially, for the endangered spoon-billed sandpiper (Aung et al. 2020). However, these areas are highly dynamic experiencing both rapid erosion and accretion (Steijn et al. 2019; Zöckler et al. 2019). Thailand has low-lying coastal plains comprising substantial areas as tidal flats; however, no proper estimate could be retrieved. Umitsu et al. (2007) pointed out that these tidal flats of Thailand, especially, in Banda Aceh are susceptible to flooding and sea-level rise, and the 2006 Boxing day tsunami led to the rampant destruction of other coastal habitats and converted them to tidal flats. Like Thailand, estimates of tidal flat extent are not available for Singapore and Indonesia too. However, carbon accumulation potential in the tidal flats of Indonesia is studied (Yamamoto et al. 2019). Zakaria (2016) indicated that the intertidal zones of Indonesia have several types of tidal flats like muddy, muddy to sandy, mixed flat, and sand flat. The presence of vegetation less tidal flats is reported from Yemen (Kürschner et al. 1998; Sagheer 2013), however, quantified estimates are missing. On the contrary, Pakhirehzan and Rahbani (2014), while working on the northwest coasts of the Gulf of Oman, reported an estimate are of 173 km^2 as flats, out of which sabkha (salt flats) covered 121 km^2 and tidal flats 52 km^2. Al-Farraj (2005) studied in detail the development process of sabkhas (supratidal flats) since the early Holocene along the north coast of UAE. He discussed the various stages through which these tidal flats took the present shape. Several studies have been carried out on the coastal areas of Iran from the perspective of vulnerability assessment (Mafi-Gholami et al. 2019), Holocene sedimentary evolution (Heyvaert and Baeteman 2007), and channel characterization (Hashemi et al. 2014); however, endeavors of quantifying flat area could not be traced. Pakistan has a long coastline of almost 970 km (Siddiqui and Maajid 2004), and several studies were carried out on the crab burrow morphology (Saher and Qureshi 2011), creek evolution pattern (Ijaz et al. 2018); however, only one study, while characterizing mangrove distribution cover, gave a quantitative estimate of tidal flats (3980 km^2) along the coastline of Pakistan (Abbas et al. 2013).

Among the African countries, South Africa has a substantial cover of tidal flats, and a lot of work has been done studying their sedimentological character (Noffke et al. 2008; Dladla et al. 2019) and paleo records (Noffke et al. 2006); however, recent quantified estimates on area coverage are hard to find. However, Ferreira et al. (2009), while mapping the coastal habitats along the Mozambique coastline observed an estimated cover of 657 km^2 as intertidal flats and 52 km^2 as supratidal flats as of the year 2005. Yap and Landoy (1986) estimated that the Kenyan coastlines encompass around 500 km^2 of tidal swamps and flats. Somalia has the longest coastline (3025 km) among the African nations and encompasses a wide variety of coastal ecosystems, which remains largely unexplored. Though the coastal setting and the geomorphology are extensively studied by Carbone and Accordi (2000), the quantified estimates of tidal flat cover could not be retrieved from any literature. Notable efforts were observed in studying the avian biodiversity in the tidal flats of Djibouti (Buechley et al. 2019); however, the tidal flat cover area is not mentioned in any of those studies.

Among the island nations, Balasuriya (2018) estimated an area of about 23.7 km^2 of tidal flats around the coastline of Sri Lanka; however, controversies exist, whether this entire area is unvegetated or occupied by salt marshes. The Maldives, being a small oceanic island occupies an area of about 300 km^2 and is mostly surrounded by coral reefs, reef flats, and sandy beaches, with almost no existence of muddy tidal flats (Clark and Edwards 1994; Khan et al. 2002). Similarly, Seychelles occupies an area of about 50,000 km^2, and mangroves, seagrass, and corals are the dominant coastal ecosystems, with very little mention of tidal flats (https://wedocs.unep.org/; Sheppard et al. 2005). Mentions of the presence of marshlands in Madagascar can be found (Lepoint et al. 2008); however, exact quantification, like many other countries, is lacking at present. Harding (2019) pointed that out of 4250 km^2 of the tidally influenced area of Madagascar, 3270 km^2 is occupied by mangroves. This implies that the remaining 980 km^2 comprises tidal flats; however, nothing is mentioned whether these are vegetated or not. A brief mention of tidal mudflats is found in some literature; however, Comoros Archipelago mostly hosts coral reef flats in its periphery (https://www.cbd.int/doc/meetings/mar/ebsa-sio-01/other/ebsa-sio-01-comoros-en.pdf). Like Seychelles and Maldives, Mauritius also encompasses mostly tidal reef flats in its periphery, and no quantitative proof of tidal flats could be retrieved from the existing literature (Arndt and Fricke 2019).

6.2.3 Mapping Saltmarshes

Compared to unvegetated tidal flats, saltmarshes have received a tad bit more attention in terms of quantifying the areal cover. Unlike unvegetated tidal flats with transient above and below water temporal changes, salt marshes are easy to map. However, like all other ecosystems, mapping this ecosystem from a synoptic scale, requires the implementation of remotely sensed technologies. Wang et al. (2007) successfully implemented the vegetation community-based Neural Network Classifier technique on air-borne spectral data to map the salt marsh cover in Venice. These marshy vegetations often remain associated with the underlying muddy substrate in such a way that it becomes difficult to differentiate the signatures of the two. Hladik et al. (2013) effectively used a combination of hyperspectral and light detection and ranging (LIDAR) data to accurately map the salt marsh habitat along with its species diversity in the Sapelo Island, Georgia, USA. Implementing remotely sensed indices like normalized difference vegetation index (NDVI) has also led to meaningful results in mapping salt marshes (Sun et al. 2016). Recently, Sentinel-1 synthetic aperture radar (SAR) images have also been utilized to map the salt marsh extents (Hu et al. 2021). Pham et al. (2019) reviewed the extensive use of remotely sensed data and imageries to map the blue carbon ecosystem. They observed that most of the remote-sensing-based studies on quantifying salt marsh spatial extent and understanding their species diversity are focused on the USA, Australia, and China. As per the latest estimates carried out by Mcowen et al. (2017), globally salt marshes occupy an area of almost 54,950 km^2, out of which major portions are concentrated in North and South America, Oceania, and China.

6.2.4 Spatial Extent of Salt Marshes in the Indian Ocean Countries

Mapping of saltmarshes received ample attention in India; however, varying estimates are observed in the different studies (Table 6.1). Garg et al. (1998) reported an area of about 1700 km^2 covered by salt marshes in India; whereas later on the estimates were revised to 1600 km^2 (SAC 2011). However, recently, Viswanathan et al. (2020) indicated that the exclusive salt marshes cover in India is only 290 km^2. They argued that the earlier studies included all the areas covered by salt marshes as well as halophytic grasslands, which led to such inconsistencies in the estimates. Whether those halophytic grasslands should be considered salt marshes or not remain an area of debate. Hossain et al. (2021) stated that the Bangladesh coastline encompasses an area of about 1115 km^2 as salt marshes. These marshes are intricately associated with the world's largest mangrove stretch of Sundarbans. Despite having a favorable landscape, an extensive study by Murray et al. (2020) observed that there is no proper information on the spatial extent of salt marshes in Myanmar. They also mentioned the possibility that the entire salt marsh areas have been anthropogenically converted to other land use classes like agriculture or aquaculture plots. Proper estimates of areal extent are almost absent in Thailand, Singapore, Indonesia, and Malaysia. Even in the western Indian Ocean, Yemen and Oman have no proper record of salt marsh extent. However, Mateos-Molina et al. (2020) measured an estimated salt marsh cover of around 13.9 km^2 along the coastline of the UAE. Though many studies characterized the species diversity of saltmarshes in Pakistan, quantified estimates of its spatial extent could not be found in the existing literature. In the western Indian Ocean, among the African countries, Adams (2020) reported an areal cover of almost 150 km^2 all along the coasts of South Africa as of the year 2018. Adams (2020) observed that between the years 1930 and 2018, the South African salt marshes witnessed a decline of almost 43% due to anthropogenic encroachments. Mozambique is known to have salt marshes interspersed with mangrove vegetation; however, many of these patches remain poorly sampled to date. Adams and Bandeira (1995) indicated that there could be 10 km^2 of salt marsh cover in the Mozambique coastline. However, the spatial extent of salt marshes in Tanzanian territory is yet to be quantified properly (Gullström et al. 2021). Besides South Africa, and partially Mozambique, none of the other African countries in the western Indian Ocean has proper documentation regarding the spatial extent of salt marsh cover (Lugendo 2016).

As mentioned above, among the island nations of the Indian Ocean, Sri Lanka has an estimated area of around 24 km^2 as tidal flats; however, to what extent these flats are vegetated with salt marshes remains a question. The Maldives has no proper documentation on the area covered by salt marshes (Shaig 2011; Patro et al. 2017). The presence of salt marshes in Seychelles could be traced in few pieces of literature (Gerlach 1993); however, exact area demarcation is missing. Harding (2019) indicated extensive salt marsh cover in Madagascar comprising almost 980 km^2; however, the demarcation between vegetated flats and unvegetated regions remains unclear. Like in the case of tidal flats, Comoros and Mauritius have no quantified estimates of salt marsh cover.

Table 6.1 Reports of areal coverage of tidal flats and salt marshes in the Indian Ocean countries

Country	Area (km^2)$^\alpha$	Place	Authors
India (AS)	290 (SM)	Entire coastline	Viswanathan et al. (2020) (SM)
	1700 (SM)	Entire coastline	Garg et al. (1998) (SM)
Bangladesh (AS)	238 (TF)	Sundarban	Payo et al. (2016) (TF)
	5 (TF)	Kalapara	Islam et al. (2020) (TF)
	1115 (SM)	Entire coastline	Hossain et al. (2021) (SM)
Myanmar (AS)	4000 (TF)	Gulf of Mottama	Zöckler et al. (2014)(TF)
Thailand (AS)	ND		
Singapore (AS)	ND		
Indonesia (AS)	14.2 (SM) 16.2 (TF)	Semarang Metropolitan Region (SMR), Central Java	Sejati et al. (2020)
Malaysia (AS)	ND		
Yemen (AS)	ND		
Oman (AS)	173 (TF)	Gulf of Oman	Pakhirehzan and Rahbani (2014) (TF)
UAE (AS)	450 (TF)	North coast of UAE	Al-Farraj (2005) (TF)
Iran (AS)			
Pakistan (AS)	3980 (TF)	Entire coastline	Abbas et al. (2013) (TF)
	23.5 (SM)	Entire coastline	Abbas et al. (2013) (SM)
South Africa (AF)	150 (SM)		Adams (2020) (SM)
Mozambique (AF)	700 (TF)	Entire coastline	Ferreira et al. (2009) (TF)
	10 (SM)	Entire coastline	Adams and Bandeira (1995) (SM)
Tanzania (AF)	ND		
Kenya (AF)	500 (TF)	Entire coastline	Yap and Landoy (1986) (TF)
Somalia (AF)	ND		
Djibouti (AF)	ND		
Sri Lanka (AS)	23.7 (TF and SM)	Entire coastline	Balasuriya (2018) (TF and SM)
Maldives (AS)	ND		
Seychelles (AF)	ND		
Madagascar (AF)	980 (TF and SM)	Entire coastline	Harding (2019) (TF and SM)
Comoros (AF)	ND		
Mauritius (AF)	ND		

ND no data; *SM* saltmarsh; *TF* tidal flats

6.3 Physicochemical and Hydrogeological Settings of the Tidal Flats and Salt Marsh Sediments

Tidal flats occur at slightly elevated soils near the coastal or estuarine periphery and saltmarshes prefer a less disturbed semi-enclosed tidal flat region. The tidal flats, as well as salt marshes, thrive in micro to meso to macro tidal environments in the periphery of the Indian Ocean (Hena et al. 2007; Saha et al. 2011; Saha and Sinha 2021). Most of the tidal flats in India are siliciclastic, whereas carbonate sediments are also seldom observed (Saha et al. 2011; Saha and Sinha 2021). Many of these tidal flats and salt marshes received untreated industrial sewage water, as observed by Kaviarasan et al. (2019) in the southern salt marsh patches of India. The salt marshes in south India had a sediment pH range of 7.4–8.3 (Kaviarasan), whereas in western India it exhibited a narrow range of 8.40–8.48 (Chaudhary et al. 2018). The electrical conductivity in western Indian salt marsh sediments varied between 10.0 and 18.5 dS m^{-2}. The salt marsh beds in south India recorded more than 80% sand followed by 12–18% silt and a very small fraction as clay (1.3–2.8%). Several studies observed that the salt marshes and the underlying sediments in Bangladesh can accumulate substantial quantities of heavy metals. Siddiqueet al. (2012) observed Pb, Cd, Zn, and Cu levels in the salt marsh sediments of Cox's Bazar adjacent to Bakkhali River estuary varying from 21–32 mg kg^{-1}, 0.16–0.36 mg kg^{-1}, 93–106 mg kg^{-1}, and 27–42 mg kg^{-1}, respectively. Hossain et al. (2021) also observed an abundance of a suite of heavy metals in the sediments of Bangladesh salt marshes. They observed that the roots of *Porteresia* sp. accumulated substantial heavy metals; however, the translocation rate to other plant parts was poor. The only exception was Pb, which had significantly higher levels in the shoots and leaves of these marshes. Choi et al. (2020), while studying the morphodynamics of the Sittaung River Estuary in Myanmar observed that tidal flats and salt marshes rapidly interconvert among themselves, and they attributed the dominance of tide along with high sediment load behind such observations. The tidal flats situated at the landward side of the salt marshes in this region exhibited varied textural composition. Most of the sites were dominated by sand (60–80%) with absolutely zero contribution of clay. However, in many of the sites silts were predominant followed by clay with a negligible proportion of sand. Studies reporting the physicochemical and hydrogeological settings of the tidal flats and salt marsh sediments are not available for Thailand and Singapore. Mokhtari et al. (2016) studied the role of crab burrows in the tidal mudflats of Indonesia. They observed that crab burrows facilitated oxygen to the anoxic layers of sediment and stimulated remineralization of organic matter and iron reduction. Brown and Rajkaran (2020) exhaustively studied the physicochemical properties of salt marsh sediments in six estuaries of South Africa. The soil moisture content and sediment pH showed a wide spectrum varying from 30.6% to 62.0% and 6.5 to 9.4, respectively. Thus, it is evident that the tidal flats and salt marshes in the Indian Ocean periphery experience a wide variety of physicochemical conditions.

6.4 Species Diversity of Salt Marshes

Figure 6.1 shows the principal locations where species diversity and carbon assessments have been done in the salt marsh and tidal flat habitats. According to Banerjee et al. (2017), almost 15 species have been identified to thrive all along the coastline of India. Viswanathan et al. (2020), in a comparatively recent study, indicated a possibility of only 14 species along the Indian coast (Table 6.2). Viswanathan et al. (2020) mainly categorized the salt marshes as succulents, grasses, and sedges. They observed that tidal flats which are regularly inundated exhibit a dominance of *Porteresia coarctata* and *Myriostachya wightiana*, whereas regions with irregular tidal activities witness a growth of *Suaeda* sp. The highly elevated regions which receive tidal waters only during the spring tide usually exhibit a predominance of *H. curassavicum*, *C. cretica*, and *Suaeda* sp. Gopi et al. (2019) observed similar species composition in the down south of peninsular India. Hena et al. (2007) reported two salt marshes species, namely *Imperata cylindrica* and *Spartina* sp. from the Cox

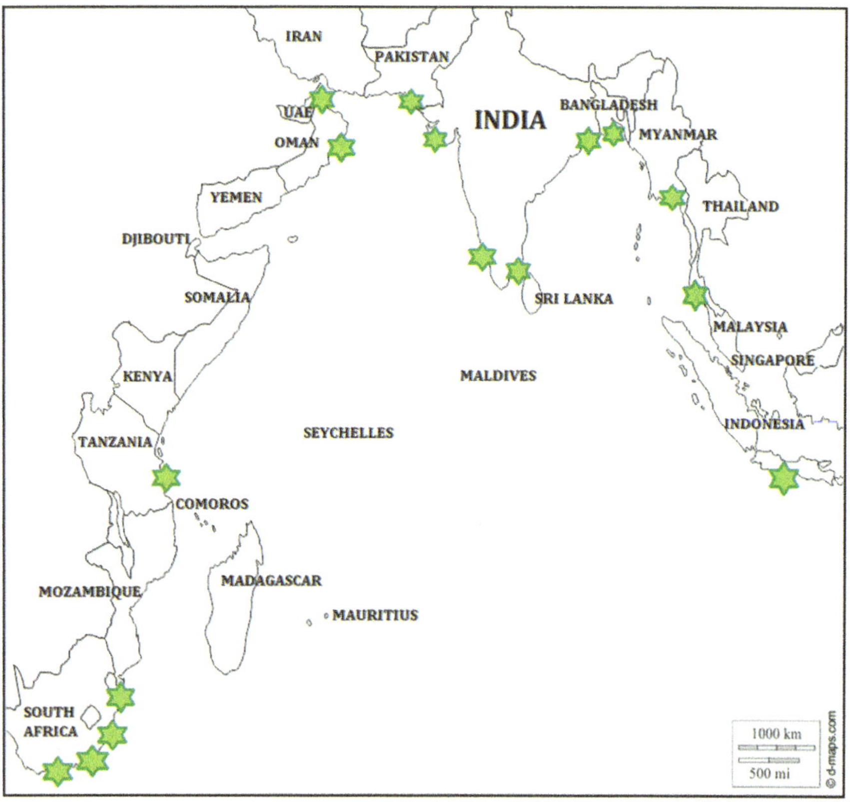

Fig. 6.1 The principal locations where species diversity or carbon stock assessment in the salt marsh and tidal flat habitats has been done

Table 6.2 Dominant salt marsh species thriving in the Indian Ocean countries

Species Name	IN	BD	MM	TH	SG	ID	MY	IR	YM	OM	UAE	SL	MD	SC	MG	CM	KN	MA	MZ	SM	SA	TZ	DJ	PK
Arthrocnemum indicum	+											+							+					+
Cressa cretica	+							+			+													+
Fimbristylis ferruginea	+																							
Heliotropium curassavicum	+																							
Myriostachya wightiana	+	+																						
Porteresia coarctata	+	+																						
Salicornia brachiata	+											+							+					
Scirpus littoralis	+																							
Sesuvium portulacastrum	+																							
Suaeda fruticosa	+							+											+					+
Suaeda maritima	+										+	+							+					
Suaeda monoica	+									+		+							+					
Suaeda nudiflora	+											+							+					
Urochondra setulosa	+								+	+														+
Spartina sp.		+																			+			
Imperata cylindrica		+																						
Eriochloa procera		+																						
Phragmites karka		+						+																
Aeluropus lagopoides								+	+		+													
Halopeplis perfoliata									+		+													
Halopyrum mucronatum									+															+

(continued)

Table 6.2 (continued)

Species Name	IN	BD	MM	TH	SG	ID	MY	IR	YM	OM	UAE	SL	MD	SC	MG	CM	KN	MA	MZ	SM	SA	TZ	DJ	PK
Sevada schimperi									+															
Anabasis ehrenbergii									+															
Arthrocnemum macrostachyum									+	+	+								+					+
Cyperus conglomeratus									+		+													
Suaeda vermiculata										+	+													
Suaeda moschata										+														
Agriophyllum minus											+													
Halocnemum strobilaceum								+			+													
Salicornia europaea											+								+					
Suaeda aegyptiaca								+			+													
Cyperus arenarius											+													
Cyperus laevigatus											+													
Cyperus rotundus								+			+													
Fimbristylis cymosa											+													
Fimbristylis ferruginea											+													
Phragmites australis								+			+								+		+			
Phyla nodiflora											+													
Aeluropus littoralis								+																
Spergula fallax								+																
Salsola jordanicola								+																
Alhagi mannifera								+																

(continued)

Table 6.2 (continued)

Species Name	IN	BD	MM	TH	SG	ID	MY	IR	YM	OM	UAE	SL	MD	SC	MG	CM	KN	MA	MZ	SM	SA	TZ	DJ	PK
melilotus indicus								+																
Atriplex stocksii												+												+
Zygophyllum simplex																								+
Salsola imbricata																								+
Juncus kraussii																					+			
Bassia diffusa																					+			
Cotula coronopifolia																					+			
Limonium linifolium																					+			
Stuckenia pectinata																								
Triglochin striata																					+			
Sporobolus virginicus																			+		+			
Stenotaphrum secundatum																					+			
Cynodon dactylon																			+					
Panicum maximum																			+					
Pennisetum sp.																			+					
Cyperus compressus																			+					
Urochloamosambicensis																			+					
Pemphis acidula																			+					
Suriana maritima																			+					
Ruppia maritima																			+					

Bazar region of Bangladesh. *Porteresia coarctata* is also common in the coastlines of Bangladesh (Hossain et al. 2021). Kamal and Khan (2009) reported 5 species altogether from Bangladesh, which did not include *Spartina* sp. Thus, a total of 6 salt marsh species reportedly exist in Bangladesh. Myanmar, as of now, has no record of salt marsh species in its coastal periphery (Murray et al. 2020). Thailand, Singapore, Indonesia, and Malaysia also have no proper record of salt marsh species distribution. Though the spatial extent of tidal salt marsh remains unknown in Yemen, Kürschner et al. (1998) reported eight salt marsh species from the south of Yemen. Similarly, Ghazanfar (1999) reported at least 5 salt marshes along with other associated salt-tolerant plants from the coasts of Oman. Gairola et al. (2019) reported as many as 18 salt marsh species from the coasts of UAE. Alaie (2001) recorded 13 plant communities on the coasts of Iran. Asri and Ghorbanli (1997) observed similar species composition in an inland salt lake of Iran. Several studies identified the salt marsh species composition along the coastline of Pakistan (Gul and Khan 1995; Khan et al. 2000, 2005; Khan and Gul 2002; Gulzar et al. 2003; Agha et al. 2009; Patro et al. 2017). Altogether these studies indicate the availability of 9 species in Pakistan. Among the African countries, the South African saltmarshes are perhaps the most widely studied (Adams et al. 2012, 2016; Tabot and Adams 2013; Brown et al. 2019; Adams 2020; Raw et al. 2020). Adams et al. (2016) observed that *Stenotaphrum secundatum*, *Juncus kraussii*, *Sporobolus virginicus*, and *Phragmites australis* are the most dominant salt marshes in South Africa; however, altogether, they reported as many as 38 species from the South African estuaries and coastline. Adams and Bandeira (1995) reported a rich salt marsh species diversity from Mozambique comprising as many as 18 species. However, even at the present date, not much is known about the salt marsh species diversity of Tanzania (Gullström et al. 2021). A similar absence of data prevails for the other western Indian Ocean African countries like Kenya, Somalia, and Djibouti. Besides a few anecdotal observations, proper documentation about species diversity is missing from these nations (Lugendo 2016).

Among the island nations of the Indian Ocean, several studies have been conducted in Sri Lanka (Pemadasa et al. 1979; Abeywickrama and Arulgnanam 1993; Mathuranthini 2014; Patro et al. 2017). A total of 6 species have been so far recorded from Sri Lanka. However, nothing is known about the salt marsh species diversity in the Maldives (Shaig 2011; Patro et al. 2017). Similarly, the scenario in Seychelles is also poorly constrained. A report by Senterreet al. (2015) indicated the presence of salt marshes belonging to Cyperaceae and Poaceae families. Adams and Bandeira (1995) and Harding (2019) indicated the possibility of extensive salt marshes in Madagascar; however, no information on species diversity is available as of the year 2021. Comoros and Mauritius have also no proper information on salt marsh species diversity.

6.5 Carbon Stock in Living Biomass of Salt Marshes

While assessment of the spatial extent and species diversity remains poorly constrained, it is obvious that studies focused on the carbon content of salt marshes are even fewer in number from the Indian Ocean rim countries. Kaviarasan et al. (2019) measured the aboveground and belowground biomass and their carbon content in four dominant salt marsh species in the Tuticorin coast, situated in southern India. They observed that the aboveground carbon content in the vegetated structures varied between 0.632 and 1.487 Mg ha^{-1}, whereas the belowground carbon content varied between 0.186 and 0.449 Mg ha^{-1} (Table 6.3). Chaudhary et al. (2018) while studying the biomass dynamics of *Salicornia brachiata* in the Gujarat coast of India, observed that the aboveground biomass carbon ranged from 0.05 to 2.1 Mg ha^{-1}, whereas the belowground biomass carbon ranged from 0.009 to 0.335 Mg ha^{-1}. These observations were in parity with earlier studies conducted in various parts of India (Jagtap et al. 2002; Das et al. 2015; Rathore et al. 2016). Recently, Kaladharan et al. (2020) reported that the salt marshes Gulf of Mannar and Palk Bay in India have a blue carbon stock of around 0.001782 Tg C and 0.043996 Tg C, respectively. Hena et al. (2007), while studying the biomass of the salt marsh species *Sporobolus alterniflorus*, alternatively known as *Spartina alterniflora*, in the Cox Bazar region of Bangladesh observed aboveground biomass varying between 0.43 and 1.66 Mg ha^{-1}, whereas the belowground biomass was 0.078 Mg ha^{-1}. Schile et al. (2017) carried out an exhaustive study across 58 sampling sites bordering the coastline of UAE. They observed that the aboveground and belowground biomass carbon in the salt marsh species varied from 1.0 to 3.8 Mg C ha^{-1} and 0.9 to 3.4 Mg C ha^{-1}, respectively. Thus, the total biomass ranged between 1.9 and 7.2 Mg C ha^{-1}, having a mean of 3.6 Mg C ha^{-1}. Similar observations were made in the Persian Gulf by Almahasheer (2021). Sejati et al. (2020) observed a total blue carbon content (including biomass and sediments) of 17 Mg C ha^{-1} in the Semarang Metropolitan Region (SMR), Central Java, Indonesia.

Among the African countries, Brown and Rajkaran (2020) observed very high carbon content in both the aboveground (5.9–31.4 Mg C ha^{-1}) and belowground (16.7–33.4 Mg C ha^{-1}) biomass of salt marshes at six prominent estuaries of South Africa, namely, Berg, Heuningnes, Kwelerha, Langebaan, Olifants, and Nahoon estuaries. However, the measurement of biomass carbon in the salt marshes of other African countries like Tanzania, Mozambique, Kenya, Somalia, and Djibouti is absent. The Indian Ocean island nations also exhibited an absence of data in this regard.

Table 6.3 Reports of carbon stock in the live biomass of the saltmarshes, bed sediments along with the ecosystem productivity observed in the Indian Ocean countries

Country	Live biomass carbon (Mg C ha^{-1})	Soil organic carbon (Mg C ha^{-1})	Ecosystem productivity	Authors
India (AS)	0.05–2.1 (AGB) 0.009–0.449 (BGB)	8.42–54.46	ND	Chaudhary et al. (2018), Kaviarasan et al. (2019)
Bangladesh (AS)	0.43–1.66 (AGB) 0.078 (BGB)	ND	5.93–8.619 g m^{-2} d^{-1}	Podder et al. (2012)
Myanmar (AS)	ND	ND	ND	
Thailand (AS)	ND	ND	40–75 mmol O_2 m^{-2} d^{-1} (TF)	Holmer et al. (2001)
Singapore (AS)	ND	ND	ND	
Indonesia (AS)	17	6 (TF)	7.4×10^3 Mg C km^{-2} yr^{-1} (TF)	Yamamoto et al. (2019), Sejati et al. (2020)
Malaysia (AS)	ND	ND	ND	
Yemen (AS)	ND	ND	ND	
Oman (AS)	ND	ND	ND	
UAE (AS)	1.9–7.2 (SM)	29.5–163.7 (SM) 51.0–120.5 (TF)	ND	Schile et al. (2017)
Iran (AS)	ND	ND	ND	
Pakistan (AS)	ND	51.7 ± 3.9 (TF)	ND	Nazim et al. (2018)
	ND	ND	ND	
South Africa (AF)	5.9–31.4 (AGB) 16.7–33.4 (BGB)	2.61–109.6	ND	Raw et al. (2019), Banda et al. (2021), Brown and Rajkaran (2020)
Mozambique (AF)	1–13 (TF)	ND	ND	Gullström et al. (2021)
Tanzania (AF)	2–8 (TF)	ND	ND	Gullström et al. (2021)
Kenya (AF)	ND	ND	ND	
Somalia (AF)	ND	ND	ND	
Djibouti (AF)	ND	ND	ND	

(continued)

Table 6.3 (continued)

Country	Live biomass carbon (Mg C ha^{-1})	Soil organic carbon (Mg C ha^{-1})	Ecosystem productivity	Authors
	ND	ND	ND	
Sri Lanka (AS)	ND	ND	ND	
Maldives (AS)	ND	ND	ND	
Seychelles (AF)	ND	ND	ND	
Madagascar (AF)	ND	ND	ND	
Comoros (AF)	ND	ND	ND	
Mauritius (AF)	ND	ND	ND	

ND no data; *SM* saltmarsh; *TF* tidal flats; *AGB* aboveground biomass; *BGB* belowground biomass

6.6 Tidal Flats and Salt Marsh Sediments: A Strong Carbon Sink

Compared to other blue carbon ecosystems like that of mangroves, tidal flats and salt marshes have received far less attention from the viewpoint of quantifying soil organic carbon. Kaviarasan et al. (2019), while studying the salt marsh dynamics in the Tuticorin coast, South India, observed a varying soil organic carbon content ranging from 4.13 ± 0.08% to 7.23 ± 0.09%, which was equivalent to 8.42 ± 0.64 Mg ha^{-1} to 54.46 ± 1.46 Mg ha^{-1} (Table 6.3). Sivakumar et al. (2014) observed an average soil organic carbon of 4.28% in the salt marshes adjoining the Muthupet mangroves of India. However, Das et al. (2015) reported a comparatively lower range of soil organic carbon (1.04–1.49%) from the eastern Indian salt marshes. Aysha et al. (2015) observed a sedimentation rate of 4.9–7.5 mg cm^{-2} d^{-1} and 48.8–135.5 mg cm^{-2} d^{-1} in the low marshes and high marshes, respectively, in the Sitakunda coast, Chittagong, Bangladesh. In the same region, they observed a sedimentation rate varying between 9.8 and 9.9 mg cm^{-2} d^{-1} in the bare mudflats. Aysha et al. (2015) further observed that the mean sediment organic carbon content in the unvegetated mudflats and salt marsh beds was 2.0 ± 0.3% and 1.8 ± 0.5%, respectively. Yamamoto et al. (2019), while studying the tidal mudflats of Bengkalis island in Indonesia observed peat formation of thick layers sandwiched between clay layers, and the carbon accumulation rate in these peat layers (7.4 Mg C km^{-2} yr^{-1}) was substantially higher than that observed in the adjacent mangrove soils (1.7 Mg C km^{-2} yr^{-1}). Sejati et al. (2020) observed a blue carbon content of 6 Mg C ha^{-1} in the Semarang Metropolitan Region (SMR), Central Java, Indonesia. Mokhtari et al. (2016) did not report carbon accumulation rates or soil organic carbon pool in the mudflats of Malaysia; however, they observed that a higher abundance of crabs, which for burrows lead to oxidation of organic matter and a lower amount of carbon compared to regions which were devoid of crab burrows. Schile et al. (2017), while studying the salt marshes and the tidal flats encompassing the entire coastline of UAE observed a similar soil organic carbon pool in the bare tidal flats (51.0–120.5 Mg C ha^{-1}; mean: 82.4 Mg C ha^{-1}) and the vegetated salt marsh sediments (29.5–163.7 Mg C ha^{-1}; mean: 80.4 Mg C ha^{-1}). However, the carbon concentrations were substantially less than that observed in the adjacent mangrove and microbial mat sediments. The total carbon stock including the carbon locked in the vegetated salt marsh patches (31.4–205.0 Mg C ha^{-1}; mean: 97.1 Mg C ha^{-1}) was higher than the bare mudflats (51.0–120.5 Mg C ha^{-1}; mean: 82.4 Mg C ha^{-1}). Similar observations were recorded by Cusack et al. (2018) and Ashok et al. (2019) in the Persian Gulf adjoining Saudia Arabia and Qatar. Nazim et al. (2018), while working in the Sandspit sediments (bare intertidal mudflats) of Pakistan observed a soil organic carbon stock of 51.7 ± 3.9 Mg C ha^{-1}; however, data on salt marsh sediments could not be retrieved.

Among the African nations, South Africa perhaps received the highest attention in quantifying carbon stocks in salt marsh ecosystems. Raw et al. (2019), while working in the Nahoon Estuary of South Africa observed similar sediment carbon content in the mangrove (110.1 Mg C ha^{-1}), salt marsh (109.6 Mg C ha^{-1}), and the intermediate ecotone (114.5 Mg C ha^{-1}) regions. However, Banda et al. (2021) reported a soil organic carbon content as low as 2.61 Mg C ha^{-1} from a warm-temperate estuary of South Africa. Anecdotal information indicates the presence of salt marshes in Mozambique and Tanzania; however, proper assessment of their carbon stock is still missing. Gullström et al. (2021) tabulated that carbon stock in unvegetated coastal mudflats varies from 2 to 8 Mg C ha^{-1} (mean: 6 Mg C ha^{-1}) and 1 to 13 Mg C ha^{-1} (mean: 5 Mg C ha^{-1}) in Tanzania and Mozambique, respectively. However, such assessments from other Western Indian Ocean African countries like Kenya, Somalia, and Djibouti are absent. The island nations in the periphery of the Indian Ocean are also devoid of any data in this regard.

6.7 Salt Marsh Ecosystem Productivity

Very few studies are conducted all over the world to quantify the ecosystem productivity in these crucial marine ecosystems. Podder et al. (2012) studied the productivity of one of the most abundant salt marsh species in Bangladesh, namely *Porteresia coarctata*. This species is also very common in the adjoining Indian Sundarban. They observed a plant growth rate varying between 5.93 and 8.62 g m^{-2} d^{-1}. They further observed that the plant growth rate increased with aging; however, the net assimilation rate in the salt marshes decreased with age. Holmer et al. (2001) observed a productivity rate of 40–75 mmol O_2 m^{-2} d^{-1} in the waters lying above the tidal flats of Thailand.

6.8 Aquatic Biogeochemistry in the Water Column Above the Tidal Flats and Saltmarshes

Indiscriminate pollutant load flows through the salt marshes and tidal flats all over the world. Mia et al. (2021) recently observed the physicochemical dynamics in the water column lying above the salt marsh beds adjoining the Karnafully River Estuary of Bangladesh. They observed a very low dissolved oxygen range varying between 0.5 and 2.7 mg l^{-1} accompanied by very high biochemical oxygen demand levels of 20.7–69.7 mg l^{-1}. The pH in the water column and the porewater salinity ranged from 6.2 to 7.3 and 1.9 to 6.9, respectively. The water temperature in this region ranged between 25.1 and 31.5 °C. However, in the Moheshkhali area of Bangladesh, water above salt marsh beds exhibited a temperature range of 24–28 °C having a substantially high salinity of 18–33 (Sarkar et al. 2018). However, the pH (6.4–7.6) and dissolved oxygen (0.9–2.0 mg $1 l^{-1}$) ranges were in parity with the observations of Mia et al. (2021). Sarkar et al. (2018) further inferred that the nutrient content in the adjoining water column and the mineral content in the salt marshes exhibited a significant positive correlation. Behera et al. (2013) extensively studied the water quality in the West Johor Strait, which flows between Singapore and Malaysia, and covers wide tidal flats on both banks (Chan et al. 2006). They observed a water temperature, salinity, chlorophyll-a, and turbidity range of 28–30.5 °C, 27–31, 4–13 mg m^{-3}, and 0.5–8 NTU, respectively, all round the year. Schile et al. (2017), while analyzing the pore waters of salt marsh sediments and tidal flats observed significantly higher salinities compared to open ocean water. The salinity levels in the salt marsh pore waters of UAE were as high as 50; whereas in the tidal flats, these values varied between 150 and 250. They also observed significantly reduced sulfate levels in the salt marsh pore waters compared to that observed in the bare tidal flats, where the dominance of seawater was prominent with much higher sulfate levels. Brown and Rajkaran (2020) observed that the pore water salinity in the salt marshes of South African Estuaries varied between 25.8 and 44.5, whereas the pore water pH ranged from 6.9 to 8.5. Thus, it is evident that adjoining water mass in the tidal flats and salt marshes exhibits a wide spectrum in terms of the range of biogeochemical parameters. Figure 6.2 shows the biogeochemical attributes in the sediment and the aquatic column adjoining the salt marshes and tidal flats in the Indian Ocean periphery.

6.9 Factors Regulating the Carbon Stock in Salt Marshes and Tidal Flats

In the case of salt marsh, root decomposition of the vegetated structures plays a crucial role in enhancing the soil organic carbon pool. Ouyang et al. (2017) indicated that root decomposition depends on ambient temperature, average annual rainfall, and the type of terrain, where these salt marshes grow. They estimated a global carbon burial rate

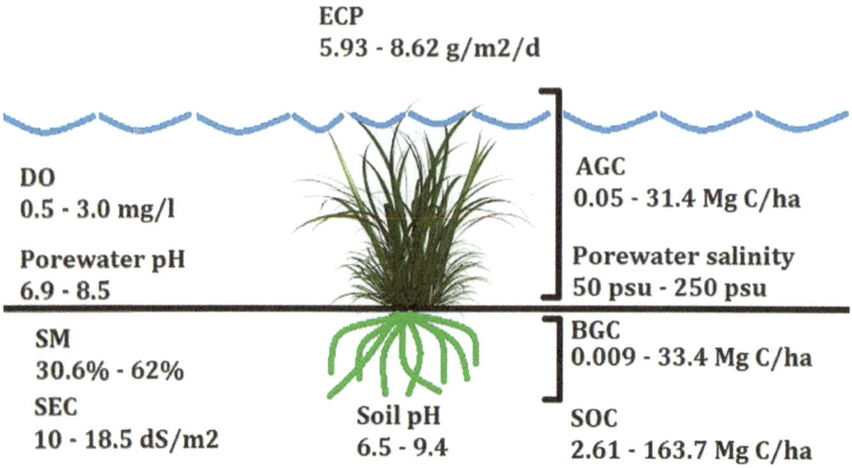

Fig. 6.2 The biogeochemical attributes observed in the sediments and the aquatic column adjoining the salt marshes and the tidal flats in the Indian Ocean periphery

of 191 g C m^{-2} yr^{-1} in the salt marsh ecosystems due to root decomposition; however, no observations from the Indian Ocean periphery were present in their study. Apple et al. (2006) also emphasized the role of ambient temperature and water temperature in regulating the microbial community respiration as well as the growth efficiency of the salt marshes. Salt marshes and tidal flats usually exist at hydrodynamically active regions of the coast. The tidal amplitudes, wave disturbances, sediment supply (both toward the flats and marshes and away from it) determine the carbon accrual in these ecosystems (Caçador et al. 2016). Considerably fast water currents often facilitate the outwelling of trapped carbon from these ecosystems to the open oceans (Townend et al. 2011), where they remineralize and transform to gaseous CO_2 and revert toward the atmosphere. Erosion and accretion dynamics also play a significant role in regulating the salt marsh regimes in the coastal periphery (Cao et al. 2018). Relative sea-level rise, an ongoing phenomenon, can significantly alter the spatial extent as well as the species diversity of the salt marshes (Sharpe and Baldwin 2012). The salt marshes in the future could undergo a reduction in species diversity due to the rising sea level and this would lead to more prominent erosion activities, thus creating a vicious circle (Valentim et al. 2013). However, changing paradigms in the coastal morphology can witness shifting extents of salt marshes into unvegetated coastal regions, and that can lead to increment in the carbon stock of such regions (as observed by Xuehui et al. 2021). Though coastal pollution can prove to be detrimental for any marine ecosystems, recent studies indicate that tidal salt marsh habitats can store substantial quantities of both autochthonous and allochthonous (anthropogenic) organic matters in their sediment column (Kumar et al. 2020), thus, acts as a carbon trapper and increases the soil organic carbon pool. Many more studies have focused on varying governing factors that regulate the carbon stock in salt marsh habitats; however, such studies are scarce from the Indian Ocean periphery.

6.10 Scope of Future Research

Ever since the coining of the term blue carbon, salt marshes, as well as unvegetated tidal flats, have received substantial attention; however, in the Indian Ocean periphery, the knowledge on these two crucial ecosystems is still in its infancy. Many of the countries still do have a proper ground estimate on the spatial extent of these ecosystems. Synoptic scale remotely sensed estimates on a global scale have come up; however, to reduce the uncertainties, meticulous ground-truthing is very essential. Collating the existing database, the present chapter indicated that the physicochemical conditions and the hydrogeological setting of these ecosystems exhibit a wide variety. All the countries should strive hard to characterize their respective morphological attributes, where these ecosystems flourish to better understand and design conservation measures to safeguard these habitats. Salt marshes are conventional blue carbon ecosystems, yet many countries have no data at all on the carbon stock of this vegetation and the underlying sediments. Keeping the ongoing climate change in mind, studies to understand the future species composition, expected variability in spatial extent, and most essentially blue carbon stock dynamics have become a dire need. Indian Ocean periphery offers tremendous potential to mitigate climate change by offsetting the anthropogenic carbon emissions; however, many of the countries in this rim are unaware of this potential or negligent in quantifying the carbon stocks of these essential ecosystems. In the future, the blue economy can play a crucial role in safeguarding the crucial blue carbon ecosystems. However to actively participate in such a carbon financing network, a base knowledge of the national scale blue carbon stock is an utmost necessity. Without proper baseline information, it is always difficult to set goals for conservation and restoration programs. Common man in many places perceives salt marshes to a large extent as wastelands being ignorant of their potential and ecosystem services. Such ignorance leads to the widespread conversion of these crucial habitats to other landforms. Local or even national governments should conduct awareness programs to educate the common mass about the utility and natural values of these ecosystems. Overall, the Indian Ocean salt marshes require much more attention from multidisciplinary researchers to uplift the knowledge base.

6.11 Summary and Conclusion

The American continents, China, and Australia along with few other European countries have carried out substantial research on the salt marshes; however, in the Indian Ocean rim, there is an acute scarcity of data from various angles concerning the tidal flats and salt marshes. Accurate demarcation of any ecosystem is a basic prerequisite to have a holistic knowledge of its habitat and characteristics. Out of the 24 Indian Ocean nations considered in this chapter, 11 nations have no proper estimation of the areal extent of salt marshes. Mapping of unvegetated tidal flats is even more difficult than mapping salt marshes. The coastal peripheries of the respective countries should pay more attention to mapping both these habitats and understand their temporal dynamics. Almost 14 out of the 24 countries have no proper record of the salt marsh species composition or of the most dominant species, except few anecdotal reports. Few countries have documented the physicochemical and hydrogeological conditions of these habitats; however, such data is missing in most of the other Indian Ocean countries. The existing literature shows substantial and at the varying carbon stock of 0.009–33.4 Mg C ha^{-1} in the biomass and 2.61–163.7 Mg C ha^{-1} in the sediment column of these ecosystems. However, only 7 out of 24 countries have measured data on salt marsh biomass, whereas only 5 countries recorded soil organic carbon data. Assessments of ecosystem productivity are even more scarce. Thus, it is evident that there is still a lot of work and scope of research in this region. These ecosystems require and most essentially deserve more attention from the scientific community. Future research can unravel many more exciting findings that can show us avenues to fight against the evil of climate change through conserving these ecosystems.

References

Abbas S, Qamer FM, Ali G, Tripathi NK, Shehzad K, Saleem R, Gilani H (2013) An assessment of status and distribution of mangrove forest cover in Pakistan. J Biodivers Environ Sci 3(6):64–78

Abeywickrama BA, Arulgnanam P (1993) The coastal plants of Sri Lanka, part I (Salt Marsh Plants). UNESCO: Man and the Biosphere National Committee for Sri Lanka. Publication No. 20, Natural Resources, Energy and Science Authority of Sri Lanka, Colombo, p 19

Adams JB (2020) Salt marsh at the tip of Africa: patterns, processes and changes in response to climate change. Estuar Coast Shelf Sci 237:106650. https://doi.org/10.1016/j.ecss.2020.106650

Adams JB, Veldkornet D, Tabot P (2016) Distribution of macrophyte species and habitats in South African estuaries. S Afr J Bot 107:5–11. https://doi.org/10.1016/j.sajb.2016.08.001

Adams J, Bandeira S (1995) Salt marshes. Nairobi Convention, Kenya

Adams JB, Grobler A, Rowe C, Riddin T, Bornman TG, Ayres DR (2012) Plant traits and spread of the invasive salt marsh grass, Spartina alterniflora Loisel., in the Great Brak Estuary, South Africa. Afr J Mar Sci 34(3):313–322. https://doi.org/10.2989/1814232X.2012.725279

Agha F, Gul B, Khan MA (2009) Seasonal variation in productivity of *Atriplex stocksii* from a coastal marsh along the Arabian Sea coast. Pak J Bot 41(3):1053–1068

Akhand A, Chanda A, Watanabe K, Das S, Tokoro T, Hazra S, Kuwae T (2021) Reduction in riverine freshwater supply changes inorganic and organic carbon dynamics and air–water CO_2 fluxes in a tropical mangrove dominated estuary. J Geophys Res Biogeosci e2020JG006144. https://doi.org/10.1029/2020JG006144

Alaie E (2001) Salt marshes and salt deserts of SW Iran. Pak J Bot 33(1):77–92

Al-Farraj A (2005) An evolutionary model for sabkha development on the north coast of the UAE. J Arid Environ 63(4):740–755. https://doi.org/10.1016/j.jaridenv.2005.01.025

Almahasheer H (2021) Assessment of coastal salt marsh plants on the Arabian Gulf Region. Saudi J Biol Sci. https://doi.org/10.1016/j.sjbs.2021.06.002

Angamuthu B, Darby SE, Nicholls RJ (2018) Impacts of natural and human drivers on the multi-decadal morphological evolution of tidally-influenced deltas. Proc R Soc A 474(2219):20180396. https://doi.org/10.1098/rspa.2018.0396

Apple JK, Del Giorgio PA, Kemp WM (2006) Temperature regulation of bacterial production, respiration, and growth efficiency in a temperate salt-marsh estuary. Aquat Microb Ecol 43(3):243–254. https://doi.org/10.3354/ame043243

Arndt E, Fricke R (2019) Intertidal fishes of Mauritius with special reference to shallow tidepools. Biodivers Data J 7. https://doi.org/10.3897/BDJ.7.e36754

Ashok A, Cusack M, Saderne V, Krishnakumar PK, Rabaoui L, Qurban MA et al (2019) Accelerated burial of petroleum hydrocarbons in Arabian Gulf blue carbon repositories. Sci Total Environ 669:205–212. https://doi.org/10.1016/j.scitotenv.2019.01.437

Asri Y, Ghorbanli M (1997) The halophilous vegetation of the Orumieh lake salt marshes, NW. Iran. Plant Ecol 132(2):155–170. https://doi.org/10.1023/A:1009790901167

Aung PP, Moses S, Clark NA, Anderson GQ, Hilton GM, Buchanan GM et al (2020) Recent changes in the number of spoon-billed sandpipers *Calidris pygmaea* wintering on the Upper Gulf of Mottama in Myanmar. Oryx 54(1):23–29. https://doi.org/10.1017/S0030605318000698

Aysha A, Abu Hena MK, Mishra M, Nesarul MH, Padhi BK, Mishra SK et al (2015) Sediment and carbon accumulation in sub-tropical salt marsh and mangrove habitats of north-eastern coast of Bay of Bengal, Indian Ocean. Int J Fish Aquat Stud 2(4):184–189

Bai J, Zhang G, Zhao Q, Lu Q, Jia J, Cui B, Liu X (2016) Depth-distribution patterns and control of soil organic carbon in coastal salt marshes with different plant covers. Sci Rep 6(1):1–12. https://doi.org/10.1038/srep34835

Balasuriya A (2018) Coastal area management: biodiversity and ecological sustainability in Sri Lankan perspective. In: Sivaperuman C, Singh AK (eds) Biodiversity and climate change adaptation in Tropical Islands. Academic Press, pp 701–724. https://doi.org/10.1016/B978-0-12-813064-3.00025-9

Banda SP, Adams JB, Rajkaran A, Johnson JL, Raw JL (2021) Blue carbon storage comparing mangroves with saltmarsh and seagrass habitats at a warm temperate continental limit. In: Sidik F, Friess DA (eds) Dynamic sedimentary environments of mangrove coasts. Elsevier, pp 447–471. https://doi.org/10.1016/B978-0-12-816437-2.00008-2

Banerjee K, Sappal SM, Ramachandran P, Ramesh R (2017) Salt marsh: ecologically important, yet least studied blue carbon ecosystems in India. J Clim Chang 3(2):59–72. https://doi.org/10.3233/JCC-170014

Behera MR, Chun C, Palani S, Tkalich P (2013) Temporal variability and climatology of hydrodynamic, water property and water quality parameters in the West Johor Strait of Singapore. Mar Pollut Bull 77(1–2):380–395. https://doi.org/10.1016/j.marpolbul.2013.09.043

Bernal B, Mitsch WJ (2012) Comparing carbon sequestration in temperate freshwater wetland communities. Glob Chang Biol 18(5):1636–1647. https://doi.org/10.1111/j.1365-2486.2011.02619.x

Botto F, Iribarne O (2000) Contrasting effects of two burrowing crabs (*Chasmagnathus granulata* and *Uca uruguayensis*) on sediment composition and transport in estuarine environments. Estuar Coast Shelf Sci 51(2):141–151. https://doi.org/10.1006/ecss.2000.0642

Brown CE, Rajkaran A (2020) Biomass partitioning in an endemic southern African salt marsh species *Salicornia tegetaria* (Chenopodiaceae). Afr J Aquat Sci 45(1–2):41–48. https://doi.org/10.2989/16085914.2019.1687419

Brown CE, Veldkornet DA, Boatwright JS, Engelbrecht A, Rajkaran A (2019) The phylogeography and genetic diversity of the salt marsh species *Salicornia tegetaria* (S. Steffen, L. Mucina and G. Kadereit) Piirainen and G. Kadereit, endemic to South Africa, Namibia and Mozambique. S Afr J Bot 123:270–277. https://doi.org/10.1016/j.sajb.2019.03.007

Buechley ER, de la Cruz MA, Roman JR, Caucal G, Rayaleh H (2019) Notable bird observations for Djibouti, including the first record of Semi-collared Flycatcher Ficedula semitorquata. Bull Afr Bird Club 26(2):179–185

Burns CJ, Alber M, Alexander CR (2021) Historical changes in the vegetated area of salt marshes. Estuar Coast 44(1):162–177. https://doi.org/10.1007/s12237-020-00781-6

Buzzi NS, Fernández EM, Severini MDF, Truchet DM, Gilabert AS, Villagran DM, Spetter CV (2021) Environmental quality assessment by multiple biogeochemical indicators of an intertidal flat under anthropogenic influence from the southwest of Buenos Aires (Argentina). Environ Earth Sci 80(7):1–17. https://doi.org/10.1007/s12665-021-09438-4

Caçador I, Duarte B, Marques JC, Sleimi N (2016) Carbon mitigation: a salt marsh ecosystem service in times of change. In: Khan MA, Ozturk M, Gul B, Ahmed MZ (eds) Halophytes for food security in dry lands. Academic Press, pp 83–110. https://doi.org/10.1016/B978-0-12-801854-5.00006-6
Cao H, Zhu Z, Balke T, Zhang L, Bouma TJ (2018) Effects of sediment disturbance regimes on Spartina seedling establishment: implications for salt marsh creation and restoration. Limnol Oceanogr 63(2):647–659. https://doi.org/10.1002/lno.10657
Carbone F, Accordi G (2000) The Indian Ocean coast of Somalia. Mar Pollut Bull 41(1–6):141–159. https://doi.org/10.1016/S0025-326X(00)00107-7
Chakrabarti A (2005) Sedimentary structures of tidal flats: a journey from coast to inner estuarine region of eastern India. J Earth Syst Science 114(3):353–368. https://doi.org/10.1007/BF02702954
Chan ES, Tkalich P, Gin KYH, Obbard JP (2006) The physical oceanography of Singapore coastal waters and its implications for oil spills. In: Wolanski E (ed) The environment in Asia Pacific Harbours. Springer, Netherlands, pp 393–412. https://doi.org/10.1007/1-4020-3655-8_23
Chaudhary DR, Rathore AP, Jha B (2018) Aboveground, belowground biomass and nutrients pool in *Salicornia brachiata* at coastal area of India: interactive effects of soil characteristics. Ecol Res 33(6):1207–1218. https://doi.org/10.1007/s11284-018-1634-9
Chen J, Wang D, Li Y, Yu Z, Chen S, Hou X et al (2020) The carbon stock and sequestration rate in tidal flats from coastal China. Glob Biogeochem Cycl 34(11):e2020GB006772. https://doi.org/10.1029/2020GB006772
Chmura GL (2013) What do we need to assess the sustainability of the tidal salt marsh carbon sink? Ocean Coast Manag 83:25–31. https://doi.org/10.1016/j.ocecoaman.2011.09.006
Chmura GL (2009) Tidal salt marshes. The management of natural coastal carbon sinks. The management of natural costal carbon sinks. IUCN, Gland, Switzerland, p 53
Choi K, Kim D, Jo J (2020) Morphodynamic evolution of the macrotidal Sittaung River estuary, Myanmar: tidal versus seasonal controls. Mar Geol 430:106367. https://doi.org/10.1016/j.margeo.2020.106367
Clark S, Edwards AJ (1994) Use of artificial reef structures to rehabilitate reef flats degraded by coral mining in the Maldives. Bull Mar Sci 55(2–3):724–744
Correll MD, Wiest WA, Hodgman TP, Shriver WG, Elphick CS, McGill BJ et al (2017) Predictors of specialist avifaunal decline in coastal marshes. Conserv Biol 31(1):172–182. https://doi.org/10.1111/cobi.12797
Costa CS, Iribarne OO, Farina JM (2009) Human impacts and threats to the conservation of South American salt marshes. In: Silliman BR, Grosholz ED, Bertness MD (eds) Human impacts on salt marshes: a global perspective. University of California Press, Berkeley, pp 337–360
Cusack M, Saderne V, Arias-Ortiz A, Masque P, Krishnakumar PK, Rabaoui L et al (2018) Organic carbon sequestration and storage in vegetated coastal habitats along the western coast of the Arabian Gulf. Environ Res Lett 13(7):074007. https://doi.org/10.1088/1748-9326/aac899
Das S, Zaman S, Pramanick P, Pal N, Mitra A (2015) *Suaeda maritima*: a potential carbon reservoir of coastal zone. Int Adv Res J Sci Eng Technol 2(5):61–65
Davidson KE, Fowler MS, Skov MW, Doerr SH, Beaumont N, Griffin JN (2017) Livestock grazing alters multiple ecosystem properties and services in salt marshes: a metaanalysis. J Appl Ecol 54(5):1395–1405. https://doi.org/10.1111/1365-2664.12892
Dladla NN, Green AN, Cooper JAG, Humphries MS (2019) Geological inheritance and its role in the geomorphological and sedimentological evolution of bedrock-hosted incised valleys, lake St Lucia, South Africa. Estuar Coast Shelf Sci 222:154–167. https://doi.org/10.1016/j.ecss.2019.04.013
Dong H, Qian L, Yan J, Wang L (2020) Evaluation of the carbon accumulation capability and carbon storage of different types of wetlands in the Nanhui tidal flat of the Yangtze River estuary. Environ Monit Assess 192(9):1–12. https://doi.org/10.1007/s10661-020-08547-0

Dontis EE, Radabaugh KR, Chappel AR, Russo CE, Moyer RP (2020) Carbon storage increases with site age as created salt marshes transition to mangrove forests in Tampa Bay, Florida (USA). Estuar Coast 43(6):1470–1488. https://doi.org/10.1007/s12237-020-00733-0
Endo T, Otani S (2019) Carbon storage in tidal flats. In: Kuwae T, Hori M (eds) Blue carbon in shallow coastal ecosystems. Springer, Singapore, pp 129–151. https://doi.org/10.1007/978-981-13-1295-3_5
Fan Y, Chen S, Zhao B, Yu S, Ji H, Jiang C (2018) Monitoring tidal flat dynamics affected by human activities along an eroded coast in the Yellow River Delta, China. Environ Monit Assess 190(7):1–17. https://doi.org/10.1007/s10661-018-6747-7
Ferreira MA, Andrade F, Cardoso P, Paula J (2009) Coastal habitat mapping along the Tanzania/Mozambique transboundary area using Landsat 5 TM imagery. West India Ocean J Mar Sci 8(1). https://doi.org/10.4314/wiojms.v8i1.56670
Flores-Verdugo FJ, Agraz-Hernández C, Benitez-Pardo D (2007) Ecosistemas acuáticos costeros: Importancia, retos y prioridades para su conservación. In: Sánchez O, Herzig M, Peters E et al (eds) Perspectivas sobre conservación de ecosistemas acuáticos en México. Secretaria de Medio Ambiente y Recursos Naturales. Instituto Nacional de Ecología, México, pp 145–147
Friese JD, Temming A, Dänhardt A (2021) Preference, avoidance, or coincidence? How fish and crustaceans use intertidal salt-marsh creeks in the German Wadden Sea. Estuar Coast Shelf Sci 255:107297. https://doi.org/10.1016/j.ecss.2021.107297
Gairola S, Mahmoud T, Shabana HA (2019) Diversity and distribution of salt-tolerant plants of the United Arab Emirates: perspectives for sustainable utilization and future research. In: Gul B, Böer B, Khan M, Clüsener-Godt M, Hameed A (eds) Sabkha Ecosystems. Springer, Cham, pp. 349–362. https://doi.org/10.1007/978-3-030-04417-6_21
Garg JK, Singh TS, Murthy TVR (1998) Wetlands of India. Project Report: RSAM/SAC/RESA/PR/01/98. Space Applications Centre (ISRO), Ahmedabad, p 239
Gerlach J (1993) The conservation of Silhouette Island, Seychelles. Phelsuma 1:18–29
Ghadeer SG, Macquaker JH (2012) The role of event beds in the preservation of organic carbon in fine-grained sediments: analyses of the sedimentological processes operating during deposition of the Whitby Mudstone Formation (Toarcian, Lower Jurassic) preserved in northeast England. Mar Petrol Geol 35(1):309–320. https://doi.org/10.1016/j.marpetgeo.2012.01.001
Ghazanfar SA (1999) Coastal vegetation of Oman. Estuar Coast Shelf Sci 49:21–27. https://doi.org/10.1016/S0272-7714(99)80004-3
Gopi M, Kumar MP, Jeevamani JJJ, Raja S, Muruganandam R, Samuel VD et al (2019) Distribution and biodiversity of tropical saltmarshes: Tamil Nadu and Puducherry, Southeast coast of India. Estuar Coast Shelf Sci 229:106393. https://doi.org/10.1016/j.ecss.2019.106393
Gul B, Khan MA (1995) Productivity of an *Arthrocnemum indicum* dominated coastal salt marsh at Karachi, Pakistan. In: Khan MA, Ungar IA (eds) Biology of salt-tolerant plants. University of Karachi, Karachi, pp 99–105
Gullström M, Dahl M, Lindén O, Vorhies F, Forsberg S, Ismail RO, Björk M (2021) Coastal blue carbon stocks in Tanzania and Mozambique: support for climate adaptation and mitigation actions. IUCN, Gland, p 80
Gulzar S, Khan MA, Ungar IA (2003) Salt tolerance of a coastal salt marsh grass. Commun Soil Sci Plant Anal 34(17&18):2595–2605. https://doi.org/10.1081/CSS-120024787
Guo L, Su N, Zhu C, He Q (2018) How have the river discharges and sediment loads changed in the Changjiang River basin downstream of the Three Gorges Dam? J Hydrol 560:259–274. https://doi.org/10.1016/j.jhydrol.2018.03.035
Harding S (2019) Madagascar. In Sheppard C (ed) World Seas: an Environmental Evaluation, Academic Press, pp 121–144. https://doi.org/10.1016/B978-0-08-100853-9.00005-1
Hashemi S, Javaherian A, Ataee-pour M, Tahmasebi P, Khoshdel H (2014) Channel characterization using multiple-point geostatistics, neural network, and modern analogy: a case study from a carbonate reservoir, southwest Iran. J Appl Geophys 111:47–58. https://doi.org/10.1016/j.jappgeo.2014.09.015

Healy T, Wang Y, Healy J (2002) Muddy coasts of the world: processes, deposits, and function. Elsevier Science, Amsterdam

Hena MA, Short FT, Sharifuzzaman SM, Hasan M, Rezowan M, Ali M (2007) Salt marsh and seagrass communities of Bakkhali Estuary, Cox's Bazar, Bangladesh. Estuar Coast Shelf Sci 75(1–2):72–78. https://doi.org/10.1016/j.ecss.2007.01.022

Heyvaert VMA, Baeteman C (2007) Holocene sedimentary evolution and palaeocoastlines of the Lower Khuzestan plain (southwest Iran). Mar Geol 242(1–3):83–108. https://doi.org/10.1016/j.margeo.2007.01.008

Hill NK, Woodworth BK, Phinn SR, Murray NJ, Fuller RA (2021) Global protected-area coverage and human pressure on tidal flats. Conserv Biol 35(3):933–943. https://doi.org/10.1111/cobi.13638

Hladik C, Schalles J, Alber M (2013) Salt marsh elevation and habitat mapping using hyperspectral and LIDAR data. Remote Sens Environ 139:318–330. https://doi.org/10.1016/j.rse.2013.08.003

Holmer M, Andersen FØ, Holmboe N, Kristensen E, Thongtham N (2001) Spatial and temporal variability in benthic processes along a mangrove-seagrass transect near the Bangrong Mangrove, Thailand. Wetl Ecol Manag 9(2):141–158. https://doi.org/10.1023/A:1011100300357

Hopkinson CS, Wolanski E, Brinson MM, Cahoon DR, Perillo GME (2018) Coastal wetlands: a synthesis. In: Perillo GME, Wolanski E, Cahoon DR, Hopkinson CS (eds) Coastal Wetlands: an integrated ecosystem approach. Elsevier, Amsterdam, pp 1–75

Hossain MB, Rakib MRJ, Jolly YN, Rahman M (2021) Metals uptake and translocation in salt marsh macrophytes, Porteresia sp. from Bangladesh coastal area. Sci Total Environ 764:144637. https://doi.org/10.1016/j.scitotenv.2020.144637

https://wedocs.unep.org/bitstream/handle/20.500.11822/25894/Seychelles_MEDA.pdf?sequence=1&isAllowed=y. Accessed on 7 June 2021

https://www.cbd.int/doc/meetings/mar/ebsa-sio-01/other/ebsa-sio-01-comoros-en.pdf. Accessed 7 June 2021

Hu Y, Tian B, Yuan L, Li X, Huang Y, Shi R et al (2021) Mapping coastal salt marshes in China using time series of Sentinel-1 SAR. ISPRS J Photogram Remote Sens 173:122–134. https://doi.org/10.1016/j.isprsjprs.2021.01.003

Ijaz MW, Mahar RB, Siyal AA, Anjum MN (2018) Geospatial analysis of creeks evolution in the Indus Delta, Pakistan using multi-sensor satellite data. Estuar Coast Shelf Sci 200:324–334. https://doi.org/10.1016/j.ecss.2017.11.025

Islam MA, Paull DJ, Griffin AL, Murshed S (2020) Assessing ecosystem resilience to a tropical cyclone based on ecosystem service supply proficiency using geospatial techniques and social responses in coastal Bangladesh. Int J Disaster Risk Reduct 49:101667. https://doi.org/10.1016/j.ijdrr.2020.101667

Jagtap TG, Bhosale SH, Nagle VL (2002) Ecological observations on major Salicornia beds from highly saline coastal wetlands of India. Wetlands 22(3):443–450. https://doi.org/10.1672/0277-5212(2002)022[0443:EOOMSB]2.0.CO;2

Kabir MA, Salauddin M, Hossain KT, Tanim IA, Saddam MMH, Ahmad AU (2020) Assessing the shoreline dynamics of Hatiya Island of Meghna estuary in Bangladesh using multiband satellite imageries and hydro-meteorological data. Reg Stud Mar Sci 35:101167. https://doi.org/10.1016/j.rsma.2020.101167

Kaladharan P, Zacharia PU, Thirumalaiselvan S, Anasu Koya A, Lavanya R, Batcha SM (2020) Blue carbon stock of the seagrass meadows of Gulf of Mannar and Palk Bay off Coromandel Coast, South India. India J Fish 67(4):149–153. http://eprints.cmfri.org.in/id/eprint/14775

Kamal AHM, Khan MAA (2009) Coastal and estuarine resources of Bangladesh: management and conservation issues. Maejo Int J Sci Technol 3(2):313–342

Kaviarasan T, Dahms HU, Gokul MS, Henciya S, Muthukumar K, Shankar S, James RA (2019) Seasonal species variation of sediment organic carbon stocks in Salt Marshes of Tuticorin Area, Southern India. Wetlands 39(3):483–494. https://doi.org/10.1007/s13157-018-1094-6

Khan MA, Gul B (2002) Salt tolerant plants of coastal Sabkhat of Pakistan. In: Barth B (ed) Sabkha ecosystems. Kluwer Academic Publishers, Amsterdam, pp 123–139

Khan MA, Ungar IA, Showalter AM (2000) The effect of salinity on the growth, water status, and ion content of a leaf succulent perennial halophyte *Suaeda fruticosa* (L.) Forssk. J Arid Environ 45:73–84. https://doi.org/10.1006/jare.1999.0617

Khan TMA, Quadir DA, Murty TS, Kabir A, Aktar F, Sarker MA (2002) Relative sea level changes in Maldives and vulnerability of land due to abnormal coastal inundation. Mar Geod 25(1–2):133–143. https://doi.org/10.1080/014904102753516787

Khan MA, Ungar IA, Showalter AM (2005) Salt stimulation and tolerance in an intertidal stem-succulent halophyte. J Plant Nutr 28:1365–1374. https://doi.org/10.1081/PLN-200067462

Kim T, Noh J, Kwon BO, Lee C, Kim B, Kwon I et al (2020) Natural purification capacity of tidal flats for organic matters and nutrients: a mesocosm study. Mar Pollut Bull 154:111046. https://doi.org/10.1016/j.marpolbul.2020.111046

Kumar M, Boski T, González-Vila FJ, de la Rosa JM, González-Pérez JA (2020) Discerning natural and anthropogenic organic matter inputs to salt marsh sediments of Ria Formosa lagoon (South Portugal). Environ Sci Pollut Res 27:28962–28985. https://doi.org/10.1007/s11356-020-09235-9

Kürschner H, Al-Gifri AN, Al-Subai MY, Rowaished AK (1998) Vegetational patterns within coastal salines in southern Yemen. Feddes Repertorium 109(1–2):147–159

Lakshminarayana G, Kumar DV, Kumar SK, Vaidyanadhan R (2020) Tidal Flat Sedimentation in Bairenkonda Formation of the Tirumala Hills, Southwestern Part of the Cuddapah Basin, Andhra Pradesh. J Geol Soc India 96(4):325–336. https://doi.org/10.1007/s12594-020-1561-6

Leonardi N, Carnacina I, Donatelli C, Ganju NK, Plater AJ, Schuerch M, Temmerman S (2018) Dynamic interactions between coastal storms and salt marshes: a review. Geomorphology 301:92–107. https://doi.org/10.1016/j.geomorph.2017.11.001

Lepoint G, Frédérich B, Gobert S, Parmentier E (2008) Isotopic ratios and elemental contents as indicators of seagrass C processing and sewage influence in a tropical macrotidal ecosystem (Madagascar, Mozambique channel). Sci Mar 72(1):109–117. http://hdl.handle.net/2268/14700

Liu J, Yong DL, Choi CY, Gibson L (2020a) Transboundary frontiers: an emerging priority for biodiversity conservation. Trend Ecol Evol 35(8):679–690. https://doi.org/10.1016/j.tree.2020.03.004

Liu SP, Shi B, Gu K, Zhang CC, Yang JL, Zhang S, Yang P (2020b) Land subsidence monitoring in sinking coastal areas using distributed fiber optic sensing: a case study. Nat Haz 103(3):3043–3061. https://doi.org/10.1007/s11069-020-04118-1

Loconsole D, Cristiano G, De Lucia B (2019) Glassworts: from wild salt marsh species to sustainable edible crops. Agriculture 9(1):14. https://doi.org/10.3390/agriculture9010014

Lugendo B (2016) Mangroves, salt marshes, and seagrass beds. In: UNEP-Nairobi convention, WIOMSA (eds) The regional state of the coast report: Western Indian Ocean, Nairobi, Kenya: UNEP and WIOMSA, pp 49–64

Macreadie PI, Hughes AR, Kimbro DL (2013) Loss of 'blue carbon' from coastal salt marshes following habitat disturbance. PLoS ONE 8(7):e69244. https://doi.org/10.1371/journal.pone.0069244

Macreadie PI, Anton A, Raven JA, Beaumont N, Connolly RM, Friess DA et al (2019) The future of Blue Carbon science. Nat Commun 10(1):1–13. https://doi.org/10.1038/s41467-019-11693-w

Mafi-Gholami D, Zenner EK, Jaafari A, Bakhtyari HRR, Bui DT (2019) Multi-hazards vulnerability assessment of southern coasts of Iran. J Environ Manag 252:109628. https://doi.org/10.1016/j.jenvman.2019.109628

Mateos-Molina D, Antonopoulou M, Baldwin R, Bejarano I, Burt JA, García-Charton JA et al (2020) Applying an integrated approach to coastal marine habitat mapping in the north-western United Arab Emirates. Mar Environ Res 161:105095. https://doi.org/10.1016/j.marenvres.2020.105095

Mathuranthini S (2014) Cyclic succession in salt marsh vegetation involving *Halosarcia indica* and *Salicornia brachiata* in Sri Lanka. Sciscitator 1:7–9

Mcowen CJ, Weatherdon LV, Van Bochove JW, Sullivan E, Blyth S, Zockler C et al (2017) A global map of saltmarshes. Biodivers Data J 5:e11764. https://doi.org/10.3897/BDJ.5.e11764

Mia MM, Sarkar MSI, Khan MAA, Hassan N (2021) Monitoring of Water Chemistry in Salt Marsh Area along Karnafully River Coast, Chittagong, Bangladesh. Int J Res Eng Technol 6(2):1–12
Miththapala S (2013) Tidal flats, vol 5. IUCN, Colombo
Mok J (2019) Creating added value for Korea's Tidal Flats: using blue carbon as an incentive for coastal conservation. University of California, California Digital Library, p 67
Mokhtari M, Abd Ghaffar M, Usup G, Che Cob Z (2016) Effects of fiddler crab burrows on sediment properties in the mangrove mudflats of Sungai Sepang, Malaysia. Biology 5(1):7. https://doi.org/10.3390/biology5010007
Murray NJ, Phinn SR, DeWitt M, Ferrari R, Johnston R, Lyons MB et al (2019) The global distribution and trajectory of tidal flats. Nature 565(7738):222–225. https://doi.org/10.1038/s41586-018-0805-8
Murray NJ, Keith DA, Tizard R, Duncan A, Htut WT, Hlaing N, Oo AH, Ya KZ, Grantham H (2020) Threatened ecosystems of Myanmar. An IUCN Red list of ecosystems assessment, version 1.0. Wildlife Conservation Society
Nazim K, Ahmed M, Khan MU (2018) Carbon sequestration potential of sandspit backwater mangrove and sediment, Pakistan. FUUAST J Biol 8(2):209–215
Noffke N, Eriksson KA, Hazen RM, Simpson EL (2006) A new window into Early Archean life: microbial mats in Earth's oldest siliciclastic tidal deposits (3.2 Ga Moodies Group, South Africa). Geology 34(4):253–256. https://doi.org/10.1130/G22246.1
Noffke N, Beukes N, Bower D, Hazen RM, Swift DJP (2008) An actualistic perspective into Archean worlds–(cyano) bacterially induced sedimentary structures in the siliciclastic Nhlazatse Section, 2.9 Ga Pongola Supergroup, South Africa. Geobiology 6(1):5–20. https://doi.org/10.1111/j.1472-4669.2007.00118.x
Ouyang X, Lee SY, Connolly RM (2017) The role of root decomposition in global mangrove and saltmarsh carbon budgets. Earth-Sci Rev 166:53–63. https://doi.org/10.1016/j.earscirev.2017.01.004
Pakhirehzan M, Rahbani M (2014) Short term monitoring the Northwest Coasts of Gulf of Oman using both satellite images and numerical model. In: 2014 IEEE/OES Baltic international symposium (BALTIC), IEEE, pp 1–8
Pascual J, Alberti J, Daleo P, Fanjul E, Rocca C, Iribarne O (2019) Herbivory and dropping effects by small mammals on saltmarsh vegetation vary across microhabitats. J Veg Sci 30(2):322–330. https://doi.org/10.1111/jvs.12735
Patro S, Krishnan P, Deepak Samuel V, Purvaja R, Ramesh R (2017) Seagrass and Salt Marsh ecosystems in South Asia: an overview of diversity, distribution, threats and conservation status. In: Prusty B, Chandra R, Azeez P (eds) Wetland science. Springer, New Delhi. https://doi.org/10.1007/978-81-322-3715-0_5
Payo A, Mukhopadhyay A, Hazra S, Ghosh T, Ghosh S, Brown S et al (2016) Projected changes in area of the Sundarban mangrove forest in Bangladesh due to SLR by 2100. Clim Chang 139(2):279–291. https://doi.org/10.1007/s10584-016-1769-z
Pemadasa MA, Balasubramaniam S, Wijewansa HG, Amarasinghe L (1979) The ecology of a saltmarsh in Sri Lanka. J Ecol 67(1):41–63. https://doi.org/10.2307/2259336
Perillo GM (2019) Geomorphology of tidal courses and depressions. In: Perillo GME, Wolanski E, Cahoon DE, Charles S. HopkinsonCoastal Wetlands, Elsevier, pp 221–261. https://doi.org/10.1016/B978-0-444-63893-9.00006-X
Pham TD, Xia J, Ha NT, Bui DT, Le NN, Tekeuchi W (2019) A review of remote sensing approaches for monitoring blue carbon ecosystems: mangroves, seagrasses and salt marshes during 2010–2018. Sensors 19(8):1933. https://doi.org/10.3390/s19081933
Podder M, Mandol S, Ali MM, Islam MS (2012) Carbon assimilation in *Porterecia coarctata* salt marsh at Faujderhat coast, Chittagong, Bangladesh. Poll Res 31(2):307–313
Rathore AP, Chaudhary DR, Jha B (2016) Biomass production, nutrient cycling, and carbon fixation by *Salicornia brachiata* Roxb.: a promising halophyte for coastal saline soil rehabilitation. Int J Phytoremed 18(8):801–811. https://doi.org/10.1080/15226514.2016.1146228

Raw JL, Julie CL, Adams JB (2019) A comparison of soil carbon pools across a mangrove-salt marsh ecotone at the southern African warm-temperate range limit. S Afr J Bot 127:301–307. https://doi.org/10.1016/j.sajb.2019.11.005

Raw JL, Riddin T, Wasserman J, Lehman TWK, Bornman TG, Adams JB (2020) Salt marsh elevation and responses to future sea-level rise in the Knysna Estuary, South Africa. Afr J Aquat Sci 45(1–2):49–64. https://doi.org/10.2989/16085914.2019.1662763

Reed D, van Wesenbeeck B, Herman PM, Meselhe E (2018) Tidal flat-wetland systems as flood defenses: understanding biogeomorphic controls. Estuar Coast Shelf Sci 213:269–282. https://doi.org/10.1016/j.ecss.2018.08.017

Reef R, Spencer T, Möller I, Lovelock CE, Christie EK, McIvor AL et al (2017) The effects of elevated CO_2 and eutrophication on surface elevation gain in a European salt marsh. Glob Chang Biol 23(2):881–890. https://doi.org/10.1111/gcb.13396

Reise K, Baptist M, Burbridge P, Dankers NMJA, Fischer L, Flemming B et al (2010) The Wadden Sea a universally outstanding tidal wetland. In: The Wadden Sea 2010. Common Wadden Sea Secretariat (CWSS); Trilateral Monitoring and Assessment Group: Wilhelmshaven. Wadden Sea Ecosystem; 29/editors, Harald Marencic and Jaap de Vlas, vol 7

Ruiz-Fernández AC, Carnero-Bravo V, Sanchez-Cabeza JA, Pérez-Bernal LH, Amaya-Monterrosa OA, Bojórquez-Sánchez S et al (2018) Carbon burial and storage in tropical salt marshes under the influence of sea level rise. Sci Total Environ 630:1628–1640. https://doi.org/10.1016/j.scitotenv.2018.02.246

Rupprecht F, Möller I, Paul M, Kudella M, Spencer T, Van Wesenbeeck BK et al (2017) Vegetation-wave interactions in salt marshes under storm surge conditions. Ecol Eng 100:301–315. https://doi.org/10.1016/j.ecoleng.2016.12.030

Ryu JH, Choi JK, Lee YK (2014) Potential of remote sensing in management of tidal flats: a case study of thematic mapping in the Korean tidal flats. Ocean Coast Manag 102:458–470. https://doi.org/10.1016/j.ocecoaman.2014.03.003

SAC (2011) National Wetland Atlas. SAC/EPSA/ABHG/NWIA/ATLAS/34/2011. Space Applications Centre, ISRO, Ahmedabad, p 310

Sagheer AA (2013) Combined use of satellite remote sensing and GIS to support land-use planning and management of At-Tif-AlFaza Coastal Area, Red Sea, Yemen. J King Abdulaziz Univ 24(1):85

Saha K, Sinha S (2021) Grain size analysis and characterization of sedimentary process in tidal flat of Chandipur region, East Coast of Odisha, India. Mar Geod 44:485–503. https://doi.org/10.1080/01490419.2021.1922554

Saha S, Ghosh A, Banerjee S, Saraswati PK, Burley SD (2011) Characteristics of an open coast tidal flat: example from Daman, west coast of India. J Geol Soc India 77(5):409–418

Saher US, Qureshi NA (2011) Burrow morphology of three species of fiddler crab (Uca) along the coast of Pakistan. Belg J Zool 142(2):114–126

Sarkar MSI, Khan AA, Bhuyan MS (2018) Physicochemical features of water collected from Salt Marsh Bed (*Spartina alterniflora*) at Moheshkhali Coastal Area, Cox's Bazar, Bangladesh. Int J Pollut Noise Cont 4(2):6–18. https://doi.org/10.37628/jpnc.v4i2.757

Schile LM, Kauffman JB, Crooks S, Fourqurean JW, Glavan J, Megonigal JP (2017) Limits on carbon sequestration in arid blue carbon ecosystems. Ecol Appl 27(3):859–874. https://doi.org/10.1002/eap.1489

Schutte CA, Ahmerkamp S, Wu CS, Seidel M, De Beer D, Cook PLM, Joye SB (2019) Biogeochemical dynamics of coastal tidal flats. In: Perillo GME, Wolanski E, Cahoon DR, Hopkinson CS (eds) Coastal Wetlands. Elsevier, pp 407–440. https://doi.org/10.1016/B978-0-444-63893-9.00012-5

Sejati AW, Buchori I, Kurniawati S, Brana YC, Fariha TI (2020) Quantifying the impact of industrialization on blue carbon storage in the coastal area of Metropolitan Semarang, Indonesia. Appl Geogr 124:102319. https://doi.org/10.1016/j.apgeog.2020.102319

Senterre B, Henriette E, Labrosse J-C, Pinchart J, Senaratne N (2015) Rapid biodiversity assessment of Anse Royale and North-East Point wetlands (Mahé, Seychelles). Consultancy Report, Government of Seychelles, United Nations Development Programme, Victoria, Seychelles, p 62

Serrano O, Kelleway JJ, Lovelock C, Lavery PS (2019) Conservation of blue carbon ecosystems for climate change mitigation and adaptation. In: Perillo GME, Wolanski E, Cahoon DR, Hopkinson CS (eds) Coastal wetlands. Elsevier, pp 965–996. https://doi.org/10.1016/B978-0-444-63893-9.00028-9

Sethi D, Butler TO, Shuhaili F, Vaidyanathan S (2020) Diatoms for carbon sequestration and bio-based manufacturing. Biology 9:217. https://doi.org/10.3390/biology9080217

Shaig A (2011) Survey of climate change adaptation measures in the Maldives. Integration of climate change risks into resilient island planning in the Maldives Project. Final Report. Prepared for the Ministry of Housing and Development

Sharpe PJ, Baldwin AH (2012) Tidal marsh plant community response to sea-level rise: a mesocosm study. Aquat Bot 101:3440. https://doi.org/10.1016/j.aquabot.2012.03.015

Sheppard C, Dixon DJ, Gourlay M, Sheppard A, Payet R (2005) Coral mortality increases wave energy reaching shores protected by reef flats: examples from the Seychelles. Estuar Coast Shelf Sci 64(2–3):223–234. https://doi.org/10.1016/j.ecss.2005.02.016

Siddique MAM, Mustafa Kamal AH, Aktar M (2012) Trace metal concentrations in salt marsh sediments from Bakkhali River estuary, Cox's Bazar, Bangladesh. Zool Ecol 22(3–4):254–259. https://doi.org/10.1080/21658005.2012.735842

Siddiqui MN, Maajid S (2004) Monitoring of geomorphological changes for planning reclamation work in coastal area of Karachi, Pakistan. Adv Space Res 33(7):1200–1205. https://doi.org/10.1016/S0273-1177(03)00373-9

Sivakumar S, Krishnakumar P, Lakshumanan C (2014) Estimation of carbon stocks in above ground biomass in Muthupet mangrove, Southeast coast of India. Int J Intellect Adv Res Eng Comput 5(2):139–150

Steijn RC, Cleveringa J, Leuven JRFW, van der Baan J, van der Zanden JJG (2019) Extreme coastal dynamics of Sittaung Estuary (Myanmar). In: Trung Viet N, Xiping D, Thanh Tung T (eds) International conference on Asian and Pacific Coasts. Springer, Singapore, pp 1193–1200. https://doi.org/10.1007/978-981-15-0291-0_163

Sun C, Liu Y, Zhao S, Zhou M, Yang Y, Li F (2016) Classification mapping and species identification of salt marshes based on a short-time interval NDVI time-series from HJ-1 optical imagery. Int J Appl Earth Observ Geoinfo 45:27–41. https://doi.org/10.1016/j.jag.2015.10.008

Tabot PT, Adams JB (2013) Ecophysiology of salt marsh plants and predicted responses to climate change in South Africa. Ocean Coast Manag 80:89–99. https://doi.org/10.1016/j.ocecoaman.2013.04.003

Tang J, Ye S, Chen X, Yang H, Sun X, Wang F et al (2018) Coastal blue carbon: concept, study method, and the application to ecological restoration. Sci China Earth Sci 61(6):637–646. https://doi.org/10.1007/s11430-017-9181-x

Toniolo MA, Seitz C, Perillo GM (2021) Origin and evolution of tidal depressions in a tidal flat and their role in carbon sequestration in the Bahía Blanca Estuary (Argentina). Mar Geol 436:106467. https://doi.org/10.1016/j.margeo.2021.106467

Townend I, Fletcher C, Knappen M, Rossington K (2011) A review of salt marsh dynamics. Water Environ J 25:477488. https://doi.org/10.1111/j.1747-6593.2010.00243.x

Tseng KH, Kuo CY, Lin TH, Huang ZC, Lin YC, Liao WH, Chen CF (2017) Reconstruction of time-varying tidal flat topography using optical remote sensing imageries. ISPRS J Photogram Remote Sens 131:92–103. https://doi.org/10.1016/j.isprsjprs.2017.07.008

Umitsu M, Tanavud C, Patanakanog B (2007) Effects of landforms on tsunami flow in the plains of Banda Aceh, Indonesia, and Nam Khem, Thailand. Mar Geol 242(1–3):141–153. https://doi.org/10.1016/j.margeo.2006.10.030

Valentim JM, Vaz N, Silva H, Duarte B, Cacador I, Dias JM (2013) Tagus estuary and Ria de aveiro salt marsh dynamics and the impact of sea level rise. Estuar Coast Shelf Sci 130:138151. https://doi.org/10.1016/j.ecss.2013.04.005

Valiela I, Lloret J, Bowyer T, Miner S, Remsen D, Elmstrom E et al (2018) Transient coastal landscapes: rising sea level threatens salt marshes. Sci Total Environ 640:1148–1156. https://doi.org/10.1016/j.scitotenv.2018.05.235

Viswanathan C, Purvaja R, Jeevamani JJJ, Samuel VD, Sankar R, Abhilash KR et al (2020) Salt marsh vegetation in India: species composition, distribution, zonation pattern and conservation implications. Estuar Coast Shelf Sci 242:106792. https://doi.org/10.1016/j.ecss.2020.106792

Wang C, Menenti M, Stoll MP, Belluco E, Marani M (2007) Mapping mixed vegetation communities in salt marshes using airborne spectral data. Remote Sens Environ 107(4):559–570. https://doi.org/10.1016/j.rse.2006.10.007

Whitfield AK (2017) The role of seagrass meadows, mangrove forests, salt marshes, and reed beds as nursery areas and food sources for fishes in estuaries. Rev Fish Biol Fish 27(1):75–110. https://doi.org/10.1007/s11160-016-9454-x

Xuehui Z, Zhongsheng Z, Zhe L, Min L, Haitao W, Ming J (2021) Impacts of *Spartina alterniflora* invasion on soil carbon contents and stability in the Yellow River Delta, China. Sci Total Environ 775:145188. https://doi.org/10.1016/j.scitotenv.2021.145188

Yamamoto K, Basir N, Sutikno S, Kanno A, Kagawa H, Suzuki M et al (2019) Tropical peat debris storage in the tidal flat in northern part of the Bengkalis island, Indonesia. In: MATEC Web of conferences. EDP Sciences, vol 276, p 06002

Yap WG, Landoy RJ (1986) Report on a survey of the coastal areas of Kenya for shrimp farm development. FAO, Rome, Italy

Zakaria R (2016) A study of open coast tidal flat sedimentology based on outcrop data in ngrayong formation, kadiwono area, central Java. In proceeding, seminar nasional kebumian ke-9 peran penelitian ilmu kebumian dalam pemberdayaan masyarakat 6–7 oktober 2016; grha sabha pramana. Departemen teknik geologi ft ugm

Zhang M, Dai Z, Bouma TJ, Bricker J, Townend I, Wen J et al (2021) Tidal-flat reclamation aggravates potential risk from storm impacts. Coast Eng 166:103868. https://doi.org/10.1016/j.coastaleng.2021.103868

Zhao C, Qin CZ, Teng J (2020) Mapping large-area tidal flats without the dependence on tidal elevations: a case study of Southern China. ISPRS J Photogramm Remote Sens 159:256–270. https://doi.org/10.1016/j.isprsjprs.2019.11.022

Zöckler C, Naing TZ, Moses S, Soe RN, Hla TH (2014) The importance of the Myanmar coast for Water Birds. Stilt 66:37–51

Zöckler C, Moses S, Lwin ST (2019) The importance of the Myeik mangroves and mudflats, Tanintharyi, Myanmar for migratory waders and other waterbirds. Wader Study 126:129–141. https://doi.org/10.18194/ws.00149

Chapter 7
An Overview of the Indian Ocean Coral Ecosystems

Abhra Chanda

Abstract Despite the debate that continues to go on, at present, corals do not qualify to be a part of the conventional blue carbon ecosystems. Nonetheless, these marine ecosystems are crucial as they provide a plethora of ecosystem services not only to humans but also to several blue carbon ecosystems. These ecosystems provide shelter to the adjacent seagrass and mangroves. Corals act as wave breakers in many parts of the world and thereby furnishes a calm and stable environment to blue carbon ecosystems lying adjacent to each other. The present chapter collated the knowledge acquired so far on the coral species count and areal cover in the different countries bordering the Indian Ocean. This chapter also discussed the various threats that make corals vulnerable on these coastlines. We observed a spatial disparity in the degree of attention paid to the corals bordering the Indian Ocean. The coral triangle region received much higher attention than that observed in the western Indian Ocean. Several countries are yet to configure the principal threats to the corals lying on their coastline. Thus, this chapter apart from detailing the present state-of-the-art identifies the arenas that require scientific attention in the Indian Ocean periphery.

Keywords Species diversity · Reefs · Fringing corals · Porites · Acropora · Coral triangle · Western Indian Ocean · Threat · Bleaching · Unsustainable fishing

7.1 Introduction

Corals have always been a point of interest for marine ecologists and conservationists as these ecosystems shelter rich biodiversity all through the world (Benkwitt et al. 2020). These calcareous structures encompass the coastal periphery in several countries. The fascinating beauty of corals has always attracted tourists and marine divers to cherish their ambiance. These marine habitats provide a plethora of ecosystem services (Mumby et al. 2018; Woodhead et al. 2019). The reef systems shelter around 0.83 million species worldwide (Fisher et al. 2015). They also provide a complicated

A. Chanda (✉)
School of Oceanographic Studies, Jadavpur University, 188 Raja S.C. Mullick Road, Kolkata, West Bengal 700084, India
e-mail: abhrachanda.sos.rs@jadavpuruniversity.in

A. Chanda et al. (eds.), *Blue Carbon Dynamics of the Indian Ocean*,
https://doi.org/10.1007/978-3-030-96558-7_7

but conducive habitat for several life forms at different life stages (Graham and Nash 2013; Gillis et al. 2014). The reef structures act as a barrier to several physical forcing events like storm surges and cyclones and provide direct protection benefit to millions of people who live very close to the mean sea level (Ferrario et al. 2014; Beck et al. 2018). Several researchers indicated that coral reefs significantly regulate the ambient water quality and nutrient levels (De Goeij et al. 2013; Hicks et al. 2014). Various fisheries sectors stand on the pillars provided by reef ecosystems in the nearshore regimes (Grafeld et al. 2017). This ecosystem also offers incredible medicinal values and biotechnological potentials (Albert et al. 2015; Motuhi et al. 2016). Coral-based tourism plays a crucial role in the economy of several countries (Spalding et al. 2017).

Despite such global significance, the ongoing climate change has long initiated a decline in the coral cover across the planet (Adam et al. 2021). Several blue carbon ecosystems exhibited a promising potential to expand in the higher latitudes due to an increase in temperature (Chen et al. 2020). However, the fate of corals in the near future is not bright in this regard. Muir et al. (2015) argued that enhanced temperature beyond temperate latitudes can facilitate coral growth but the lack of optimum photosynthetically active radiation would restrict their extent only in shallower depths. The steady increase of atmospheric CO_2 due to multifarious anthropogenic activities has accelerated ocean acidification. This phenomenon is detrimental to the coralline structures, especially in the warm tropics (Hoegh-Guldberg et al. 2017; Mollica et al. 2018; Comeau et al. 2019; Guo et al. 2020). The marine heatwaves produced out of climatic phenomena like El Nino Southern Oscillation (ENSO) led to considerable bleaching of corals throughout the planet (Stuart-Smith et al. 2018; Sully et al. 2019; Suggett and Smith 2020; De et al. 2021). A range of coastal polluting activities severely affect the corals (Dubinsky and Stambler 1996). Untreated sewage discharge often leads to eutrophication in the coral-adjacent waters and proves to be deleterious to the reef structures (Reopanichkul et al. 2009). Several coastal construction processes introduce a substantial sediment load that compromises the light availability in the coralline water masses (Weber et al. 2012). Oil pollution can be detrimental too for the scleractinian corals (Loya and Rinkevich 1980; Haapkylä et al. 2007). Recent pieces of research indicate that thermal and microplastic pollution can affect these ecosystems in near future (Reichert et al. 2021). Toxic materials in the nearshore environment like heavy metals often hamper the physiological functions of the corals (Patterson et al. 2020). Disasters like tsunami can also lead to the large-scale devastation of corals (Datta Majumdar et al. 2018). Thus, it is unequivocal that coral ecosystems are facing several threats of both natural and anthropogenic nature.

Despite being a marine ecosystem, corals do not qualify to be a blue carbon ecosystem. Debates are ongoing on whether to include this ecosystem within the blue carbon arena. However, to date, corals remain excluded from the umbrella of the carbon system (Lovelock and Duarte 2019). Kinsey and Hopley (1991) argued that corals sequester a substantial amount of carbon in the form of calcareous structures. Several other researchers also went on to agree to this concept. However, the very mechanism through which calcium carbonate deposition takes place in the oceanic water column involves the generation of CO_2. This mechanism is one of the most fundamental reasons why corals remain outside the blue carbon ecosystems. The readers must be wondering that if corals are not a part of the blue carbon ecosystems, why is there a chapter altogether on this topic. We have to nurture several aspects of corals to get an answer to this pertinent question. Though the corals are not a part of the conventional blue carbon ecosystem, they provide an array of ecosystem services to the mangroves, seagrasses, and salt marshes that grow adjacent to the corals. Seagrasses prefer a calm and sheltered habitat. Corals often furnish such a protected environment to the seagrasses. The corals bear the primary brunt of the wave action and reduce the disturbance in the ambiance of the seagrasses and allow them to flourish and sequester substantial quantities of carbon (Guerra-Vargas et al. 2020). Coral ecosystems also accentuate wave tolerance in the salt marsh ecosystems (Meyer et al. 1997; Shaver and Silliman 2017). Several studies indicated that these blue carbon ecosystems exhibit a higher rate of growth and recovery in the presence of nearshore coralline structures (Silliman et al. 2015). Thus, the service provided by corals to the blue carbon ecosystems makes them equally important as the blue carbon system themselves. Hence, a thorough understanding of this ecosystem has become an urgent necessity even for blue carbon scientists. The synergistic effect of their coexistence has emerged as a point of interest and an avenue to enhance the blue carbon stock.

7.2 The State-of-the-Art of Corals in the Asian Countries Bordering the Eastern Indian Ocean

7.2.1 India

India has a long coastline of around 7500 km. However, corals and reef structures remain concentrated only in few places like the Gulfs of Kutch and Mannar, Andaman and Nicobar Islands, Lakshadweep, and Malvan region in Maharashtra (Venkataraman 2006). Corals require a suitable substrate and turbidity-free water column, only available in the mentioned regions (Raghuraman et al. 2013). According to the latest pieces of research, corals in India occupy an area of about 2400 km^2 (Datta Majumdar et al. 2018) (Fig. 7.1). More than 80% of this area remains located in the Andaman and Nicobar Islands and Lakshadweep. The surveys conducted till the end of the twentieth century indicated that India had around 200 coral species. However, Raghuraman et al. (2013) highlighted that India has as many as 478 coral species in the coastal periphery (Fig. 7.2). According to the present understanding, nineteen families of coral species exist on this planet. India has corals belonging to all these nineteen families (Raghuraman et al. 2013). India has a profound diversity of scleractinian corals, and the Acroporidae family corals are the most abundant in this country. It is worth mentioning that the regions where corals thrive also host substantial cover of other blue carbon ecosystems like mangroves, seagrass, and salt marshes. Studies linking the relation between their coexistence are scarce from India.

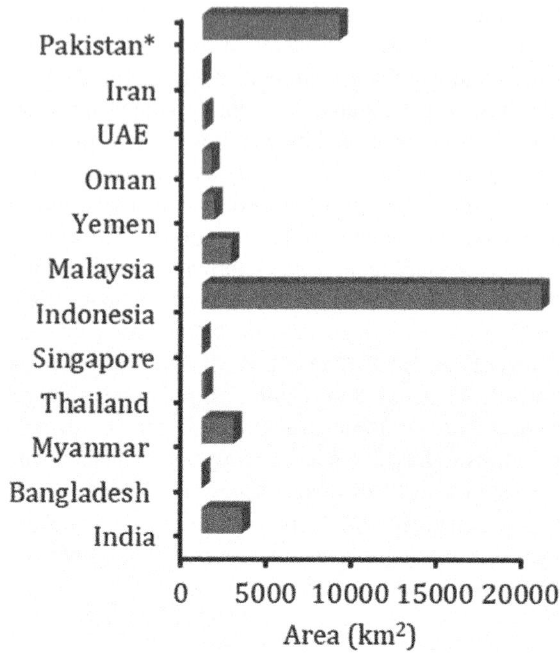

Fig. 7.1 Area of coral cover reported from the Asian countries in the eastern Indian Ocean (*the magnitude for Pakistan is an estimated figure)

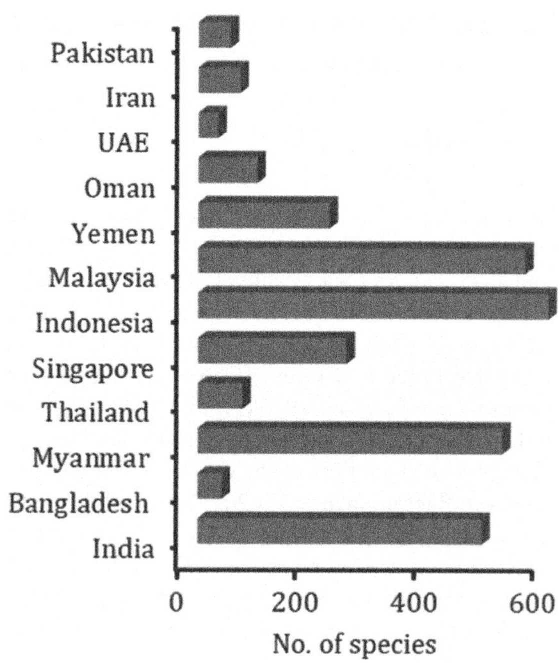

Fig. 7.2 Number of coral species reported the Asian countries in the eastern Indian Ocean

However, I believe that it would not be wrong to speculate their positive feedback to the blue carbon repositories in India. However, the corals in India, like other regions of the world, are under severe threat. Natural factors like increasing sea surface temperature led to severe coral bleaching in the Andaman Islands (Datta Majumdar et al. 2018). Records of similar bleaching events are available from Lakshadweep Archipelago (Harithsa et al. 2005; Vinoth et al. 2012) and Gulf of Mannar (Edward et al. 2008). The Indian Ocean tsunami destroyed substantial coral cover in the Indian periphery (Wilkinson et al. 2005). Datta Majumdar et al. (2018) indicated that tourism-driven trampling and sedimentation took a heavy toll on the corals of India. They also observed that branching coral was more susceptible than the massive corals in the Andaman region (Rahman and Alam 2020).

7.2.2 *Bangladesh*

Most of the Bangladesh coastline experiences a turbid water column, which does not facilitate the growth of corals. However, there happens to be only one island named Saint Martin (in Bangladesh), which has corals in the coastal periphery (Rahman and Alam 2020). The coral cover is less than 8 km^2 in Bangladesh (Alam et al. 2015). This island attracts thousands of tourists all over the year due to its aesthetic beauty, but the same reason has led to the significant destruction of corals in the Saint Martin Islands (Rani et al. 2020; Ara et al. 2021). Despite being a small area, this region shelters some threatened marine turtles (Billah et al. 2018). However, coral biodiversity has been steadily decreasing throughout the last three to four decades. English et al. (1997) recorded 141 species in 1980, whereas Rajasuriya et al. (2002) observed 66 species in 2000. Lately, Ahammed et al. (2016) recorded only 41 species in 2016, and they predicted a possibility of coral extinction from Bangladesh by 2045. Fishing activities, anchoring of boats, and tourist littering are the primary points of concern for the corals of this region. Several researchers urged for stringent law enforcement and proactive initiatives from the Bangladesh Government to save the corals of this island (Rahman and Alam 2020; Rani et al. 2020).

7.2.3 *Myanmar*

Myanmar has a long coastline, and anecdotal information indicates vast areas of pristine coral reefs. However, these corals have perhaps received the least scientific attention (Tun and Heiss 2005). Recently, some studies focused on the corals of Myanmar. However, much more research remains the need of the hour as the Myanmar government intends to cash upon the reef-based tourism (Wongthong et al. 2015). According to the report of BOBLME (2012), the corals in Myanmar cover about 1870 km^2. The corals in Myanmar remain concentrated in the Myeik Archipelago, Tanintharyi, and Rakhine regions (Anelli et al. 2019). The best possible estimates indicate 512 species from the coastlines of Myanmar. However, Wongthong et al. (2015) emphasized the possibility of overestimation. *Acropora*, *Porites*, *Fungia*, and *Psammocora* are the dominant coral species on the coastline of this country. Like in the neighboring countries, the corals of Myanmar are also under stress. Wongthong et al. (2015) observed poor health in the reefs of several regions adjoining the periphery of Myeik Archipelago. Burke et al. (2002) indicated that unsustainable fishing techniques, infrastructural development along the coastline, coastal pollution, and trading of corals are the fundamental threats in Myanmar.

7.2.4 Thailand

Thailand has a coastline of almost 2600 km and offers potential habitats for the corals. According to Yeemin et al. (2006), the corals of Thailand encompass around 153 km^2. The Gulf of Thailand (75 km^2) and the Andaman Sea coastline (78 km^2) equally share the corals of this country (Yeemin et al. 2013). The corals of Thailand bared the brunt of anomalous heat waves in 1998 and 2010 (Yeemin et al. 1998; 2010). Such recurrent enhancement of sea surface corals led to significant coral damage all around the coastline of Thailand (Yeemin et al. 2009). The dominant *Acropora* species in the Thai coastlines underwent massive bleaching due to anomalous warming events (Hoeksema et al. 2013). Several pieces of research characterized the coral species diversity in Thailand. Phongsuwan and Chansang (2012) recorded 47 species from the coastline adjacent to the Andaman Sea, whereas Sutthacheep et al. (2012; 2013) reported 44 species from the Gulf of Thailand. However, a recent study (Yucharoen et al. 2020) identified 76 species in the Losin Pinnacle region. This region usually experiences a dynamic wave system that arguably saved the corals from bleaching (Schmidt et al. 2016). Besides sea surface temperature anomalies, the Thai corals suffer from tourism-related activities, indiscriminate fishing, coastal eutrophication, and a wide range of marine pollution.

7.2.5 Singapore

Singapore is one of the smallest nations on the periphery of the Indian Ocean. However, it shelters several crucial marine ecosystems. Lately, Guest et al. (2016) furnished a comprehensive overview of the corals of this country. Singapore witnessed a drastic transformation in the past two centuries; a mere forested land to a highly urbanized metropolitan setup (Hilton and Manning 1995). The coral reefs of Singapore remained concentrated in the southern portion, and the type of corals found there are mainly patchy and fringing. Many of these calcareous structures have degraded due to the enhanced turbidity levels in the coastal waters of Singapore (Dikou and Van Woesik 2006). Infrastructural development, coastal dredging, and industrial plants together compromised the light availability in the coastal regime of

Singapore (Guest et al. 2016). Sixty years ago, corals of Singapore used to thrive at a depth of around 10 m. However, at present, the corals delimit themselves to a depth of only 2 m (Chou 1996). Despite so many constraints, the corals of Singapore exhibit a rich diversity of more than 250 species in the fringing structures of the southern part (Huang et al. 2009). According to Chou et al. (2012), Singapore still maintains an area of 10 km^2 under coral cover.

7.2.6 Indonesia

Indonesia is the epitome of coral biodiversity. The junction between the Indian Ocean and the Pacific, popularly known as the coral triangle, exhibits the richest marine biodiversity on this planet (Asaad et al. 2018). Indonesia shelters almost 590 species of corals in its coastal periphery (Veron 1995). Globally, Acropora corals constitute 114 species, out of which Indonesia shelters 91. The Tomini Bay in Indonesia has the highest species diversity of Acropora (Wallace et al. 2001). The spatial extent of the corals in the Indonesian coastline has always remained debatable (Fig. 7.3). Several authors came up with varying estimates. Dahuri (1995) was perhaps the first to comment on the nationwide coral cover of Indonesia (17,500 km^2). However, the very next year Cesar (1996) mentioned an estimate of 75,000 km^2. Spalding et al. (2001) and Burke et al. (2002) furnished an arguable estimate of 51,000 km^2. However, estimations of Tomascik et al. (1997) and Pet-Soede et al. (2002) indicate

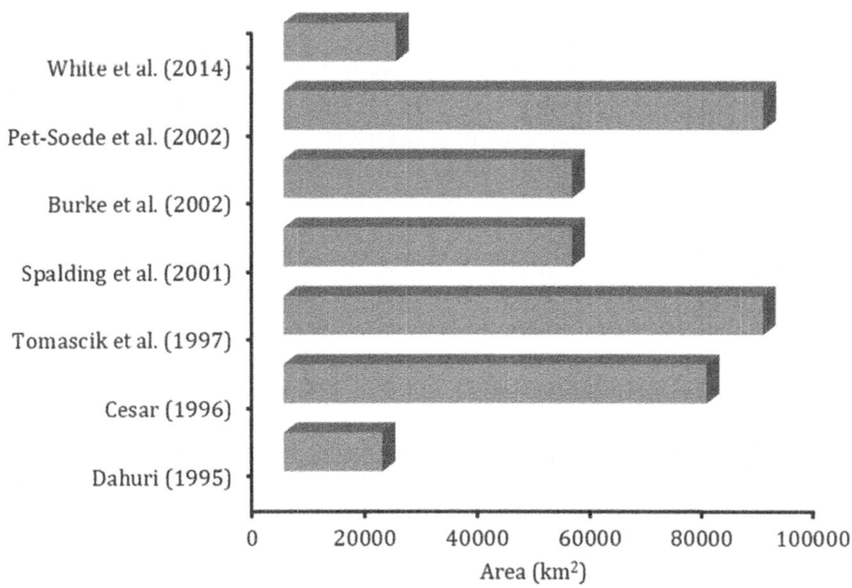

Fig. 7.3 Area of coral cover in Indonesia reported by several authors over the last three decades

that Indonesia encompasses a coralline area of almost 85,000 km^2. However, the estimates provided by White et al. (2014) are more realistic and based on ground-truthed data (19,868 km^2). According to this estimate, Indonesia comprises almost 16% of the global coral cover. Corals occupy the entire coastal stretch of Indonesia. However, the quality of corals is in excellent condition in the eastern region (Asian Development Bank Report 2014). The absence of rivers flowing into the coral inhabited regions in eastern Indonesia facilitates turbidity-free waters (Hutomo et al. 2009). Types of coral reefs in this country range from fringing to patchy to barrier. There are some of the most spectacular atolls as well. The Raja Ampat Islands record the highest species count of stony and soft corals (Veron et al. 2009). These reefs shelter a wide spectrum of fauna, especially fishes (Limmon et al. 2020). Despite being an epicenter of marine biodiversity, Indonesian corals face severe threats due to multifarious anthropogenic activities (Edinger et al. 1998). Rapid changes in land use and destructive fishing techniques continue to be the major threats to corals of this region (Quevedo et al. 2021). Plastic pollution has also become a point of concern for the marine biodiversity of Indonesia (Lasut et al. 2018).

7.2.7 Malaysia

Malaysia is also one of the six coral triangle countries like Indonesia. It encompasses a coralline area of almost 1700 km^2, equivalent to 1% of the global cover (White et al. 2014). Corals are available in both the eastern and the peninsular parts of Malaysia. According to the Coral Reef Monitoring Report (2008), Malaysia hosts an area of 4000 km^2 comprising 550 species (as mentioned in Praveena et al. 2012). Porites, Montipora, and Acropora are the most abundant corals of Malaysia (Akmal et al. 2019). The Semporna region in eastern Sabah exhibits the highest species diversity of corals within the Malaysian coastal waters (Waheed and Hoeksema 2013). The mushroom corals are available in plenty in the Malaysian periphery. Waheed and Hoeksema (2013) recorded 44 mushroom or plate coral species from the Semporna coastline and claimed the highest count within the coral triangle. Threats to corals are prevalent all through the world. However, the corals of this region are much intact compared to the neighboring countries. Blast fishing happens to be one of the significant concerns for the corals of Malaysia (Waheed and Hoeksema 2013). The scleractinian coral diversity is also significantly high in the coastal waters of Sabah (Waheed and Hoeksema 2014). The prevalence of different varieties of reef structure enabled numerous species to acclimatize within a small area (Waheed and Hoeksema 2013; 2014). Though few places in Malaysia show good quality corals, several threats have continued to degrade the corals of Malaysia (Praveena et al. 2012). The 1998 and 2010 bleaching events left a disastrous signature in many Malaysian coral patches (Tan and Heron 2011). Overall, Praveena et al. (2012) observed that unsustainable fishing technique is the primary cause of coral degradation in the Sabah region, whereas the sand mining activities and consequent reduction in water column transparency are worrisome for the Sarawak region.

7.3 The State-of-the-Art of Corals in the Asian Countries Bordering the Western Indian Ocean

7.3.1 Yemen

Yemeni coastal waters shelter around 700 km^2 of coralline area (Spalding et al. 2001). Turak et al. (2007) reported almost 221 coral species from this region. Hassan et al. (2002) observed that half of the corals were alive in the Yemeni coastline, and the percentage share of live corals was almost 70% in the Red Sea region. The hard coral cover exhibited significant spatial variation in this region. In places like Mukalla, the hard coral cover was only 15%, whereas, in Belhaf, it was close to 70%. According to Hassan et al. (2002), the Kamaran Island region witnessed massive degradation of corals. Destructive collection techniques of ornamental fishes have principally led to such consequences. Benzoni et al. (2003) observed that substrate characteristics, upwelling variability, and degree of exposure to sunlight were the primary factors in the Yemeni coastline that governed the coral dynamics. Turak et al. (2007) highlighted that the 1990s witnessed extensive coral bleaching events and the outbreak of crown-of-thorns sea stars in the Yemen coastline. The coral population in Yemen has an overall dominance of Porites. However, sponge infestations led to significant damage to *Porites* colonies in South Yemen (Benzoni et al. 2008). However, notable conservation endeavors also exist in this part of the world. Lately, almost 1500 coral colonies underwent replacement to facilitate a natural gas plant in the offshore regions of Yemen (Seguin et al. 2008). The Socotra Island, situated away from the Yemen coastline, comes under the purview of the marine protected area, and the archipelagic corals of this region are one of the finest (DeVantier et al. 2004).

7.3.2 Oman

Oman has a long coastline, and coral reefs occur at various locations adjoining the coastal periphery. Dhofar, Muscat, Musandam, and Masirah Gulf record the most prolific occurrences of corals in Oman (Sheppard and Salm 1988). Like Yemen, *Porites* is the dominant coral form in the coastal waters of Oman (Salm 1993). Burt et al. (2016) mentioned that the corals of Oman grow under three different water masses, namely the Sea of Oman, Persian Gulf waters, and the Arabian Sea. However, the species diversity (106 species) recorded so far is far less than that of Yemen (Claereboudt 2006). The reef-based fisheries play a crucial role in the economy of Oman (Grandcourt 2012). Though reef surveys in Oman date back to the 1980s (Glynn 1993), these reefs are one of the least studied in the world (Vaughan and Burt 2016). Starfish *Acanthaster planci*, tropical cyclones, and algal blooms are the principal threats to the corals of these regions (Burt et al. 2016). Al-Azri et al. (2016) also indicated that monsoon-induced upwelling in the Oman coast delimits the coral growth. The upwelled water has a low temperature which disfavors coral flourishment, and at the same time, the nutrient-rich waters facilitate fungal spread over the coral bodies (Savidge et al. 1990; Sheppard et al. 1992).

7.3.3 UAE

The United Arab Emirates (UAE) used to shelter one of the largest stretches of coral reefs in the Arabian Sea region during the mid-twentieth century (Purser and Evans 1973; Shinn 1976). The UAE witnessed a rampant change in the land-use-land-cover in the last three to four decades that took a serious toll on marine habitats like reefs (Sale et al. 2011). According to the estimates of Spalding et al. (2001), corals encompass around 1200 km^2 in the coastal waters of UAE. However, the recent study by Grizzle et al. (2016) indicates only 132 km^2 coral cover bordering the UAE, most of which lies in the outskirts of Abu Dhabi. The species diversity also counts to only 34 hard corals (Riegl and Purkis 2012). Natural and anthropogenic factors both led to significant damages to the corals of the UAE. The corals of this region experienced one of the deadliest bleaching events during the late 1990s (Palumbi et al. 2014; Hume et al. 2015). The urban infrastructural development, as well as industrial activities ranging from tourism to petrochemical mining, has also degraded the corals of the UAE (Rezai et al. 2004). Despite such widespread destruction, the corals of the UAE have become a point of interest for present-day marine scientists. Lately, researchers have observed temperature-resistant microalgae (Symbiodinium thermophilum) in many corals of the UAE, which enhanced their tolerance against climate-extreme events (Ben-Romdhane et al. 2018; 2020a; 2020b). Five decades ago, the dominant species used to be *Acropora*; however, in the present day, the species diversity has significantly reduced to only *Porites* dominance observed all through the UAE coral cover (Grizzle et al. 2016).

7.3.4 Iran

Iran has a long coastline of more than 2800 km; however, this region shelters coral reefs occupying an area of only 10.5 km^2. The corals of Iran remain restricted to only sixteen islands (Shokri et al., 2005). Fringing reefs are the most dominant among the available reef structures (Shojae et al. 2010). Several studies indicated that the Acropora is the dominant genus in the Iranian reefs (Sheppard and Sheppard 1991). The reef-oriented research enhanced substantially in Iran during the last two decades. However, the coral species count made by different authors was localized area-specific (Rezai and Jahangard 1994; Rezai 1996; Sadeghi 1997; Fatemi and Shokri 2001). None of them provided a nationwide estimate. Lately, Riegl and Purkis (2011) reported a total of 71 coral species in the Iran adjacent Persian Gulf region. Anthropogenic threats in the Iranian coastline led to a significant reduction of the predominant Acropora species (Fatemi and Shokri 2001). Upwelling-induced low temperatures have always bothered the Gulf corals (Coles and Fadlallah 1991). The prevalence of petroleum industries is responsible for metal pollution in the coral habitats of Iran (Jafarabadi et al. 2018; 2020). Polybrominated diphenyl ethers and chlorinated paraffin have shown substantial toxic effects on the corals of Iran (Jafarabadi et al. 2021a; 2021b).

7.3.5 *Pakistan*

Coral reef research has gained impetus in Pakistan since the onset of the twenty-first century (Raza et al. 2014). Lately, Raza (2013) identified 54 coral species belonging to 37 genera from the coastline of Pakistan. In 2006, a survey conducted by Pakistan Wetland Program led to several discoveries of reef structures in the nearshore waters of Pakistan (Iqbal et al. 2018). The corals of the Astola Islands have special significance in Pakistan. Pakistan Government declared this region as their first marine protected area due to the coral and associated biodiversity. Two decades ago, the scientific community perceived that Pakistan's coastal water is devoid of corals. However, inspired by the early observations, Pakistan Navy undertook several cruises and made the landmark discovery of corals in the Gwader region that shelters 17 coral species and several coral fishes (Pakistan Wetlands Programme 2011). Iqbal et al. (2018) implemented advanced remote sensing techniques to map the coral biodiversity in Astola Islands. Ali et al. (2014) carried out a comprehensive survey in the coastal waters of Pakistan. They reported that Favites and Porites are the most dominant in this country's coastal periphery. Ghaffar (2015) estimated a potential habitat of almost 8000 km^2 where corals can be present; routine surveys should verify the extent of coral spread. Ghaffar (2015) further indicated the possibility of extensive coral patches in the deltaic offshore waters of Indus. These regions are yet to receive survey-based attention. Lately, Ali et al. (2020) identified 50 live coral species from the Sindh region. Churna Island exhibited potential coral habitats. They suggested that wave dynamics, elevated nutrient levels, rapidly changing temperature, and indiscriminate fishing activities are the principal threats to the corals of Pakistan.

7.4 The State-of-the-Art of Corals in the African Countries Bordering the Indian Ocean

The western Indian Ocean adjacent to the African coastline received the least attention with regards to identifying and documenting the coral biodiversity (Obura 2012). South Africa marks the end of the latitudinal spread of corals. Very few studies like

Boshoff (1981) and Riegl (1995) recorded the species diversity in the nearshore environment of South Africa. Several authors indicated that the species diversity is much less in South Africa. However, the bleaching events that destroyed substantial coral cover in the western Indian Ocean did not leave any such impact in South Africa (Riegl 2003). Smit and Glassom (2017) showed that intertidal corals, though exhibit less species diversity, can acclimatize well in temperate conditions. Thus, the corals of South Africa might perform better under the climate change scenario, unlike the corals in the northern Indian Ocean. Overall, Schleyer et al. (2008) reported 132 species from the St Lucia Wetland Park, South Africa, that accounts for most of this country's coral cover. Schleyer (2000) observed that almost three-fourth of the species diversity is of Scleractinia types. Schleyer et al. (2008) also indicated that the sea surface temperature is steadily rising in the South African periphery, which has led to more prevalence of hard corals replacing soft ones. However, the overall coral cover has reduced too. They estimated a coral area of around 50 km^2 in this country's coastline.

Mozambique has a long coastline of 2700 km. The northern and the southern stretch of this coastline shelters corals (Costa et al. 2005). The central swampy sector of the coastline does not shelter any coral due to the high turbidity caused by the drainage of several estuaries into the Indian Ocean (Pereira et al. 2004). The northern part has some continuous reef structures, whereas the south mostly displays patchy reefs (Motta et al. 2000). However, Obura (2012) reported very many species from the Mozambique channel that includes the reef structures of western Madagascar, Seychelles, and Comoros. They observed that the species diversity of this region is comparable to that observed in the coral triangle. Extensive reef structures are available in the Tanzanian coastline (Wagner 2004). However, these regions have become highly susceptible to marine heatwaves (Magel et al. 2020; Mawren et al. 2021). Among the anthropogenic reasons, coral mining has been a concern for Tanzania (Dulvy et al. 1995). In addition, sedimentation has been a major problem for the Kenyan corals (McClanahan and Obura 1997). The estuarine sediment load is critical in regulating coral growth throughout several places in the eastern boundary of Africa (Van Katwijk et al. 1993).

Much less is known about the corals of Somalia. Carbone et al. (1994) identified 63 species on the Somalia coast. However, their study was limited to few study sites. They reported a comparatively poor species diversity. Studies focusing on the threats of Somalia corals are scarce to date. The coastal periphery of Djibouti has a substantial coral cover. However, Hassan et al. (2002) indicated that out of the total coralline area, only one-third encompasses live corals. The site-specific spatial variability of live coral cover was significant in the coastline of Djibouti. The Gulf of Tadjoura had almost 90% dead corals, whereas the magnitude reduced to only 40% Sable Blanc. Besides the starfish, coral-consuming gastropods and sea urchins cause substantial damage to the corals of this country. Tourism-related activities are also increasing hand-in-hand with a lesser degree of environmental awareness (Hassan et al. 2002).

7.5 The State-of-the-Art of Corals in the Island Countries of the Indian Ocean

Sri Lanka has a coastline of almost 1600 km and shelters fringing to patchy reefs in its coastline (Berg et al. 1998). According to Rajasuriya and White (1995), only 2% of the entire coastline acts as an abode for corals in Sri Lanka. Most of the reefs are near the coastline, whereas few exist at substantial distances from the shoreline (Rajasuriya et al. 2002). Rajasuriya and White (1995) reported 171 species from the coastal sectors of this country. Turbidity enhancement and mining of corals for trade and commerce are the two bothering concerns of the Sri Lankan corals. The nearshore corals are susceptible to more damage than the offshore ones. However, indiscriminate fishing techniques and the use of explosives have made the offshore corals equally vulnerable (Ellepola et al. 2021). The 30-year-long civil war also took a serious toll on the corals of this country. However, Arulananthan et al. (2021) recently identified 36 new coral species not reported earlier from the Sri Lankan coastline. Ellepola et al. (2021) indicated a possibility of 2739 km^2 coral cover in the coastal waters of Sri Lanka. They also predicted that a substantial part of this cover would cease to thrive in the coming years unless strict government intervention comes forward.

The Maldives is a paradise for tourists, and the prime attraction has always been the coralline beauty (Kundur 2012). However, it is the tourism-related activities that have become a point of concern for the corals of this archipelago (Cowburn et al. 2018). Naseer and Hatcher (2004) estimated 4500 km^2 coral cover in and around the Maldives. Overall, the oceanic reef structures are predominant in the Maldives, with few regions having patchy reefs, especially in the atolls. Unlike many other countries, Maldives value the coral reefs as they offer protection from strong physical forcing events (Rajasuriya et al. 2002). However, apart from tourism-related activities, marine heat waves-driven coral bleaching and mining of corals have emerged as threats to this crucial heritage (Brown and Dunne 1988; McClanahan 2000; Jaleel 2013). McClanahan (2000) indicated that Porites and Astreopora dominance flourished at the cost of preexisting Acropora colonies in the post-bleaching phases after 1998. Similar threats exist in the coral communities of Seychelles. Payet and Agricole (2006) indicated that corals of Seychelles are paying a heavy price due to the ongoing climate change. Marine heatwaves have led to significant degradation of corals in the Seychelles coastline (Spencer et al. 2000). Chong-Seng et al. (2014) observed that a limited supply of larvae and premature deaths of juvenile corals are preventing a sustainable recovery of bleached corals in Seychelles. Coral degradation has made several parts of Seychelles vulnerable to wave dynamics (Sheppard et al. 2005). Similar observations are prevalent in another tourist-dominated country like Mauritius. Lately, Van der Schyff et al. (2020) indicated that shipping and navigation in shallow waters pose significant problems to the well-being of the coral colonies. Incidentally, the corals of Mauritius suffered less from bleaching events. However, tourism-related activities and fishing techniques are incessantly posing threats to the Mauritian corals (Turner et al. 2000). Thus, it is evident that corals are in dire need of conservation, restoration, and management interventions, especially in the small islands of the Indian Ocean.

7.6 Conclusion

Collating the available information on the corals of the Indian Ocean, it is evident that this unique ecosystem received substantial attention. However, their role in providing positive or negative feedback to the conventional blue carbon ecosystem requires a lot more study. Unlike the blue carbon ecosystems, corals are more susceptible to episodic events like bleaching led by marine heatwaves. Hence the species count for corals always varies substantially as many species often fail to cope up with the natural and anthropogenic alterations caused in the coral environment. Several countries are yet to generate updated databases on the species count and the areal spread. The publishing dates of the scientific literature show that steady monitoring is also missing in many of the countries. However, it was unanimous that corals bordering all the countries are under severe threat, except for those nations, where such research endeavors are absent altogether. Indiscriminate fishing, turbidity enhancement, and reckless tourism are the major threats to the corals of this region. Since marine conservationists have little to do in combatting heatwaves, rigorous restoration and conservation measures are in dire need to prevent the corals from further degradation.

References

Adam AA, Garcia RA, Galaiduk R, Tomlinson S, Radford B, Thomas L, Richards ZT (2021) Diminishing potential for tropical reefs to function as coral diversity strongholds under climate change conditions. Divers Distrib. https://doi.org/10.1111/ddi.13400

Ahammed SS, Hossain MA, Abedin MZ, Khaleque MA (2016) A study of environmental impacts on the coral resources in the vicinity of the Saint Martin Island. Bangladesh. Int J Sci Technol Res 5(1):37–39

Akmal KF, Shahbudin S, Faiz MHM, Hamizan YM (2019) Diversity and abundance of scleractinian corals in the East Coast of peninsular Malaysia: a case study of Redang and Tioman Islands. Ocean Sci J 54(3):435–456. https://doi.org/10.1007/s12601-019-0018-6

Alam O, Deng TL, Uddin MN, Alamgir M (2015) Application of environmental ethics for sustainable development and conservation of Saint Martin's Island in Bangladesh. J Environ Sci Nat Resour 8(1):19–27. https://doi.org/10.3329/jesnr.v8i1.24628

Al-Azri AR, Piontkovski SA, Al-Hashmi KA, Goes JI, Do Gomes HR (2010) Chlorophyll a as a measure of seasonal coupling between phytoplankton and the monsoon periods in the Gulf of Oman. Aquat Ecol 44(2):449–461. https://doi.org/10.1007/s10452-009-9303-2

Albert JA, Olds AD, Albert S, Cruz-Trinidad A, Schwarz A-M (2015) Reaping the reef: provisioning services from coral reefs in Solomon Islands. Mar Pol 62:244–251. https://doi.org/10.1016/j.marpol.2015.09.023

Ali A, Ormond R, Leujak W, Siddiqui PJA (2014) Distribution, diversity and abundance of coral communities in the coastal waters of Pakistan. J Mar Biol Assoc UK 94(1):75–84. https://doi.org/10.1017/S0025315413001203

Ali A, Siddiqui PJA, Rasheed M, Ahmad N, Shafique S, Khokhar FN (2020) Status of corals along the Sindh coast of Pakistan: prevailing environmental conditions, their impacts on community structure and conservation approaches. Reg Stud Mar Sci 39:101391. https://doi.org/10.1016/j.rsma.2020.101391

Anelli M, Julitta T, Fallati L, Galli P, Rossini M, Colombo R (2019) Towards new applications of underwater photogrammetry for investigating coral reef morphology and habitat complexity in the Myeik Archipelago, Myanmar. Geocarto Int 34(5):459–472. https://doi.org/10.1080/10106049.2017.1408703

Ara S, Alif MAUJ, Islam KA (2021) Impact of tourism on LULC and LST in a coastal island of Bangladesh: a geospatial approach on St. Martin's Island of Bay of Bengal. J India Soc Remote Sens. https://doi.org/10.1007/s12524-021-01389-4

Arulananthan A, Herath V, Kuganathan S, Upasanta A, Harishchandra A (2021) The status of the coral reefs of the Jaffna Peninsula (Northern Sri Lanka), with 36 coral species new to Sri Lanka confirmed by DNA bar-coding. Oceans 2(3):509–529. https://doi.org/10.3390/oceans2030029

Asaad I, Lundquist CJ, Erdmann MV, Costello MJ (2018) Delineating priority areas for marine biodiversity conservation in the Coral Triangle. Biol Conserv 222:198–211. https://doi.org/10.1016/j.biocon.2018.03.037

Asian Development Bank Report (2014) State of the coral triangle: Indonesia. Asian Development Bank. Mandaluyong City, Philippines

Beck MW, Losada IJ, Menendez P, Reguero BG, Díaz-Simal P, Fernandez F (2018) The global flood protection savings provided by coral reefs. Nat Commun 9(1):2186. https://doi.org/10.1038/s41467-018-04568-z

Benkwitt CE, Wilson SK, Graham NA (2020) Biodiversity increases ecosystem functions despite multiple stressors on coral reefs. Nat Ecol Evolution 4(7):919–926. https://doi.org/10.1038/s41559-020-1203-9

Ben-Romdhane H, Al-Musallami M, Marpu PR, Ouarda TB, Ghedira H (2018) Change detection using remote sensing in a reef environment of the UAE during the extreme event of El Niño 2015–2016. Int J Remote Sens 39:6358–6382. https://doi.org/10.1080/01431161.2018.1460502

Ben-Romdhane H, Jabado RW, Grandcourt EM, Perry RJO, Al Blooshi AY, Marpu PR et al (2020) Coral reefs of Abu Dhabi, United Arab Emirates: analysis of management approaches in light of international best practices and a changing climate. Front Mar Sci 7:541. https://doi.org/10.3389/fmars.2020.00541

Ben-Romdhane H, Ouarda TB, Marpu P, Rajan A, Bugla I, Perry RJ et al (2020a) Studying coral reef patterns in UAE waters using panel data analysis and multinomial logit and probit models. Ecol Indic 112106050. https://doi.org/10.1016/j.ecolind.2019.106050

Benzoni F, Bianchi CN, Morri C (2003) Coral communities of the northwestern Gulf of Aden (Yemen): variation in framework building related to environmental factors and biotic conditions. Coral Reefs 22(4):475–484. https://doi.org/10.1007/s00338-003-0342-1

Benzoni F, Calcinai B, Eisinger M, Klaus R (2008) Coral disease mimic: sponge attacks Porites lutea in Yemen. Coral Reefs 27(3):695. https://doi.org/10.1007/s00338-008-0371-x

Berg H, Öhman MC, Troëng S, Lindén O (1998) Environmental economics of coral reef destruction in Sri Lanka. Ambio 27(8):627–634. https://www.jstor.org/stable/4314808

Billah MM, Kader MA, Mahmud SS, Asif AA, Siddiqui AAM (2018) Diversity and distribution of seaweeds in Saint Martin Island, Bangladesh. Int J Fish Aquat Stud 6(6):166–169

BOBLME (2012) Transboundary diagnostic analysis vol 2. Background and environmental assessment Phuket, Thailand: Bay of Bengal Large Marine Ecosystem Project, p 108

Boshoff P (1981) An annotated checklist of Southern African scleractinia. Oceanographic Research Institute. Durban Invest Rep 49:1–45

Brown BE, Dunne RP (1988) The environmental impact of coral mining on coral reefs in the Maldives. Environ Conserv 15(2):159–165. https://doi.org/10.1017/S0376892900028976

Burke L, Selig E, Spalding M (2002) Reefs at risk in Southeast Asia. World Resources Institute, Washington

Burt JA (2013) The growth of coral reef science in the Gulf: a historical perspective. Mar Pollut Bull 72(2):289–301. https://doi.org/10.1016/j.marpolbul.2013.05.016

Burt JA, Coles S, Van Lavieren H, Taylor O, Looker E, Samimi-Namin K (2016) Oman's coral reefs: a unique ecosystem challenged by natural and man-related stresses and in need of conservation. Mar Pollut Bull 105(2):498–506. https://doi.org/10.1016/j.marpolbul.2015.11.010

Carbone F, Matteucci R, Rosen BR, Russo A (1994) Recent coral facies of the Indian Ocean coast of Somalia with an interim check list of corals. Facies 30(1):1–13. https://doi.org/10.1007/BF02536886

Cesar H (1996) Economic analysis of Indonesian coral reefs. Environment Department, The World Bank, Washington

Chen E, Blaze JA, Smith RS, Peng S, Byers JE (2020) Freeze tolerance of poleward-spreading mangrove species weakened by soil properties of resident salt marsh competitor. J Ecol 108(4):1725–1737. https://doi.org/10.1111/1365-2745.13350

Chong-Seng KM, Graham NAJ, Pratchett MS (2014) Bottlenecks to coral recovery in the Seychelles. Coral Reefs 33(2):449–461. https://doi.org/10.1007/s00338-014-1137-2

Chou LM (1996) Response of Singapore reefs to land reclamation. Galaxea 13:85–92

Chou LM, Toh KB, Tay YC, Phang VXH (2012) Coral reefs in Singapore: past, present and future. In: The Asian conference on sustainability, energy and the environment: conference proceedings, pp 431–436

Claereboudt MR (2006) Coral reefs and reef corals of the Gulf of Oman. Al-Roya Publishing, Muscat

Coles SL, Fadlallah H (1991) Reef coral survival and mortality at low temperatures in the Arabian Gulf: new species-specific lower temperature limits. Coral Reefs 9:231–237. https://doi.org/10.1007/BF00290427

Comeau S, Cornwall CE, DeCarlo TM, Doo SS, Carpenter RC, McCulloch MT (2019) Resistance to ocean acidification in coral reef taxa is not gained by acclimatization. Nat Clim Chang 9(6):477–483. https://doi.org/10.1038/s41558-019-0486-9

Coral Reef Monitoring Report (2008) Annual report reef check Malaysia

Costa A, Pereira MA, Motta H, Schleyer M (2005) Status of coral reefs of Mozambique: 2004. In: Souter D, Linden O (eds) Coral reef degradation in the Indian Ocean: status report. CORDIO Publishers, Sweden, pp 54–60

Cowburn B, Moritz C, Birrell C, Grimsditch G, Abdulla A (2018) Can luxury and environmental sustainability co-exist? Assessing the environmental impact of resort tourism on coral reefs in the Maldives. Ocean Coast Manag 158:120–127. https://doi.org/10.1016/j.ocecoaman.2018.03.025

Dahuri R (1995) Indonesia: national status and approaches to coastal management. In: Hotta K, Dutton IM (eds) Coastal management in the Asia-Pacific region: issues and approaches. Japan International Marine Science and Technology Federation (JIMSTEF), Tokyo, pp 277–290

Datta Majumdar S, Hazra S, Giri S, Chanda A, Gupta K, Mukhopadhyay A, Roy SD (2018) Threats to coral reef diversity of Andaman Islands, India: a review. Reg Stud Mar Sci 24:237–250. https://doi.org/10.1016/j.rsma.2018.08.011

De Goeij JM, van Oevelen D, Vermeij MJA, Osinga R, Middelburg JJ, de Goeij AFPM, Admiraal W (2013) Surviving in a marine desert: the sponge loop retains resources within coral reefs. Science 342(6154):108–110. https://doi.org/10.1126/science.1241981

De K, Nanajkar M, Arora M, Nithyanandan M, Mote S, Ingole B (2021) Application of remotely sensed sea surface temperature for assessment of recurrent coral bleaching (2014–2019) impact on a marginal coral ecosystem. Geocarto Int 1–25. https://doi.org/10.1080/10106049.2021.1886345

DeVantier L, De Ath G, Klaus R, Al-Moghrabi S, Abdulaziz M, Reinicke GB, Cheung C (2004) Reef-building corals and coral communities of the Socotra Archipelago, a zoogeographic crossroads' in the Arabian Sea. Fauna Arabia 20:117–168

Dikou A, Van Woesik R (2006) Survival under chronic stress from sediment load: spatial patterns of hard coral communities in the southern islands of Singapore. Mar Pollut Bull 52(11):1340–1354. https://doi.org/10.1016/j.marpolbul.2006.02.011

Dubinsky ZVY, Stambler N (1996) Marine pollution and coral reefs. Glob Chang Biol 2(6):511–526. https://doi.org/10.1111/j.1365-2486.1996.tb00064.x
Dulvy NK, Stanwell-Smith D, Darwall WR, Horrill CJ (1995) Coral mining at Mafia Island, Tanzania: a management dilemma. Ambio 24(6):358–365. https://www.jstor.org/stable/4314367
Edinger EN, Jompa J, Limmon GV, Widjatmoko W, Risk MJ (1998) Reef degradation and coral biodiversity in Indonesia: effects of land-based pollution, destructive fishing practices and changes over time. Mar Pollut Bull 36(8):617–630. https://doi.org/10.1016/S0025-326X(98)00047-2
Edward JP, Mathews G, Patterson J, Ramkumar R, Wilhelmsson D, Tamelander J, Linden O (2008) Status of coral reefs of the Gulf of Mannar, southeastern India. Coastal Oceans Research and Development in the Indian Ocean, East Africa, Kenya, p 489
Ellepola G, Harischandra S, Ranawana KB (2021) Spatial analysis of geophysical and environmental factors characterize distinct coral reef habitats around Sri Lanka; implications for management. Ocean Coast Manag 210:105667. https://doi.org/10.1016/j.ocecoaman.2021.105667
English S, Wilkinson C, Baker V (1997) Survey manual for tropical marine resources, 2nd edn. Australian Institute of Marine Science, Townsville
Fatemi SMR, Shokri MR (2001) Iranian coral reefs status with particular reference to Kish Island, Persian Gulf. In: International coral reef initiative Indian Ocean regional workshop, Maputo, Mozambique, 26–28 November 2001
Ferrario F, Beck MW, Storlazzi CD, Micheli F, Shepard CC, Airoldi L (2014) The effectiveness of coral reefs for coastal hazard risk reduction and adaptation. Nat Commun 5:3794. https://doi.org/10.1038/ncomms4794
Fisher R, O'Leary RA, Low-Choy S, Mengersen K, Knowlton N, Brainard RE, Caley MJ (2015) Species richness on coral reefs and the pursuit of convergent global estimates. Curr Biol 25(4):500–505. https://doi.org/10.1016/j.cub.2014.12.022
George JD, John DM (2000). The coral reefs of Abu Dhabi, United Arab Emirates: past, present and future. In George J, John D (eds) Proceedings of the 2nd Arab international conference and exhibition on environmental biotechnology (Coastal Habitats), Abu Dhabi, pp 142–156
Ghaffar B (2015) Estimating the potential habitat of coral reefs along the coastline of Pakistan using GIS and RS techniques. Int J Fauna Biol Stud 2(3):92–95
Gillis LG, Bouma TJ, Jones CG, van Katwijk MM, Nagelkerken I, Jeuken C et al (2014) Potential for landscape-scale positive interactions among tropical marine ecosystems. Mar Ecol Prog Ser 503:289–303. https://doi.org/10.3354/meps10716
Glynn PW (1993) Monsoonal upwelling and episode ic Acanthaster predation as probable controls of coral reef distribution and community structure in Oman, Indian Ocean. Atoll Res Bull 379:1–66
Grafeld S, Oleson KLL, Teneva L, Kittinger JN (2017) Follow that fish: Uncovering the hidden blue economy in coral reef fisheries. PLoS One 12(8):e0182104. https://doi.org/10.1371/journal.pone.0182104
Graham NAJ, Nash KL (2013) The importance of structural complexity in coral reef ecosystems. Coral Reefs 32(2):315–326. https://doi.org/10.1007/s00338-012-0984-y
Grandcourt E (2012) Reef fish and fisheries in the Gulf. In: Riegl B, Purkis S (eds) Coral reefs of the Gulf. Springer, Dordrecht, pp 127–161. https://doi.org/10.1007/978-94-007-3008-3_8
Grizzle RE, Ward KM, AlShihi RM, Burt JA (2016) Current status of coral reefs in the United Arab Emirates: distribution, extent, and community structure with implications for management. Mar Pollut Bull 105(2):515–523. https://doi.org/10.1016/j.marpolbul.2015.10.005
Guerra-Vargas LA, Gillis LG, Mancera-Pineda JE (2020) Stronger together: do coral reefs enhance seagrass meadows "blue carbon" potential? Front Mar Sci 7:628. https://doi.org/10.3389/fmars.2020.00628
Guest JR, Tun K, Low J, Vergés A, Marzinelli EM, Campbell AH, Bauman AG, Feary DA, Chou LM, Steinberg PD (2016) 27 years of benthic and coral community dynamics on turbid, highly urbanised reefs off Singapore. Sci Rep 6(1):1–10. https://doi.org/10.1038/srep36260
Guo W, Bokade R, Cohen AL, Mollica NR, Leung M, Brainard RE (2020) Ocean acidification has impacted coral growth on the great barrier reef. Geophys Res Lett 47(19):e2019GL086761. https://doi.org/10.1029/2019GL086761

Haapkylä J, Ramade F, Salvat B (2007) Oil pollution on coral reefs: a review of the state of knowledge and management needs. Vie et Milieu/Life Environ 2007:91–107. https://hal.sorbonne-universite.fr/hal-03234945

Harithsa S, Raghukumar C, Dalal SG (2005) Stress response of two coral species in the Kavaratti atoll of the Lakshadweep Archipelago, India. Coral Reefs 24(3):463–474. https://doi.org/10.1007/s00338-005-0008-2

Hassan M, Kotb MM, Al-Sofyani AA (2002) Status of coral reefs in the Red Sea-Gulf of Aden. In: Clive Wilkinson (ed) Status of coral reefs of the world. Australian Institute of Marine Science, pp 45–52. http://hdl.handle.net/1834/389

Hicks CC, Stoeckl N, Cinner JE, Robinson J (2014) Fishery benefits and stakeholder priorities associated with a coral reef fishery and their implications for management. Environ Sci Pol 44:258–270. https://doi.org/10.1016/j.envsci.2014.04.016

Hilton MJ, Manning SS (1995) Conversion of coastal habitats in Singapore: indications of unsustainable development. Environ Conserv 22:307–322. https://doi.org/10.1017/S0376892900034883

Hoegh-Guldberg O, Poloczanska ES, Skirving W, Dove S (2017) Coral reef ecosystems under climate change and ocean acidification. Front Mar Sci 4:158. https://doi.org/10.3389/fmars.2017.00158

Hoeksema BW, Scott C, True JD (2013) Dietary shift in corallivorous Drupella snails following a major bleaching event at Koh Tao, Gulf of Thailand. Coral Reefs 32:423–428. https://doi.org/10.1007/s00338-012-1005-x

Huang DW, Tun KPP, Chou LM, Todd PA (2009) An Inventory of zooxanthellate scleractinian corals in Singapore including 33 new records. Raffles Bull Zool 22:69–80

Hume BC, D'Angelo C, Smith EG, Stevens JR, Burt J, Wiedenmann J (2015) Symbiodinium thermophilum sp. nov., a thermotolerant symbiotic alga prevalent in corals of the world's hottest sea, the Persian/Arabian Gulf. Sci Rep 5:8562. https://doi.org/10.1038/srep08562

Hutomo M, Nontji N, Dirhamsyah (2009) Regional coastal resources and environment profile for ASEAN Region. Research Centre for Oceanography, Indonesian Institute of Sciences, Jakarta

Iqbal A, Qazi WA, Shahzad N, Nazeer M (2018) Identification and mapping of coral reefs using Landsat 8 OLI in Astola Island, Pakistan coastal ocean. In: 14th International Conference on Emerging Technologies (ICET) IEEE, pp 1–6

Jafarabadi AR, Bakhtiari AR, Spanò N, Cappello T (2018) First report of geochemical fractionation distribution, bioavailability and risk assessment of potentially toxic inorganic elements in sediments of coral reef Islands of the Persian Gulf, Iran. Mar Pollut Bull 137:185–197. https://doi.org/10.1016/j.marpolbul.2018.09.052

Jafarabadi AR, Dashtbozorg M, Raudonytė-Svirbutavičienė E, Bakhtiari AR (2021a) A potential threat to the coral reef environments: polybrominated diphenyl ethers and phthalate esters in the corals and their ambient environment (Persian Gulf, Iran). Sci Total Environ 775:145822. https://doi.org/10.1016/j.scitotenv.2021.145822

Jafarabadi AR, Dashtbozorg M, Raudonytė-Svirbutavičienė E, Bakhtiari AR (2021b) Chlorinated paraffins (SCCPs and MCCPs) in corals and water-SPM-sediment system in the Persian Gulf, Iran: a potential global threat for coral reefs. Environ Pollut 275:116531. https://doi.org/10.1016/j.envpol.2021.116531

Jafarabadi AR, Mitra S, Raudonytė-Svirbutavičienė E, Bakhtiari AR (2020) Large-scale evaluation of deposition, bioavailability and ecological risks of the potentially toxic metals in the sediment cores of the hotspot coral reef ecosystems (Persian Gulf, Iran). J Haz Mat 400:122988. https://doi.org/10.1016/j.jhazmat.2020.122988

Jaleel A (2013) The status of the coral reefs and the management approaches: the case of the Maldives. Ocean Coast Manag 82:104–118. https://doi.org/10.1016/j.ocecoaman.2013.05.009

Kinsey DW, Hopley D (1991) The significance of coral reefs as global carbon sinks—response to greenhouse. Palaeogeogr Palaeoclim Palaeoecol 89(4):363–377. https://doi.org/10.1016/0031-0182(91)90172-N

Kundur SK (2012) Development of tourism in Maldives. Int J Sci Res Pub 2(4):1–5

Lasut MT, Weber M, Pangalila F, Rumampuk ND, Rimper JR, Warouw V et al (2018) From Coral Triangle to Trash Triangle—How the Hot spot of Global Marine Biodiversity Is Threatened by Plastic Waste. In: Cocca M, Di Pace E, Errico M, Gentile G, Montarsolo A, Mossotti R (eds) Proceedings of the international conference on microplastic pollution in the Mediterranean sea. Springer Water, Springer, Cham. https://doi.org/10.1007/978-3-319-71279-6_15

Limmon G, Delrieu-Trottin E, Patikawa J, Rijoly F, Dahruddin H, Busson F et al (2020) Assessing species diversity of Coral Triangle artisanal fisheries: a DNA barcode reference library for the shore fishes retailed at Ambon harbor (Indonesia). Ecol Evol 10(7):3356–3366. https://doi.org/10.1002/ece3.6128

Lovelock CE, Duarte CM (2019) Dimensions of blue carbon and emerging perspectives. Biol Lett 15(3):20180781. https://doi.org/10.1098/rsbl.2018.0781

Loya Y, Rinkevich B (1980) Effects of oil pollution on coral reef communities. Mar Ecol Prog Ser 3(16):180. https://www.jstor.org/stable/24813665

Magel JM, Dimoff SA, Baum JK (2020) Direct and indirect effects of climate change-amplified pulse heat stress events on coral reef fish communities. Ecol Appl 30(6):e02124. https://doi.org/10.1002/eap.2124

Mawren D, Hermes J, Reason CJC (2021) Marine heatwaves in the Mozambique channel. Clim Dyn. https://doi.org/10.1007/s00382-021-05909-3

McClanahan TR (2000) Bleaching damage and recovery potential of Maldivian coral reefs. Mar Pollut Bull 40(7):587–597. https://doi.org/10.1016/S0025-326X(00)00064-3

McClanahan TR, Obura D (1997) Sedimentation effects on shallow coral communities in Kenya. J Exp Mar Biol Ecol 209(1–2):103–122. https://doi.org/10.1016/S0022-0981(96)02663-9

Meyer DL, Townsend EC, Thayer GW (1997) Stabilization and erosion control value of oyster cultch for intertidal marsh. Restor Ecol 5(1):93–99. https://doi.org/10.1046/j.1526-100X.1997.09710.x

Mollica NR, Guo W, Cohen AL, Huang KF, Foster GL, Donald HK, Solow AR (2018) Ocean acidification affects coral growth by reducing skeletal density. Proc Natl Acad Sci 115(8):1754–1759. https://doi.org/10.1073/pnas.1712806115

Motta H, Rodrigues MJ, Schleyer M (2000) Coral reef monitoring and management in Mozambique. Aqua Docs. http://hdl.handle.net/1834/471

Motuhi SE, Mehiri M, Payri CE, La Barre S, Bach S (2016) Marine natural products from New Caledonia—a review. Mar Drug 14(3):58. https://doi.org/10.3390/md14030058

Muir PR, Wallace CC, Done T, Aguirre JD (2015) Limited scope for latitudinal extension of reef corals. Science 348(6239):1135–1138. https://doi.org/10.1126/science.1259911

Mumby PJ, Broad K, Brumbaugh DR, Dahlgren CP, Harborne AR, Hastings A et al (2008) Coral reef habitats as surrogates of species, ecological functions, and ecosystem services. Conserv Biol 22(4):941–951. https://doi.org/10.1111/j.1523-1739.2008.00933.x

Naseer A, Hatcher BG (2004) Inventory of the Maldives' coral reefs using morphometrics generated from Landsat ETM+ imagery. Coral Reefs 23(1):161–168. https://doi.org/10.1007/s00338-003-0366-6

Obura D (2012) The diversity and biogeography of Western Indian Ocean reef-building corals. PLoS One 7(9):e45013. https://doi.org/10.1371/journal.pone.0045013

Pakistan Wetlands Programme (2011) Report on discovery of Corel reefs in Gwader. The Ministry of Environment's Pakistan Wetlands Program

Palumbi SR, Barshis DJ, Traylor-Knowles N, Bay RA (2014) Mechanisms of reef coral resistance to future climate change. Science 344:895–898. https://doi.org/10.1126/science.1251336

Patterson J, Jeyasanta KI, Sathish N, Edward JP, Booth AM (2020) Microplastic and heavy metal distributions in an Indian coral reef ecosystem. Sci Total Environ 744:140706. https://doi.org/10.1016/j.scitotenv.2020.140706

Payet R, Agricole W (2006) Climate change in the Seychelles: implications for water and coral reefs. Ambio 35(4):182–189. https://www.jstor.org/stable/4315717

Pereira MA, Gonçalves PMB (2004) Effects of the 2000 southern Mozambique floods on a marginal coral community: the case at Xai-Xai. Afr J Aquat Sci 29(1):113–116. https://doi.org/10.2989/16085910409503800

Phongsuwan N, Chansang H (2012) Repeated coral bleaching in the Andaman Sea, Thailand, during the last two decades. Phuket Mar Biol Cent Res Bull 71:19–41

Praveena SM, Siraj SS, Aris AZ (2012) Coral reefs studies and threats in Malaysia: a mini review. Rev Environ Sci Biotechnol 11(1):27–39. https://doi.org/10.1007/s11157-011-9261-8

Purser BH, Evans G (1973) Regional sedimentation along the Trucial coast, SE Persian Gulf. In: Purser BH (ed) The Persian Gulf, Springer, Heidelberg, pp 211–231. https://doi.org/10.1007/978-3-642-65545-6_13

Quevedo JMD, Uchiyama Y, Muhamad Lukman K, Kohsaka R (2021) How blue carbon ecosystems are perceived by local communities in the coral triangle: comparative and empirical examinations in the Philippines and Indonesia. Sustainability 13(1):127. https://doi.org/10.3390/su13010127

Raghuraman R, Raghunathan C, Venkataraman K (2013) Present status of coral reefs in India. In: Venkataraman K, Sivaperuman C, Raghunathan C (eds) Ecology and conservation of tropical marine faunal communities, Springer, Heidelberg, pp 351–379. https://doi.org/10.1007/978-3-642-38200-0_23

Rahman MM, Alam MA (2020) Regulatory and Institutional framework for the conservation of coral reefs in Bangladesh: a critical review. In: Alam MA, Alam F, Begum D (eds) Knowledge Management, Governance and Sustainable Development: Lessons and Insights from Developing Countries. India: Routledge, pp 231–244. http://dx.doi.org/https://doi.org/10.2139/ssrn.3794550

Rajasuriya A, Zahir H, Muley EV, Subramanian K, Venkataraman MVM, Wafar SM, Khan MH, Whittingham E (2002) Status of coral reefs in South Asia: Bangladesh, India, Maldives, Sri Lanka, in Wilkinson, C (ed) Status of coral reefs of the world, Australian Institute of Marine Science:Townsville, pp. 101–121.

Rajasuriya A, White AT (1995) Coral reefs of Sri Lanka: review of their extent, condition, and management status. Coast Manag 23(1):77–90. https://doi.org/10.1080/08920759509362257

Rani S, Ahmed MK, Xiongzhi X, Yuhuan J, Keliang C, Islam MM (2020) Economic valuation and conservation, restoration & management strategies of Saint Martin's coral island, Bangladesh. Ocean Coast Manag 183:105024. https://doi.org/10.1016/j.ocecoaman.2019.105024

Raza A, Shaukat SS, Perveen R, Hussain R (2014) Current status of coral research in Pakistan. Int J Fauna Biol Stud 1(6):105–107

Raza AA (2013) Taxonomic Study of Collections of Scleractinians Corals (Cnidaria; Anthozoa; Zoantharia) in the Zoological and Geological Museum of Karachi University, Karachi. Ph.D. Thesis, Department of Zoology, University of Karachi, Karachi, Pakistan.

Reichert J, Tirpitz V, Anand R, Bach K, Knopp J, Schubert P et al (2021) Interactive effects of microplastic pollution and heat stress on reef-building corals. Environ Pollut 118010. https://doi.org/10.1016/j.envpol.2021.118010

Reopanichkul P, Schlacher TA, Carter RW, Worachananant S (2009) Sewage impacts coral reefs at multiple levels of ecological organization. Mar Pollut Bull 58(9):1356–1362. https://doi.org/10.1016/j.marpolbul.2009.04.024

Rezai H (1996) Observation of some corals in shallow waters of several remote Iranian islands in the Persian Gulf. Abzeeyan Mon Mag 7:4–10

Rezai H, Jahangard A (1994) Observation of the sea star *Acanthaster planci* (L.) in shallow water of Small Tunb Island (a report). Bandar Lengeh, Iran: Persian Gulf Molluscs Fisheries Research Center.

Rezai H, Wilson S, Claereboudt M, Riegl B (2004) Coral reef status in the ROPME sea area: Arabian/Persian Gulf, Gulf of Oman and Arabian Sea. Status Coral Reefs World 1:155–170

Riegl B (1995) Description of four new species in the hard coral genus Acropora Oken, 1815 (Scleractinia: Astrocoeniina: Acroporidae) from south-east Africa. Zool J Linn Soc 113:229–247. https://doi.org/10.1111/j.1096-3642.1995.tb00934.x

Riegl B (2003) Climate change and coral reefs: different effects in two high-latitude areas (Arabian Gulf, South Africa). Coral Reefs 22:433–446. https://doi.org/10.1007/s00338-003-0335-0

Riegl B, Purkis S (2011) Persian/Arabian Gulf Coral Reefs. In: Hopley D. (eds) Encyclopedia of Modern Coral Reefs. Encyclopedia of Earth Sciences Series, Springer, Dordrecht. https://doi.org/10.1007/978-90-481-2639-2_123

Riegl BM, Purkis SJ (2012) Coral reefs of the Gulf: adaptation to climatic extremes in the world's hottest sea, In Riegl BM, Purkis SJ (eds) Coral Reefs of the Gulf, Springer, Dordrecht, pp. 1–4. https://doi.org/10.1007/978-94-007-3008-3_1

Sadeghi SM (1997) Identification and distribution survey of stone corals on Kish Island. MSc dissertation. Tehran Azad University, Iran.

Sale PF, Feary DA, Burt JA, Bauman AG, Cavalcante GH, Drouillard KG et al (2011) The growing need for sustainable ecological management of marine communities of the Persian Gulf. Ambio 40(1):4–17. https://doi.org/10.1007/s13280-010-0092-6

Salm RV (1993) Coral reefs of the Sultanate of Oman. Atoll Res Bull 380:1–85

Savidge G, Lennon J, Matthews AJ (1990) A shore-based survey of upwelling along the coast of Dhofar region, southern Oman. Cont Shelf Res 10(3):259–275. https://doi.org/10.1016/0278-4343(90)90022-E

Schleyer MH (2000) South African coral communities. Coral reefs of the Indian ocean: their ecology and conservation. Oxford University Press, New York, pp. 83–105.

Schleyer MH, Kruger A, Celliers L (2008) Long-term community changes on a high-latitude coral reef in the Greater St Lucia Wetland Park. South Africa. Mar Pollut Bull 56(3):493–502. https://doi.org/10.1016/j.marpolbul.2007.11.011

Schmidt GM, Wall M, Taylor M, Jantzen C, Richter C (2016) Large-amplitude internal waves sustain coral health during thermal stress. Coral Reef 35:1–13. https://doi.org/10.1007/s00338-016-1450-z

Seguin F, Le Brun O, Hirst R, Al-Thary I, Dutrieux E (2008) Large coral transplantation in Bal Haf (Yemen): an opportunity to save corals during the construction of a Liquefied Natural Gas plant using innovative techniques. In Proc. 11th Int. Coral Reef Symp. Ft. Lauderdale, Florida, pp. 1267–1270.

Shaver EC, Silliman BR (2017) Time to cash in on positive interactions for coral restoration. PeerJ 5:e3499. https://doi.org/10.7717/peerj.3499

Sheppard CRC, Sheppard ALS (1991) Coral and coral communities of Arabia. Fauna Saudi Arabia 12:3–170

Sheppard CRC, Salm RV (1988) Reef and coral communities of Oman, with a description of a new coral species (Order Scleractinia, genus Acanthastrea). J Nat Hist 22(1):263–279. https://doi.org/10.1080/00222938800770201

Sheppard C, Dixon DJ, Gourlay M, Sheppard A, Payet R (2005) Coral mortality increases wave energy reaching shores protected by reef flats: examples from the Seychelles. Estuar Coast Shelf Sci 64(2–3):223–234. https://doi.org/10.1016/j.ecss.2005.02.016

Sheppard C, Price A, Roberts C (1992) Marine ecology of the Arabian region: patterns and processes in extreme tropical environments. FAO, Rome, Italy.

Shinn EA (1976) Coral reef recovery in Florida and the Persian Gulf. Environ Geol 1(4):241

Shojae F, Kamrani E, Shokri MR, Ranjbar MS, Moradi M, Hesni MA (2010) New records of three hard coral species from north-east of Larak Island (Persian Gulf, Iran). Mar Biodivers Rec e65. https://doi.org/10.1017/S1755267210000515

Shokri MR, Fatemi SMR, Crosby MP (2005) The status of butterflyfishes (Chaetodontidae) in the northern Persian Gulf, I.R. Iran. Aquat Conserv Mar Freshw Ecosyst 15:91–99. https://doi.org/10.1002/aqc.714

Silliman BR, Schrack E, He Q, Cope R, Santoni A, Van Der Heide T et al (2015) Facilitation shifts paradigms and can amplify coastal restoration efforts. Proc Natl Acad Sci 112(46):14295–14300. https://doi.org/10.1073/pnas.1515297112

Smit K, Glassom D (2017) Large fluctuations but constant mean temperatures allow corals to persist in intertidal rock pools on the east coast of South Africa. Helgol Mar Res 71:3. https://doi.org/10.1186/s10152-017-0482-2

Spalding M, Ravilious C, Green EP (2001) World Atlas of Coral Reefs. University of California Press, Berkeley

Spalding M, Burke L, Wood SA, Ashpole J, Hutchison J, zu Ermgassen P (2017) Mapping the global value and distribution of coral reef tourism. Mar Pol 82:104–113. https://doi.org/10.1016/j.marpol.2017.05.014

Spencer T, Teleki KA, Bradshaw C, Spalding MD (2000) Coral bleaching in the southern Seychelles during the 1997–1998 Indian Ocean warm event. Mar Pollut Bull 40(7):569–586. https://doi.org/10.1016/S0025-326X(00)00026-6

Stuart-Smith RD, Brown CJ, Ceccarelli DM, Edgar GJ (2018) Ecosystem restructuring along the Great Barrier Reef following mass coral bleaching. Nature 560(7716):92–96. https://doi.org/10.1038/s41586-018-0359-9

Suggett DJ, Smith DJ (2020) Coral bleaching patterns are the outcome of complex biological and environmental networking. Glob Chang Biol 26(1):68–79. https://doi.org/10.1111/gcb.14871

Sully S, Burkepile DE, Donovan MK, Hodgson G, Van Woesik R (2019) A global analysis of coral bleaching over the past two decades. Nat Commun 10(1):1–5. https://doi.org/10.1038/s41467-019-09238-2

Sutthacheep M, Yucharoen M, Klinthong W, Pengsakun S, Sangmanee K, Yeemin T (2012) Coral mortality following the 2010 mass bleaching event at Kut Island, Thailand. Phuket Mar Biol Cent Res Bull 71:83–92

Sutthacheep M, Yucharoen M, Klinthong W, Pengsakun S, Sangmanee K, Yeemin T (2013) Impacts of the 1998 and 2010 mass coral bleaching events on the Western Gulf of Thailand. Deep Res Part II Top Stud Oceanogr 96:25–31. https://doi.org/10.1016/j.dsr2.2013.04.018

Tan CH, Heron SF (2011) First observed severe mass bleaching in Malaysia, Greater Coral Triangle. Galaxea J Coral Reef Stud 13(1):27–28. https://doi.org/10.3755/galaxea.13.27

Tun K, Heiss G (2005) Status of coral reefs in Myanmar: Post-tsunami Assessment. Australian Institute of Marine Science and Global Coral Reef Monitoring Network.

Turak E, Brodie J, DeVantier L (2007) Reef-building corals and coral communities of the Yemen Red Sea. Fauna Arabia 23:1–40

Turner JR, Hardman E, Klaus R, Fagoonee I, Daby D, Baghooli R, Persands S (2000) The reefs of Mauritius. Coral reef degradation in the Indian Ocean: status report. CORDIO Publishers, Sweden.

Van der Schyff V, Du Preez M, Blom K, Kylin H, Yive NSCK, Merven J et al (2020) Impacts of a shallow shipwreck on a coral reef: A case study from St. Brandon's Atoll, Mauritius, Indian Ocean. Mar Environ Res 156:104916. https://doi.org/10.1016/j.marenvres.2020.104916

Van Katwijk MM, Meier NF, Van Loon R, Van Hove EM, Giesen WBJT, Van der Velde G, Den Hartog C (1993) Sabaki River sediment load and coral stress: correlation between sediments and condition of the Malindi-Watamu reefs in Kenya (Indian Ocean). Mar Biol 117(4):675–683. https://doi.org/10.1007/BF00349780

Vaughan GO, Burt JA (2016) The changing dynamics of coral reef science in Arabia. Mar Pollut Bull 105(2):441–458. https://doi.org/10.1016/j.marpolbul.2015.10.052

Venkataraman K (2006) Coral reefs in India. National Biodiversity Authority.

Veron JEN (1995) Corals in space and time: the biogeography and evolution of the Scleractinia. Cornell University Press, Sydney

Veron JE, Devantier LM, Turak E, Green AL, Kininmonth S, Stafford-Smith M, Peterson N (2009) Delineating the coral triangle. Galaxea J Coral Reef Stud 11(2):91–100. https://doi.org/10.3755/galaxea.11.91

Vinoth R, Gopi M, Kumar TTA, Thangaradjou T, Balasubramanian T (2012) Coral reef bleaching at Agatti Island of Lakshadweep atolls. India. J Ocean Univ China 11(1):105–110. https://doi.org/10.1007/s11802-012-1822-0

Wagner GM (2004) Coral reefs and their management in Tanzania. West India Ocean J Mar Sci 3(2):227–243

Waheed Z, Hoeksema BW (2013) A tale of two winds: species richness patterns of reef corals around the Semporna peninsula. Malaysia. Mar Biodivers 43(1):37–51. https://doi.org/10.1007/s12526-012-0130-7

Waheed Z, Hoeksema BW (2014) Diversity patterns of Scleractinian corals at Kota Kinabalu, Malaysia, in relation to exposure and depth. Raffles Bull Zool 62:66–82. http://zoobank.org/urn:lsid:zoobank.org:pub:10140B8E-BD9C-4883-A384-63893CDEDBF

Wallace CC, Richard Z, Suharsono (2001) Regional Distribution Patterns of Acropora and Their Use in the Conservation of Coral Reefs in Indonesia. Jurnal Pesisir & Lautan (Indones J Coast Mar Resour) 4(1):40–58.

Weber M, De Beer D, Lott C, Polerecky L, Kohls K, Abed RM et al (2012) Mechanisms of damage to corals exposed to sedimentation. Proc Natl Acad Sci 109(24):E1558–E1567. https://doi.org/10.1073/pnas.1100715109

White AT, Aliño PM, Cros A, Fatan NA, Green AL, Teoh SJ et al (2014) Marine protected areas in the Coral Triangle: progress, issues, and options. Coast Manag 42(2):87–106. https://doi.org/10.1080/08920753.2014.878177

Wilkinson C, Souter D, Goldberg J (2005) Status of coral reefs in tsunami-affected countries: 2005. Australian Institute of Marine Science; Global Coral Reef Monitoring Network.

Wongthong P, True J, Manopawitr P (2015) Situation analysis of the Myeik Archipelago. BOBLME-2015-Ecology-36.

Woodhead AJ, Hicks CC, Norström AV, Williams GJ, Graham NA (2019) Coral reef ecosystem services in the Anthropocene. Funct Ecol 33(6):1023–1034. https://doi.org/10.1111/1365-2435.13331

Yeemin T, Ruengsawang N, Buaruang J, Nopchinwong P, Asa S, Panthewee W (1998) Coral bleaching in the Gulf of Thailand: a significant natural disturbance. Proc 24th Congr Sci Technol Thailand, Bangkok, pp 698–699.

Yeemin T, Saenghaisuk C, Pengsakun S, Donsomjit W, Klinthong W, Sangmanee K, Sutthacheep M (2010) Patterns of coral bleaching at Koh Yak, Mu Koh Chang National Park. Proc 36th Congr Sci Technol Thailand, Bangkok, pp 1–5.

Yeemin T, Saenghaisuk C, Sutthacheep M, Pengsakun S, Klinthong W, Saengmanee K (2009) Conditions of coral communities in the Gulf of Thailand: a decade after the 1998 severe bleaching event. Galaxea J Coral Reef Stud 11:207–217. https://doi.org/10.3755/galaxea.11.207

Yeemin T, Pengsakun S, Yucharoen M et al (2013) Long-term decline in Acropora species at Kut Island, Thailand, in relation to coral bleaching events. Mar Biodiv 43:23–29. https://doi.org/10.1007/s12526-012-0138-z

Yeemin T, Sutthacheep M, Pettongma R (2006) Coral reef restoration projects in Thailand. Ocean Coast Manag 49(9–10):562–575. https://doi.org/10.1016/j.ocecoaman.2006.06.002

Yucharoen M, Chankong A, Phongsuwan N, Ninwat S, Darumas U, Thongkao S, Phasombun S (2020) Coral Diversity at Losin Pinnacle, an Offshore Reef in the Gulf of Thailand: Toward a Future MPA. Front Mar Sci 7:130. https://doi.org/10.3389/fmars.2020.00130

Chapter 8
Blue Carbon Dynamics of Australia

Anirban Akhand

Abstract Australia has a long coastline and faces the Indian Ocean and the Pacific Ocean. This continent has a typical tropical and temperate regime along the coastline, which shelters varied blue carbon ecosystems. The coastal periphery shelters all three conventional blue carbon habitats like mangroves, seagrasses, and salt marshes. Australian Government took several initiatives to conserve and rehabilitate several blue carbon habitats, and they incorporated these crucial ecosystems within the framework of the national greenhouse gas inventory. The present chapter focused exclusively on the blue carbon habitats of Australia. The chapter collated the significant observations made so far by the researchers who worked on these three ecosystems. This chapter speaks about the spatial distribution and species diversity of the three ecosystems. However, the carbon content characterization holds the prime emphasis. We discussed the aboveground, belowground, and pedosphere carbon assimilation and some specialized observations from all the Australian blue carbon habitats. Overall, this chapter gives a brief overview of the state-of-the-art knowledge acquired on the blue carbon ecosystems of this continent.

Keywords New South Wales · Western Australia · Northern Territory · Queensland · Victoria · Mangroves · Seagrass · Salt marsh · Biomass carbon · Soil organic carbon

8.1 Introduction

Australia is the sixth-largest country in the world and encompasses an area of almost 7,692,000 km^2. Including the fragmented islands, this continent has a coastline of around 59,700 km. The latitudinal extent of the Commonwealth of Australia ranges between 10° 41′ S and 43° 38′ S. The entire continental landmass experiences a typical tropical to sub-tropical climate. The long coastline and a suitable environment make Australia one of the blue carbon giants of the present day. This country holds

A. Akhand (✉)
Coastal and Estuarine Environment Research Group, Port and Airport Research Institute, 3-1-1, Nagase, Yokosuka 239-0826, Kanagawa, Japan

A. Chanda et al. (eds.), *Blue Carbon Dynamics of the Indian Ocean*,
https://doi.org/10.1007/978-3-030-96558-7_8

the credit of sheltering the second-largest mangrove (after Indonesia) (Friess and Webb 2014) and the largest seagrass cover of the world (Green et al. 2003). Salt marshes are also available in continuity with mangrove and seagrass meadows on the Australian coastlines (Brown et al. 2016). The coastal carbon-rich ecosystems of this country have received substantial attention from the scientific community in the recent past (Duarte et al. 2010; Duarte 2017). Australian Government realizes their blue carbon strength and appreciates it as a tool to combat climate change (Bell-James 2016). Australia has formulated several in-house and international strategies to fight out the evil of climate change. Australia's the-then Environment minister officially declared in the UNCCC conference held in Paris in 2015 that they have incorporated the blue carbon ecosystems in their legislative framework of the National GHG Inventory. This initiative is a welcome step from a developed country like Australia. Such endeavor would promote the significance of blue carbon ecosystems in many other countries. A sincere attempt like this can help the world rejuvenate the lost blue carbon stretches and conserve them for the future.

8.2 Australian Mangroves

According to Köppen's climate classification, Australia experiences varied climatic conditions throughout the country (Stern et al. 2000), and the mangroves also thrive under different categories. Arid climate is the predominant type, followed by the temperate climate. Typical tropical and cold climatic conditions are also prevalent in the country. Crosbie et al. (2012) argued that the arid and tropical climatic features would enhance in Australia with time, and the temperate and cold conditions would gradually diminish. Lymburner et al. (2020) pointed out that the highly productive mangroves of Australia thrive under monsoon, savannah, warm oceanic, and humid subtropical climates. These mangrove stands exhibit high species diversity. The dispersed mangroves with low productivity and species diversity grow in the temperate, Mediterranean, and semi-arid to desert climates. Mangroves situated in the northern border, Queensland, New South Wales, and the western margin of this continent belong to the former category, whereas the ones that grow in the southern part of New South Wales, Victoria, and the southern boundary come in the latter class.

8.2.1 Spatial Distribution of Mangroves

The Indian Ocean and the Pacific Ocean surround the Australian continent. Mainland Australia has five subdivisions. Western Australia is one of them, which remains encompassed by the Indian Ocean. The other regions like the Northern Territory, Queensland, New South Wales, and South Australia face the Pacific Ocean. Mangroves are available in the entire coastal stretch. However, their characteristic features vary spatially. Clough (1982) detailed the Australian mangroves and their specialty in a vivid way. In the same volume, Galloway (1982) elaborated on the spatial distribution and the physiographic patterns of the mangrove floristic stands in the Australian periphery. Galloway (1982) reported a modest mangrove cover of around 11,500 km^2, and according to this estimation, almost 6000 km of coastal stretches in mainland Australia and 1000 km in the shelter of the fringing islands mangrove vegetations. The southern part of Western Australia and the Tasmanian coastlines are almost devoid of any mangrove cover. The remaining continental margin shelters sparse to dense mangroves. The mangroves of Australia are very thin at some places, at best one or two trees in the coastal periphery, whereas some locations have extensive mangrove cover of few kilometers wide. According to Galloway (1982), almost 40% of mangroves of the entire country thrive in Queensland and 36% in Northern Territory. 21% of mangroves grow in Western Australia. The remaining 3% remains scattered in South Australia, New South Wales, and Victoria (Fig. 8.1).

Australian Government has taken the initiative of protecting the mangroves under the umbrella of state and federal laws of this country. More than 19% of the nation's mangrove cover falls under the IUCN protected forest reserves. Asbridge et al. (2016) suggested that such protective measures prevented the Australian mangroves from direct anthropogenic impacts. However, the natural factors like the strong-physical-forcing events (Asbridge et al. 2018), rise in regional sea level (Woodroffe 2018; Sandi et al. 2021), fluctuating ambient temperature (Osland et al. 2017; Hickey et al. 2021), and precipitation patterns continue to threaten the mangroves of Australia. Despite having a substantial mangrove cover, mapping endeavors have accelerated

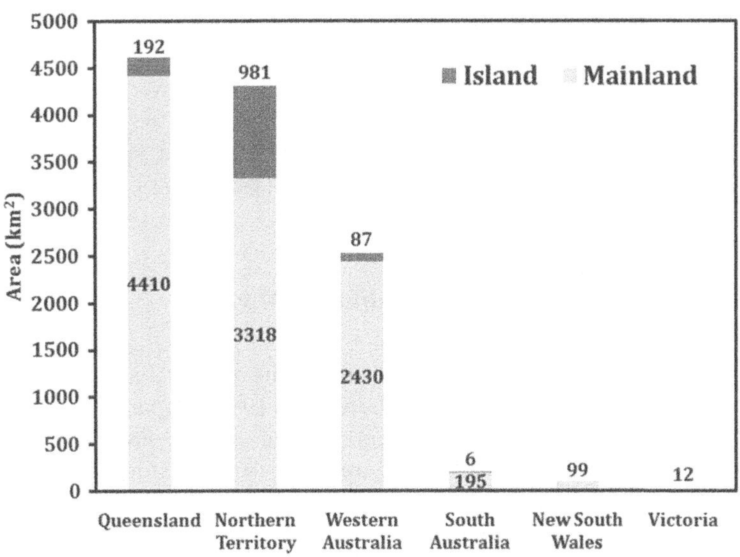

Fig. 8.1 The spatial distribution of mangroves in the different sectors of Australia (according to Galloway (1982))

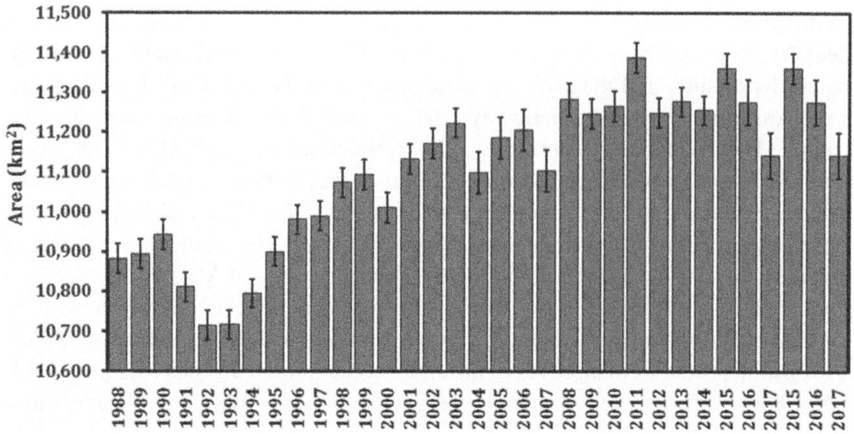

Fig. 8.2 The changes in mangrove area observed in Australian periphery between the years 1988 and 2017 (after Lymburner et al. (2020))

in this continent only in the past two decades. The National Forest Inventory of Australia put together the ground-level observations at the state and territory levels to generate the nationwide mangrove maps (Lymburner et al. 2020). The worldwide mangrove mapping missions (Spalding et al. 1997, 2010; Giri et al. 2011; Bunting et al. 2018) generated first-hand knowledge of Australian mangroves. However, most of the national reports of Australia relied on estimates made in discrete years. Lately, Lymburner et al. (2020) analyzed the country's mangrove cover dynamics from 1988 to 2017. They observed that mangroves recorded a minimum of around 10,700 km^2 during 1992–93, whereas a maximum of approximately 11,400 km^2 in 2011 (Fig. 8.2). Lymburner et al. (2020) reported that the woodland mangroves, open canopy, and closed canopy structures encompassed 12%, 49%, and 39% of Australia's total mangrove cover. They further inferred that most of the spatial changes in mangrove areas occurred in the northern periphery of the continent.

8.2.2 Species Diversity in Australian Mangroves

Several authors studied the mangrove species diversity in Australia. The most notable endeavors date back to the 1970s and 1980s (Clough 1982). Disparities exist between the different reports, and the species diversity varied spatially throughout the continent. Queensland and Northern Territory reported the highest number of species. Western Australia also records a substantial number of true mangroves. However, the biodiversity in the southern states is substantially poor than its northern counterparts. Dowling and McDonald (1982) reported discontinuous mangrove patches all along the Queensland coastline and the islands situated at least 1 km away from the mainland. The Queensland sector has many low-lying coastal plains having silty and muddy substrates, which favors the growth of luxuriant mangrove cover. Dowling and McDonald (1982) pointed out that the sediments of this region often tend to become hypersaline due to irregular tidal flushing activity. Macnae (1966), Jones (1971), and Lear and Turner (1977) reported a varying number of mangrove species from Queensland. However, Dowling and McDonald (1982) compiled those observations and enlisted 32 species from this region. The definition of true mangroves varied among authors, and so did the distinction between woody plants and shrubs, which led to the discrepancies in the number of species reported. Lately, Cresswell and Semeniuk (2018) reviewed the observations on species diversity. The review indicates that there hasn't been any significant change in the mangrove floristic distribution in the last three to four decades, except for the area covered by mangroves. *Aegialitis annulata*, *Aegiceras corniculatum*, *Avicennia marina*, *Bruguiera sp.*, *Ceriops sp.*, *Sonneratia sp.*, and *Xylocarpus sp.* are some of the dominant mangroves of the Queensland region (Dowling and McDonald 1982). Wells (1982) reported 25 species from the Northern Territory. *Avicennia marina*, *Lumnitzera racemosa*, *Excoecaria agallocha*, *Aegiceras corniculatum*, *Xylocarpus sp.*, *Bruguiera exaristata*, *Ceriops australis*, and *Rhizophora stylosa* are widely abundant in this region. However, the mangroves of Western Australia experience an arid climate and exhibit a sparse vegetation cover (Kenneally 1982). The geomorphic setting in this region is substantially different from the eastern and northern counterparts of Australia. Before the initiative of Semeniuk et al. (1978), information on Western Australian mangroves was scarce. Kenneally (1982) reported a list of 17 species from the Western Australian sector, and a substantial portion of this study followed the footsteps of Semeniuk et al. (1978). Bridgewater (1982) indicated that the southern Australian coastlines had the least species diversity. *Avicennia marina* was the dominant most in this sector. Several places had mono species stands of *Avicennia marina*. However, *Aegialitis annulata*, *Aegiceras corniculatum*, *Rhizophora stylosa*, *Bruguiera exaristata*, and *Ceriops tagal* frequently appeared in minor quantities in association with *Avicennia marina*. Among all the mangrove floras in Australia, two are endemic to this continent, namely *Avicennia integra* and *Ceriops australis* (Duke 2017). Collating all the observations, a modest count of 33 species of true mangroves prevail in Australia. Table 8.1 shows that *Aegialitis annulata*, *Aegiceras corniculatum*, *Avicennia marina*, *Bruguiera exaristata*, *Ceriops tagal*, and *Rhizophora stylosa* are the most dominant species and are available in all the nooks and corners of this continent.

8.2.3 Aboveground Biomass Carbon (AGC) Stock

Mangrove vegetations store substantial quantities of carbon in the different compartments of the ecosystem. Aboveground biomass carbon (AGC), belowground biomass carbon (BGC), and soil organic carbon (SOC) are the most significant storehouses in this regard (Donato et al. 2011). The mangroves of Australia received substantial attention from the scientific community in characterizing these estimates. The researchers also examined the factors that regulate the magnitude of these carbon

Table 8.1 The list of mangrove species that thrives in the Australian continent (collated from Clough 1982)

Sr. No	Species	Queensland	Northern Territory	Western Australia	Southern Australia
1	*Acanthus ilicifolius*	+	+		
2	*Aegialitis annulata*[a]	+	+	+	+
3	*Aegiceras corniculatum*[a]	+	+	+	+
4	*Avicennia eucalyptifolia*	+		+	
5	*Avicennia marina*[a]	+	+	+	+
6	*Avicennia officinalis*		+		
7	*Bruguiera cylindrica*	+			
8	*Bruguiera exaristata*[a]	+	+	+	+
9	*Bruguiera gymnorhiza*	+	+		
10	*Bruguiera parviflora*	+	+	+	
11	*Bruguiera sexangula*	+	+		
12	*Camptostemon schultzii*	+	+	+	
13	*Ceriops australis*	+	+		
14	*Ceriops decandra*	+	+		
15	*Ceriops tagal*[a]	+	+	+	+
16	*Cynometra iripa*	+			
17	*Excoecaria agallocha*	+	+	+	
18	*Heritiera littoralis*	+			
19	*Hibiscus tiliaceaus*	+			
20	*Lumnitzera littorea*	+	+		
21	*Lumnitzera racemosa*	+	+	+	
22	*Nypa fruticans*	+			
23	*Osbornia octodonta*	+	+	+	
24	*Pemphis acidula*	+		+	

(continued)

Table 8.1 (continued)

Sr. No	Species	Queensland	Northern Territory	Western Australia	Southern Australia
25	*Rhizophora apiculata*	+	+		
26	*Rhizophora mucronata*	+			
27	*Rhizophora stylosa*[a]	+	+	+	+
28	*Scyphiphora hydrophyllacea*	+	+	+	
29	*Sonneratia alba*	+	+	+	
30	*Sonneratia caseolaris*	+	+		
31	*Thespesia populnea*	+			
32	*Xylocarpus australasicus*	+	+	+	
33	*Xylocarpus granatum*	+	+	+	

[a] Available in all sectors of Australia

stocks and the spatial variability. Serrano et al. (2019) compiled a primary dataset with secondary unpublished literature to derive an estimate of carbon stocks for all the coastal vegetated ecosystems of Australia. According to them, Australian mangroves exhibit a mean AGC of 125±90 Mg C ha^{-1} (Fig. 8.3). This estimate stands on obser

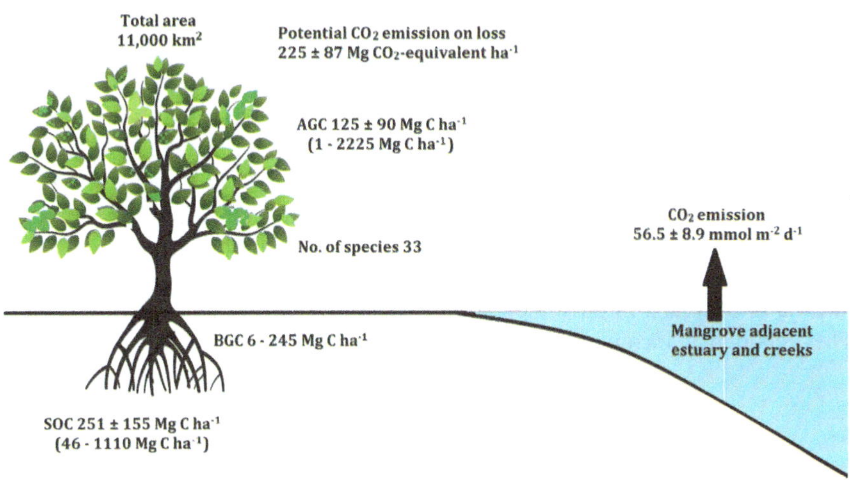

Fig. 8.3 A snapshot of the carbon dynamics in the mangroves of Australia

vations from 37 mangrove sites in the Australian periphery. Extrapolating this figure for the nationwide scale, Australia holds approximately 158 Tg C in the aboveground compartments of the mangrove ecosystems. Hickey et al. (2018) measured the AGC in the Mangrove Bay situated in north-western Australia. This region experiences a typical semi-arid climate. They coupled remotely sensed data with ground-based estimates to deduce a mean of AGC of 35 Mg C ha^{-1} from this area. Carbon estimation took place in other mangrove patches of Australia. Clough (1998) recorded an aboveground biomass range of 143–619 Mg ha^{-1} (equivalent to an AGC range of 72–310 Mg C ha^{-1}) in the mangroves of Hinchinbrook Channel, Queensland, whereas Heenkenda et al. (2016) reported a wide range of AGC from 1 to 2225 Mg C ha^{-1} in Rapid Creek, Northern Territory. The AGC range (22.9–141.4 Mg C ha^{-1}) observed by Alongi et al. (2000) in the semi-arid mangroves of NW Australia was a bit higher than the reports of Hickey et al. (2018). Lamont et al. (2020) measured the AGC in the mangroves of Hawkesbury River, New South Wales, at an interval of 30 years. They compared their estimate in 2018 with the observations of Saintilan (1997). In 2018 they recorded an AGC range of 11–125 Mg C ha^{-1}. They observed a significant difference between the tall and stunted patches.

8.2.4 Belowground Biomass Carbon (BGC) Stock

Mangroves have a typical root structure that extends horizontally and vertically over long distances from the main stem of a tree. These luxuriant root structures store anomalously higher carbon than many other terrestrial vegetated ecosystems (Alongi 2020). Various studies indicated that the contribution of belowground root structure often exceeds the aboveground counterparts (He et al. 2018). However, quantifying the BGC stock is a challenging endeavor (Adame et al. 2017). Destructive measurements lead to the most accurate measurements; however, it is not advisable. Serrano et al. (2019) conducted a meta-analysis of carbon stocks in all the blue carbon ecosystems of Australia. However, they did not report the mean BGC separately. Hickey et al. (2018), while working in the semi-arid mangroves of NW Australia, observed a mean BGC of 20 Mg C ha^{-1} (Table 8.2). They combined destructive protocols and allometric approaches to derive the results. The observations of Alongi et al. (2000) from different sites within the semi-arid mangroves of NW Australia (5.8–27.9 Mg C ha^{-1}) were in parity with Hickey et al. (2018). Lamont et al. (2020) compared the BGC of the Hawkesbury River mangroves in New South Wales at an interval of 30 years. In 2018 they observed a BGC range of 194–245 Mg C ha^{-1}. They reported an increment rate in the BGC of 5.1–6.6 Mg C ha^{-1} yr^{-1}.

Table 8.2 Aboveground, belowground, and soil carbon stock observed in the mangroves of Australia

Place	AGC (Mg C ha^{-1})	BGC (Mg C ha^{-1})	SOC (Mg C ha^{-1})	Authors
Entire continent of Australia	125±90		251±155	Serrano et al. (2019)
Mangrove Bay, NW Australia	35	10		Hickey et al. (2018)
Hinchinbrook Channel, Queensland	72 to 310			Clough (1998)
Rapid Creek, Northern Territory	1 to 2225			Heenkenda et al. (2016)
Semi-arid mangroves, NW Australia	22.9 to 141.4	5.8 to 27.9	45.9 to 183	Alongi et al. (2000)
Hawkesbury River, New South Wales	11 to 125	194 to 245	753 to 1110	Lamont et al. (2020)

8.2.5 Soil Organic Carbon (SOC) Stock

The dead remains of the aboveground and belowground plant structures and the incessant litter fall contribute to substantial quantities of carbon in the mangrove soils (Lovelock 2008). The SOC accumulates in the mangrove soils over millennial scales. Mangrove soils are one of the richest in carbon. The organic matter in the pedosphere usually maintains a proportional relationship with SOC (Chaikaew and Chavanich 2017). Researchers often found substantially higher SOC than the combined estimates of AGC and BGC in several mangrove forests of the world (Sitoe et al. 2014; Friess et al. 2016; Schile et al. 2017). However, the factors that govern the SOC magnitudes vary across the planet, and the principle behind such variations needs more study (Kida et al. 2017). Serrano et al. (2019) estimated that the mangrove soils of Australia stores a mean SOC of 251±155 Mg C ha^{-1}, equivalent to 257 Tg C for the entire continent. However, Alongi et al. (2000), while working in the semi-arid mangroves of NW Australia, observed a lower range of 45.9–183 Mg C ha^{-1}. Lamont et al. (2020), while working in the Hawkesbury River, New South Wales, estimated the SOC levels in a tall and stunted mangrove cover at an interval of 30 years. In 2018 they observed a massive SOC range of 753–1110 Mg C ha^{-1}, and they recorded that SOC increased at a rate of 2.2 to 4.3 Mg C ha^{-1} yr^{-1}.

8.2.6 Significant Observations Related to Carbon Stock in the Australian Mangrove Ecosystems

Several biogeochemical, geomorphic, hydrological, and climatic factors govern the carbon dynamics within the mangrove ecosystems. Hickey et al. (2018) observed that the distance of the mangrove trees from the nearest hydrological feature played a critical role in the carbon sequestration potential. The distant locations had less carbon stock in the biomass, whereas the nearer ones had substantially high carbon content. Clough (1998) observed that only 12% of the total fixed carbon through photosynthesis ends up in the living biomass. He inferred that the age of the plant plays a crucial role in governing the sequestration potential. Heenkenda et al. (2016) implemented Worldview-2 satellite data to derive the AGB, and they stressed transect sampling to enhance the quality of assessments due to the widely varying range of biomass within a small area. Law and Garnett (2011) indicated that coastal development in Australian coastlines occurs at the expense of degrading the blue carbon ecosystems. According to their estimate, one hectare of mangrove loss could lead to 225±87 Mg CO_2-equivalent ha^{-1}. Alongi et al. (2000) researched the semi-arid mangroves of NW Australia. They observed a directly proportional relationship between soil temperature and the rate of degradation of soil organic matter. The metabolic pathways of organic matter degradation varied among the species stands. Oxic respiration and sulfate reduction are the dominant pathways observed in the semi-arid mangroves of Australia. Lamont et al. (2020) inferred that species composition and the above-ground structure play a decisive role in enhancing the blue carbon stock over the years when left undisturbed. They pinpointed that tall stands of *Avicennia marina* enhanced SOC at a much faster rate than the mixed stunted patches of *Avicennia marina* and *Aegiceras corniculatum*.

Sanders et al. (2016) studied the carbon and nutrient burial rate in the Moreton Bay mangrove sediments. They observed that anthropogenic phosphorus enrichment in the creek systems facilitates the enhancement in carbon burial rate. Sediment accretion rate and anthropogenic discharge of nutrients play a crucial role in governing the long-term carbon burial dynamics. Several researchers emphasized that benthic fauna controls the carbon burial in mangrove sediments. However, Mazumder and Saintilan (2010) inferred that crabs do not draw their nutrition from the mangrove leaves in the temperate regions of Australia. Southern Australia has limited mangrove cover and species diversity. However, lately, Foster et al. (2019) indicated an 8 km^2 increase in mangrove cover between 1987 and 2015. Mangrove adjacent estuaries all over the world act as prominent sources of CO_2. The Australian estuaries and creeks are no exception. Rosentreter et al. (2018) reported a mean CO_2 flux of 56.5±8.9 mmol m^{-2} d^{-1} from three river estuaries of the Queensland region. Sippo et al. (2019) indicated that mangrove dieback leads to reduced outwelling of carbon through the adjacent shelf waters. They observed that an event of mangrove death across an area of almost 70 km^2 in the Gulf of Carpentaria led to a reduction in cross-shelf transport of dissolved inorganic carbon, dissolved organic carbon, and total alkalinity compared to the regions where mangroves are alive. Lovelock et al.

(2014), while studying the mangroves of South East Queensland, observed that sediment accretion rate plays a decisive role in regulating the carbon sequestration potential of the mangrove forests. Maher et al. (2013) inferred that groundwater discharge plays a crucial role in the carbon sink mechanism.

8.3 Australian Seagrasses

Like the mangroves, Australia shelters extensive seagrass meadows in the entire coastal periphery of this continent. Unlike the mangroves, which are almost absent in the southern part of this continent, seagrasses spread across the coastal fringes of Tasmania, New South Wales, and Victoria (Macreadie et al. 2018). The coastal geomorphology of Australia has a gentle slope, except for the eastern temperate boundary, where the continental slopes are much steeper than the rest of the periphery (Kilminster et al. 2015). These gentle seascapes allow a luxuriant growth of seagrasses. Seagrasses usually thrive in less disturbed calm environments (Tecchiato et al. 2019). The extensive reef system encircling the continent of Australia lays a foundation for seagrass abundance in this continental margin. The Great Barrier Reef in Australia is famous, which guards the northeastern margin of this country (McGowan and Theobald 2017). However, very few people know that an even more extensive temperate reef system exists in the southern parts of Australia, which facilitates ideal environments for the seagrasses to thrive (Gallop et al. 2015). The estuaries in the Australian boundary also host substantial seagrass meadows, especially in the eastern sector, where reef structures are almost absent. In this region, most seagrass meadows spread in the semi-enclosed bays and the eutrophic zones of the river estuaries.

8.3.1 Spatial Distribution of Seagrasses

Estimations carried out in the last decade indicate that global seagrass cover could be between 300,000 km^2 and 600,000 km^2 (Durate et al. 2010; Durate 2017). However, the recent compilation of data and subsequent analysis by McKenzie et al. (2020) suggests that seagrasses occupy nearly 160,000 km^2. They observed that 103 countries shelter seagrass meadows in their coastal periphery. The estimate of 160,000 km^2 has a moderate to high confidence level. They additionally reported the possibility of seagrass presence in 33 other countries, which can encompass 106,000 km^2. However, the confidence level for this figure is substantially low. The total comes around 266,000 km^2, which is much lower than the lower range of the earlier estimates. Such decreasing trend in global seagrass cover has become a point of concern (Unsworth et al. 2018; 2019; Dunic et al. 2021; Green et al. 2021). McKenzie et al. (2020) confirmed that Australia hosts the largest share of global seagrass meadows. More than 31% of the seagrass that thrives on this planet are available in the Australian coastal belt. Seagrasses usually occupy the sheltered habitats in the intertidal regimes. However, Carter et al. (2021) lately reported seagrass meadows at depths of more than 70 m in the Great Barrier Reef-protected coastal periphery. McKenzie et al. (2020) documented a total seagrass area of 83,000 km^2 from Australia, out of which close to 74,500 km^2 is available in the tropical bioregion. More than 35,000 km^2 of seagrass thrives in the Great Barrier Reef region.

8.3.2 Species Diversity in Australian Seagrasses

Australia exhibits very high seagrass species diversity. The tropical and temperate regions display high species richness in Australia (Carruthers et al. 2002; 2007). Short et al. (2007) demarcated the world into several bioregions to classify the seagrass habitats and species distribution. A further amendment to this classification led to three distinct biogeographic regions in Australia where the seagrass population varies in terms of species diversity (Fig. 8.4). The tropical northern, southwestern, and southeastern sectors are the principal subdivisions of Australia from the viewpoint of seagrass distribution. Lately, Kuo et al. (2018) reported 33 seagrass species from the coastlines of Australia. However, Kilminster et al. (2015) observed 37 species (Table 8.3). Many of the seagrass species that thrive in the temperate regions of this continent are endemic to this landmass. However, the species available in the tropical setup of Australia are also available in the Indo-Pacific sectors. Seagrass species can be colonizing, opportunistic, and persistent. All three types of seagrass species are available in the three biogeographic zones (Kilminster et al. 2015). The southwestern sectors exhibit the highest species diversity (26), including the maximum number of persistent species. The tropical north and the southeastern sector recorded 19 and 15 seagrass species, respectively. Both of these sectors have two persistent species. Four to five opportunistic species and nine to eleven colonizing species are available in the three biogeographic regions. *Halophila ovalis*, *Posidonia australis*, *Zostera muelleri*, and *Posidonia coriacea* are the dominant species that thrive in all the

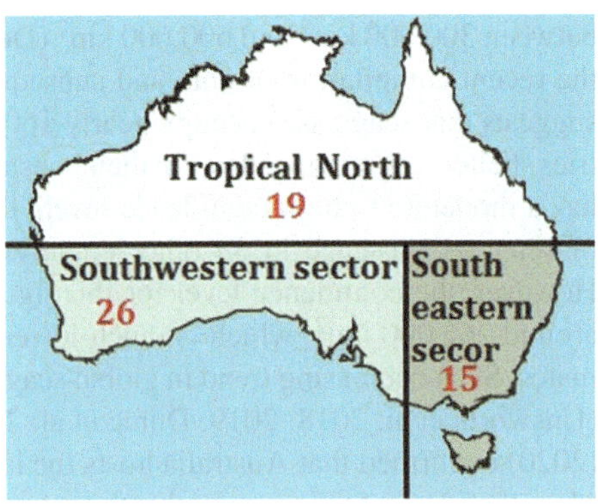

Fig. 8.4 The three distinct seagrass biogeographic zones of Australia and their species count (in red font)

Table 8.3 The seagrass species diversity observed in the three distinct biogeographic regions of Australia (after Kilminster et al. 2015)

Sr. No	Genus	Species	Temperate southwest	Temperate southeast	Tropical north
1	*Amphibolis*	*A. antarctica*	+	+	
2		*A. griffitthi*	+		
3	*Cymodocea*	*C. angustata*			+
4		*C. serrulata*			+
5		*C. rotundata*			+
6	*Enhalus*	*E. acoroides*			+
7	*Halodule*	*H. uninervis*	+		+
8		*H. pinifolia*			+
9	*Halophila*	*H. australis*	+	+	
10		*H. decipiens*	+	+	+
11		*H. ovalis*	+	+	+
12		*H. spinulosa*	+		+
13		*H. minor*			+
14		*H. capricorni*			+
15		*H. tricostata*			+
16	*Lepilaena*	*L. cylindrocarpa*	+	+	
17		*L. marina*	+	+	
18		*L. preisii*	+	+	
19	*Posidonia*	*P. angustifolia*	+	+	
20		*P. australis*	+	+	
21		*P. sinuosa*	+		
22		*P. ostenfeldii*	+		
23		*P. coriacea*	+		
24		*P. denhartogii*	+		
25		*P. kirkmanii*	+		
26		*P. robertsoniae*	+		
27	*Ruppia*	*R. megacarpa*	+	+	+
28		*R. tuberosa*	+	+	+
29		*R. maritima*	+	+	+
30	*Syringodium*	*S. isoetifolium*	+		+
31	*Thalassia*	*T. hemprichii*			+
32	*Thalassodendron*	*T. pachyrhizum*	+		
33		*T. ciliatum*			+
34	*Zostera*	*Z. muelleri*	+	+	+
35		*Z. nigricaulis*	+	+	
36		*Z. polychlamus*	+		

(continued)

Table 8.3 (continued)

Sr. No	Genus	Species	Temperate southwest	Temperate southeast	Tropical north
37		*Z. tasmanica*		+	

Table 8.4 Biomass and soil carbon stock observed in the seagrass meadows of Australia

Place	Biomass C (Mg C ha^{-1})	SOC (Mg C ha^{-1})	Authors
Entire continent of Australia	1.9±2.0	112±88	Serrano et al. (2019)
West Australia and Queensland		2.6–48.3	Lavery et al. (2013)
Coffs Harbour, New South Wales		365±225	Brown et al. (2016)
Jervis Bay, New South Wales		25 to 90	Macreadie et al. (2015)
Shark Bay, Western Australia		128±7	Arias-Ortiz et al. (2018)

bioregions. According to Kilminster et al. (2015), genera *Cymodocea*, *Enhalus*, and *Thalassia* are available only in the Tropical Northern sector in Australia.

8.3.3 Seagrass Biomass Carbon Stock

Seagrasses are one of the blue carbon giants of this planet. They store substantial quantities of carbon, which often supersedes the carbon content per unit area observed in mangrove and saltmarsh ecosystems (Fourqurean et al. 2012; Miyajima et al. 2015; Rahmawati et al. 2020). However, the carbon stored in the seagrass biomass is much less than that in mangroves and saltmarshes. Serrano et al. (2019) reported a mean carbon stock of 1.9±2.0 Mg C ha^{-1} from the Australian seagrasses, equivalent to 16–22 Tg C for the entire continent (Table 8.4).

8.3.4 Seagrass Soil Carbon Stock

Seagrass beds are active depocenters of carbon in the continental margins and coastal peripheries (Fourqurean et al. 2012). The sediment accumulation rate usually remains substantially high in the seagrass adjoining areas (Serrano et al. 2016), which facilitates a steady organic carbon burial (Carnell et al. 2020). Serrano et al. (2019) reported a mean seagrass bed carbon stock of 112±88 Mg C ha^{-1} from the Australian

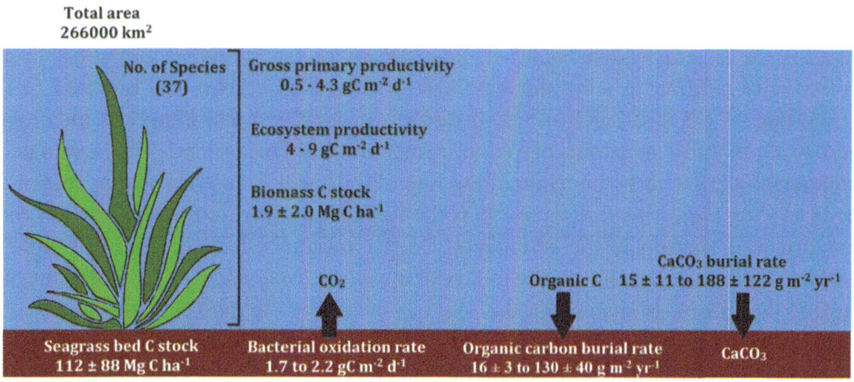

Fig. 8.5 A snapshot of carbon dynamics observed in the seagrass meadows of Australia

periphery (Fig. 8.5). Extrapolating this figure for the entire continent leads to an estimated total of 762 to 1051 Tg C. Lavery et al. (2013) sampled across 17 sites spread in Western Australia and Queensland and measured the soil carbon pool in mono-species stands and mixed meadows comprising ten seagrass species. They observed a wide range of 2.6 Mg C ha^{-1} to 48.3 Mg C ha^{-1} across the various types of habitat and species. The SOC range was much less than that reported by Serrano et al. (2019). However, Lavery et al. (2013) considered only the top 25 cm in the seagrass bed. Brown et al. (2016) observed a modest mean sedimentary carbon stock of 365±225 Mg C ha^{-1} in the seagrass patches of Coffs Harbour, New South Wales. Macreadie et al. (2015) researched the Jervis Bay seagrasses in New South Wales and examined the role of disturbances in the sedimentary carbon pools. They observed a SOC range of 25–90 Mg C ha^{-1}. Arias-Ortiz et al. (2018) studied the impact of marine heatwaves in the seagrasses of Shark Bay in Western Australia. The seagrass stretch of Shark Bay is the largest single contiguous stretch of the world, which stores 1.3% of the carbon stored in global seagrass meadows (Walker et al. 1988). Arias-Ortiz et al. (2018) reported a mean carbon stock of 128 ± 7 Mg C ha^{-1} from this region, close to the global median estimates of 140 Mg C ha^{-1} (Fourqurean et al. 2012).

8.3.5 Significant Observations Related to Carbon Stock in Australian Seagrass Meadows

The seagrasses in Australia received substantial attention from the scientific community from multiple viewpoints. Pollard and Moriarty (1991) implemented a lacunal gas technique to measure the primary productivity of the seagrasses in the Gulf of Carpentaria. The gross primary production rate varied between 0.5 and 4.3 gC m^{-2} d^{-1}, whereas the ecosystem productivity rate ranged between 4 and

9 gC m^{-2} d^{-1}. A diverse composition of algae enhanced the net ecosystem productivity in the seagrass adjacent water mass. However, the sulfate-reducing bacterial community oxidized a substantial portion of this fixed carbon and returned CO_2 at the rate of 1.7 to 2.2 gC m^{-2} d^{-1}. Lavery et al. (2013) researched 17 locations across the seagrass meadows of Western Australia and Queensland. They estimated an annual organic carbon accumulation rate of 0.93 metric tons across the seagrass beds of Australia. They observed that the depths of seagrass habitats govern the magnitude of carbon in the sediments. Deeper meadows had significantly higher carbon stocks than the subtidal or intertidal meadows. Macreadie et al. (2015) investigated the effects of disturbing the seagrass meadows on the sedimentary carbon pools. They inferred that more than 70% of the carbon that took more than a millennium to accumulate in the seagrass soils washed away due to an event of a disturbance. However, recovered regions showed substantially high carbon content than the disturbed areas. This observation is quite encouraging to promote the restoration of seagrass beds. Arias-Ortiz et al. (2018) observed that marine heatwaves can also lead to a significant destruction of seagrass meadows, which exposes the sedimentary carbon pool to oxic conditions that favor the emission of the stored carbon as CO_2. Sanders et al. (2019) researched three temperate estuaries of the southeastern sector of Australia having seagrass meadows. They observed a varying burial rate of organic carbon in these estuaries ranging from 16 ± 3 to 130 ± 40 g m^{-2} yr^{-1}. They inferred that seagrass beds are capable of trapping autochthonous organic carbon produced within the seagrass ecosystems and allochthonous inputs from adjacent mangrove ecosystems and algal communities. These seagrasses also store substantial quantities of terrestrial organic carbon that ends up in the seagrass water column. Sanders et al. (2019) also recorded a varying calcium carbonate burial rate of 15 ± 11 to 188 ± 122 g m^{-2} yr^{-1}. This result indicates that calcifying organisms can substantially offset the organic carbon burial in the seagrass beds. Thomson et al. (2019) carried out an ex-situ experiment by collecting sediment and water samples from the seagrass meadows of Fagans Bay, Australia. They examined the role of bioturbating faunas in remineralizing sediment-trapped organic carbon. They inferred that bottom-dwelling shrimps like *Trypaea australiensis* enhance the CO_2 emission from the seagrass beds by disturbing the substrate. O'Brien et al. (2018) studied the effect of light deprivation on the seagrass meadows of Australia. Light is one of the most essential factors that regulate the autotrophic potential of the seagrasses and their wellbeing. They observed that seagrass's resilience toward light deprivation varies across species diversity. The persistent species can tolerate light deficiency for longer time scales than the colonizing ones. They also discussed the adaptation mechanisms of seagrass to combat the varying degree of light deprivation. Macreadie et al. (2019) researched the seagrass population of Fagans Bay, New South Wales, and inferred that warming and oxic environments facilitate the microbial attack on the stored blue carbon in the sediment substrates.

8.4 Saltmarshes of Australia

Lack of awareness about saltmarshes played a crucial role in the degradation of this ecosystem throughout the world. The perception of people to consider these valuable habitats as wastelands and even the neglect from the scientific community were responsible for the diminishing of saltmarsh habitats. Australia is also no exception in this regard. Since the 1970s, several pieces of research pointed out this fact (Kratochvil et al. 1972; Bucher and Saenger 1991; Finlayson and Rea 1999; Prahalad 2014). The New South Wales region witnessed the most alarming loss in the past decades (Bowen et al. 1995). However, Australia is one of the very few nations, which recognized the potential of this crucial blue carbon system and acted accordingly to safeguard this habitat by including them as part of the endangered communities (Laegdsgaard 2006). Saltmarshes, like seagrasses, prefer a calm habitat where the straightforward brunt of oceanic waves is absent. Sheltered bays and estuarine mouths are the most suitable locations for salt marshes. The Australian coastal periphery has many such regions. However, the local climatic factors and salinity regime also play a pivotal role in regulating the spatial distribution of salt marshes (Underwood and Chapman 1993). Australia is also one of the pioneers in restoring salt marshes and rehabilitating them in suitable habitats (Adam 2019). Australian Government has put on substantial efforts to conserve saltmarsh habitats, especially in the southeastern sector (Gulliver et al. 2020). Restricting cattle grazing, deviating the flow of polluted waters from these habitats, and removing unwanted weeds are some basic activities that help rehabilitate and conserve saltmarshes (Howitt et al.2014; Hurst and Boon 2016).

8.4.1 Spatial Distribution of Saltmarshes

Ambient temperature, rainfall, and salinity regimes principally regulate the spatial distribution of the saltmarshes. Saintilan (2009) indicated that these factors play a crucial role in governing the saltmarsh distribution in Australia. There is a clear north–south divide in terms of the spatial distribution of saltmarshes in Australia. The 23° S latitude divides the entire continent into two halves. The northern portion of this line has a poor species diversity with less area. The southern part shelters the biodiverse salt marsh patches of this country. According to Saintilan (2009), the average minimum ambient temperature is the decisive factor that governs the extent of this vegetation in Australia. The lower the minimum temperature, the higher was the abundance of salt marshes. Rainfall also plays a decisive role in governing the salt marsh dynamics. Laegdsgaard (2006) pointed out that precipitation volume is directly proportional to the degree of salinity stress alleviation. The salt marshes of Australia exhibit abundant growth under oligohaline conditions. Thus, the salt marsh extent mimics the average total precipitation observed in the coastlines of Australia (Fig. 8.6). Overall, the conditions in terms of both temperature and rainfall

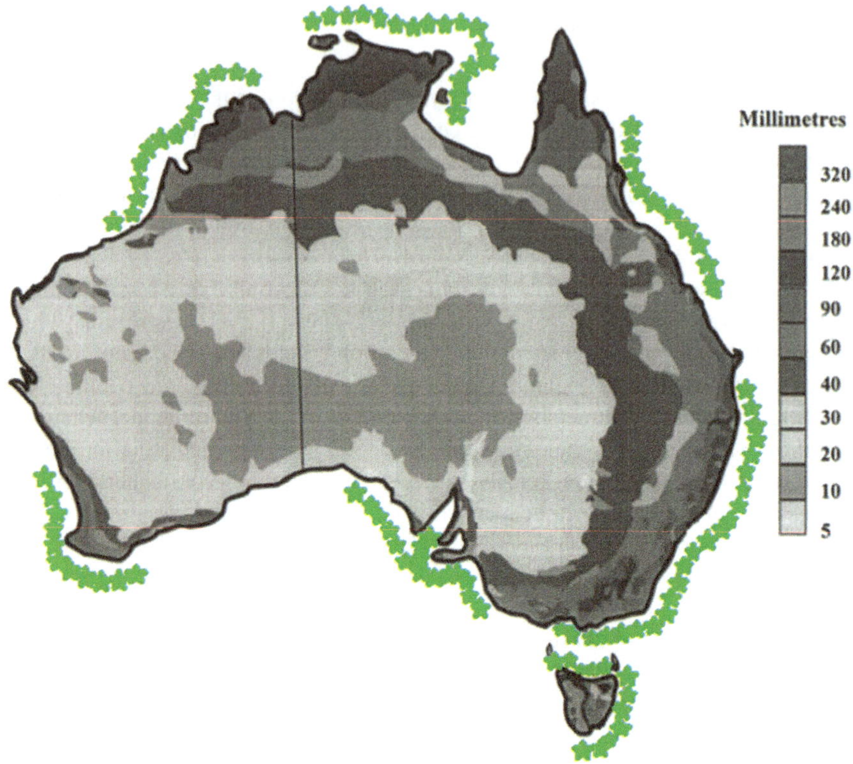

Fig. 8.6 The spatial distribution of seagrass vis-à-vis the average rainfall observed along the Australian coastal periphery (recreated after Laegdsgaard 2006)

are favorable in the southern part of the continent, which exhibits a substantially better abundance of salt marshes in Australia.

8.4.2 Species Diversity in Australian Saltmarshes

Unlike the mangroves (50–70) and seagrasses (<75), salt marshes have a long list of species that thrive throughout the world (>500). Several studies on Australian salt marshes documented their species diversity in the different bioregions of this country. Saintilan (2009) reported a modest total of 93 salt marsh species from this continent. However, in the same paper, the author mentioned the possibility of more than 100 species thriving in this landmass (Adam 1990). Almost 70% of these species belong to the Asteraceae, Aizoaceae, Cyperaceae, Chenopodiaceae, and Poaceae families. Many of these species are endemic to Australia. The northern sector of this continent shelters the typical tropical salt marshes, whereas to the south of 23° S the temperate species have predominance. Overall the species diversity is substantially higher among the temperate group compared to the tropical group of species (Specht 1981). Several studies recognized four dominant salt marsh species among the temperate cluster, namely, *Sporobolus virginicus*, *Sarcocornia quinqueflora*, *Samolus repens*, and *Triglochin striatum* (Clarke and Jacoby 1994; Turner and Streever 1997). Australia also has several rare species which thrive in a limited number of places all over the world. *Wilsonia backhousei* is one such species that is available in only two locations within Australia (Adam and Hutchings 1987). The other rare salt marsh species that are available in Australia are *Sclerostegia arbuscula*, *Halosarcia* spp., and *Agrostis limitanea*. Overall, the species abundance throughout Australia is well-documented. However, recent observations have focused on other matters rather than a temporal assessment of species diversity change all along the Australian coastal fronts.

8.4.3 Saltmarsh Biomass Carbon Stock

Attempts of measuring the carbon stock in the living biomass of salt marshes (of Australia) go back to the 1980s. Congdon and McComb (1981) observed a wide range of biomass (3.0–33 Mg ha^{-1}) in the salt marshes of Blackwood River in Molloy Island (Table 8.5). Clarke and Jacoby (1994) measured the biomass of three salt marsh species in three estuaries of southeast Australia. They observed the highest mean biomass in *Juncus kraussii* (11.2 Mg ha^{-1}), followed by *Sporobolus virginicus* (3.5 Mg ha^{-1}) and *Sarcocornia quinqueflora* (3.2 Mg ha^{-1}). Owers et al. (2018) researched four rivers and creeks in the SE Australian coastline. They observed an aboveground biomass with a range of 7.5–16.0 Mg ha^{-1}. They also inferred that the structural differences among the various types of salt marshes led to varying biomasses. The rush type had substantially higher biomass than the herb and sedge types.

8.4.4 Saltmarsh Soil Carbon Stock

Clarke and Jacoby (1994) measured the soil organic carbon in the three estuaries of southeast Australia; however, they did not report it as unit carbon per unit area. The SOC% ranged between 3.7 and 5.9%. They observed that deeper samples had higher SOC than the surface strata. Lately, Brown et al. (2016) reported a mean SOC of 858±190 Mg C ha^{-1} from the Coffs Harbour, New South Wales, slightly lower than the adjacent mangrove SOC. However, Howe et al. (2009) reported a much lower SOC range of 81.2–129.7 Mg C ha^{-1} from the salt marshes of Kooragang Island. Livesley and Andrusiak (2012) observed a similar mean SOC (168 Mg C ha^{-1}) in the salt marshes of Victoria, Australia. Gorham et al. (2021) researched the temperate tidal marshes of Western Australia. They observed a mean SOC range of 156–211 Mg C ha^{-1}. They inferred that the estuarine vegetations stored higher quantities

Table 8.5 Aboveground, belowground, and soil carbon stock observed in the salt marshes of Australia

Place	AGC (Mg C ha^{-1})	BGC (Mg C ha^{-1})	SOC (Mg C ha^{-1})	Authors
Blackwood River	1.5–11.5			Congdon and McComb (1981)
Jervis Bay, Port Hacking, Batemans Bay, SE Australia	1.59–5.58	0.94–2.68		Clarke and Jacoby (1994)
Hunter and Minnamurra Rivers, Currambene and Cararma Inlets	3.76–7.98			Owers et al. (2018)
Coffs Harbour, New South Wales			858±190	Brown et al. (2016)
Kooragang Island			81.2–129.7	Howe et al. (2009)
Westernport Bay, Victoria			168	Livesley and Andrusiak (2012)
Temperate Western Australia			156–211	Gorham et al. (2021)
Entire Australia			165±7	Macreadie et al. (2017)

of carbon in the sediment profile than their marine counterparts. They derived that Western Australian marshes hold almost 57 Tg C, equivalent to 2% of the global salt marsh's sediment carbon content. In this regard, Macreadie et al. (2017) carried out a holistic compilation of SOC data from the Australian tidal marshes. They reported a mean SOC of 165±7 Mg C ha^{-1} for the entire continent.

8.4.5 *Significant Observations Related to Carbon Stock in Australian Saltmarshes*

Saltmarshes have received substantial attention from the scientific community in Australia from varying viewpoints. Lately, Kelleway et al. (2016) studied the transition from salt marsh to the mangrove in Botany Bay, Southeast Australia. They observed that a steady mangrove encroachment into the salt marsh habitat enhanced the carbon content in the aboveground and belowground compartments. This observation indicates that salt marshes act as a foundation for the initial colonization of mangroves. Saintilan et al. (2013) monitored that the carbon accumulation rate in the salt marsh habitats of Australia ranged between 0.46 and 2.07 Mg C ha^{-1} yr^{-1}. They commented that accretion rate and regional sea level play crucial roles in regulating the carbon assimilation in these habitats. Santos et al. (2019) studied the fate of carbon buried in the salt marsh and mangrove sediments of the Evans Head region. They observed a SOC burial rate of 0.1 to 0.6 Mg C ha^{-1} yr^{-1}. They also inferred that CH_4 flux offsets around 5% of the carbon buried in these soils, and lateral export of dissolved inorganic carbon acts as a long-term sink. Livesley and Andrusiak (2012) observed that the salt marsh sediments emit CH_4 and NO_2. However, the emission rate was negligible compared to the carbon stored in the sediment column. They observed that salt marsh sediments had a higher mean SOC than the adjacent mangrove soils. Macreadie et al. (2017) derived a mean SOC accumulation rate of 0.55 ± 0.02 Mg C ha^{-1} yr^{-1} for the entire continent. They inferred that the geomorphologic setting played the most decisive role in regulating the carbon stocks of the sediment profile in salt marsh ecosystems.

8.5 Conclusion

Analyzing the significant outcomes from the studies carried out in the last three to four decades, we can infer confidently that Australian blue carbon ecosystems received substantial attention. Australian mangroves, seagrasses, and salt marshes exhibit very high species diversity. The presence of different climatic regimes enabled spatial variability in the species distribution. Several studies documented the carbon content in the various blue carbon habitats of this country. Many pieces of research focused on the factors that regulate the carbon concentration in the aboveground, belowground, and sediment of these ecosystems. On the whole, this continent can act as a blueprint for conservation endeavors and, at the same time, scientific studies for other countries having these crucial coastal habitats. Initiatives taken by the Australian Government and the scientific community are laudable and can show the other blue carbon nations the right path to walk upon to safeguard the interests of these vegetated ecosystems along the coast.

References

Adam P (1990) Saltmarsh ecology. Cambridge University Press, Cambridge
Adam P, Hutchings P (1987) The saltmarshes and mangroves of Jervis Bay. Wetl Austr 6(2):5864
Adam P (2019) Salt marsh restoration. In Perillo GME, Wolanski E, Cahoon DR, Hopkinson (eds) Coastal Wetlands: an integrated ecosystem approach coastal Wetlands, Elsevier pp 817–861. https://doi.org/10.1016/B978-0-444-63893-9.00023-X

Adame MF, Cherian S, Reef R, Stewart-Koster B (2017) Mangrove root biomass and the uncertainty of belowground carbon estimations. Forest Ecol Manag 403:52–60. https://doi.org/10.1016/j.foreco.2017.08.016

Alongi DM (2020) Global significance of mangrove blue carbon in climate change mitigation. Sci 2(3):67. https://doi.org/10.3390/sci2030067

Alongi DM, Tirendi F, Clough BF (2000) Below-ground decomposition of organic matter in forests of the mangroves *Rhizophora stylosa* and *Avicennia marina* along the arid coast of Western Australia. Aquat Bot 68(2):97–122. https://doi.org/10.1016/S0304-3770(00)00110-8

Arias-Ortiz A, Serrano O, Masqué P, Lavery PS, Mueller U, Kendrick GA, Duarte CM (2018) A marine heatwave drives massive losses from the world's largest seagrass carbon stocks. Nat Clim Chang 8(4):338–344. https://doi.org/10.1038/s41558-018-0096-y

Asbridge E, Lucas R, Ticehurst C, Bunting P (2016) Mangrove response to environmental change in Australia's Gulf of Carpentaria. Ecol Evol 6:3523–3539. https://doi.org/10.1002/ece3.2140

Asbridge E, Lucas R, Rogers K, Accad A (2018) The extent of mangrove change and potential for recovery following severe Tropical Cyclone Yasi, Hinchinbrook Island, Queensland, Australia. Ecol Evol 8(21):10416–10434. https://doi.org/10.1002/ece3.4485

Bell-James J (2016) Developing a framework for'blue carbon'in Australia: legal and policy considerations. Univ New S Wales Law J 39(4):1583–1611. https://search.informit.org/doi/10.3316/ielapa.481381256681194

Bowen R, Stephens N, Donnelly P (1995) SEPP-14 wetland protection and the role of mitigation. Wetl Austral 14:2

Bridgewater PB (1982) Mangrove vegetation of the Southern and Western Australian coastline. In Clough BF (ed) Mangrove ecosystems in Australia: structure, function and management, Proceedings of the Australian National Mangrove Workshop: Australian Institute of Marine Science, Australian National University Press, pp 111–120

Brown DR, Conrad S, Akkerman K, Fairfax S, Fredericks J, Hanrio E, Sanders LM, Scott E, Skillington A, Tucker J, van Santen ML, Sanders CJ (2016) Seagrass, mangrove and saltmarsh sedimentary carbon stocks in an urban estuary; Coffs Harbour, Australia. Reg Stud Mar Sci 8(1):1–6. https://doi.org/10.1016/j.rsma.2016.08.005

Bucher D, Saenger P (1991) An inventory of Australian estuaries and enclosed marine waters: an overview of results. Austr Geogr Stud 29:370381. https://doi.org/10.1111/j.1467-8470.1991.tb00726.x

Bunting P, Rosenqvist A, Lucas R, Rebelo L-M, Hilarides L, Thomas N, Hardy A, Itoh T, Shimada M, Finlayson M (2018) The global mangrove watch—a new 2010 global baseline of mangrove extent. Remote Sens 10(10):1669. https://doi.org/10.3390/rs10101669

Carnell PE, Ierodiaconou D, Atwood TB, Macreadie PI (2020) Overgrazing of seagrass by sea urchins diminishes blue carbon stocks. Ecosystems 23(7):1437–1448. https://doi.org/10.1007/s10021-020-00479-7

Carruthers TJB, Dennison WC, Longstaff BJ, Waycott M, Abal EG, McKenzie LJ, Long WJ (2002) Seagrass habitats of northeast Australia: models of key processes and controls. Bull Mar Sci 71(3):1153–1169

Carruthers TJB, Dennison WC, Kendrick GA, Waycott M, Walker DI, Cambridge ML (2007) Seagrasses of south–west Australia: a conceptual synthesis of the world's most diverse and extensive seagrass meadows. J Exp Mar Biol Ecol 350(1–2):21–45. https://doi.org/10.1016/j.jembe.2007.05.036

Carter AB, McKenna SA, Rasheed MA, Collier C, McKenzie L, Pitcher R, Coles R (2021) Synthesizing 35 years of seagrass spatial data from the Great Barrier Reef World Heritage Area, Queensland, Australia. Limnol Oceanogr Lett https://doi.org/10.1002/lol2.10193

Chaikaew P, Chavanich S (2017) Spatial variability and relationship of mangrove soil organic matter to organic carbon. Appl Environ Soil Sci. Article ID 4010381, https://doi.org/10.1155/2017/4010381

Clarke PJ, Jacoby CA (1994) Biomass and above-ground productivity of salt marsh plants in south-eastern Australia. Austr J Mar Freshw Res 45:15211528. https://doi.org/10.1071/MF9941521

Clough B (1998) Mangrove forest productivity and biomass accumulation in Hinchinbrook Channel, Australia. Mangroves Salt Marshes 2(4):191–198. https://doi.org/10.1023/A:1009979610871
Clough BF (ed) (1982) Mangrove ecosystems in Australia: structure, function, and management. In: Proceedings of the Australian national mangrove workshop: Australian Institute of Marine Science, Australian National University Press
Congdon RA, McComb AJ (1981) The vegetation of the Blackwood River estuary South-West Australia. J Ecol 69:116. https://doi.org/10.2307/2259812
Cresswell ID, Semeniuk V (2018) Australian mangroves: Their distribution and protection. In: Makowski C, Finkl C (eds) Threats to mangrove forests. Coastal Research Library, vol 25. Springer, Cham. https://doi.org/10.1007/978-3-319-73016-5_1
Crosbie RS, Pollock DW, Mpelasoka FS, Barron OV, Charles SP, Donn MJ (2012) Changes in Köppen-Geiger climate types under a future climate for Australia: hydrological implications. Hydrol Earth Syst Sci 16(9):3341–3349. https://doi.org/10.5194/hess-16-3341-2012
Donato DC, Kauffman JB, Murdiyarso D, Kurnianto S, Stidham M, Kanninen M (2011) Mangroves among the most carbon-rich forests in the tropics. Nat Geosci 4(5):293–297. https://doi.org/10.1038/ngeo1123
Dowling RM, McDonald TJ (1982) Mangrove communities of Queensland. In: Clough BF (ed) Mangrove ecosystems in Australia: structure, function and management. In: Proceedings of the Australian national mangrove workshop: Australian Institute of Marine Science, Australian National University Press, pp 79–93
Duarte CM (2017) Reviews and syntheses: hidden forests, the role of vegetated coastal habitats in the ocean carbon budget. Biogeosciences 14:301–310. https://doi.org/10.1038/ngeo1123
Duarte CM, Marba N, Gacia E, Fourqurean JW, Beggins J, Barron C, Apostolaki ET (2010) Seagrass community metabolism: assessing the carbon sink capacity of seagrass meadows. Global Biogeochem Cycles 24(4). https://doi.org/10.1029/2010GB003793
Duke NC (2017) Mangrove floristics and biogeography revisited: further deductions from biodiversity hot spots, ancestral discontinuities, and common evolutionary processes. In: Rivera-Monroy V, Lee S, Kristensen E, Twilley R (eds) Mangrove ecosystems: a global biogeographic perspective. Springer, Cham. https://doi.org/10.1007/978-3-319-62206-4_2
Dunic JC, Brown CJ, Connolly RM, Turschwell MP, Côté IM (2021) Long-term declines and recovery of meadow area across the world's seagrass bioregions. Glob Chang Biol. https://doi.org/10.1111/gcb.15684
Finlayson CM, Rea N (1999) Reasons for the loss and degradation of Australian wetlands. Wetl Ecol Manag 7:111. https://doi.org/10.1023/A:1008495619951
Foster N, Jones AR, Waycott M, Gillanders BM (2019) Coastal carbon opportunities: technical report on changes in the distribution of mangrove and saltmarsh across South Australia (1987–2015). Goyder Institute for Water Research Technical Report Series, (19/23)
Fourqurean JW, Duarte CM, Kennedy H, Marbà N, Holmer M, Mateo MA et al (2012) Seagrass ecosystems as a globally significant carbon stock. Nat Geosci 5(7):505–509. https://doi.org/10.1038/ngeo1477
Friess DA, Webb EL (2014) Variability in mangrove change estimates and implications for the assessment of ecosystem service provision. Glob Ecol Biogeogr 23(7):715–725. https://doi.org/10.1111/geb.12140
Friess DA, Richards DR, Phang VX (2016) Mangrove forests store high densities of carbon across the tropical urban landscape of Singapore. Urban Ecosyst 19(2):795–810. https://doi.org/10.1007/s11252-015-0511-3
Gallop SL, Bosserelle C, Haigh ID, Wadey MP, Pattiaratchi CB, Eliot I (2015) The impact of temperate reefs on 34 years of shoreline and vegetation line stability at Yanchep, southwestern Australia and implications for coastal setback. Mar Geol 369:224–232. https://doi.org/10.1016/j.margeo.2015.09.001
Galloway RW (1982) Distribution and physiographic patterns of Australian mangroves. In: Clough BF (ed) Mangrove ecosystems in Australia: structure, function and management, proceedings of

the Australian national mangrove workshop: Australian institute of marine science. Australian National University Press, pp 31–54

Giri C, Ochieng E, Tieszen LL, Zhu Z, Singh A, Loveland T, Masek J, Duke N (2011) Status and distribution of mangrove forests of the world using earth observation satellite data. Glob Ecol Biogeogr 20:154–159. https://doi.org/10.1111/j.1466-8238.2010.00584.x

Gorham C, Lavery P, Kelleway JJ, Salinas C, Serrano O (2021) Soil carbon stocks vary across geomorphic settings in Australian temperate tidal marsh ecosystems. Ecosystems 24(2):319–334. https://doi.org/10.1007/s10021-020-00520-9

Green EP, Short FT, Frederick T (2003) World atlas of seagrasses. University of California Press

Green AE, Unsworth RK, Chadwick MA, Jones PJ (2021) Historical analysis exposes catastrophic seagrass loss for the United Kingdom. Front Plant Sci 12:261. https://doi.org/10.3389/fpls.2021.629962

Gulliver A, Carnell PE, Trevathan-Tackett SM, de Paula D, Costa M, Masqué P, Macreadie PI (2020) Estimating the potential blue carbon gains from tidal marsh rehabilitation: a case study from south eastern Australia. Front Mar Sci 7:403. https://doi.org/10.3389/fmars.2020.00403

He Z, Peng Y, Guan D, Hu Z, Chen Y, Lee SY (2018) Appearance can be deceptive: shrubby native mangrove species contributes more to soil carbon sequestration than fast-growing exotic species. Plant Soil 432(1):425–436. https://doi.org/10.1007/s11104-018-3821-4

Heenkenda MK, Maier SW, Joyce KE (2016) Estimating mangrove biophysical variables using Worldview-2 satellite data: rapid creek, Northern Territory. Australia. J Imag 2(3):24. https://doi.org/10.3390/jimaging2030024

Hickey SM, Callow NJ, Phinn S, Lovelock CE, Duarte CM (2018) Spatial complexities in aboveground carbon stocks of a semi-arid mangrove community: a remote sensing height-biomass-carbon approach. Estuar Coast Shelf Sci 200:194–201. https://doi.org/10.1016/j.ecss.2017.11.004

Hickey SM, Radford B, Callow JN, Phinn SR, Duarte CM, Lovelock CE (2021) ENSO feedback drives variations in dieback at a marginal mangrove site. Sci Rep 11(1):1–9. https://doi.org/10.1038/s41598-021-87341-5

Howe AJ, Rodríguez JF, Saco PM (2009) Surface evolution and carbon sequestration in disturbed and undisturbed wetland soils of the Hunter estuary, southeast Australia. Estuar Coast Shelf Sci 84(1):75–83. https://doi.org/10.1016/j.ecss.2009.06.006

Howitt JA, Mondon J, Mitchell BD, Kidd T, Eshelman B (2014) Urban stormwater inputs to an adapted coastal wetland: role in water treatment and impacts on wetland biota. Sci Total Environ 485:534–544. https://doi.org/10.1016/j.scitotenv.2014.03.101

Hurst T, Boon PI (2016) Agricultural weeds and coastal saltmarsh in south-eastern Australia: an insurmountable problem? Austr J Bot 64(4):308–324. https://doi.org/10.1071/BT16027

Jones WT (1971) The field identification and distribution of mangroves in eastern Australia. Queensl Nut 20:35–51

Kelleway JJ, Saintilan N, Macreadie PI, Skilbeck CG, Zawadzki A, Ralph PJ (2016) Seventy years of continuous encroachment substantially increases 'blue carbon'capacity as mangroves replace intertidal salt marshes. Glob Chang Biol 22(3):1097–1109. https://doi.org/10.1111/gcb.13158

Kenneally KF (1982) Mangroves of Western Australia. In: Clough BF (ed) Mangrove ecosystems in Australia: structure, function and management, proceedings of the Australian national mangrove workshop: Australian institute of marine science. Australian National University Press, pp 94–110

Kida M, Tomotsune M, Iimura Y, Kinjo K, Ohtsuka T, Fujitake N (2017) High salinity leads to accumulation of soil organic carbon in mangrove soil. Chemosphere 177:51–55. https://doi.org/10.1016/j.chemosphere.2017.02.074

Kilminster K, McMahon K, Waycott M, Kendrick GA, Scanes P, McKenzie L, O'Brien KR, Lyons M, Ferguson A, Maxwell P, Glasby T, Udy J (2015) Unravelling complexity in seagrass systems for management: Australia as a microcosm. Sci Total Environ 534:97–109. https://doi.org/10.1016/j.scitotenv.2015.04.061

Kratochvil M, Hannon NJ, Clarke LD (1972) Mangrove swamp andsaltmarsh communities in southeastern Australia. Proc Linnaen Soc NSW 97:262274

Kuo J, Cambridge ML, McKenzie LJ, Coles RG (2018) Taxonomy of Australian seagrasses. In: Larkum A, Kendrick G, Ralph P (eds) Seagrasses of Australia. Springer, Cham. https://doi.org/10.1007/978-3-319-71354-0_23

Laegdsgaard P (2006) Ecology, disturbance and restoration of coastal saltmarsh in Australia: a review. Wetl Ecol Manag 14(5):379–399. https://doi.org/10.1007/s11273-005-8827-z

Lamont K, Saintilan N, Kelleway JJ, Mazumder D, Zawadzki A (2020) Thirty-year repeat measures of mangrove above-and below-ground biomass reveals unexpectedly high carbon sequestration. Ecosystems 23(2):370–382. https://doi.org/10.1007/s10021-019-00408-3

Lavery PS, Mateo MÁ, Serrano O, Rozaimi M (2013) Variability in the carbon storage of seagrass habitats and its implications for global estimates of blue carbon ecosystem service. PloS One 8(9):e73748. https://doi.org/10.1371/journal.pone.0073748

Law R, Garnett ST (2011) Mapping carbon in tropical Australia: estimates of carbon stocks and fluxes in the Northern Territory using the national carbon accounting toolbox. Ecol Manag Restor 12(1):61–68. https://doi.org/10.1111/j.1442-8903.2011.00566.x

Lear R, Turner T (1977) Mangroves of Australia. University of Queensland Press, St Lucia, pp 84

Livesley SJ, Andrusiak SM (2012) Temperate mangrove and salt marsh sediments are a small methane and nitrous oxide source but important carbon store. Estuar Coast Shelf Sci 97:19–27. https://doi.org/10.1016/j.ecss.2011.11.002

Lovelock CE (2008) Soil respiration and belowground carbon allocation in mangrove forests. Ecosystems 11(2):342–354. https://doi.org/10.1007/s10021-008-9125-4

Lovelock CE, Adame MF, Bennion V, Hayes M, O'Mara J, Reef R, Santini NS (2014) Contemporary rates of carbon sequestration through vertical accretion of sediments in mangrove forests and saltmarshes of South East Queensland, Autralia. Estuar Coasts 37(3):763–771. https://doi.org/10.1007/s12237-013-9702-4

Lymburner L, Bunting P, Lucas R, Scarth P, Alam I, Phillips C, Ticehurst C, Held A (2020) Mapping the multi-decadal mangrove dynamics of the Australian coastline. Remote Sens Environ 238:111185. https://doi.org/10.1016/j.rse.2019.05.004

Macreadie PI, Trevathan-Tackett SM, Skilbeck CG, Sanderman J, Curlevski N, Jacobsen G, Seymour JR (2015) Losses and recovery of organic carbon from a seagrass ecosystem following disturbance. Proc R Soc B Biol Sci 282(1817):20151537. https://doi.org/10.1098/rspb.2015.1537

Macreadie PI, Ollivier QR, Kelleway JJ, Serrano O, Carnell PE, Lewis CE et al (2017) Carbon sequestration by Australian tidal marshes. Sci Rep 7(1):1–10. https://doi.org/10.1038/srep44071

Macreadie PI, Atwood TB, Seymour JR, Fontes MS, Sanderman J, Nielsen DA, Connolly RM (2019) Vulnerability of seagrass blue carbon to microbial attack following exposure to warming and oxygen. Sci Total Environ 686:264–275. https://doi.org/10.1016/j.scitotenv.2019.05.462

Macreadie PI, Sullivan B, Evans SM, Smith TM (2018) Biogeography of Australian seagrasses: NSW, Victoria, Tasmania and Temperate Queensland. In: Larkum A, Kendrick G, Ralph P (eds) Seagrasses of Australia. Springer, Cham. https://doi.org/10.1007/978-3-319-71354-0_2

Maher DT, Santos IR, Golsby-Smith L, Gleeson J, Eyre BD (2013) Groundwater-derived dissolved inorganic and organic carbon exports from a mangrove tidal creek: the missing mangrove carbon sink? Limnol Oceanogr 58(2):475–488. https://doi.org/10.4319/lo.2013.58.2.0475

Mazumder D, Saintilan N (2010) Mangrove leaves are not an important source of dietary carbon and nitrogen for crabs in temperate Australian mangroves. Wetlands 30(2):375–380. https://doi.org/10.1007/s13157-010-0021-2

McGowan H, Theobald A (2017) ENSO weather and coral bleaching on the Great Barrier Reef, Austrlia. Geophys Res Lett 44(20):10–601. https://doi.org/10.1002/2017GL074877

Macnae W (1966) Mangroves in eastern and southern Australia. Aust J Bot 14(1):67–104. https://doi.org/10.1071/BT9660067

Miyajima T, Hori M, Hamaguchi M, Shimabukuro H, Adachi H, Yamano H, Nakaoka M (2015) Geographic variability in organic carbon stock and accumulation rate in sediments of East and Southeast Asian seagrass meadows. Glob Biogeochem Cycl 29(4):397–415. https://doi.org/10.1002/2014GB004979

O'Brien KR, Adams MP, Ferguson AJP, Samper-Villarreal J, Maxwell PS, Baird ME, Collier C (2018) Seagrass resistance to light deprivation: implications for resilience. In: Larkum A., Kendrick G, Ralph P (eds) Seagrasses of Australia. Springer, Cham. https://doi.org/10.1007/978-3-319-71354-0_10

Osland MJ, Feher LC, Griffith KT, Cavanaugh KC, Enwright NM, Day RH, Stagg CL, Krauss KW, Howard RJ, Grace JB, Rogers K (2017) Climatic controls on the global distribution, abundance, and species richness of mangrove forests. Ecol Monogr 87(2):341–359. https://doi.org/10.1002/ecm.1248

Owers CJ, Rogers K, Woodroffe CD (2018) Spatial variation of above-ground carbon storage in temperate coastal wetlands. Estuar Coast Shelf Sci 210:55–67. https://doi.org/10.1016/j.ecss.2018.06.002

Pollard PC, Moriarty DJW (1991) Organic carbon decomposition, primary and bacterial productivity, and sulphate reduction, in tropical seagrass beds of the Gulf of Carpentaria, Australia. Mar Ecol Prog Ser Oldendorf 69(1):149–159

Prahalad VN (2014) Human impacts and saltmarsh loss in the Circular Head coast, north-west Tasmania, 1952–2006: implications for management. Pac Conserv Biol 20(3):272–285. https://doi.org/10.1071/PC140272

Rahmawati S, Irawan A, Hadiyanto H, Prayudha B, Hafizt M, Afdal A et al (2020) Assessing carbon stock and sequestration of the tropical seagrass meadows in Indonesia. Ocean Sci J 55(1):85–97. https://doi.org/10.1007/s12601-020-0003-0

Rosentreter JA, Maher DT, Erler DV, Murray R, Eyre BD (2018) Seasonal and temporal CO_2 dynamics in three tropical mangrove creeks–a revision of global mangrove CO_2 emissions. Geochim Cosmochim Acta 222:729–745. https://doi.org/10.1016/j.gca.2017.11.026

Saintilan N (1997) Above and below-ground biomasses of two species of mangrove on the Hawkesbury River estuary, New South Wales. Mar Freshw Res 48:147–152. https://doi.org/10.1071/MF96079

Saintilan N (2009) Biogeography of Australian saltmarsh plants. Austr Ecol 34(8):929–937. https://doi.org/10.1111/j.1442-9993.2009.02001.x

Saintilan N, Rogers K, Mazumder D, Woodroffe C (2013) Allochthonous and autochthonous contributions to carbon accumulation and carbon store in southeastern Australian coastal wetlands. Estuar Coast Shelf Sci 128:84–92. https://doi.org/10.1016/j.ecss.2013.05.010

Sanders CJ, Santos IR, Maher DT, Breithaupt JL, Smoak JM, Ketterer M, Call M, Sanders L, Eyre BD (2016) Examining 239+ 240Pu, 210Pb and historical events to determine carbon, nitrogen and phosphorus burial in mangrove sediments of Moreton Bay, Australia. J Environ Radioact 151:623–629. https://doi.org/10.1016/j.jenvrad.2015.04.018

Sanders CJ, Maher DT, Smoak JM, Eyre BD (2019) Large variability in organic carbon and CaCO3 burial in seagrass meadows: a case study from three Australian estuaries. Mar Ecol Prog Seri 616:211–218. https://doi.org/10.3354/meps12955

Sandi SG, Rodriguez JF, Saco PM, Saintilan N, Riccardi G (2021) Accelerated sea-level rise limits vegetation capacity to sequester soil carbon in coastal Wetlands: a study case in South-Eastern Australia. Earth Future e2020EF001901. https://doi.org/10.1029/2020EF001901

Santos IR, Maher DT, Larkin R, Webb JR, Sanders CJ (2019) Carbon outwelling and outgassing versus burial in an estuarine tidal creek surrounded by mangrove and saltmarsh wetlands. Limnol Oceanogr 64(3):996–1013. https://doi.org/10.1002/lno.11090

Schile LM, Kauffman JB, Crooks S, Fourqurean JW, Glavan J, Megonigal JP (2017) Limits on carbon sequestration in arid blue carbon ecosystems. Ecol Appl 27(3):859–874. https://doi.org/10.1002/eap.1489

Semeniuk V, Kenneally KF, Wilson PG (1978) Mangroves of Western Australia. Western Australian Naturalists' Club Handbook No. 11, Perth, W.A

Serrano O, Ruhon R, Lavery PS, Kendrick GA, Hickey S, Masqué P, Arias-Ortiz A, Steven A, Duarte CM (2016) Impact of mooring activities on carbon stocks in seagrass meadows. Sci Rep 6:23193. https://doi.org/10.1038/srep23193

Serrano O, Lovelock CE, Atwood TB, Macreadie PI, Canto R, Phinn S et al (2019) Australian vegetated coastal ecosystems as global hotspots for climate change mitigation. Nat Commun 10(1):1–10. https://doi.org/10.1038/s41467-019-12176-8

Short F, Carruthers T, Dennison W, Waycott M (2007) Global seagrass distribution and diversity: a bioregional model. J Exp Mar Biol Ecol 350(1–2):3–20. https://doi.org/10.1016/j.jembe.2007.06.012

Sippo JZ, Maher DT, Schulz KG, Sanders CJ, McMahon A, Tucker J, Santos IR (2019) Carbon outwelling across the shelf following a massive mangrove dieback in Australia: insights from radium isotopes. Geochim Cosmochim Acta 253:142–158. https://doi.org/10.1016/j.gca.2019.03.003

Sitoe AA, Mandlate LJC, Guedes BS (2014) Biomass and carbon stocks of Sofala bay mangrove forests. Forests 5(8):1967–1981. https://doi.org/10.3390/f5081967

Spalding MD, Blasco F, Field CD (eds) (1997) World mangrove atlas. Okinawa (Japan): International society for mangrove ecosystems. Compiled by UNEPWCMC, in collaboration with the International Society for Mangrove Ecosystems (ISME), pp 178

Spalding M, Kainuma M, Collins L (2010) World Atlas of Mangroves (Version 1.1). A Collaborative Project of ITTO, ISME, FAO, UNEP-WCMC, UNESCO-MAB, UNUINWEH and TNC. Earthscan, London, UK, pp 319

Specht RL (1981) Biogeography of halophytic angiosperms (salt-marshes, mangrove and sea-grass). In: Keast A (ed) Ecological biogeography of Australia. Dr Junk, The Hague

Stern H, De Hoedt G, Ernst J (2000) Objective classification of Australian climates. Austr Met Mag 49(2):87–96

Tecchiato S, Buosi C, Ibba A, Del Deo C, Parnum I, O'Leary M, De Muro S (2019) Geomorphological and sedimentological surrogates for the understanding of seagrass distribution within a temperate nearshore setting (Esperance Western Australia). Geo-Mar Lett 39(3):249–264. https://doi.org/10.1007/s00367-019-00571-5

Thomson AC, Trevathan-Tackett SM, Maher DT, Ralph PJ, Macreadie PI (2019) Bioturbator-stimulated loss of seagrass sediment carbon stocks. Limnol Oceanogr 64(1):342–356. https://doi.org/10.1002/lno.11044

Turner PA, Streever WJ (1997) The relationship between the density of Aedes vigilax (Diptera, Culicidae) eggshells and environmental factors on Kooragang Island, New South Wales, Australia. J Am Mosq Contr Assoc 13(4):361367

Underwood AJ, Chapman MG (1993) Seashores a beachcomber's guide. NSW University Press, Australia

Unsworth RK, Ambo-Rappe R, Jones BL, La Nafie YA, Irawan A, Hernawan UE et al (2018) Indonesia's globally significant seagrass meadows are under widespread threat. Sci Total Environ 634:279–286. https://doi.org/10.1016/j.scitotenv.2018.03.315

Unsworth RK, McKenzie LJ, Collier CJ, Cullen-Unsworth LC, Duarte CM, Eklöf JS et al (2019) Global challenges for seagrass conservation. Ambio 48(8):801–815. https://doi.org/10.1007/s13280-018-1115-y

Unsworth LC, Roelfsema C, Unsworth RK (2020) The global distribution of seagrass meadows. Environ Res Lett 15(7):074041. https://doi.org/10.1088/1748-9326/ab7d06

Walker D, Kendrick G, McComb A (1988) The distribution of seagrass species in Shark Bay, Western Australia, with notes on their ecology. Aquat Bot 30:305–317. https://doi.org/10.1016/0304-3770(88)90063-0

Wells AG (1982) Mangrove vegetation of Northern Australia. In: Clough BF (ed) Mangrove ecosystems in Australia: structure, function and management, proceedings of the Australian national mangrove workshop: Australian institute of marine science. Australian National University Press, pp 57–79

Woodroffe CD (2018) Mangrove response to sea level rise: palaeoecological insights from macrotidal systems in northern Australia. Mar Freshw Res 69(6):917–932. https://doi.org/10.1071/MF17252

Chapter 9
Threats to the Blue Carbon Ecosystems Adjoining the Indian Ocean

Abhra Chanda

Abstract The scientific community and the common mass have unequivocally realized the significance of blue carbon habitats in mitigating climate change and providing livelihood options to the marginalized section of coastal people. Even then, these ecosystems witnessed a rapid decline in the last few decades, and several natural and anthropogenic factors pose threats to these crucial habitats. Mangroves are mostly susceptible to land-use changes like conversion to agriculture or aquaculture plots, whereas seagrasses suffer from coastal discharges that enhance the sedimentation rate and turbidity of the water column. Salt marshes, on the other hand, have received much less attention that what it deserves. The present chapter discussed the country-wise variability in the type of threats and the significant observations. Thus, this chapter can act as a snapshot for the present understanding of threats to blue carbon ecosystems in the periphery of the Indian Ocean. The collated findings can be helpful for early career researchers striving to look for solutions and policy managers to prioritize management action plans.

Keywords Land conversion · Cattle grazing · Aquaculture · Sea-level rise · Boating · Unsustainable fishing · Sand mining · Shell collection · Lime production · Indiscriminate tourism

9.1 Introduction

The scientific community all around the globe unequivocally recognizes the importance of blue carbon ecosystems. However, these ecosystems are under severe anthropogenic pressure in almost all corners of the world. A suite of natural as well as human-driven phenomena is posing immense threats to these ecosystems (DasGupta and Shaw 2017). Several studies suggested that the areal extent of mangroves in the coastal peripheries was manifold higher than what it is at present. Several countries

A. Chanda (✉)
School of Oceanographic Studies, Jadavpur University, 188 Raja S.C. Mullick Road, Kolkata, West Bengal 700084, India
e-mail: abhrachanda.sos.rs@jadavpuruniversity.in

A. Chanda et al. (eds.), *Blue Carbon Dynamics of the Indian Ocean*,
https://doi.org/10.1007/978-3-030-96558-7_9

during the past centuries went for massive deforestation to harvest wood for multifarious activities (Selvam 2021). The other intents were to expand human settlements and agricultural fields in the coast-adjoining hinterlands. The last century has witnessed an exponential growth in human population in the coastal margins throughout the world (Sengupta et al. 2018), especially in the tropics and the subtropics, where the major chunks of blue carbon ecosystems thrive. The ever-increasing population in the land-sea transition zones has significantly changed the land use and land cover in the adjacency of coastlines (Kaliraj et al. 2017; Etemadi et al. 2018; Lei 2021; Qiu et al. 2021). In some ways or the other, the blue carbon ecosystems, like mangroves, seagrass, and salt marshes, get compromised due to such land conversions. Transformation of substantial mangrove cover to agricultural plots occurred in the past decades (Fauzi et al. 2019). Coastal aquaculture farming has emerged as a new trend (Prasad et al. 2019; Ren et al. 2019; Stiller et al. 2019) to complement the uncertainties in open ocean capture fisheries and meet the ever-increasing food demand (Sainsbury et al. 2018; Plagányi 2019; Lima et al. 2020). Many of these aquaculture ponds came into existence at the cost of clearing mangroves (Boone Kauffman et al. 2017). The reclamation pressure and coastal pollution due to various anthropogenic activities have taken a toll on the areal extent of salt marshes throughout the world (Zedler et al. 2018; Gu et al. 2018; Giuliani and Bellucci 2019). Unsustainable fishing techniques, indiscriminate waste disposal, and activities like seaweed farming have degraded many of the seagrass patches in the world (Unsworth et al. 2018; Griffiths et al. 2020).

Besides anthropogenic disturbances, various natural and climatic phenomena, which might be partially responsible in the first place due to human-driven impacts, pose severe threats to all the blue carbon ecosystems. The ongoing global warming is consistently elevating the regional sea levels in all the corners of the world (Hoozemans and Hulsbergen 2021). This rising mean sea level could engulf almost 30% of the coastal wetlands, and under an aggravated scenario, the estimated loss could be even more (Schuerch et al. 2018). Several pieces of research suggested that the mangroves are highly vulnerable to rising sea levels as this phenomenon can significantly alter the salinity and inundation dynamics (Saintilan et al. 2020). The construction of dams and barrages in many of the riverine and estuarine stretches has substantially reduced the sediment flow into the mangrove and salt marsh habitats (Xie et al. 2017; Ladd et al. 2019; Yuan et al. 2020). Adequate sedimentation helps prograde the deltaic regions and allows the mangroves to cope up with the rise in sea level. Thus, an absence of sediment flow would make the scenario worse for mangroves (Rogers 2021). Regional sea-level rise and increased intensity of physical forcing events enhanced the rate of soil erosion in the coastal margins. Such eroding coastlines wash away substantial mangrove areas in several locations (Besset et al. 2019; Sánchez-Núñez et al. 2019; Das 2020; van Bijsterveldt et al. 2020). Salt marshes are equally susceptible to regional sea-level rise and the change in salinity regime, which can make the blue carbon stock of these vegetated regions highly vulnerable (Li et al. 2018; Gonneea et al. 2019; Langston et al. 2020). Tropical cyclones, hurricanes, and tornados often make landfall in mangrove fringes and lead to severe destruction (Asbridge et al. 2018; Armitage et al. 2020; Rogers 2019; Payton et al. 2020). Several studies indicate that the frequency and intensity of these atmospheric events have increased in the past decades and are likely to keep on increasing (Klotzbach et al. 2018; Lindner 2019; Knutson et al. 2019; Kossin et al. 2020). Most of the mangrove communities have exhibited a regenerating capability following these types of strong physical forcing events (Azad et al. 2020; Fickert2020; Halder et al. 2021). However, given the recurrency and increased vigors of these storms, the natural regeneration potential of this coastal vegetation could get significantly compromised (Krauss and Osland 2020). The seagrass communities are also susceptible to multiple natural threats like storm surges, tsunamis, and coastal erosion (Moseby et al. 2020; Twomey et al. 2021; Xu et al. 2021). Thus, there are several concerns, which require considerable attention from scientists and conservationists to protect and safeguard and protect these crucial blue carbon ecosystems. To device proper management strategies and restoration and rehabilitation programs, a holistic idea of the present threats and their plausible adverse effects on these vegetation communities is extremely necessary.

9.2 Threats to Mangroves, Seagrasses, and Salt Marshes of the Indian Ocean

The periphery of the Indian Ocean shelters some of the largest mangrove forests in the world (Vanderklift et al. 2019). However, the mangrove cover in these regions has witnessed a drastic reduction in the recent past and experiences threats of a wide spectrum (DasGupta and Shaw 2017; Romañach et al. 2018; Hamad et al. 2019). Felling of mangrove trees for wood harvest, conversion of mangrove habitats to agriculture and aquaculture plots, coastal pollution, sea-level rise, erosion, cyclonic disasters, and storm surges are some of the common issues that have become a point of concern for these mangroves. Several researchers have characterized the different threats on the mangroves of this region. Though the spatial extent of seagrass and salt marsh in the Indian Ocean periphery requires much more attention, several studies pinpointed the existing threats to these crucial ecosystems. The present chapter collated all such findings to develop a holistic understanding of this issue.

9.3 Threats in the Asian Countries Bordering the Eastern Indian Ocean

9.3.1 India

India has a long coastline of around 7500 km. According to the latest estimates, it shelters a mangrove area of almost 4975 km^2 (India State of Forest Report 2019). Almost 40% of Sundarbans, the world's largest mangrove forest (shared by India and Bangladesh) lies in the Indian coastline, along with many other prominent mangrove patches throughout the east and west coasts facing the Bay of Bengal and the Arabian Sea, respectively (Kathiresan 2010). Chaudhuri et al. (2015) characterized the threats to the mangrove habitats on the east coast of India. They observed massive mangrove losses in many of the deltaic regions in the Indian Peninsula. However, a net increase in mangrove area also took place in many of these regions, as the Krishna basin (Ravishankar et al. 2004), Godavari basin (Ramasubramanian et al. 2006), and Mahanadi estuary (Reddy et al. 2007). Chaudhari et al. (2015) also noticed that though dense mangrove areas increased in places, a reduction in open or sparse mangroves concomitantly took place in many of these regions. The Sundarbans, which comprise almost 40% of the country's total mangrove cover witnessed massive destruction in the past. During the colonial era, hired laborers cleared thousands of square kilometers of this area for developmental activities (Ghosh et al. 2015). At present, this region shelters a population of almost 4.4 million people, which imposes various types of threats to this delicate ecosystem (Hajra et al. 2017; DasGupta et al. 2019). Akhand et al. (2017) observed that between the years 1975 and 2013, this forest encountered a net loss of 107 km^2 mangrove area (due to a combined effect of erosion, sea-level rise, and anthropogenic encroachment), which led to an estimated loss of 1570 Gg of blue carbon (Table 9.1). The huge population pressure and the emerging trend of coastal aquaculture for better economic return have increased the vulnerability of this mangrove forest (Dubey et al. 2017). The Sundarban is the abode of the majestic Royal Bengal Tiger, and man-animal conflicts (crocodile and tiger attack, snake bite) lead to frequent casualties in this region (Das 2015; Das and Jana 2018). Despite such life risks, thousands of local inhabitant's ventures into the forest local meet their livelihood demands. Moreover, many of the estuaries in this forest have stopped receiving direct freshwater input from the upstream regions leading to acute water scarcity and salinization of the mangrove soils (Chatterjee et al. 2013). Such freshwater scarcity also leads to nutrient limitation and deteriorates the health of the standing mangroves (Chowdhury et al. 2019). Situated downstream to a peri-urban setup and with major industrial hubs and metropolises lying in the north of Sundarban, this region experiences severe coastal pollution. The sediments and floral stand of these mangroves exhibit signatures of heavy metals (Chowdhury et al. 2017; Mondal et al. 2018; Chanda et al. 2021) and persistent organic pollutants (Zanardi-Lamardo et al. 2019; Basu et al. 2021; Henderson et al. 2021) throughout the spatial extent of Sundarbans. The tropical cyclones often form in the Bay of Bengal and leave disastrous signatures on the mangroves all along the east coast of India (Pal and Ghosh 2018; Paul and Chatterjee 2019; Mitra 2020; Pramanik et al. 2021). Thus, several factors govern the threat levels in the mangroves of India; however, all these issues have started receiving adequate attention from the government authorities, NGOs, scientific, and local communities.

The seagrasses and salt marshes have received much less attention in India compared to mangroves. Thangaradjou and Bhatt (2018) observed that both natural and anthropogenic factors have led to considerable loss of seagrass beds in the intertidal zones of India. Anthropogenic disturbances led to a significant destruction of seagrass beds in Palk Bay (Sridhar et al. 2010). The Gulf of Mannar also witnessed a

Table 9.1 Some of the major observations related to threats to the blue carbon habitats in the Asian countries of the eastern Indian Ocean

Country		Major threats	Observations	Authors
India	M	• Sediment reduction • Erosion • Sea-level rise • Salinization	• Cauvery river mangroves—a net loss of 24 km^2 between 1976 and 2013 • Indian Sundarban mangroves—a net loss of 107 km^2 between 1975 and 2013	Chaudhuri et al. (2015) Akhand et al. (2017)
	SG	• Anthropogenic pressure • Tsunami • Subsidence	• Palk Bay seagrass cover—a net loss of 2.64 km^2 between 1996 and 2004 • Gulf of Mannar seagrass cover—a net loss of 28.6 km^2 between 1998 and 2005 • Lakshadweep seagrass cover—a net loss of 0.73 km^2 between 2000 and 2008 • Andaman and Nicobar Islands seagrass cover—a net loss of 28.6 km^2 between 2004 and 2007	Thangaradjou et al. (2008, 2009) Nobi and Thangaradjou (2012) Thangaradjou and Bhatt (2018)
	SM	Industrial waste Agricultural use Sea walls	• 9 out of 14 salt marsh species are vulnerable	Gopi et al. (2019) Viswanathan et al. (2020)
Bangladesh	M	Salinity stress Sea-level rise	• Sundarban mangroves may witness a net loss of 81 to 1393 km^2 by the year 2010 • Oligohaline species would be rapidly replaced by salt-tolerant species • The total blue carbon stock of the forest would be significantly compromised in near future due to this change in species composition	Mukhopadhyay et al. (2015) Payo et al. (2016) Chanda et al. (2016) Ghosh et al. (2016, 2017)
	SG	Sea-level rise Pollution Rising temperature	• Sea-level rise, increase in salinity, temperature, and turbidity are the major concerns for seagrass communities	Kamal (2009) Kamal and Khan (2009) Akther (2016)
	SM	Heavy metal pollution Shipbreaking Grazing	• *Porteresia* sp. roots accumulated substantial magnitudes of heavy metals • Shipbreaking industries, overgrazing, oil spills, sand mining, and walkways pose a significant threat to salt marshes	Hossain et al. (2021) Meah and Barua (2020)
Myanmar	M	Illegal tree felling Population growth	• Myanmar mangroves—a net loss of 559 km^2 between the years 1954 and 1998 • Myanmar mangroves a net loss of 1911 km^2 between the years 2000 and 2014	Oo (2002) Estoque et al. (2018)

(continued)

Table 9.1 (continued)

Country		Major threats	Observations	Authors
	SG	Hazardous waste and sediment flow Sand mining Bottom trawling	• Toxic waste effluent runoffs from the industrial and urban sector, unsustainable fishing techniques, increase in turbidity due to disturbance in the shallow aquatic column are the major threats	Soe-Htun et al. (2015) Thu-Rein and Murray-Jones (2019)
	SM	Conversion to agricultural plots Sedimentation	• The tidal flats and salt marshes interchange their habitats in a temporally dynamic way. Sedimentation, erosion, and utilization as agricultural plots are the major threats	Choi et al. (2020) Murray et al. (2020)
Thailand	M	Shrimp farming Agriculture Lack of knowledge on mangrove ecology Coastal erosion	• Mangrove cover in Thailand – a net loss of 1420 km^2 between the years 1961 and 2007 • Shrimp farming is the primary factor leading to >50% destruction of mangrove habitats	Plathong and Plathong (2004) Thampanya et al. (2006) Pumijumnong (2014)
	SG	Sedimentation Tsunami	• Sedimentation in the coastline due to developmental activities is the chief threat • Libong Island witnessed a net loss of 2 km^2 seagrass bed and 5% of the entire south Andaman coast seagrass was affected by the 2004 Indian Ocean tsunami	Adulyanukosol and Poovachiranon (2006) Stankovic et al. (2021)
	SM	NS	• Local people of the Samut Songkhram region use salt marsh species *Suaedamaritima* for various cuisines • A direct report of threats could not be retrieved from the existing literature	Pornpitakdamrong and Sudjaroen (2014)
Singapore	M	Dam construction Prawn ponds	• The mangrove area reduced from 75 km^2 (in the year 1819) to 6.59 km^2 (in the year 2010) • Singapore might lose 33% of its present mangrove due to the proposed future development projects	Yee et al. (2010) Friess et al. (2012, 2016) Lai et al. (2015)
	SG	Land reclamation Sedimentation Artificial beach and lagoon construction	• Continuous beds are almost absent, with few patchy occurrences of seagrass meadows • Besides sedimentation, anchoring boats and cage culture poses significant threats to the seagrasses	Loo et al. (1996) , Yaakub et al. (2013) McKenzie et al. (2016)

(continued)

Table 9.1 (continued)

Country		Major threats	Observations	Authors
	SM	NS	NS	NS
Indonesia	M	Aquafarming Oil palm cultivation	• Net mangrove area loss of 13,000 km^2 between the years 1980 and 2005 • The largest mangrove forest of Central Java reduced by 85 km^2 between the years 1978 and 2003	Ardli and Wolff (2009) Murdiyarso et al. (2015) Richards and Friess (2016)
	SG	Coastal development Sand and coral mining	• 30,000 km^2 of seagrass cover (measured during the year 2000) has reduced to a large extent; however, there are no quantified estimates	Riani et al. (2012) Unsworth et al. (2018)
	SM	Mangrove degradation	• Mangrove degradation can indirectly hamper the growth of other associated blue carbon habitats like salt marshes	Sejati et al. (2020)
Malaysia	M	Coastal erosion Unsustainable tree felling Illegitimate agriculture	• Matang mangrove forest lost 80 km^2 area between the years 1993 and 2011 • Iskandar region lost 67 km^2 area between the years 1989 and 2014	Ibrahim et al. (2015) Kanniah et al. (2015)
	SG	Sand shifting Sand mining Excessive sedimentation	• A net loss of 0.94 km^2 and 0.34 km^2 seagrass cover in the Sungai Pulai estuary and the Tanjung Adang Darat, respectively, between 1994 and 2017	Hossain et al. (2019)
	SM	NS	NS	NS

massive reduction in seagrass cover due to submergence and erosion (Thangaradjou et al. 2008). Lakshadweep, Andaman, and Nicobar witnessed a seagrass area loss of 0.73, 1.46, and 14.72 km^2, during 2000–2008, 2004–2007, and 2004–2007, respectively (Nobi and Thangaradjou 2012). The 2004 Indian Ocean tsunami affected most of the seagrass cover in the Andaman and Nicobar Islands. Shell collection led to considerable damage of seagrasses in Lakshadweep. Thangaradjou et al. (2009) detailed various threats to the Indian seagrass communities. Lately, Mishra and Apte (2021) suggested that the seagrass species *Halophilabeccarii*, which is abundant in the Indian coastlines, has become vulnerable due to coastal development and consequent pollution. Similarly, many of the salt marsh species and stretches in the Indian coastline have become vulnerable and threatened (Gopi et al. 2019; Viswanathan et al. 2020). Urbanization and industrial waste disposal are the main concerns of salt marshes along the Indian coastline. Several studies indicated that the degree of vulnerability is much higher on the east coast compared to the west coast (Jagtap et al. 2002).

9.3.2 Bangladesh

The lion's share of Sundarban mangrove forest comprises a substantial part of the Bangladesh coastline. Emch and Peterson (2006) observed drastic changes in mangrove cover between 1989 and 2000 and identified salinity stress as one of the principal threats to the mangroves of Bangladesh Sundarbans. Unlike the Indian counterpart of Sundarbans, Iftekhar and Saenger (2008) observed a predominance of Heritiera fomes across the Bangladesh Sundarbans. However, Mukhopadhyay et al. (2015) predicted a radical shift in the mangrove species composition of this area considering the business-as-usual scenario. Many of the freshwater-loving species are under severe threat in this forest, whereas the salt-tolerant ones are gradually replacing them. They predicted that many of the oligohaline mangroves would substantially reduce in numbers followed by a dominance of hypersaline mangroves. Chanda et al. (2016) predicted that such changes in future species assemblage would significantly compromise the blue carbon stock of this forest. Giri et al. (2007) observed that the areal cover in Bangladesh Sundarban remained stable between the years 1973 and 2000. However, erratic and irregular rainfall coupled with increasing ambient temperature is the principal factor that could lead to a drastic change in the species composition of this forest (Ghosh et al. 2016, 2017). Dasgupta et al. (2017) anticipated virtual extinction of *Heritiera fomes* from this forest due to ongoing salinization, and that would be a big loss owing to its high carbon sequestration potential and timber value. Payo et al. (2016) estimated that Bangladesh Sundarbans may witness a mangrove area loss of almost 81 to 1393 km^2 by the year 2100, due to the rising sea level and subsidence (depending on various climate change scenarios). Besides, Bangladesh is prone to cyclonic disasters. In the past decades, several cyclones have hard-hit Bangladesh resulting in significant loss of life and property (Chowdhury et al. 1993; Alam and Collins 2010; Hoque et al. 2019). The mangroves of Sundarban bear the brunt of almost every cyclone that makes landfall in Bangladesh. Many of these cyclones destroyed substantial mangrove cover in Bangladesh (Bhowmik and Cabral 2013; Halder et al. 2021); however, their regeneration potential enabled these trees to cope up during the post-disaster phases. Like Indian Sundarbans, significant metal contamination prevails in the Bangladesh counterpart too (Rahman et al. 2009; Islam et al. 2017). Thus, multiple threats can bother this unique mangrove stand in the future unless we come up with an adequate conservation strategy and implement proper management interventions.

The seagrass and salt marsh patches are much smaller compared to the mangroves of Sundarbans in Bangladesh. Kanal and Short (2009) reported the existence of seagrass species Halophilabeccarii from the coastline of Bangladesh; however, they could not identify any threats. Hena et al. (2007) studied the physiological traits of two salt marsh and one seagrass species in Bangladesh, and they also did not report any significant threats to these communities. Kamal (2009) observed that all the macrophyte communities in the intertidal habitats of Bangladesh would suffer from regional sea-level rise including the seagrasses. Akthter (2016) mentioned that seagrasses in the Bangladesh coastline are vulnerable to tropical cyclones, rise in temperature and mean sea level, pollution, and unsustainable extraction. Kamal and Khan (2009) observed that changes in salinity and turbidity levels due to alterations in hydrodynamics can affect the seagrass communities. Lately, Hossain et al. (2021) observed significant bioaccumulation of heavy metals in the salt marshes thriving on the Bangladesh coastline. Meah and Barua (2020) identified several threats to the *Porteresia* sp. communities in Bangladesh. Flourishing of shipbreaking industries, overgrazing, oil spills, sand mining, and walkways are some of the potential factors that lead to considerable damage to the salt marshes of Bangladesh.

9.3.3 Myanmar

Myanmar has a coastline of approximately 3000 km and experiences a typical tropical environment, which is very conducive for the coastal vegetations to flourish (Zöckler and Aung 2019). The coastlines of Myanmar shelter some of the largest patches of mangroves in the world. However, the mangroves of Myanmar are at the same time one of the worst-affected regions of the world. Oo (2002) pointed out that in the year 1954, Myanmar had a mangrove cover of around 2345 km^2, which reduced to 1786 km^2 in 1984. The population growth in coastal Myanmar along with the illegal feeling to harvest mangrove woods led to massive deforestation in the mid of twentieth century and has continued to date (De Alban et al. 2020). The Ayeyarwady Delta had dense mangrove cover. The Second World War led to rampant tree felling in the mangrove forests of this region to serve war demands. At present, the densely vegetated mangroves of the Ayeyarwady region have transformed into sparse open forest (Webb et al. 2014). Not only the mangroves, the entire country of Myanmar, which had thick forest patches witnessed massive deforestation in the last century and the recent past as well (Wang and Myint 2016). Estoque et al. (2018) suggested that in the last two decades, the rate of mangrove deforestation attained extreme highs, with almost 146 km^2 mangrove area loss per year. Their estimates indicate that Myanmar has lost 1911 km^2 mangroves between the years 2000 and 2014. They attributed improper forest management and anthropogenic pressure like conversion to agriculture and aquaculture behind such a rapid rate of deforestation. The Wunbaik Reserved Mangrove Forest in the Rakhine state and the Tanintharyi area shelter some of the largest mangrove forests of Myanmar, besides the Ayeyarwady Delta. These two sites have also witnessed rampant deforestation in the recent past (Saw and Kanzaki 2015; Gaw et al. 2018). Poor enforcement of land-use policies at the national and local levels has resulted in such deforestation. The local population switched to aquaculture ponds for higher income (Saw and Kanzaki 2015), which led to this catastrophic reduction in mangrove forest cover. Lately, Veettil et al. (2018) also recognized the rapidly diminishing state of the Myanmar mangroves; however, they also mentioned that since the year 2010, significant efforts for the restoration of degraded habitats have begun.

Unlike the mangroves, the seagrass beds in Myanmar cover small areas and are mostly concentrated in few coastal margins. Lampi Island and surrounding areas like

Pub Nala, Wa-Ale Kyun, Gregory, and Kanzagyi Islands shelter the most prominent seagrass beds of Myanmar (Novak et al. 2009). However, recent studies discovered more areas where seagrasses are available on the Myanmar coastline, like Baw Di, TharthanarDauk, Chan Pyin, NgweSaung, Yay MyetTaung, Wet Thay, and GyaingKauk (Soe-Htun et al. 2018). Soe-Htun et al. (2015) reported that all these seagrass beds are under severe threats from multifarious factors. Toxic waste discharged through urban runoffs is one of the chief concerns. The bottom trawling fishing technique damages the seagrass beds. The sediment runoff often buries the seagrass vegetation in the Myanmar coastline. Sand mining practiced in many of the sectors along Myanmar coastline increases the turbidity of the water column, which can indirectly hamper the productivity of seagrasses in Myanmar. Information on salt marshes of Myanmar is scarce. Murray et al. (2020), while characterizing the threatened ecosystems of Myanmar, emphasized that salt marshes in Myanmar are data deficient, and nothing conclusive can be stated about the possible threats that this ecosystem faces at present. However, the morphodynamic study by Choi et al. (2020) on the Sittaung River estuary, Myanmar, indicated that conversion to agricultural lands, sedimentation, and erosion are the major threats to salt marsh ecosystems of Myanmar.

9.3.4 Thailand

Thailand has a coastline of around 2815 km. The tidal flats, especially in the Gulf of Thailand, shelter large patches of mangroves. Like many other countries, Thailand has also witnessed substantial changes in the mangrove cover in the last few decades. Pumijumnong (2014) detailed a zone-wise analysis of mangrove cover dynamics from the year 1961 to 2007. Pumijumnong (2014) observed that the coastlines of Thailand in 1961 encompassed almost an area of 3720 km^2 under dense mangrove patches. However, this areal cover reduced to 1675 km^2 in the year 1996. However, significant conservation measures led to an increase in the mangrove area to 2300 km^2 in the year 2007. They observed that the threats to Thailand mangroves arise from the local inhabitants in many places (Krabi, Phuket, Trang, and Ranong) where they continually encroach on the mangrove patches for intensive aquaculture and tree harvesting. Shrimp farming-induced mangrove land conversions are most prominent in the eastern part of the Thai Gulf and the Nakhon Si Thammarat and Surat Thani provinces (Pumijumnong 2014). Imai et al. (2006) observed that disturbances in the mangrove patches of the Ranong region led to the formation of gaps in the canopy cover. Giri et al. (2008) mentioned that mining activities also led to significant degradation of mangroves in Thailand. Plathong and Plathong (2004) observed that knowledge on mangrove ecological aspects among the local community coupled with anthropogenic pressure from the inland fishing sector is jointly responsible for the widespread destruction of mangrove habitats in Thailand. Loopholes in law enforcement and the lack of active participation of the local inhabitants in conservation programs have worsened the scenario. The perception of local fisher communities and the fear of compromising financial benefits to safeguard mangrove habitats are some of the other factors that increase the threats to mangroves in this part of the world (Jumnongsong et al. 2015). Aksornkoae (2012) pointed out that shrimp farming led to the destruction of more than 50% of Thailand's mangrove cover. Moreover, Thampanya et al. (2006) indicated an increased coastal erosion in the Thai coastline, which led to substantial loss of mangroves.

Like mangroves, seagrass beds also exist in the nooks and corners of the coastline. The Andaman sea coast in Thailand shelters most of the seagrass patches. The Indian Ocean tsunami in the year 2004 destroyed almost 5% of the total seagrass cover in this region (Adulyanukosol and Poovachiranon 2006). Nakaoka et al. (2007) also observed a decline in the seagrass biomass in the post-tsunami phase all along the coastal regions of Thailand. Prathep et al. (2010), while working in the KohTha Rai seagrass beds of Nakhon Si Thammarat Province, observed that sediment deposition from a nearby development project caused a significant reduction in the biomasses of seagrasses. The sediment-laden runoffs during the monsoon season can also pose a natural threat to the existing seagrass meadows of Thailand. Khogkhaoa et al. (2017) observed significantly less biomass and lesser species diversity in the wet season compared to the dry season in the Trang River seagrass communities of Thailand. Recent baseline studies in the Thailand coast also mentioned similar threats and emphasized the requirement of public awareness (Rattanachot et al. 2020). Stankovic et al. (2021) observed a decrease of almost 2 km^2 between the years 1999 and 2019 in the Libong Island of Trang Province and attributed urbanization as the prime cause behind such reduction in areal extent. Several pieces of literature indicate the presence of salt marshes in the Thai coastlines (Yuvaniyama and Arunin 1993; Ueda et al. 2000; Choowong 2002). However, accurate quantification of the areal extent and identifying threats to this ecosystem are still lacking. Lately, Pornpitakdamrong and Sudjaroen (2014) reported that local inhabitants of the Samut Songkhram region use salt marsh species *Suaedamaritima* for cooking.

9.3.5 Singapore

Singapore encompasses a small land surface area of only 710 km^2, yet shelters many of the crucial marine ecosystems. During the early nineteenth century, mangrove forests surrounding Singapore covered an area of almost 75 km^2 (Corlett 1987, 1991). During the British colonial era, the mangrove area in Singapore diminished to a large extent due to charcoal and timber extraction (Yee et al. 2010). Following their separation from Malaysia (in the year 1965), Singapore witnessed a series of developmental projects like the construction of dams and a flourish in prawn ponds. In the late nineteenth century, Singapore had to compromise a substantial part of the pre-existing mangrove cover due to all these activities. In the last decade, Yee et al. (2010) quantified the present mangrove area to be 6.59 km^2. However, Yang et al. (2011) reported a contemporary mangrove cover of around 7.35 km^2. Friess et al. (2012) extensively studied the Mandai mangroves of Singapore and identified all the representative threats of this nation's mangrove cover. They observed that poor access to many of the mangrove locations facilitated less disturbance and effectively conserved the mangroves. However, development like railway tracks, which enhanced the accessibility, led to the degradation of mangroves. Friess et al. (2012) also identified that sea-level rise could be of severe threat to patchy mangroves as observed in the coastal regions of Singapore. Besides, aquafarming operations led to the reduction of many of the old mangrove forest stands of mainland Singapore (Bird et al. 2004). Due to the proposed development projects that are to initiate soon, almost 33% of the country's total mangroves are at stake (Lai et al. 2015; Friess et al. 2016). Due to urbanization, pollutants like heavy metals (Cuong et al. 2005) and polybrominated diphenyl ethers (Bayen et al. 2005) accumulate substantially in the mangrove habitats of Singapore, which can pose severe threats to these floral stands.

Yaakub et al. (2013) identified that Cyrene, Pulau Semakau, and Chek Jawa are the three locations with prominent seagrass meadows on the coastline of Singapore. They mentioned that Singapore used to shelter large seagrass beds in the mid-twentieth century (Chuang 1961); however, at present, these sites encompass only 0.337 km^2 of seagrass cover. They feared that two seagrass species namely *Halophila ovata* and *Halodule pinifolia* might have become extinct from the coastal periphery of Singapore. Loo et al. (1996) pointed out that rampant land cover change, land reclamation, and enhanced rate of sedimentation substantially destroyed the seagrass beds of Singapore. Artificial beach and lagoon construction hampered the growth rate of seagrass species. They also reported that many of the seagrass species ceased to thrive along the Singapore coastline. Yaakub et al. (2014) and McKenzie et al. (2016) observed that light availability and temperature are two of the most crucial environmental drivers that regulate the health of seagrass in Singapore. The near-shore construction activities significantly compromise the light attenuation coefficient of the water column due to both bottom churning and sediment influx. Seagrass prefers calm and less disturbed habitats, and these sites are often used for anchoring boats and cage culture. A thorough study on the potential threats of seagrass meadows and formulating suitable options to alleviate these threats is still due for Singapore. No information is available at present on the salt marshes of Singapore.

9.3.6 Indonesia

Indonesia shelters more than 20% of the global mangrove cover. However, agricultural expansion, aquafarming, and urban encroachment have degraded substantially in the last few decades (DasGupta and Shaw 2017). Giri et al. (2008) pointed out that almost 200 km^2 of shrimp aquaculture farms came into existence in Indonesia at the cost of mangrove deforestation, since the year 1975. Dalimunthe and Putri (2017) mentioned that ing activities, forest clearing for timber and construction materials, ship-based pollution, and land conversion to aquafarming are the major threats to Indonesian mangroves. Mulyadi and Amin (2016), while working in the Dumai city of Sumatra, observed several threats to the adjacent mangroves. The threats ranged from domestic and industrial effluents, construction activities, water-based navigation, tourism activities, to encroachment of human settlements and forest felling for timber extraction. Richards and Friess (2016) mentioned that a steady conversion of mangrove forests to oil palm estates could pose a severe threat in the future. According to them, Sulawesi, Borneo, and Sumatra are the hotspots of mangrove deforestation in Indonesia. Sulawesi and Kalimantan witnessed rapid mangrove conversion to aqua-farming plots (Richards and Friess 2016). According to Murdiyarso et al. (2015), the Indonesian coastline of 95,000 km, used to shelter 42,000 km^2 of mangroves in the year 1980, which reduced to ^{29,000} km^2 by the year 2005. They argued that the expansion of aquaculture is primarily responsible for this destruction. Several studies indicated that the national government at some point in time promoted the growth of aquaculture. Ardli and Wolff (2009) recorded similar observations from the largest mangrove forest of Central Java, Segara Anakan. They quantified that this forest lost an area of about 85 km^2 between the years 1978 and 2003. Hermon et al. (2018) observed that the Belitung Island of Indonesia lost more than 50% of its blue carbon stock between the years 1995 and 2015.

Besides mangroves, Indonesia has the pride of sheltering almost 5% of the global seagrass meadows encompassing an area of about 30,000 km^2 (Kuriandewa et al. 2003). Several studies indicate that the seagrass meadows underwent rapid degradation due to anthropogenic activities, which included infrastructural development along the coastline, land renovation, sand mining, and mooring boats (Unsworth et al. 2018). Riani et al. (2012) documented similar threats while characterizing the challenges of seagrass management in Indonesian coastlines. However, there are no quantitative data on the reduction of seagrass areal cover from the coastlines of Indonesia. Some studies indicate the possibility of underestimating the seagrass extent as thousands of islands remain unexplored (Ooiet al. 2011; Alongi et al. 2016). Salt marshes of Indonesia are yet to receive adequate attention in terms of identifying threats. However, Sejati et al. (2020) in a recent study commented that mangrove destruction can leave a disastrous signature on the other associated blue carbon habitats, which includes the coastal marshes.

9.3.7 Malaysia

Malaysian (with more than 3700 km coastline) mangroves are under severe threat like many other countries in the periphery of the Indian Ocean (Latiff and Faridah-Hanum2014). This country shelters 3.7% of the global mangrove cover, which is close to 5800 km^2; however, several studies indicated a massive rate of deforestation from the mangrove patches of Malaysia. The Matang mangrove forest is the largest in Malaysia, encompassing an area of about 400 km^2. Ibrahim et al. (2015) reported a net mangrove area loss of 80 km^2 from this forest between 1993 and 2011. They identified coastal erosion, unsustainable tree felling, shrimp aquaculture, illegitimate agriculture, and trespassing are the major threats that the mangroves experience. Kanniah et al. (2015) observed a mangrove area loss of 67 km^2 in the Iskandar region between 1989 and 2014. Selangor, Pahang, and Johor states witnessed rampant mangrove degradation due to urbanization and industrial spread in the mangrove peripheries (Azian and Mubarak 2012; Omar 2012). Sarmin et al. (2016a) implemented the Driver-Pressure-State-Impact-Response (DPSIR) framework to characterize the plausible threats to mangroves of Johor state. Sarmin et al. (2016b) suggested that, like Indonesia, Malaysia also has a growing sector of manufacturing oil palm, which could lead to substantial loss of mangroves in the future. Reports of mangrove clearing for building residential areas also exist from Langkawi Island, Malaysia (Shahbudin et al. 2012). Jusoff and Taha (2008) and Jusoff (2013) attributed population boom as one of the main drivers of mangrove clearing. Several workers noted that rice cultivation also continues to be one of the significant reasons behind the rapid deforestation of mangrove cover in the populated coastlines of Malaysia (Ghazali et al. 2016).

Besides mangroves, Malaysian is an abode of a spectacular seagrass cover. However, a combination of natural and anthropogenic factors has led to substantial degradation of seagrass cover in this country. Hossain et al. (2015) observed substantial loss of seagrass cover in the Paka and Pengkalan Nangka lagoons and Punag-Sari estuary in Malaysia between the years 1998 and 2014. Natural sand shifting and coastal flooding led to considerable loss of seagrass meadows. Among the anthropogenic factors, dredging, construction activities, and sand mining substantially disturbed the seagrass habitats. Hossain et al. (2019) observed a loss of 0.94 and 0.34 km^2 seagrass cover in the Sungai Pulai estuary and the Tanjung Adang Darat, respectively, between 1994 and 2017. Coastal reclamation and unwarranted sedimentation led to such a reduction in area cover. Freeman et al. (2008) observed short-term seagrass loss in the Sabah region of Malaysia mainly due to light-limited conditions induced by enhanced turbidity in the water column. Bujang et al. (2016) pointed out that rough sea conditions during the northeast monsoon season bring in particulate matter that buries the seagrass beds leading to their destruction. However, these natural threats are negligible compared to the threats caused by population growth in the coastal sector and enhanced anthropogenic activities (Bujang et al. 2006). Though seagrasses of Malaysia received substantial attention from the scientific community, information on the threats to salt marshes is absent to date.

9.4 Threats in the Asian Countries Bordering the Western Indian Ocean

9.4.1 Yemen

Yemen comprises a long coastline of about 2200 km, which shelters some of the luxuriant mangrove patches of the west coast of the Indian Ocean. However, in the recent past, these mangroves have become vulnerable due to urban sprawl, aquafarming, tourism industry, and timber harvesting (Nagi and Abubakr 2013). Oil spills, indiscriminate waste disposal, domestic and industrial sewage, unsustainable fishing activities, and heavy metal accumulation (Al Hagibi et al. 2018) have also become points of concern. Al-Hodeidah and Al-Luhayyah are some of the cities, which witnessed a significant reduction in mangrove cover due to construction activities related to urban development. A large section of the rural people in Yemen depends on the mangroves to feed their cattle, which led to the disappearance of green leaves in the mangrove plants and eventually their death (Al-Hemiari 2004). The tourism activities in the Bir Ali and Buhees villages led to the considerable loss of mangroves (Nagi and Abubakr 2013). Besides, industrial pollution, clearing mangroves for creating salt pans, and aquafarming have led to a significant destruction of mangroves in the Yemeni coastlines (Khalil 2002). However, camel grazing and timber collection are the two most crucial reasons that led to the widespread destruction of mangroves throughout the coast of Yemen (Almahasheer et al. 2016). Compared to the continuous stretches of mangroves, the seagrasses occur in patches along the coastlines of Yemen. The survey by Barratt et al. (1987) still serves as the baseline for the state of the art of seagrass vegetation on the Yemeni coast. Rouphael et al. (1998) mentioned that, unlike many other coastal peripheries, sedimentation and the associated increase in the turbidity levels are not the prime concerns for the seagrasses of Yemen. The water column in the Yemen near-shore waters has substantial water clarity. According to Rouphael et al. (1998), coastal industrial pollution (especially oil pollution), urbanization, and shipping activities impose severe threats to the seagrasses of Yemen. However, Mahdy et al. (2021) observed that landfill activities and coastal engineering enhance the turbidity in the seagrass overlying water column. Thus, full-proof characterization of threats to seagrasses of Yemen requires more attention. Several studies indicated the presence of salt marshes along the coasts of Yemen (Kürschner et al. 1998; Al-Hawshabi and El-Naggar 2015; Świerczewski et al. 2017). However, no proper report on the threats to these ecosystems exists at present (Table 9.2).

9.4.2 Oman

Oman has a long coastline and shelters typical mangroves of arid environments. Very few studies focused on their spatial extent (Choudri et al. 2016). However, many of the observations showed a periodic increase and decrease in mangrove cover in recent decades (Al-Afifi 2018). Al-Afifi (2018) further observed that the mangroves of Oman suffer due to cyclonic storms and associated surges. Flash floods from the terrestrial sectors also lead to substantial damage to mangroves. Besides, urban developmental activities also narrow down the mangrove lines in the coastal periphery of Oman. Paleo research in the coastline of Oman indicated that the mangroves of this region vanished 6000 years ago (Decker et al. 2021). The reason was neither the rise in regional sea level nor anthropogenic pressure. The shift in precipitation patterns led to acute water scarcity in the arid mangroves of Oman. Enhanced salinity was the main reason behind the absence of mangroves. Such observations indicate that

Table 9.2 Some of the major observations related to threats to the blue carbon habitats in the Asian countries of the western Indian Ocean

Country		Major threats	Observations	Authors
Yemen	M	• Camel grazing • Timber harvesting	• Al-Hodeidah, Al-Luhayyah, and Midi cities witnessed significant mangrove destruction in the peripheries • Tourism degraded mangrove habitats in the Bir Ali and Buhees villages	Nagi and Abubakr (2013) Almahasheer et al. (2016)
	SG	• Shipping industry • Oil pollution • Landfill activities	• Industrial and oil pollution is the primary threat	Rouphael et al. (1998) Mahdy et al. (2021)
	SM	NS	• NS	NS
Oman	M	• Storm surges • Coastal flooding • Urbanization	• Cyclones Gonu and Phet in the year 2007 and 2010 led to considerable destruction of mangroves in Oman	Al-Afifi (2018)
	SG	• Overgrazing • Environmental stress • Monsoon rain-induced low temperature and salinity	• Physical isolation of small patches from the larger patches leads to considerable damage to seagrasses	Jupp et al. (1996) Ferreira et al. (2006)
	SM	NS	• NS	NS
UAE	M	• Cement industry • Coastal development • Dredging activities • Regional sea-level rise	Total mangrove area decreased in Abu Dhabi by 2.9 km^2 between 1900 and 2000, and 2.2 km^2 between 2000 and 2006	Elmahdy and Mohamed (2013) Alsumaiti et al. (2017) Paleologos et al. (2019)
	SG	• Desalination plants • Power plants • Boating activities	• Regional sea-level rise can hamper the functioning of seagrasses	Howari et al. (2009) Schile et al. (2017) Schile-Beers et al. (2018)
	SM	-DO-	-DO-	-DO-
Iran	M	• Regional sea-level rise • Recurrent droughts	• Mangroves of Khamir, Tiab, and Jask regions witnessed a net decrease of 12 km^2 between 1998 and 2017	Etemadi et al. (2018) Mafi-Gholami et al. (2019, 2020)

(continued)

Table 9.2 (continued)

Country		Major threats	Observations	Authors
	SG	• Shrimp capture by bottom trawling • Unstable political conditions	• Oil pollution and wars involving the marine sector takes a heavy toll on the benthic flora	Sharifinia et al. (2019) Keshavarzifard et al. (2021)
	SM	NS	• NS	NS
Pakistan	M	• Coastal erosion • Regional sea-level rise • Cattle grazing • Freshwater supply reduction	• The Indus Delta mangroves witnessed a net loss of almost 2000 km^2 between 1980 and 2006	Qureshi (1993) WWF-Pakistan (2006) Mukhtar and Hannan (2012)
	SG	NS	• NS	NS
	SM	NS	• NS	NS

natural factors like temperature and rainfall play a more crucial role in governing the mangrove dynamics than the anthropogenic factors in the coastline of Oman. Like mangroves, seagrasses exist in several regions along the coastline of Oman. Jupp et al. (1996), while working on the seagrasses of Masirah Island, observed that detachment of patchy meadows from the larger seagrass bodies, environmental stress, and overgrazing are the concurrent threats to the seagrasses of Oman. Ferreira et al. (2006) indicated that the monsoon-induced dip in temperature and low saline waters are not suitable for the seagrass beds to flourish on the coast of Oman. However, proper delineation of anthropogenic threats remains a task for future researchers. Regarding salt marshes, the research in Oman is still in its infancy and requires substantial attention.

9.4.3 Uae

UAE (having a 1300 km coastline) hosts patchy mangroves mainly concentrated in Abu Dhabi (Habshi et al. 2007). Khor Umm Al-Quwain, Al Dabbiya, and Khor Kalba are some of the most crucial mangrove sites of the UAE. Besides all these regions, the UAE shelters a thin mangrove line in the fringes of the tidal waterways encircling the country (Habshi et al. 2007). Several studies quantified the total mangrove cover in UAE; however, the estimates remain debatable to date. Elmahdy et al. (2020) observed temporal increase and decrease in sparse and dense mangrove vegetated areas in selected mangrove sites of UAE in the last two decades. The total estimated mangrove cover ranges between 30 km^2 (Rabanal and Beuschel 1978) and 40 km^2 (Habshi et al. 2007). Elmahdy and Mohamed (2013) recorded a net decrease in Abu Dhabi's forest area by 2.9 km^2 between 1900 and 2000 and 2.2 km^2 between 2000 and 2006. Paleologos et al. (2019) recently reported a 155 km^2 of Avicennia marina stretch in Abu Dhabi. According to them, Abu Dhabi has been witnessing rapid coastal development activities. Cement and heavy metal industries have rampantly flourished in the coastal sector. Tourism activities also went up, which led to many artificial coastal constructions. Such constructions along with the increased waterway traffic significantly jeopardized the hydrodynamic setting of this region. Instead of settling in the coastal peripheries, particulate matters have shown a tendency to wash away toward the sea, which poses a severe threat to the existing mangrove stands as they are bereft of essential substrates. The widespread dredging activities all along the coastline are making these crucial vegetations in the UAE more vulnerable (Yagoup and Kolan 2006). Alsumaiti et al. (2017) also indicated that oil spills and regional sea-level rise could pose severe threats to the mangroves of Abu Dhabi in the long run. Seagrasses in the Abu Dhabi coastline also exist in patches. Howari et al. (2009) observed that diversion of freshwater runoff and boat movements related to tourism and navigation activities are the major threats to the seagrass meadows of this region. Schile et al. (2017) pointed out that the seagrass functioning in the UAE coastline is highly susceptible to small changes in regional sea level. Erftemeijer and Shuail (2012) observed that desalination units and power plants in the coastal sectors are posing severe threats to the blue carbon habitats of the UAE. Though explicit indicators of threat are not available for the salt marshes, the same for seagrasses are applicable for the coastal marshes (Schile-Beers et al. 2018).

9.4.4 Iran

Iranian coastlines shelter one of the prominent mangrove patches of the entire Middle East, covering about 192 km^2 in the present date (Mafi-Gholami et al. 2020). The mangroves of this region are typically arid. Regional sea-level rise and recurrent droughts are the points of severe concern in these coastlines. The mangrove lines in the Persian Gulf and Gulf of Oman sectors have substantially retreated due to the ongoing sea-level rise coupled with enhanced occurrences of droughts between the years 1998 and 2017 (Mafi-Gholami et al. 2020). They observed a collective net decrease of 12 km^2 mangrove area in the Khamir, Tiab, and Jask regions. They further predicted that under the business-as-usual scenario of the dry spells, more than 60% of the mangrove cover could cease to exist in the Iranian coastlines by the end of 2100. Etemadi et al. (2018), while working in the protected areas of Hormozgan Province, observed that sea-level rise would soon engulf a substantial portion of the seaward mangroves of this region. The regional sea-level rise in these coastal stretches varied between 2.9 and 5.4 mm/year. Mafi-Gholami et al. (2019a) also pointed out that dry spells leading to droughts significantly reduce the leaf area index in the mangrove forests, which takes a toll on the overall health of these floral stands. Besides these natural factors, anthropogenic threats include heavy metals (Keshavarz et al. 2012; Ghasemi et al. 2018) and microplastic pollution (Maghsodian et al. 2021). The Gabrik Wetland in Iran hosts mangrove stands that are of international importance. Parvaresh (2011) identified that dune movements, camel grazing, and the construction of dams and ports are the major threats to the mangroves of this region. Ghodrati Shojaei et al.

(2021) seconded these observations. Maghsoodloo and Eghtesadi (2008) mentioned that seagrass habitats are visible throughout the Iranian coastline in patches. However, their proper area demarcation is still missing in the available literature. Erftemeijer and Shuail (2012) reported a 160 km^2 seagrass bed in the Shidvar Wildlife Refuge of Iran. Keshavarzifard et al. (2021) indicated that bottom trawling activities for shrimp capture are harmful to the seagrasses of this region. Sharifinia et al. (2019) further observed that oil pollution, population growth, and an unstable political paradigm, which often leads to wars through the marine fronts, leave an irreparable scar in the blue carbon habitats of this coastal margin. Iran has richer biodiversity of salt marshes (Asri and Ghorbanli 1997; Alaie 2001) than many of the other Middle East countries. However, documents on threats to salt marshes of this region are scarce. Mafi-Gholami et al. (2019b) mentioned that mangroves and salt marshes exchange their habitats over temporal scales due to a combination of factors like anthropogenic land-use change, local hydrogeological changes, and climatic change.

9.4.5 Pakistan

Pakistan has a long coastline of about 1000 km interspersed by three principal estuaries, namely the Indus, Hingol, and Dasht, giving rise to an optimum environment for blue carbon habitats to flourish in the coastal sectors. Besides the Indus Delta, mangroves are prevalent in the Gawatar, Kalmat, and Sonmiani regions. However, numerous anthropogenic stress has made these mangroves significantly vulnerable (Mukhtar and Hannan 2012). Acute scarcity of freshwater is one of the potent threats to the mangroves of Pakistan. The Indus Delta comprises more than 97% of the country's total mangrove cover (Amjad and Jusoff 2007). Kidwai et al. (2019) observed that scarcity of freshwater in the Indus Delta potentially hamper the health of the mangroves. Coastal erosion, cattle grazing, regional sea-level rise, tree felling for timber, and firewood are the primary concerns for the mangroves of Pakistan (Adhikari et al. 2010). The mangroves of the Indus Delta consistently reduced from 2800 km^2 in 1980 to 1600 km^2 in 1990 to only 730 km^2 in 2006 (Qureshi 1993; WWF-Pakistan 2006; Mukhtar and Hannan 2012). An increase in industrial pollutant load in the deltaic regimes has also led to the widespread destruction of mangroves in Pakistan (Amjad et al. 2007). Export of mangrove seedlings for reforestation purposes within the country and abroad has also emerged as a threat to the natural seed banks of the existing vegetated stands (Rasool 1996). Though the mangroves of Pakistan have received attention, there is no information on this country's seagrass cover (Patro et al. 2017). On the contrary, several studies focused on the salt marshes of Pakistan (Gul and Khan 1995; Khan et al. 2000, 2005; Gulzar et al. 2003; Agha et al. 2009); however, none of them indicated the plausible threats to this ecosystem.

9.5 Threats in the African Countries Bordering the Western Indian Ocean

9.5.1 *South Africa*

The Nahoon estuary on the east coast of South Africa marks the country's southern limit of mangrove spread. Being at a junction of sub-tropical to temperate climate, the mangroves of South Africa display unique characteristics (Naidoo 2016). Mangroves approximately encompass around 19 km2 in this country (Hoppe-Speer 2012). The species diversity declines as one moves south (Naidoo 2016). Though South Africa witnessed significant efforts on enhancing mangrove cover throughout the country (Steinke 1999), the existing mangroves face several natural and anthropogenic threats. The mangroves of South Africa grow in patches (Quisthoudt et al. 2013), which makes them more vulnerable than those in large continuous stretches. Exploiting mangroves for timber and cattle grazing has significantly affected the seed bank in many of these mangrove patches (Hoppe-Speer et al. 2015). Sustainable mangrove harvest action has become a dire need to safeguard the future of this vegetation in South Africa (Rajkaran and Adams 2012). An enhancement in salinity, which the mangroves of South Africa experience in patches, is a chief point of concern. Hoppe-Speer et al. (2011) observed that acute salinity stress shortens the tree structure of the mangroves, and the floral stands experience leaf necrosis. Yang et al. (2014), while studying the mangroves of Mngazana estuary, observed that sea-level rise and the increased frequency of storm events lead to substantial loss of mangrove cover by the end of this century. The mangroves of South Africa have a limited phylogenetic diversity, unlike many other tropical countries (Daru et al. 2015). The conservationists argue that the lack of biodiversity makes these forests prone to diseases or pest attacks that proliferate among the same type of species (Yessoufou and Stoffberg 2016). Many of the mangrove pockets in South Africa cease to experience tides due to changes in land use and geomorphology, which induces drought-like conditions (Hoppe-Speer et al. 2013). Such situations are difficult altogether for the mangroves to acclimatize. Besides, signatures of anthropogenic pollution in the form of microplastics and coal dust from the adjacent mines are prevalent in the mangrove estuaries of South Africa (Naidoo and Chirkoot 2004; Naidoo and Naidoo 2005; Naidoo et al. 2020).

Besides mangroves, South African coastlines shelter seagrass beds of varying dimensions (Adams 2016). *Zostera capensis* is the most dominant variety available in this country. The seagrasses and saltmarshes of this country are highly susceptible to the regional sea-level rise (Bornman et al. 2016). The seagrass beds in many of the South African estuaries experience flooding. The deposition of sediment, drastic changes in salinity, and reduction in light penetrability induced by floods leave a disastrous signature on the health of the seagrass (Hanekom and Baird 1988). Anthropogenic activities like trampling, sand mining, and digging irreversibly degrades the seagrass habitats of this region (Pillay et al. 2010). Interrupted tidal flow due to clogging of river mouths and dam construction coupled with subsequent changes in nutrient level leading to eutrophication are some of the other threats to seagrasses of South African estuaries (Nunes and Adams 2014; Human et al. 2016). The seashores of South Africa experience high waves and incessant disturbances, which restricts the suitable habitat for salt marsh growth to few pockets of backwaters (Adams et al. 2016). Raw et al. (2020) indicated that the salt marshes of South Africa are susceptible to future sea-level rise. The salt marsh would migrate toward land with changing sea levels. However, the absence of a proper area to flourish would squeeze this ecosystem into competition for space (Raw et al. 2021).

9.5.2 Mozambique

Mozambique shelters a substantial share of the total mangrove cover in the African coastline adjoining the western Indian Ocean. According to Barbosa et al. (2001), Mozambique encompasses about 4000 km^2 of mangrove in its coastal periphery. However, these mangroves are susceptible to several threats. Local inhabitants use mangrove timber for construction purposes (Charrua et al. 2020). Lack of electricity and high population density in many of the coastal sectors of Mozambique compels the local inhabitants to harvest mangrove trees like Avicennia sp. and Rhizophora sp. to meet the demand for fuel wood (Macamo et al. 2016) (Table 9.3). The Maputo city in Mozambique shelters one of the largest mangrove forest reserves in adjacency to a growing urban settlement. Urbanization has led to substantial loss of mangroves in and around this area (Beja da Costa and Ribeiro 2017). Cabral et al. (2017) indicated that physical forcing events and coastal erosion pose severe threats to the Mozambique mangroves. Asante et al. (2009) observed an overall low elevation in the Mozambique coastline where the mangroves flourish, which makes them vulnerable to tropical cyclones and storm surges. Recently, the super cyclone Idai (March 2019) devastated the Mozambique mangroves (Devi 2019). Anthropogenic sewage pollution is another point of concern for the fringe mangroves lying in the vicinity to discharge points along the creeks of Mozambique (Penha-Lopes et al. 2011). Fatoyinbo et al. (2008) summed up that land-use change to agricultural plots, salt production units, marine pollution, and reduction in freshwater supply are some of the crucial factors that lead to mangrove degradation in Mozambique. Earlier observations of Saket and Matusse (1994) and Doddema (1997) were in parity with these observations. Fatoyinbo et al. (2008) also challenged the previous estimates of around 4000 km^2 mangrove area and reported only 2900 km^2 of present area coverage in Mozambique. Mozambique has extensive seagrass cover. Several pieces of research focused on this crucial ecosystem. However, characterizing threats is at present a lacuna. Among the few studies, Bandeira (1995) and Ruwa (1996) indicated that mangrove felling, invertebrates collection, unsustainable fishing, heavy siltation, sewage discharge, coastal constructions, and alteration of river flow by the construction of dams are the principal challenges for conserving seagrass beds. Similarly, the salt marshes of Mozambique have received much less attention despite their reported existence adjoining several mangrove patches. Adams and Bandeira (1995) observed

Table 9.3 Some of the major observations related to threats to the blue carbon habitats in the African countries of the western Indian Ocean

Country		Major threats	Observations	Authors
South Africa	M	• Non-tidal conditions Droughts Cattle grazing Harvesting	• South Africa losses 11% of their mangrove cover every decade	Hoppe-Speer (2012) Yessoufou and Stoffberg (2016)
	SG	• Clogging f river mouth • Eutrophication	• Flooding poses severe threat to the seagrass patches and it takes few years for the seagrasses to recover	Nunes and Adams (2014) Adams (2016) Bornman et al. (2016)
	SM	• Sea-level rise • Less space for future landward migration	• Only two species can cope with the changing rate of sea level • 40% of salt marsh cover in Thesen's Island would likely disappear by 2100	Raw et al. (2020 2021)
Mozambique	M	• Overexploitation • Solar salt production	• Mozambique experiences mangrove deforestation at a rate of 18 km^2 per year	Barbosa et al. (2001)
	SG	• Dam construction • Alteration of river flow • Coastal development • Unsustainable fishing	• Proper characterization of threats to seagrass communities is yet to be done	Bandeira (1995) Ruwa (1996)
	SM	• Urbanization Salt pans • Coastal development	• Maputo and Beira cities in Mozambique witnessed large-scale degradation of salt marshes	Adams and Bandeira (1995)
Tanzania	M	• Flooding • Drought • Human settlement	• Tanzanian coastline lost 17 km^2 mangrove between 1988 and 2000	Wang et al. (2003) Ferreira et al. (2009)
	SG	• Fishing activity • Seaweed farming • Overgrazing by sea urchins	• Zanzibar witnessed a net seagrass area loss of 3.4 km2 between 1987 and 2003	Gullström et al. (2006)
	SM	• NS	• NS	S
Kenya	M	• Fuel wood and collection • Tourism • Aquaculture	• Kenyan coastline lost 97 km^2 mangrove between 1985 and 2010	Kirui et al. (2013)

(continued)

Table 9.3 (continued)

Country		Major threats	Observations	Authors
	•		• Kenya lost around 100 km2 seagrass meadows between 1986 and 2016	Harcourt et al. (2018)
	SM	• NS	• NS	NS
Somalia	M	• Pastoralism	• Low biodiversity and sporadic occurrences	Diops et al. (2002) Bosire et al. (2015)
	SG	• Port and fishing activities	• The pre-colonial and colonial era witnessed significant destruction of coral and adjoining seagrass beds	Carbone and Accordi (2000)
	SM	• NS	NS	NS
Djibouti	M	• Sand dune shifting • Camel grazing • Pastoralism	• Lack of tidal activity and industrial pollution led to significant destruction of mangroves	Witsen (2012)
	SG	• Oil pollution		Sougueh and Jonasson (2018)
	SM	• Oil pollution		Sougueh and Jonasson (2018)

that urbanization, agricultural land conversion, salt pans, road construction, and port development are some of the factors that are making the salt marshes of Mozambique vulnerable.

9.5.3 *Tanzania*

Tanzanian coastlines shelter mangroves in small and large patches. Recent evidences indicate that the ongoing climate change is taking a toll on the mangroves of Tanzania. Hamad et al. (2019) observed that the Tanzanian coastlines are experiencing a steady increase in sea level. They also observed a decline in precipitation and an increase in ambient temperature over the last three decades, which hampered the productivity of these floral stands. Wang et al. (2003) observed a loss of almost 17 km^2 mangrove cover in the Tanzanian coastlines between the years 1988 and 2000. Flooding and droughts are two of the concerns that threaten the mangroves of Tanzania. Encroachment of settlements, industrial pollution, and tree felling are some of the additional threats (Ferreira et al. 2009). Agriculture and aquaculture activities have increased significantly over the last few decades in the Tanzanian coastlines (Semesi 1992; Lugendo 2016). Some of the rivers (Msimbazi River) and ports (Dar es Salaam) adjoining the mangrove stands suffer from intense metal and organochlorine pesticides pollution (Kruitwagen et al. 2008). Mainoya et al. (1986) reported three decades ago that conversion of mangrove habitat to salt evaporation pans is one of the prime threats in Tanzania. Besides mangroves, Tanzanian coastlines have luxuriant seagrass meadows, especially in Zanzibar. Gullström et al. (2006) studied the changes in these submerged aquatic vegetation between 1987 and 2003. They had mixed observations of both gains and losses in seagrass patches over the spatial stretch, but reported a net loss of around 3.4 km^2 between the same time period. They further reported that interlinked natural and anthropogenic factors together pose threats to this submerged ecosystem. The preference of local fisher community to carry out fishing in seagrass adjoining water column, seaweed farming, and overgrazing by sea urchins are the principal factors that degrade the seagrasses of Tanzania (Gullström et al. 2006). These threats on the seagrass communities of Tanzania make the dugongs of this region highly vulnerable (Muir et al. 2003). de la Torre-Castro and Rönnbäck (2004) indicated that the local fisher community benefits substantially from the seagrass-associated fishing and has substantial traditional knowledge. However, a nation-wide seagrass management strategy could have better ensured the well-being of this vegetation and the related ecosystem services. Adams and Bandeira (1995) reported that some salt marsh species are available in the Tanzanian coastlines and they grow in hypersaline soils where mangroves usually do not thrive. However, no proper information is available on the threats that the salt marshes of this region face if any.

9.5.4 Kenya

Kenya hosts a comparatively less mangrove cover than the other east African countries; however, it provides several essential services to a large population. Kirui et al. (2013) observed a net loss of around 97 km^2 mangrove area in the Kenyan coastline between 1985 and 2010. According to them, Kenya lost almost 18% of its pre-existing mangrove cover in these twenty-five years. Forest clearing for fuel wood and timber coupled with anthropogenic land conversion for salt production, aqua-farming, and tourism are taking a heavy toll on the Kenyan mangroves (Kirui et al. 2013). Owuor et al. (2019a), while working in the Mida creek, observed that almost half of the interviewed local people considered that the mangroves have degraded. Plastic pollution and night soil in mangrove habitats, oil spills, and encroachment of human settlement are the primary factors behind mangrove degradation. In the same location, Owuor et al. (2019b) observed that mangrove degradation is robbing the local community from several crucial ecosystem services and that they are willing to invest substantial time to rejuvenate the state of mangroves in this region. Mungai et al. (2019) further observed that the transboundary region between Kenya and Tanzania witnessed severe mangrove loss in the recent past. Mangrove harvesting was the principal reason for this loss. Kenya has a substantial seagrass cover along the entire coastline except for the Tana Delta. Harcourt et al. (2018) argued that the previous seagrass covers reported from Kenya were underestimates. At present, it constitutes almost 317 km^2; however, between 1986 and 2016, Kenya witnessed a loss of around ~100 km^2. Harcourt et al. (2018) mentioned that unsustainable fishing practice and mangrove clearing are the main factors that led to the degradation of seagrass habitats of Kenya. Githaiga et al. (2019) emphasized that the seagrass bed losses in Kenyan coastlines make several epifauna vulnerable and lead to significant loss of sequestered blue carbon. Musembi et al. (2019) pointed out that unsustainable gears used for fishing substantially degrade the seagrass beds around the Gazi Bay, Kenya. Uku et al. (2021) observed that the over-explosion of sea urchins and developmental projects adds to the vulnerability of the Kenyan seagrasses. Information on Kenyan salt marshes is not available to comment on their threats.

9.5.5 Somalia

Very little is known about the mangroves of Somalia even in the present date. Pieces of existing literature indicate that Somalia has a low biodiversity with only three mangrove species available in this country's coastline (Diops et al. 2002). Kathiresan and Rajendran (2005), while giving an overview of the Indian Ocean mangroves, mentioned that Somalia hosts a mangrove cover of only 20 km^2 and that too with sporadic occurrences. Bosire et al. (2015) also mentioned that Somalia requires baseline studies and mangrove area demarcation to assess the present state and thereby characterize the threats. However, pastoralism is one of the practices of local communities that poses threat to the mangroves of this country. Carbone and Accordi (2000) mentioned about the presence of mangroves, seagrasses, and corals in the Somali coastlines. However, there is no proper legislature to conserve or monitor these crucial ecosystems. The pre-colonial and colonial eras witnessed significant destruction of the coastal stretches on the whole due to port-related activities. However, quantified estimates of these vegetation covers and their temporal change are literally absent as of now. Ravage by civil war, this country is trying to gear up and develop slowly and steadily. In days to come, characterizing the ecosystem services of this arid vegetations should be taken up from the government level. Adams and Bandeira (1995) mentioned about the presence of few salt marsh species in Somalia; however, the knowledge acquired so far is not adequate to comment on the threats that this vegetation experiences in this country.

9.5.6 Djibouti

Like Somalia, Djibouti also faces several crisis like unstable political environment, poverty, food scarcity, and financial constraints among the general population. According to Witsen (2012), Djibouti shelters a mangrove cover of around 10 km^2. Witsen (2012) identified several natural and anthropogenic threats to the mangroves of Djibouti. The natural threats comprise encroachment of sand dunes, tide deficient conditions in many patches, pest attack, and top-dying disease. On the other hand, the anthropogenic factors include camel grazing, pastoralism, and industrial pollution. Appelquist and Balstrøm (2014) also observed similar threats to the entire mangrove communities of Djibouti. Lack of sedimentation is one of the fundamental problems in this region. However, climate change had no detectable impact on these floral stands. Lately, heavy metal and polyaromatic hydrocarbon pollution have emerged as points of concern for Djibouti mangroves (Ahmed et al. 2017). Very little is known about the seagrass and saltmarshes of Djibouti. However, Sougueh and Jonasson (2018) indicated that both these ecosystems exist in the periphery of Djibouti and oil pollution is the principal threat in this region. Djibouti acts as a corridor of oil transport that takes place through the Red Sea and Gulf of Aden. The coastal ecosystems of this region largely suffer due to the oil spills.

9.6 Threats in the Island Countries of the Indian Ocean

9.6.1 Sri Lanka

Sri Lanka has a luxuriant mangrove cover all around the country's coastline. Keeping in view the degradation of mangroves occurring worldwide, Sri Lanka witnessed large-scale mangrove restoration projects. However, only 20% of attempted endeavors observed successful restoration (Kodikara et al. 2017). Cattle grazing, insect attacks, and choosing wrong sites for particular species were some of the fundamental factors that led to undesirable outcomes from these restoration projects. An enhancement in nature-based tourism in mangrove areas of this country can lead to negative impacts on both the floras and faunas of this habitat (Marasinghe et al. 2021). Gunawardena and Rowan (2005) observed that land conversion to aquaculture comes at the cost of mangrove clearing in many sectors of Sri Lanka. Ofori Samuel et al. (2021) indicated that shrimp farming has emerged as principal threat to the mangroves of Sri Lanka. DasGupta and Shaw (2017) mentioned that the Indian Ocean tsunami has led to considerable damage to the mangroves of Sri Lanka. Besides mangroves, several studies reported various seagrass species in the Sri Lankan coastlines (Jayasuriya 1991; Udagedara and Dahanayaka 2020). However, very few studies reported the existing threats to this ecosystem. Recently, Udagedara et al. (2017) reported the first record of *Halodule pinifolia* in the Lankan coastline near the Valaichchenai Lagoon. They observed various types of activities, which make the seagrasses vulnerable in Sri Lanka. Dumping of soil wastes, shell collection, and locomotion of fishing boats are harming the state of seagrasses. Pernetta and Wells (1993) reported similar threats from the Negombo and Puttalam Lagoons. Naveenan et al. (2015) observed substantial seagrass cover in the Sri Lankan side of the Gulf of Mannar, which is suffering from organic pollution, high rate of sedimentation, and vessel movement. Udagedara and Dahanayaka (2020) envisaged further research on Sri Lankan seagrasses keeping in view the rising sea level and changing precipitation pattern, which can significantly affect the seagrasses in the long run. Very few studies commented on the present threats on the salt marshes of Sri Lanka. Lowry and Wickremeratne (1987) indicated that cattle grazing and land-use change to salt pans and aquafarming plots used to be the prominent threats to the salt marshes. Ahalya and Suresh (2020) further observed that local people are least interested in conserving this crucial ecosystem, and the rising salinity and sea level are making this vegetation more vulnerable than ever in the Sri Lankan periphery (Table 9.4).

9.6.2 *Maldives*

Mangroves grow in the atolls found in the Indian Ocean, and the Maldives host one of the richest biodiversity among all the other atolls of this region (Sivakumar et al. 2018). However, the patchy mangroves of this country underwent rapid destruction and degradation. Saleem and Nileysha (2003) indicated that almost 150 islands of this country had mangrove cover. However, a recent study (MEE 2015) indicated that only 74 islands at present have mangrove cover in Maldives. Coastal infrastructural development and population explosion in this small country coupled with tourism pressure are the prime threats to the mangroves (Dhunya et al. 2017). The coastline of Maldives has patchy seagrass beds. Erftemeijer and Lewis (2006) indicated that dredging activities in seagrass beds for construction of resorts are the principal threat to this vegetation in Maldives. The information on the salt marshes and the threats on them is scarce from the Maldives.

9.6.3 *Seychelles*

Seychelles is an archipelago of 115 islands comprising an area of only 459 km^2 and having a population of nearly 0.1 million people. This country is a popular tourist destination where people travel from all round the world to enjoy the beaches and rain forests. Clearing of mangroves for timber and land to allow settlements in Seychelles dates back to almost late eighteenth century. The entire twentieth century witnessed a consistent decline in mangrove forests in this country. However, it has remained more or less the same since 1980 (25 km^2) (Government of Seychelles 2011). This archipelago has substantial seagrass cover in areas like Astove, Cosmoledo, and Aldabra lagoons. However, land recuperation, marine pollution, coastal development, and regional sea-level rise, especially in the populated islands, led to substantial reduction in seagrass cover (Government of Seychelles 2011). Ingram and Dawson (2001) also observed that the effluents from populated areas like Mahé enhance the rate of sedimentation and deteriorate the water quality in the seagrass adjoining water column. Senterre et al. (2015) mentioned about the presence of salt marshes in Seychelles coastline; however, nothing conclusive is present on the threats to this ecosystem.

Table 9.4 Some of the major observations related to threats to the blue carbon habitats in the Island countries of the Indian Ocean

Country		Major threats	Observations	Authors
Sri Lanka	M	• Cattle trampling • Insect attacks • Tsunami • Shrimp farming	• Indian Ocean tsunami 2004 destroyed substantial mangrove covers of Sri Lanka	DasGupta and Shaw (2017) Ofori Samuel et al. (2021)
	SG	• Shell collection • Boat movement • Sedimentation • Marine pollution	• Traditional fishing technique is taking a heavy toll on the seagrasses	Jayasuriya (1991)
	SM	• Salinization • Sea-level rise • Cattle grazing	• Lack of interest in conserving this ecosystem is the prime reason behind its destruction	Lowry and Wickremeratne (1987) Ahalya and Suresh (2020)
Maldives	M	• Coastal infrastructure • Tourism • Population increase	• 150 islands used to have mangroves, which now reduced to 74 islands only	Saleem and Nileysha (2003) Dhunya et al. (2017)
	SG	• Dredging • Fishing boats	• Construction of resorts requires dredging, which destroys seagrass beds	Erftemeijer and Lewis (2006)
	SM	• NS	• NS	NS
Seychelles	M	• Coastal Development • Sea-level rise	• Mangrove cover reduced to only 25 km2 by the end of 1980	Government of Seychelles (2011)
	SG	• Marine pollution • Coastal development • Sea-level rise	• The populated islands are witnessing enhanced reduction in seagrass cover	Government of Seychelles (2011)
	SM	• NS	• NS	NS
Madagascar	M	• Agriculture • Aquaculture • Tree felling • Lime production	• Madagascar lost most 210 km^2 mangrove cover between 1975 and 2005	Giri and Muhlhausen (2008) Scales and Friess (2019)
	SG	• Shell collection • Unsustainable fishing • Tropical cyclones	• Cyclone Haruna affected the species *Syringodium isoetifolium* and *Thalassia hemprichii*	Hantanirina and Benbow (2013) Côté-Laurin et al. (2017)
	SM	• NS	• NS	NS

(continued)

Table 9.4 (continued)

Country		Major threats	Observations	Authors
Comoros	M	• Unsustainable fishing practices • Indiscriminate coastal development • Overutilization of marine resources	• Comoros lost almost 8% of its pre-existing mangrove cover in the last few decades	Appadoo et al. (2016)
	SG	• Coastal runoff • Sedimentation • Soil erosion	• Population pressure, especially in Moheli Island, plays a crucial role in seagrass degradation	Poonian et al. (2016)
	SM	• NS	• NS	NS
Mauritius	M	• Land conversion • Resort building • Agriculture	• Marine debris deposition in mangrove environment is posing a threat	Ramessur (2002) Appadoo (2003) Seeruttun et al. (2021)
	SG	• Tuna fishing • Tourism-induced seagrass removal	• Removal of seagrasses for coastal development is the main threat	Daby (2003) Lalljee et al. (2018)
	SM	NS	NS	NS

9.6.4 Madagascar

Giri and Muhlhausen (2008) indicated that mangroves of Madagascar are declining at a much lower rate than that observed throughout the world. However, the principal reasons are land conversions and deterioration of mangrove health. Mangroves of Madagascar reduced from 3007 km^2 in 1975 to 2797 km^2 in 2005. Conversion to agriculture and aquaculture plots and tree felling are the principal threats to the Madagascar mangroves. Lime production at the cost of clearing mangrove covers has emerged as a point of concern for the mangroves of Madagascar (Scales et al. 2018). The Bay of Assassins in southwest Madagascar witnessed widespread mangrove clearing for rendering lime from sea shells. In this process, mangrove woods serve as fuels (Scales and Friess 2019). Rasolofo (1997) indicated that the traditional fishing in Madagascar used several mangrove parts to build cage-like structures and small boats, which led to considerable destruction to some of the selected mangrove species. Benson et al. (2017) emphasized that the mangroves of southwest Madagascar hold substantial quantities of carbon. However, land-use change led to almost 3.2% reduction in carbon content between the years 2002 and 2014. Madagascar also hosts a large area of seagrasses in its coastal periphery. Hantanirina and Benbow (2013) observed that shell collectors trample and dig the seagrass beds, which leads to loss of biodiversity. Fishing with small mesh size nets along the seagrass adjacent water column is another threat. Côté-Laurin et al. (2017) observed that tropical cyclones can take a heavy toll on the seagrasses of Madagascar. Adams and Bandeira (1995) reported the presence of salt marshes in Madagascar; however, information is scarce to comment on this crucial vegetation.

9.6.5 Comoros

Comoros occupies around 1 km^2 mangrove (Kathiresan and Rajendran 2005) principally concentrated in areas like Grande Comose, Moheli, and Anjuona Islands (Ahamada 1997). Seagrass beds also exist in this small country's periphery. However, limited studies are available on the biotic strength of the blue carbon ecosystems. Granek and Brown (2005) observed that unsustainable fishing practices, indiscriminate coastal development, and overutilization of marine resources are the primary threats to the blue carbon habitats of Comoros. Appadoo et al. (2016) further indicated that 8% of pre-existing mangroves declined from the coastal periphery of Comoros in the last few decades. Poonian et al. (2016) pointed out that the heavy rainfall-induced coastal runoff and erosion make the seagrasses of Comoros highly vulnerable, especially in Mohéli Island. However, nothing is known about salt marshes in Comoros.

9.6.6 Mauritius

Mauritius occupies an area of more than 2000 km^2 having a coastline length of 322 km. The mangroves of this country have low species diversity, with only two species reported in the beginning of this century (Appadoo 2003). Rampadarath et al. (2018) also observed such low mangrove diversity in the recent days. Appadoo (2003) emphasized on loss of biodiversity due to clearing of wetlands for the flourishment of tourism. Ramessur (2002) observed a tendency of local people to convert mangroves to resorts and agricultural and aquafarming plots in Mauritius. Tuna fishing in and around the seagrass beds of Mauritius leads to considerable destruction of this vegetation (Lalljee et al. 2018). Marine debris mostly originating from tourism and nearshore anthropogenic activities has emerged as a threat to the Mauritian mangroves (Seeruttun et al. 2021). Hammond et al. (2019) showed that the population explosion in Mauritius is taking a toll on the environmentally sensitive areas of the country. Removal of seagrasses for promoting tourism has been one of the principal threats to this vegetation type (Daby 2003). Hammond et al. (2015) also observed similar trends in the Mauritian coastline. However, conclusive information on salt marshes is still missing from this country.

9.7 Conclusion

The present chapter summarizes the threats to all the blue carbon habitats in the Indian Ocean periphery countries. Mangroves mostly witness anthropogenic land-use transformation as it provides a suitable land area for so many purposes like agriculture, aquaculture, settlements, tourism infrastructures, salt pans, and cattle grazing ground. Natural threats like regional sea-level rise and tsunamis are not ignorable in many countries. Marine pollution has become a point of concern in almost all the countries discussed in this chapter. Similarly, seagrasses mostly suffer from sedimentation, indiscriminate fishing, sewage effluents, enhancement in turbidity in the water column, and coastal development. Nutrient discharge and heavy rainfall-induced runoffs are also common problems in most countries. Salt marshes received much less attention altogether compared to mangroves and seagrasses. Thus, characterizing the threats indicate that plenty of work is waiting for the upcoming researchers to nullify these threats to secure a sustainable future for these crucial vegetated ecosystems.

References

Adams JB (2016) Distribution and status of *Zostera capensis* in South African estuaries—a review. S Afr J Bot 107:63–73. https://doi.org/10.1016/j.sajb.2016.07.007

Adams JB, Veldkornet D, Tabot P (2016) Distribution of macrophyte species and habitats in South African estuaries. S Afr J Bot 107:5–11. https://doi.org/10.1016/j.sajb.2016.08.001

Adams J, Bandeira S (1995) Salt marshes. Available at https://nairobiconvention.org/clearinghouse/sites/default/files/Chapter%2011_editorial%20check.pdf

Adhikari B, Baig SP, Iftikhar UA (2010) The use and management of mangrove ecosystems in Pakistan. J Environ Dev 19(4):446–467. https://doi.org/10.1177/1070496510384392

Adulyanukosol K, Poovachiranon S (2006) Dugong (Dugong dugon) and seagrass in Thailand: present status and future challenges. In: Proceedings of the 3rd international symposium on SEASTAR2000 and Asian bio-logging science, The 7th SEASTAR2000 workshop, Graduate school of Informatics, Kyoto University, pp 41–50. http://hdl.handle.net/2433/49742

Agha F, Gul B, Khan MA (2009) Seasonal variation in productivity of *Atriplex stocksii* from a coastal marsh along the Arabian Sea coast. Pak J Bot 41(3):1053–1068

Ahalya A, Suresh K (2020) Salt Marsh Ecology in Karainagar, Sri Lanka. Sci Res J 8(7):23–29

Ahamada S (1997) Situation de la mangroves des Comores. SAREC-SIDA Regional workshop on mangrove ecology, physiology, and management, Zanzibar

Ahmed MM, Doumenq P, Awaleh MO, Syakti AD, Asia L, Chiron S (2017) Levels and sources of heavy metals and PAHs in sediment of Djibouti-city (Republic of Djibouti). Mar Pollut Bull 120(1–2):340–346. https://doi.org/10.1016/j.marpolbul.2017.05.055

Akhand A, Mukhopadhyay A, Chanda A, Mukherjee S, Das A, Das S et al (2017) Potential CO_2 emission due to loss of above-ground biomass from the Indian Sundarban mangroves during the last four decades. J Indian Soc Remote Sens 45(1):147–154. https://doi.org/10.1007/s12524-016-0567-4

Aksornkoae S (2012) Mangroves... coastal treasure of Thailand. J R Inst Thail 4:59–77

Akther S (2016) Marine biodiversity and ecosystem of the St. Martin's island in Bangladesh: perceived threats and vulnerabilities to environmental management. Orient Geogr 60(1&2): 61–74

Al Hagibi HA, Al-Selwi KM, Nagi HM, Al-Shwafi NA (2018) Study of heavy metals contamination in mangrove sediments of the Red Sea Coast of Yemen from Al-Salif to Bab-el-Mandeb Strait. J Ecol Nat Resour 2(1):121

Al-Afifi ZMM (2018) The conservation, sustainable use and management of mangrove habitats in Oman. Department of Environment and Geography, University of York

Alaie E (2001) Salt marshes and salt deserts of SW Iran. Pak J Bot 33(1):77–92

Alam E, Collins AE (2010) Cyclone disaster vulnerability and response experiences in coastal Bangladesh. Disasters 34(4):931–954. https://doi.org/10.1111/j.1467-7717.2010.01176.x

De Alban JDT, Jamaludin J, de Wen DW, Than MM, Webb EL (2020) Improved estimates of mangrove cover and change reveal catastrophic deforestation in Myanmar. Environ Res Lett 15(3):034034. https://doi.org/10.1088/1748-9326/ab666d

Al-Hawshabi OSS, El-Naggar SM (2015) Vegetation patterns and floristic composition of Yemen. Curr Life Sci 1(3):103–111

Al-Hemiari AF (2004) Study of mangrove located between Maidi and Al Luhayyah. Unpublished report submitted to Environmental Protection Agency, Ministry of Water and Environment, Yemen, pp 51. (Arabic language)

Almahasheer H, Aljowair A, Duarte CM, Irigoien X (2016) Decadal stability of Red Sea mangroves. Estuar Coast Shelf Sci 169:164–172. https://doi.org/10.1016/j.ecss.2015.11.027

Alongi DM, Murdiyarso D, Fourqurean JW, Kauffman JB, Hutahaean A, Crooks S et al (2016) Indonesia's blue carbon: a globally significant and vulnerable sink for seagrass and mangrove carbon. Wetl Ecol Manag 24(1):3–13. https://doi.org/10.1007/s11273-015-9446-y

Alsumaiti TS, Hussein K, Al-Sumaiti AS (2017) Mangroves of Abu Dhabi Emirate, UAE, in a global context: a review. Int J Environ Sci 6(4):110–121

Amjad AS, Jusoff K (2007) Mangrove conservation through community participation in Pakistan: the case of Sonmiani Bay. 5th WSEAS International conference on environment, ecosystems and development, 14–16 Dec. Tenerife, Spain, pp 433–439

Amjad AS, Kasawani I, Kamaruzaman J (2007) Degradation of indus delta mangroves in Pakistan. Int J Geolo 3:27–34

Appadoo C (2003) Status of mangroves in Mauritius. J Coast Dev 7(1):1–4

Appadoo C, Rabi FA, Antha S, Bastienne LJ, Hurbungs M, Mougal, J, Rogoonaden R, Soogun N, Vel T (2016) Mangroves of small Islands. In: Bosire JO, Mangora MM, Bandeira S, Rajkaran A, Ratsimbazafy R, Appadoo C and Kairo JG (eds) Mangroves of the Western Indian Ocean: status and management. WIOMSA, Zanzibar Town, pp 115–135

Appelquist LR, Balstrøm T (2014) Application of the Coastal Hazard Wheel methodology for coastal multi-hazard assessment and management in the state of Djibouti. Clim Risk Manag 3:79–95. https://doi.org/10.1016/j.crm.2014.06.002

Ardli ER, Wolff M (2009) Land use and land cover change affecting habitat distribution in the Segara Anakan Lagoon, Java, Indonesia. Reg Environ Chang 9(4):235. https://doi.org/10.1007/s10113-008-0072-6

Armitage AR, Weaver CA, Kominoski JS, Pennings SC (2020) Resistance to hurricane effects varies among wetland vegetation types in the marsh–mangrove ecotone. Estuar Coast 43:960–970. https://doi.org/10.1007/s12237-019-00577-3

Asante K, Brito R, Brundrit G, Epstein P, Nussbaumer P, Patt A (2009) Study on the impact of climate change on disaster risk in mozambique. INGC synthesis report on climate change—First Draft. National Institute for Disaster Management, Maputo, Mozambique

Asbridge E, Lucas R, Rogers K, Accad A (2018) The extent of mangrove change and potential for recovery following severe Tropical Cyclone Yasi, Hinchinbrook Island, Queensland, Australia. Ecol Evol 8(21):10416–10434. https://doi.org/10.1002/ece3.4485

Asri Y, Ghorbanli M (1997) The halophilous vegetation of the Orumieh lake salt marshes, NW. Iran. Plant Ecol 132(2):155–170. https://doi.org/10.1023/A:1009790901167

Azad MS, Kamruzzaman M, Kanzaki M (2020) Canopy gaps influence regeneration dynamics in cyclone affected mangrove stands in medium saline zone of the Sundarbans, Bangladesh. Acta Ecol Sin 41(4):296–303. https://doi.org/10.1016/j.chnaes.2020.03.002

Azian M, Mubarak HT (2012) Functions and values of mangroves. In: Omar M, Aziz K, Shamsudin I, Raja Barizian RS (eds) Status of Mangroves in Peninsular Malaysia, Forest Research Institute Malaysia: Kuala Lumpur, Malaysia, pp 12–25

Bandeira SO (1995) Marine botanical communities in southern Mozambique: Seagrass and seaweed diversity and conservation. Ambio 24:506–509

Barbosa FM, Cuambe CC, Bandeira SO (2001) Status and distribution of mangroves in Mozambique. S Afr J Bot 67(3):393–398. https://doi.org/10.1016/S0254-6299(15)31155-8

Barratt L, Dawson-Shephard A, Ormand R, McDowell R (1987) Yemen Arab Republic marine conservation survey, vol 1, Distribution of habitats and species along the YAR coastline, IUCN, Red Sea and Gulf of Aden Environmental Programme/TMRU, UK

Basu S, Chanda A, Gogoi P, Bhattacharyya S (2021) Organochlorine pesticides and heavy metals in the zooplankton, fishes, and shrimps of tropical shallow tidal creeks and the associated human health risk. Mar Pollut Bull 165:112170. https://doi.org/10.1016/j.marpolbul.2021.112170

Bayen S, Wurl O, Karuppiah S, Sivasothi N, Lee HK, Obbard JP (2005) Persistent organic pollutants in mangrove food webs in Singapore. Chemosphere 61(3):303–313. https://doi.org/10.1016/j.chemosphere.2005.02.097

Beja da Costa A, Ribeiro LPF (2017) Mangroves of Maputo, Mozambique: from threatened to thriving? Plan J 2(2):629–651

Benson L, Glass L, Jones TG, Ravaoarinorotsihoarana L, Rakotomahazo C (2017) Mangrove carbon stocks and ecosystem cover dynamics in southwest Madagascar and the implications for local management. Forests 8(6):190. https://doi.org/10.3390/f8060190

Besset M, Gratiot N, Anthony EJ, Bouchette F, Goichot M, Marchesiello P (2019) Mangroves and shoreline erosion in the Mekong River delta, Viet Nam. Estuar Coast Shelf Sci 226:106263. https://doi.org/10.1016/j.ecss.2019.106263

Bhowmik AK, Cabral P (2013) Cyclone Sidrimpacts on the Sundarbans floristic diversity. Earth Sci Res 2(2):62

Van Bijsterveldt CE, van Wesenbeeck BK, van der Wal D, Afiati N, Pribadi R, Brown B, Bouma TJ (2020) How to restore mangroves for greenbelt creation along eroding coasts with abandoned aquaculture ponds. Estuar Coast Shelf Sci 235:106576. https://doi.org/10.1016/j.ecss.2019.106576

Bird M, Chua S, Fifield LK, Teh TS, Lai J (2004) Evolution of the Sungei Buloh–Kranji mangrove coast, Singapore. Appl Geogr 24(3):181–198. https://doi.org/10.1016/j.apgeog.2004.04.002

Boone Kauffman J, Arifanti VB, Hernández Trejo H, del Carmen JM, Norfolk J, Cifuentes M et al (2017) The jumbo carbon footprint of a shrimp: carbon losses from mangrove deforestation. Front Ecol Environ 15(4):183–188. https://doi.org/10.1002/fee.1482

Bornman TG, Schmidt J, Adams JB, Mfikili AN, Farre RE, Smit AJ (2016) Relative sea-level rise and the potential for subsidence of the Swartkops Estuary intertidal salt marshes, South Africa. S Afr J Bot 107:91–100. https://doi.org/10.1016/j.sajb.2016.05.003

Bosire JO, Mangora MM, Bandeira SO, Rajkaran A, Ratsimbazafy R, Appadoo C, Kairo JG (2015) Mangroves of the Western Indian Ocean: status and management. WIOMSA publications

Bujang JS, Zakaria MH, Arshad A (2006) Distribution and significance of seagrass ecosystems in Malaysia. Aquat Ecosyst Health Manag 9(2):203–214. https://doi.org/10.1080/14634980600705576

Bujang JS, Zakaria MH, Short FT (2016) Seagrass in Malaysia: Issues and challenges ahead. In: Finlayson C, Milton G, Prentice R, Davidson N (eds) The Wetland book II: Distribution, description and conservation, Springer Science+Business Media Dordrecht, pp 1875–1883. https://doi.org/10.1007/978-94-007-6173-5_268-1

Cabral P, Augusto G, Akande A, Costa A, Amade N, Niquisse S et al (2017) Assessing Mozambique's exposure to coastal climate hazards and erosion. Int J Disaster Risk Reduct 23:45–52. https://doi.org/10.1016/j.ijdrr.2017.04.002

Carbone F, Accordi G (2000) The Indian Ocean coast of Somalia. Mar Pollut Bull 41(1–6):141–159. https://doi.org/10.1016/S0025-326X(00)00107-7

Chanda A, Mukhopadhyay A, Ghosh T, Akhand A, Mondal P, Ghosh S et al (2016) Blue carbon stock of the Bangladesh Sundarban mangroves: what could be the scenario after a century? Wetlands 36(6):1033–1045. https://doi.org/10.1007/s13157-016-0819-7
Chanda A, Swain S, Das S, Seal A, Ghosh T, Hazra S (2021) Spatial variation of heavy metal accumulation in the sediments adjacent to different mangrove plant species within a mixed mangrove stand. Soil Sedim Contam Int J 30(5):548–568. https://doi.org/10.1080/15320383.2020.1867505
Charrua AB, Bandeira SO, Catarino S, Cabral P, Romeiras MM (2020) Assessment of the vulnerability of coastal mangrove ecosystems in Mozambique. Ocean Coast Manag 189:105145. https://doi.org/10.1016/j.ocecoaman.2020.105145
Chatterjee M, Shankar D, Sen GK, Sanyal P, Sundar D, Michael GS et al (2013) Tidal variations in the Sundarbans estuarine system, India. J Earth Syst Sci 122(4):899–933. https://doi.org/10.1007/s12040-013-0314-y
Chaudhuri P, Ghosh S, Bakshi M, Bhattacharyya S, Nath B (2015) A review of threats and vulnerabilities to mangrove habitats: with special emphasis on east coast of India. J Earth Sci Clim Chang 6(4). http://dx.doi.org/10.4172/2157-7617.1000270
Choi K, Kim D, Jo J (2020) Morphodynamic evolution of the macrotidalSittaung River estuary, Myanmar: Tidal versus seasonal controls. Mar Geol 430:106367. https://doi.org/10.1016/j.margeo.2020.106367
Choowong M (2002) The geomorphology and assessment of indicators of sea-level changes to study coastal evolution from the Gulf of Thailand. In: Proceedings of international symposium on "Geology of Thailand". Department of Mineral Resources, Thailand, pp 207–220
Choudri BS, Baawain M, Ahmed M (2016) An overview of coastal and marine resources and their management in sultanate of Oman. J Environ Manag Tourism, VII, 1(13):21–32. https://doi.org/10.14505/jemt.7.1(13).02. Available from: http://www.asers.eu/journals/jemt/curent-issue
Chowdhury AMR, Bhuyia AU, Choudhury AY, Sen R (1993) The Bangladesh cyclone of 1991: why so many people died. Disasters 17(4):291–304. https://doi.org/10.1111/j.1467-7717.1993.tb00503.x
Chowdhury R, Favas PJ, Jonathan MP, Venkatachalam P, Raja P, Sarkar SK (2017) Bioremoval of trace metals from rhizosediment by mangrove plants in Indian Sundarban Wetland. Mar Pollut Bull 124(2):1078–1088. https://doi.org/10.1016/j.marpolbul.2017.01.047
Chowdhury R, Sutradhar T, Begam MM, Mukherjee C, Chatterjee K, Basak SK, Ray K (2019) Effects of nutrient limitation, salinity increase, and associated stressors on mangrove forest cover, structure, and zonation across Indian Sundarbans. Hydrobiologia 842(1):191–217. https://doi.org/10.1007/s10750-019-04036-9
Chuang SH (1961) On Malayan shores. MuwuShosa, Singapore, pp 225
Corlett RT (1991) Vegetation. In: Chia LS, Rahman A, Tay BH (eds) The biophysical environment of Singapore. Singapore University Press, Singapore, pp 134–154
Corlett RT (1987) Singapore. In: Umali RM, Zamora PM, Gotera RR, Jara RS, Camacho AS, Vannucci M (eds) Mangroves of Asia and the Pacific: status and management. Natural Resources Management Center and National Mangrove Committee, Ministry of Natural Resources, Manila, pp 211–218
Côté-Laurin MC, Benbow S, Erzini K (2017) The short-term impacts of a cyclone on seagrass communities in Southwest Madagascar. Cont Shelf Res 138:132–141. https://doi.org/10.1016/j.csr.2017.03.005
Cuong DT, Bayen S, Wurl O, Subramanian K, Wong KKS, Sivasothi N, Obbard JP (2005) Heavy metal contamination in mangrove habitats of Singapore. Mar Pollut Bull 50(12):1732–1738. https://doi.org/10.1016/j.marpolbul.2005.09.008
Daby D (2003) Effects of seagrass bed removal for tourism purposes in a Mauritian bay. Environ Pollut 125:313–324. https://doi.org/10.1016/S0269-7491(03)00125-8
Dalimunthe SA, Putri IAP (2017) Mangrove rehabilitation in Seribu Islands at the crossroad of awareness and Tokenism. In: DasGupta R, Shaw R (eds) Participatory mangrove management in a changing climate. Springer, Tokyo, pp 229–245. https://doi.org/10.1007/978-4-431-56481-2_15

Daru BH, van der Bank M, Davies TJ (2015) Spatial incongruence among hotspots and complementary areas of tree diversity in southern Africa. Divers Distrib 21(7):769–780. https://doi.org/10.1111/ddi.12290

Das CS (2015) Causes, consequences, and cost-benefit analysis of the conflicts caused by tiger straying incidents in Sundarban, India. Proc Zool Soc 68(2):120–130. https://doi.org/10.1007/s12595-014-0105-8

Das S (2020) Does mangrove plantation reduce coastal erosion? Assessment from the west coast of India. Reg Environ Chang 20:1–11

Das CS, Jana R (2018) Human–crocodile conflict in the Indian Sundarban: an analysis of spatiotemporal incidences in relation to people's livelihood. Oryx 52(4):661–668. https://doi.org/10.1017/S0030605316001502

Dasgupta S, Sobhan I, Wheeler D (2017) The impact of climate change and aquatic salinization on mangrove species in the Bangladesh Sundarbans. Ambio 46(6):680–694. https://doi.org/10.1007/s13280-017-0911-0

DasGupta R, Hashimoto S, Okuro T, Basu M (2019) Scenario-based land change modeling in the Indian Sundarban delta: an exploratory analysis of plausible alternative regional futures. Sustain Sci 14(1):221–240. https://doi.org/10.1007/s11625-018-0642-6

DasGupta R, Shaw R (2017) mangroves in Asia-pacific: a review of threats and responses. In: DasGupta R, Shaw R (eds) Participatory mangrove management in a changing climate. Disaster risk reduction (Methods, Approaches, and Practices). Springer, Tokyo, pp 1–16. https://doi.org/10.1007/978-4-431-56481-2_1

de la Torre-Castro M, Rönnbäck P (2004) Links between humans and seagrasses—an example from tropical East Africa. Ocean Coast Manag 47(7–8):361–387. https://doi.org/10.1016/j.ocecoaman.2004.07.005

Decker V, Falkenroth M, Lindauer S, Landgraf J, Al-Lawati Z, Al-Rahbi H, Franz SO, Hoffmann G (2021) Collapse of Holocene mangrove ecosystems along the coastline of Oman. Quat Res 10052-76. https://doi.org/10.1017/qua.2020.96

Devi S (2019) Cyclone Idai: 1 month later, devastation persists. The Lancet 393(10181):1585. https://doi.org/10.1016/S0140-6736(19)30892-X

Dhunya A, Huang Q, Aslam A (2017) Coastal habitats of Maldives: status, trends, threats, and potential conservation strategies. Int J Sci Eng Res 8:47–62

Diop ES, Gordon C, Semesi AK, Soumaré A, Diallo N, Guissé A, Diouf M, Ayivor JS (2002) Mangroves of Africa. In: de Lacerda LD (ed) Mangrove ecosystems. Springer, Berlin, pp 63–121

Doddema M (1997) A preliminary exploratory survey of mangrove vegetation in Northern Part of Sofala, Mozambique: Methods and information requirement for planning and management purposes. SAREC-Sida Regional Workshop on Mangrove Ecology, Physiology and Management, Zanzibar

Dubey SK, Trivedi RK, Chand BK, Mandal B, Rout SK (2017) Farmers' perceptions of climate change, impacts on freshwater aquaculture and adaptation strategies in climatic change hotspots: a case of the Indian Sundarban delta. Environ Dev 21:38–51. https://doi.org/10.1016/j.envdev.2016.12.002

Elmahdy SI, Mohamed MM (2013) Change detection and mapping of mangrove using multi-temporal remote sensing data: a case study of Abu Dhabi. UAE J Geomat 7(1):41–45

Elmahdy SI, Ali TA, Mohamed MM, Howari FM, Abouleish M, Simonet D (2020) Spatiotemporal mapping and monitoring of mangrove forests changes From 1990 to 2019 in the Northern Emirates, UAE using Random forest, Kernel logistic regression and naive bayes tree models. Front Environ Sci 8:102. https://doi.org/10.3389/fenvs.2020.00102

Emch M, Peterson M (2006) Mangrove forest cover change in the Bangladesh Sundarbans from 1989–2000: a remote sensing approach. Geocarto Int 21(1):5–12. https://doi.org/10.1080/10106040608542368

Erftemeijer PLA, Lewis RRR (2006) Environmental impacts of dredging on seagrasses: a review. Mar Pollut Bull 52:1553–1572. https://doi.org/10.1016/j.marpolbul.2006.09.006

Erftemeijer PL, Shuail DA (2012) Seagrass habitats in the Arabian Gulf: distribution, tolerance thresholds and threats. Aquat Ecosyst Health Manag 15(sup1):73–83. https://doi.org/10.1080/14634988.2012.668479

Estoque RC, Myint SW, Wang C, Ishtiaque A, Aung TT, Emerton L et al (2018) Assessing environmental impacts and change in Myanmar's mangrove ecosystem service value due to deforestation (2000–2014). Glob Chang Biol 24(11):5391–5410. https://doi.org/10.1111/gcb.14409

Etemadi H, Smoak JM, Karami J (2018) Land use change assessment in coastal mangrove forests of Iran utilizing satellite imagery and CA–Markov algorithms to monitor and predict future change. Environ Earth Sci 77(5):1–13. https://doi.org/10.1007/s12665-018-7392-8

Fatoyinbo TE, Simard M, Washington-Allen RA, Shugart HH (2008) Landscape-scale extent, height, biomass, and carbon estimation of Mozambique's mangrove forests with Landsat ETM+ and shuttle radar topography mission elevation data. J Geophys Res Biogeosci 113(G2). https://doi.org/10.1029/2007JG000551

Fauzi A, Sakti A, Yayusman L, Harto A, Prasetyo L, Irawan B et al (2019) Contextualizing mangrove forest deforestation in southeast Asia using environmental and socio-economic data products. Forests 10(11):952. https://doi.org/10.3390/f10110952

Ferreira B, Garcia M, Jupp BP, Al-Kiyumi A (2006) Diet of the green turtle (*Chelonia mydas*) at Ra's Al Hadd, Sultanate of Oman. Chelonian Conserv Biol 5(1):141–146. https://doi.org/10.2744/1071-8443(2006)5[141:DOTGTC]2.0.CO;2

Ferreira MA, Andrade F, Bandeira SO, Cardoso P, Mendes RN, Paula J (2009) Analysis of cover change (1995–2005) of Tanzania/Mozambique transboundary mangroves using Landsat imagery. Aquat Conserv Mar Freshw Ecosyst 19(S1):38–45. https://doi.org/10.1002/aqc.1042

Fickert T (2020) To plant or not to plant, that is the question: Reforestation versus natural regeneration of hurricane-disturbed mangrove forests in Guanaja (Honduras). Forests 11(10):1068. https://doi.org/10.3390/f11101068

Freeman AS, Short FT, Isnain I, Razak FA, Coles RG (2008) Seagrass on the edge: land-use practices threaten coastal seagrass communities in Sabah, Malaysia. Biol Conserv 141(12):2993–3005. https://doi.org/10.1016/j.biocon.2008.09.018

Friess DA, Phelps J, Leong RC, Lee WK, Wee AKS, Sivasothi N et al (2012) Mandai mangrove, Singapore: lessons for the conservation of Southeast Asia's mangroves. Raffles Bull Zool 25:55–65

Friess DA, Richards DR, Phang VX (2016) Mangrove forests store high densities of carbon across the tropical urban landscape of Singapore. Urban Ecosyst 19(2):795–810. https://doi.org/10.1007/s11252-015-0511-3

Gaw LY, Linkie M, Friess DA (2018) Mangrove forest dynamics in Tanintharyi, Myanmar from 1989–2014, and the role of future economic and political developments. Singap J Trop Geogr 39(2):224–243. https://doi.org/10.1111/sjtg.12228

Ghasemi S, Moghaddam SS, Rahimi A, Damalas CA, Naji A (2018) Phytomanagement of trace metals in mangrove sediments of Hormozgan, Iran, using gray mangrove (*Avicennia marina*). Environ Sci Pollut Res 25(28):28195–28205. https://doi.org/10.1007/s11356-018-2684-9

Ghazali N, Zainuddin K, Zainal MZ, Dali HM, Mahmud MR (2016). The potential of mangrove forest as a bioshield in Malaysia. In: 2016 IEEE 12th international colloquium on signal processing & its applications (CSPA) IEEE, pp 322–327. https://doi.org/10.1109/CSPA.2016.7515854

Ghodrati Shojaei M, Delfan N, Naderloo R (2021) Mangrove ecosystems in Iran; current status, threats, and management. Iran J Biology 5(9):111–125

Ghosh A, Schmidt S, Fickert T, Nüsser M (2015) The Indian Sundarban mangrove forests: history, utilization, conservation strategies, and local perception. Diversity 7(2):149–169. https://doi.org/10.3390/d7020149

Ghosh MK, Kumar L, Roy C (2016) Mapping long-term changes in mangrove species composition and distribution in the Sundarbans. Forests 7(12):305. https://doi.org/10.3390/f7120305

Ghosh MK, Kumar L, Roy C (2017) Climate variability and mangrove cover dynamics at species level in the Sundarbans, Bangladesh. Sustainability 9(5):805. https://doi.org/10.3390/su9050805

Giri C, Muhlhausen J (2008) Mangrove forest distributions and dynamics in Madagascar (1975–2005). Sensors 8(4):2104–2117. https://doi.org/10.3390/s8042104
Giri C, Pengra B, Zhu Z, Singh A, Tieszen LL (2007) Monitoring mangrove forest dynamics of the Sundarbans in Bangladesh and India using multi-temporal satellite data from 1973 to 2000. Estuar Coast Shelf Sci 73(1–2):91–100. https://doi.org/10.1016/j.ecss.2006.12.019
Giri C, Zhu Z, Tieszen LL, Singh A, Gillette S, Kelmelis JA (2008) Mangrove forest distributions and dynamics (1975–2005) of the tsunami-affected region of Asia. J Biogeogr 35(3):519–528. https://doi.org/10.1111/j.1365-2699.2007.01806.x
Githaiga MN, Frouws AM, Kairo JG, Huxham M (2019) Seagrass removal leads to rapid changes in fauna and loss of carbon. Front Ecol Evol 7:62. https://doi.org/10.3389/fevo.2019.00062
Giuliani S, Bellucci LG (2019) Salt marshes: their role in our society and threats posed to their existence. In: Sheppard C (ed) World seas: an environmental evaluation, Academic Press, pp 79–101. https://doi.org/10.1016/B978-0-12-805052-1.00004-8
Gonneea ME, Maio CV, Kroeger KD, Hawkes AD, Mora J, Sullivan R et al (2019) Salt marsh ecosystem restructuring enhances elevation resilience and carbon storage during accelerating relative sea-level rise. Estuar Coast Shelf Sci 217:56–68. https://doi.org/10.1016/j.ecss.2018.11.003
Gopi M, Kumar MP, Jeevamani JJJ, Raja S, Muruganandam R, Samuel VD et al (2019) Distribution and biodiversity of tropical saltmarshes: Tamil Nadu and Puducherry, Southeast coast of India. Estuar Coast Shelf Sci 229:106393. https://doi.org/10.1016/j.ecss.2019.106393
Government of Seychelles (2011) Fourth national report to the United Nations Convention on Biological Diversity. Environment Department, P.O. Box 445, Botanical Gardens, Mont Fleuri, Victoria, Republic of Seychelles
Granek EF, Brown MA (2005) Co-management approach to marine conservation in Mohéli, Comoros Islands. Conserv Biol 19(6):1724–1732. https://doi.org/10.1111/j.1523-1739.2005.00301.x
Griffiths LL, Connolly RM, Brown CJ (2020) Critical gaps in seagrass protection reveal the need to address multiple pressures and cumulative impacts. Ocean Coast Manag 183:104946. https://doi.org/10.1016/j.ocecoaman.2019.104946
Gu J, Luo M, Zhang X, Christakos G, Agusti S, Duarte CM, Wu J (2018) Losses of salt marsh in China: trends, threats, and management. Estuar Coast Shelf Sci 214:98–109. https://doi.org/10.1016/j.ecss.2018.09.015
Gul B, Khan MA (1995) Productivity of an *Arthrocnemum indicum* dominated coastal salt marsh at Karachi, Pakistan. In: Khan MA, Ungar IA (eds) Biology of salt tolerant plants. University of Karachi, Karachi, pp 99–105
Gullström M, Lundén B, Bodin M, Kangwe J, Öhman MC, Mtolera MS, Björk M (2006) Assessment of changes in the seagrass-dominated submerged vegetation of tropical Chwaka Bay (Zanzibar) using satellite remote sensing. Estuar Coast Shelf Sci 67(3):399–408. https://doi.org/10.1016/j.ecss.2005.11.020
Gulzar S, Khan MA, Ungar IA (2003) Salt tolerance of a coastal salt marsh grass. Commun Soil Sci Plant Anal 34(17 & 18):2595–2605. https://doi.org/10.1081/CSS-120024787
Gunawardena M, Rowan JS (2005) Economic valuation of a mangrove ecosystem threatened by shrimp aquaculture in Sri Lanka. Environ Manag 36(4):535–550. https://doi.org/10.1007/s00267-003-0286-9
Habshi AA, Youssef T, Aizpuru M, Blasco F (2007) New mangrove ecosystem data along the UAE coast using remote sensing. Aquat Ecosyst Health Manag 10(3):309–319. https://doi.org/10.1080/14634980701512525
Hajra R, Ghosh A, Ghosh T (2017) Comparative assessment of morphological and land use/landcover change pattern of Sagar, Ghoramara, and Mousani island of Indian Sundarban delta through remote sensing. In: Environment and earth observation. Springer, Cham, pp 153–172. https://doi.org/10.1007/978-3-319-46010-9_11

Halder NK, Merchant A, Misbahuzzaman K, Wagner S, Mukul SA (2021) Why some trees are more vulnerable during catastrophic cyclone events in the Sundarbans mangrove forest of Bangladesh? For Ecol Manag 490:119117. https://doi.org/10.1016/j.foreco.2021.119117

Hamad H, Mchenga I, Hamisi M (2019) Climate change increasing threats on non-conserved mangroves forests of Micheweni, Zanzibar-Tanzania. Tanzania J Sci 45(3):527–538

Hammond DS, Gond V, Baider C, Florens FBV, Persand S, Laurance SGW (2015) Threats to environmentally sensitive areas from peri-urban expansion in Mauritius. Environ Conserv 42(3):256–267. https://doi.org/10.1017/S0376892914000411

Hammond DS, Gond V, Baider C, Florens VFB, Persand S, Laurance SGW (2019) Threats to environmentally sensitive areas from Peri-urban expansion in mauritius. In: Tang V, Shaw T, Holden M (eds) Development and sustainable growth of mauritius. Contemporary African Political Economy. Palgrave Macmillan, Cham. https://doi.org/10.1007/978-3-319-96166-8_9

Hanekom N, Baird D (1988) Distribution and variations in seasonal biomass of eelgrass Zostera capensis in the Kromme Estuary, St Francis Bay, South Africa. S Afr J Mar Sci 7(1):51–59. https://doi.org/10.2989/025776188784379099

Hantanirina JMO, Benbow S (2013) Diversity and coverage of seagrass ecosystems in south-west Madagascar. Afr J Mar Sci 35(2):291–297. https://doi.org/10.2989/1814232X.2013.800580

Harcourt WD, Briers RA, Huxham M (2018) The thin(ning) green line? Investigating changes in Kenya's seagrass coverage. Biol Lett 14(11):20180227. https://doi.org/10.1098/rsbl.2018.0227

Hena MA, Short FT, Sharifuzzaman SM, Hasan M, Rezowan M, Ali M (2007) Salt marsh and seagrass communities of Bakkhali Estuary, Cox's Bazar, Bangladesh. Estuar Coast Shelf Sci 75(1–2):72–78. https://doi.org/10.1016/j.ecss.2007.01.022

Henderson AC, Das S, Ghosh T, Panizzo VN, Moorhouse HL, Roberts LR et al (2021) The Indian Sundarbans: biogeochemical dynamics and anthropogenic impacts. In: Das S, Ghosh T (eds) Estuarine biogeochemical dynamics of the east coast of India. Springer, Cham. pp 239–260 https://doi.org/10.1007/978-3-030-68980-3_15

Hermon D, Ganefri PA, Oktorie O (2018) The model of mangrove land cover change for the estimation of blue carbon stock change in Belitung Island-Indonesia. Int J Appl Environ Sci 13(2):191–202

Hoozemans FM, Hulsbergen CH (2021) Sea level rise: a worldwide assessment of risk and protection costs. In: Eisma D (ed) Climate change. CRC Press, pp 137–161. https://doi.org/10.1201/9781003069935

Hoppe-Speer SCL, Adams JB, Rajkaran A (2013) Response of mangroves to drought and non-tidal conditions in St Lucia Estuary, South Africa. Afr J Aquat Sci 38(2):153–162. https://doi.org/10.2989/16085914.2012.759095

Hoppe-Speer SC, Adams JB, Bailey D (2015) Present state of mangrove forests along the Eastern Cape coast, South Africa. Wetl Ecol Manag 23(3):371–383. https://doi.org/10.1007/s11273-014-9387-x

Hoppe-Speer SC, Adams JB, Rajkaran A, Bailey D (2011) The response of the red mangrove *Rhizophora mucronata* Lam. to salinity and inundation in South Africa. Aquat Bot 95(2):71–76. https://doi.org/10.1016/j.aquabot.2011.03.006

Hoppe-Speer SCL (2012) Response of mangroves in South Africa to anthropogenic and natural impacts. Doctoral dissertation, Nelson Mandela Metropolitan University

Hoque MAA, Pradhan B, Ahmed N, Roy S (2019) Tropical cyclone risk assessment using geospatial techniques for the eastern coastal region of Bangladesh. Sci Total Environ 692:10–22. https://doi.org/10.1016/j.scitotenv.2019.07.132

Hossain MS, Bujang JS, Zakaria MH, Hashim M (2015) Application of Landsat images to seagrass areal cover change analysis for Lawas, Terengganu and Kelantan of Malaysia. Cont Shelf Res 110:124–148. https://doi.org/10.1016/j.csr.2015.10.009

Hossain MS, Hashim M, Bujang JS, Zakaria MH, Muslim AM (2019) Assessment of the impact of coastal reclamation activities on seagrass meadows in Sungai Pulai estuary, Malaysia, using Landsat data (1994–2017). Int J Remote Sens 40(9):3571–3605. https://doi.org/10.1080/01431161.2018.1547931

Hossain MB, Rakib MRJ, Jolly YN, Rahman M (2021) Metals uptake and translocation in salt marsh macrophytes, *Porteresia* sp. from Bangladesh coastal area. Sci Total Environ 764:144637. https://doi.org/10.1016/j.scitotenv.2020.144637

Howari FM, Jordan BR, Bouhouche N, Wyllie-Echeverria S (2009) Field and remote-sensing assessment of mangrove forests and seagrass beds in the northwestern part of the United Arab Emirates. J Coast Res 25(1):48–56. https://doi.org/10.2112/07-0867.1

Human LRD, Snow GC, Adams JB (2016) Responses in a temporarily open/closed estuary to natural and artificial mouth breaching. S Afr J Bot 107:39–48. https://doi.org/10.1016/j.sajb.2015.12.002

Ibharim NA, Mustapha MA, Lihan T, Mazlan AG (2015) Mapping mangrove changes in the Matang Mangrove Forest using multi-temporal satellite imageries. Ocean Coastal Manag 114:64–76. https://doi.org/10.1016/j.ocecoaman.2015.06.005

Iftekhar MS, Saenger P (2008) Vegetation dynamics in the Bangladesh Sundarbans mangroves: a review of forest inventories. Wetl Ecol Manag 16(4):291–312. https://doi.org/10.1007/s11273-007-9063-5

Imai N, Takyu M, Nakamura Y, Nakamura T (2006) Gap formation and regeneration of tropical mangrove forests in Ranong, Thailand. Plant Ecol 186(1):37–46. https://doi.org/10.1007/s11258-006-9110-y

India State of Forest Report (2019) Ministry of environment and forest, Government of India

Ingram JC, Dawson TP (2001) The impacts of a river effluent on the coastal seagrass habitats of Mahé, Seychelles. S Afr J Bot 67(3):483–487. https://doi.org/10.1016/S0254-6299(15)31167-4

Islam MA, Al-Mamun A, Hossain F, Quraishi SB, Naher K, Khan R et al (2017) Contamination and ecological risk assessment of trace elements in sediments of the rivers of Sundarban mangrove forest, Bangladesh. Mar Pollut Bull 124(1):356–366. https://doi.org/10.1016/j.marpolbul.2017.07.059

Jagtap TG, Bhosale SH, Nagle VL (2002) Ecological observations on major Salicornia beds from highly saline coastal wetlands of India. Wetlands 22(3):443–450. https://doi.org/10.1672/0277-5212(2002)022[0443:EOOMSB]2.0.CO;2

Jayasuriya PMA (1991) The species composition, abundance and the distribution of seagrass communities in Puttalam Lagoon. Vidyodaya J Sci 3(2):93–102

Jumnongsong S, Gallardo WG, Ikejima K, Cochard R (2015) Factors affecting fishers' perceptions of benefits, threats, and state, and participation in mangrove management in Pak Phanang Bay, Thailand. J Coast Res 31(1):95–106. https://doi.org/10.2112/JCOASTRES-D-12-00257.1

Jupp BP, Durako MJ, Kenworthy WJ, Thayer GW, Schillak L (1996) Distribution, abundance, and species composition of seagrasses at several sites in Oman. Aquat Bot 53(3–4):199–213. https://doi.org/10.1016/0304-3770(96)01023-6

Jusoff K, Taha D (2008) Managing sustainable mangrove forests in Peninsular Malaysia. J Sustain Dev 1(1):88–96

Jusoff K (2013) Malaysian mangrove forests and their significance to the coastal marine environment. Pol J Environ Stud 22(4). http://www.pjoes.com/Malaysian-Mangrove-Forests-and-their-r-nSignificance-to-the-Coastal-Marine-Environment,89055,0,2.html

Kaliraj S, Chandrasekar N, Ramachandran KK, Srinivas Y, Saravanan S (2017) Coastal landuse and land cover change and transformations of Kanyakumari coast, India using remote sensing and GIS. Egypt J Remote Sens Space Sci 20(2):169–185. https://doi.org/10.1016/j.ejrs.2017.04.003

Kamal AHM, Khan MAA (2009) Coastal and estuarine resources of Bangladesh: management and conservation issues. Maejo Int J Sci Technol 3(2):313–342

Kamal AHM (2009) Vulnerability of intertidal macrophytes diversity in Bangladesh to climate change and sea level rise. In IOP conference series, earth and environmental science, IOP Publishing, vol 6(30). https://doi.org/10.1088/1755-1307/6/30/302029

Kanal AHM, Short F (2009) A new record of seagrass *Halophila beccarii* Ascherson in Bangladesh. CMU J Nat Sci 8(2):201–206

Kanniah KD, Sheikhi A, Cracknell AP, Goh HC, Tan KP, Ho CS, Rasli FN (2015) Satellite images for monitoring mangrove cover changes in a fast-growing economic region in Southern Peninsular Malaysia. Remote Sens 7(11):14360–14385. https://doi.org/10.3390/rs71114360

Kathiresan K (2010) Importance of mangrove forests of India. J Coast Environ 1(1):11–26
Kathiresan K, Rajendran N (2005) Mangrove ecosystems of the Indian Ocean region. Indian J Mar Sci 34(1):104–113
Keshavarz M, Mohammadikia D, Gharibpour F, Dabbagh AR (2012) Accumulation of heavy metals (Pb, Cd, V) in sediment, roots and leaves of mangrove species in Sirik Creek along the seacoasts of Oman, Iran. J Appl Sci Environ Manag 16(4)
Keshavarzifard M, Niamaimandi N, Sharifinia M, Yeganeh V, Mobarrezi A (2021) Analytical review of the impacts of boat shrimp trawling on marine plants (Algae and Seagrass). Iran J Mar Sci Technol 25(1):64–71. https://doi.org/10.22034/IJMST.2021.244034
Khalil AS (2002) Status of Mangrove in Republic of Yemen. PERSGA, p 62
Khan MA, Ungar IA, Showalter AM (2000) The effect of salinity on the growth, water status, and ion content of a leaf succulent perennial halophyte *Suaeda fruticosa* (L.) Forssk. J Arid Environ 45:73–84. https://doi.org/10.1006/jare.1999.0617
Khan MA, Ungar IA, Showalter (2005) Salt stimulation and tolerance in an intertidal stem succulent halophyte. J Plant Nutr 28:1365–1374. https://doi.org/10.1081/PLN-200067462
Khogkhaoa C, Hayashizakib KI, Tuntiprapasa P, Prathepa A (2017) Changes in seagrass communities along the runoff gradient of the Trangriver, Thailand. Sci Asia 43(6):339–346. https://doi.org/10.2306/scienceasia1513-1874.2017.43.339
Kidwai S, Ahmed W, Tabrez SM, Zhang J, Giosan L, Clift P, Inam A (2019) The Indus delta—catchment, river, coast, and people. In: Wolanski E, Day JW, Elliott M, Ramachandran R (eds) Coasts and estuaries. Elsevier, pp 213–232. https://doi.org/10.1016/B978-0-12-814003-1.00012-5
Kirui KB, Kairo JG, Bosire J, Viergever KM, Rudra S, Huxham M, Briers RA (2013) Mapping of mangrove forest land cover change along the Kenya coastline using Landsat imagery. Ocean Coast Manag 83:19–24. https://doi.org/10.1016/j.ocecoaman.2011.12.004
Klotzbach PJ, Bowen SG, Pielke R, Bell M (2018) Continental US hurricane landfall frequency and associated damage: observations and future risks. Bull Am Met Soc 99(7):1359–1376. https://doi.org/10.1175/BAMS-D-17-0184.1
Knutson T, Camargo SJ, Chan JC, Emanuel K, Ho CH, Kossin J et al (2019) Tropical cyclones and climate change assessment: Part I: detection and attribution. Bull Am Met Soc 100(10):1987–2007. https://doi.org/10.1175/BAMS-D-18-0189.1
Kodikara KAS, Mukherjee N, Jayatissa LP, Dahdouh-Guebas F, Koedam N (2017) Have mangrove restoration projects worked? An in-depth study in Sri Lanka. Restor Ecol 25(5):705–716. https://doi.org/10.1111/rec.12492
Kossin JP, Knapp KR, Olander TL, Velden CS (2020) Global increase in major tropical cyclone exceedance probability over the past four decades. Proc Natl Acad Sci 117(22):11975–11980. https://doi.org/10.1073/pnas.1920849117
Krauss KW, Osland MJ (2020) Tropical cyclones and the organization of mangrove forests: a review. Ann Bot 125(2):213–234. https://doi.org/10.1093/aob/mcz161
Kruitwagen G, Pratap HB, Covaci A, Wendelaar Bonga SE (2008) Status of pollution in mangrove ecosystems along the coast of Tanzania. Mar Pollut Bull 56(5):1022–1030. https://doi.org/10.1016/j.marpolbul.2008.02.018
Kuriandewa TE, Kiswara W, Hutomo M, Soemodihardjo S (2003) The seagrasses of Indonesia. World Atlas of Seagrasses, University of California Press, Barkeley, pp 172-182
Kürschner H, Al-Gifri AN, Al-Subai MY, Rowaished AK (1998) Vegetational patterns within coastal salines in southern Yemen. Feddes Repertorium 109(1–2):147–159
Ladd CJ, Duggan-Edwards MF, Bouma TJ, Pagès JF, Skov MW (2019) Sediment supply explains long-term and large-scale patterns in salt marsh lateral expansion and erosion. Geophys Res Lett 46(20):11178–11187. https://doi.org/10.1029/2019GL083315
Lai S, Loke LH, Hilton MJ, Bouma TJ, Todd PA (2015) The effects of urbanisation on coastal habitats and the potential for ecological engineering: a Singapore case study. Ocean Coast Manag 103:78–85. https://doi.org/10.1016/j.ocecoaman.2014.11.006
Lalljee B, Velmurugan A, Singh AK (2018) Climate resilient and livelihood security–perspectives for Mauritius Island. In: Sivaperuman C, Velmurugan A, Singh AK, Jaisankar I (eds) Biodiversity

and climate change adaptation in tropical Islands. Academic Press, pp 403–431. https://doi.org/10.1016/B978-0-12-813064-3.00014-4
Langston AK, DuránVinent O, Herbert ER, Kirwan ML (2020) Modeling long-term salt marsh response to sea level rise in the sediment-deficient Plum Island Estuary, MA. Limnol Oceanogr 65(9):2142–2157. https://doi.org/10.1002/lno.11444
Latiff A, Faridah-Hanum I (2014) Mangrove ecosystem of Malaysia: status, challenges and management strategies. In: Faridah-Hanum I, Latiff A, Hakeem K, Ozturk M (eds) Mangrove ecosystems of Asia. Springer, New York, NY, pp 1–22. https://doi.org/10.1007/978-1-4614-8582-7_1
Lei X (2021) Land use planning in coastal areas based on remote sensing images and big data education resources extraction. Arab J Geosci 14(11):1–19. https://doi.org/10.1007/s12517-021-07421-8
Li SH, Ge ZM, Xie LN, Chen W, Yuan L, Wang DQ et al (2018) Ecophysiological response of native and exotic salt marsh vegetation to waterlogging and salinity: implications for the effects of sea-level rise. Sci Rep 8(1):1–13. https://doi.org/10.1038/s41598-017-18721-z
Declining fisheries and increasing prices: the economic cost of tropical rivers impoundment. Fish Res 221:105399. https://doi.org/10.1016/j.fishres.2019.105399
Lindner BL (2019) Climatology and variability of tropical cyclones affecting charleston, South Carolina, from 1670 to 1850. J Coast Res 35(2):397–409. https://doi.org/10.2112/JCOASTRES-D-18-00036
Loo MG, Tun KP, Chou LM (1996) Environmental status of seagrass communities in Singapore. J Singap Natl Acad Sci 22(24):97–102
Lowry K, Wickremeratne HJM (1987) Coastal area management in Sri Lanka. University of Chicago, Chicago, pp 263–293
Lugendo B (2016) Mangroves, salt marshes and seagrass beds. In: Paula J (ed) Regional state of the coast report: Western Indian Ocean, UNEP-nairobi convention and WIOMSA, UNEP and WIOMSA, Nairobi, Kenya, pp 53–68
Macamo C, Bandeira S, Muando S, Abreu D, Mabilana H (2016) Mangroves of mozambique. In: Bosire JO, Mangora MM, Bandeira SO, Rajkaran A, Appadoo C, Kairo JG (eds) Mangroves of the Western Indian Ocean: status and management, WIOMSA, Zanzibar Town, pp 51–73
Mafi-Gholami D, Zenner EK, Jaafari A, Ward RD (2019a) Modeling multi-decadal mangrove leaf area index in response to drought along the semi-arid southern coasts of Iran. Sci Total Environ 656:1326–1336. https://doi.org/10.1016/j.scitotenv.2018.11.462
Mafi-Gholami D, Baharlouii M, Mahmoudi B (2019b) Investigation of long-term mangrove and saltmarsh area changes in the Iranian mangrove habitats (Case study: mangroves of Hormozgan province). J Plant Ecosyst Conserv 6(13):121–140. http://pec.gonbad.ac.ir/article-1-296-en.html
Mafi-Gholami D, Zenner EK, Jaafari A, Bui DT (2020) Spatially explicit predictions of changes in the extent of mangroves of Iran at the end of the 21st century. Estuar Coast Shelf Sci 237:106644. https://doi.org/10.1016/j.ecss.2020.106644
Maghsodian Z, Sanati AM, Ramavandi B, Ghasemi A, Sorial GA (2021) Microplastics accumulation in sediments and *Periophthalmus waltoni* fish, mangrove forests in southern Iran. Chemosphere 264:128543. https://doi.org/10.1016/j.chemosphere.2020.128543
Maghsoodloo A, Eghtesadi P (2008) Coral reefs and related ecosystems in Iran: monitoring and management. Country report of the Islamic Republic of Iran. Powerpoint presentation, Marine Living Sciences Department, Iranian National Center for Oceanography (INCO)
Mahdy A, Ghallab A, Madkour H, Osman A (2021) Status of seagrass community in Northern protected Islands, Hurghada, Red Sea, Egypt. Aquat Sci Fish Resour. https://doi.org/10.21608/asfr.2021.55440.1011
Mainoya JR, Mesaki S, Banyikwa FF (1986) The distribution and socio-economic aspects of mangrove forests in Tanzania. In: Kunstadter P et al (eds) Man in the mangroves: the socio-economic situation of human settlements in mangrove forests. The United Nations University, Tokyo, Japan, pp 87–95

Marasinghe S, Perera P, Simpson GD, Newsome D (2021) Nature-based tourism development in coastal wetlands of Sri Lanka: an importance–performance analysis at Maduganga Mangrove Estuary. J Outdoor Recreat Tour 33:100345. https://doi.org/10.1016/j.jort.2020.100345

McKenzie LJ, Yaakub SM, Tan R, Seymour J, Yoshida RL (2016) Seagrass habitats of Singapore: environmental drivers and key processes. Raffles Bull Zool 34:60–77. http://zoobank.org/urn:lsid:zoobank.org:pub:7F125428-8891-48FC-B315-8D13981B200D

Meah M, Barua P (2020) Influence of soil organic content on salt marsh growth in the natural habitat of South-Eastern coast of Bangladesh. Environ Contam Rev 3(2):71–76. http://doi.org/10.26480/ecr.02.2020.50.55

MEE (2015) Fifth National Report of Maldives to the Convention on Biological Diversity. Ministry of Environment and Energy, Maldives

Mishra AK, Apte D (2021) The current status of Halophilabeccarii: An ecologically significant, yet Vulnerable seagrass of India. Ocean Coast Manag 200:105484. https://doi.org/10.1016/j.ocecoaman.2020.105484

Mitra A (2020) Mangrove Forests in India. Springer Nature, Switzerland AG. https://doi.org/10.1007/978-3-030-20595-9

Mondal P, Reichelt-Brushett AJ, Jonathan MP, Sujitha SB, Sarkar SK (2018) Pollution evaluation of total and acid-leachable trace elements in surface sediments of Hooghly River Estuary and Sundarban Mangrove Wetland (India). Environ Sci Pollut Res 25(6):5681–5699. https://doi.org/10.1007/s11356-017-0915-0

Moseby KE, Daniels A, Duri V, Tropa W, Welma S, Bero A, Soapi K (2020) Community-based monitoring detects catastrophic earthquake and tsunami impacts on seagrass beds in the Solomon Islands. Marine Pollut Bull 150:110444. https://doi.org/10.1016/j.marpolbul.2019.07.032

Muir CE, Sallema A, Abdallah O, De Luca D, Davenport TR (2003) The dugong (Dugong dugon) in Tanzania: a national assessment of status, distribution and threat. Wildlife Conservation Society, p 31

Mukhopadhyay A, Mondal P, Barik J, Chowdhury SM, Ghosh T, Hazra S (2015) Changes in mangrove species assemblages and future prediction of the Bangladesh Sundarbans using Markov chain model and cellular automata. Environ Sci Proc Impact 17(6):1111–1117. https://doi.org/10.1039/C4EM00611A

Mukhtar I, Hannan A (2012) Constrains on mangrove forests and conservation projects in Pakistan. J Coastal Conserv 16(1):51–62. https://doi.org/10.1007/s11852-011-0168-x

Mulyadi A, Amin B (2016) Vegetation structure and mangrove ecosystem threats in the coastal zone of Dumai, Riau, Indonesia. Int J Appl Environ Sci 11(3):785–798

Mungai F, Kairo J, Mironga J, Kirui B, Mangora M, Koedam N (2019) Mangrove cover and cover change analysis in the transboundary area of Kenya and Tanzania during 1986–2016. J India Ocean Reg 15(2):157–176. https://doi.org/10.1080/19480881.2019.1613868

Murdiyarso D, Purbopuspito J, Kauffman JB, Warren MW, Sasmito SD, Donato DC et al (2015) The potential of Indonesian mangrove forests for global climate change mitigation. Nat Clim Chang 5(12):1089–1092. https://doi.org/10.1038/nclimate2734

Murray NJ, Keith DA, Tizard R, Duncan A, Htut WT, Hlaing N, Oo AH, Ya KZ, Grantham H (2020) Threatened ecosystems of myanmar. An IUCN red list of ecosystems assessment. Version 1.0. Wildlife Conservation Society. https://doi.org/10.19121/2019.Report.37457

Musembi P, Fulanda B, Kairo J, Githaiga M (2019) Species composition, abundance and fishing methods of small-scale fisheries in the seagrass meadows of Gazi Bay, Kenya. J India Ocean Reg 15(2):139–156. https://doi.org/10.1080/19480881.2019.1603608

Nagi HM, Abubakr MM (2013) Threats status to the mangrove ecosystem along the coastal zone of Yemen. J King Abdulaziz Univ 24(1):101–116. https://doi.org/10.4197/Mar.24-1.8

Naidoo G (2016) The mangroves of South Africa: An ecophysiological review. S Afr J Bot 107:101–113. https://doi.org/10.1016/j.sajb.2016.04.014

Naidoo G, Chirkoot D (2004) The effects of coal dust on photosynthetic performance of the mangrove, *Avicennia marina* in Richards Bay, South Africa. Environ Pollut 127(3):359–366. https://doi.org/10.1016/j.envpol.2003.08.018

Naidoo G, Naidoo Y (2005) Coal dust pollution effects on wetland tree species in Richards Bay, South Africa. Wetl Ecol Manag 13(5):509–515. https://doi.org/10.1016/j.envpol.2003.08.018

Naidoo T, Thompson RC, Rajkaran A (2020) Quantification and characterisation of microplastics ingested by selected juvenile fish species associated with mangroves in KwaZulu-Natal, South Africa. Environ Pollut 257:113635. https://doi.org/10.1016/j.envpol.2019.113635

Nakaoka M, Tanaka Y, Mukai H, Suzuki T, Aryuthaka C (2007) Tsunami impacts on biodiversity of seagrass communities in the Andaman Sea, Thailand:(1) Seagrass abundance and diversity. Publ Seto Mar Biol Lab Spec Publ Ser 8:49–56. http://hdl.handle.net/2433/70911

Naveenan M, Weragodatenna DDD, Gunaratne ABAK, Rathnayake RMCWM, Jayamanne SC (2015) Remote sensing and GIS approach in assessment and conservation of seagrass beds in the coastal stretch along gulf of mannar. In: Proceedings of the national aquatic resources research and development agency (NARA). Scientific Sessions

Nobi EP, Thangaradjou T (2012) Evaluation of the spatial changes in seagrass cover in the lagoons of Lakshadweep islands, India, using IRS LISS III satellite images. Geocarto Int 27(8):647–660. https://doi.org/10.1080/10106049.2012.665501

Novak AB, Hines E, Kwan D, Parr L, Tun MT, Win H, Short FT (2009) Revised ranges of seagrass species in the Myeik Archipelago, Myanmar. Aquat Bot 91(3):250–252. https://doi.org/10.1016/j.aquabot.2009.07.002

Nunes M, Adams JB (2014) Responses of primary producers to mouth closure in the temporarily open/closed Great Brak Estuary in the warm-temperate region of South Africa. Afr J Aquat Sci 39(4):387–394. https://doi.org/10.2989/16085914.2014.980773

Ofori Samuel A, Arachchilage Sunanda K, Jayatissa Loku P, Madarasinghe Sanduni K, Gunathilaka Udara V, Wijesundara Isuru TD (2021) Shrimp farming, a major threat to mangrove and lagoon ecosystem in the Pambala-Chilaw lagoon complex, Sri Lanka. In: Mees J, Seys J (eds) Book of abstracts—VLIZ marine science day, VLIZ special publication 85, vlaams instituut voor de zee—flanders Marine Institute (VLIZ). Oostende, Belgium, p 129

Omar H (2012) Mangroves threats and changes. In: Omar M, Aziz K, Shamsudin I, Raja Barizian RS (eds) Status of mangroves in Peninsular Malaysia. Forest Research Institute of Malaysia, Kuala Lumpur, Malaysia, pp 89–110

Oo N (2002) Present state and problems of mangrove management in Myanmar. Trees 16(2–3):218–223. https://doi.org/10.1007/s00468-001-0150-6

Ooi JLS, Kendrick GA, Van Niel KP, Affendi YA (2011) Knowledge gaps in tropical Southeast Asian seagrass systems. Estuar Coast Shelf Sci 92:118–131. https://doi.org/10.1016/j.ecss.2010.12.021

Owuor MA, Icely J, Newton A (2019a) Community perceptions of the status and threats facing mangroves of Mida Creek, Kenya: Implications for community-based management. Ocean Coastal Manag 175:172–179. https://doi.org/10.1016/j.ocecoaman.2019.03.027

Owuor MA, Mulwa R, Otieno P, Icely J, Newton A (2019b) Valuing mangrove biodiversity and ecosystem services: a deliberative choice experiment in Mida Creek, Kenya. Ecosyst Serv 40:101040. https://doi.org/10.1016/j.ecoser.2019.101040

Pal I, Ghosh T (2018) Risk governance measures and actions in sundarbans delta (India): a holistic analysis of post-disaster situations of cyclone Aila. In: Pal I, Shaw R (eds) Disaster risk governance in India and cross cutting issues. Springer, Singapore, pp 225–243. https://doi.org/10.1007/978-981-10-3310-0_12

Paleologos EK, Welling BA, Amrousi ME, Masalmeh HA (2019) Coastal development and mangroves in Abu Dhabi, UAE. In: IOP conference series: earth and environmental science, IOP Publishing, vol 344(1), p 012020. https://doi.org/10.1088/1755-1315/344/1/012020

Parvaresh H (2011) Identification of threats on mangrove forests in Gabrik International Wetland for sustainable management. In: International conference on biology, environment and chemistry IPCBEE, vol. 24

Patro S, Krishnan P, Deepak Samuel V, Purvaja R, Ramesh R (2017) Seagrass and salt marsh ecosystems in South Asia: an overview of diversity, distribution, threats and conservation status.

In: Prusty B, Chandra R, Azeez P (eds) Wetland science. Springer, New Delhi. https://doi.org/10.1007/978-81-322-3715-0_5

Paul BK, Chatterjee S (2019) Climate change-induced environmental hazards and aila relief measures Undertaken to Sundarbans in Bangladesh and India. In: Sen H (eds) The sundarbans: a disaster-prone eco-region. Springer, Cham, pp 469–490. https://doi.org/10.1007/978-3-030-00680-8_16

Payo A, Mukhopadhyay A, Hazra S, Ghosh T, Ghosh S, Brown S et al (2016) Projected changes in area of the Sundarban mangrove forest in Bangladesh due to SLR by 2100. Clim Chang 139(2):279–291. https://doi.org/10.1007/s10584-016-1769-z

Payton A, Lagomasino D, Roman-Cuesta RM, Leach A, Moreno S, Taillie PJ, Fatoyinbo T (2020) Thirty year record of hurricane disturbance and recovery in mangrove forests across the caribbean basin. In: AGU fall meeting abstracts, pp EP050–05

Penha-Lopes G, Torres P, Cannicci S, Narciso L, Paula J (2011) Monitoring anthropogenic sewage pollution on mangrove creeks in southern Mozambique: A test of Palaemon concinnus Dana, 1852 (Palaemonidae) as a biological indicator. Environ Pollut 159(2):636–645. https://doi.org/10.1016/j.envpol.2010.09.029

Pernetta J, Wells S (eds) (1993) Marine Protected area needs in the South Asian Seas Region: Sri Lanka, vol 5, IUCN

Pillay D, Branch GM, Griffiths CL, Williams C, Prinsloo A (2010) Ecosystem change in a South African marine reserve (1960–2009): role of seagrass loss and anthropogenic disturbance. Mar Ecol Progr Ser 415:35–48. https://doi.org/10.3354/meps08733

Plagányi É (2019) Climate change impacts on fisheries. Science 363(6430):930–931. https://doi.org/10.1126/science.aaw5824

Plathong S, Plathong J (2004) Past and present threats on mangrove ecosystem in peninsular Thailand. coastal biodiversity in mangrove ecosystems: Paper presented in UNU-INWEH-UNESCO International Training Course, held at Centre of Advanced Studies, Annamalai University, pp 1–13

Poonian CNS, Tuharska K, Hauzer MD (2016) Diversity and distribution of seagrasses in the Union of the Comoros. Afr J Mar Sci 38(2):263–268. https://doi.org/10.2989/1814232X.2016.1181103

Pornpitakdamrong A, Sudjaroen Y (2014) Seablite (Suaedamaritima) product for cooking, Samut-Songkram province, Thailand. Food Nutr Sci 5(9):45247. https://doi.org/10.4236/fns.2014.59094

Pramanik M, Szabo S, Pal I, Udmale P, Pongsiri M, Chilton S (2021) Population health risks in multi-hazard environments: action needed in the Cyclone Amphan and COVID-19–hit Sundarbans region. Clim Dev, India. https://doi.org/10.1080/17565529.2021.1889948

Prasad KA, Ottinger M, Wei C, Leinenkugel P (2019) Assessment of coastal aquaculture for India from Sentinel-1 SAR time series. Remote Sens 11(3):357. https://doi.org/10.3390/rs11030357

Prathep A, Rattanachot E, Tuntiprapas P (2010) Seasonal variations in seagrass percentage cover and biomass at KohTha Rai, Nakhon Si Thammarat Province, Gulf of Thailand. Sonklanakarin J Sci Technol 32(5):497

Pumijumnong N (2014) Mangrove forests in Thailand. In: Faridah-Hanum I, Latiff A, Hakeem K, Ozturk M (eds) Mangrove ecosystems of Asia. Springer, New York, NY, pp 61–79. https://doi.org/10.1007/978-1-4614-8582-7_4

Qiu L, Zhang M, Zhou B, Cui Y, Yu Z, Liu T, Wu S (2021) Economic and ecological trade-offs of coastal reclamation in the Hangzhou Bay, China. Ecol Indic 125:107477. https://doi.org/10.1016/j.ecolind.2021.107477

Quisthoudt K, Adams J, Rajkaran A, Dahdouh-Guebas F, Koedam N, Randin CF (2013) Disentangling the effects of global climate and regional land-use change on the current and future distribution of mangroves in South Africa. Biodivers Conserv 22(6):1369–1390. https://doi.org/10.1007/s10531-013-0478-4

Qureshi MT (1993) Rehabilitation and management of mangrove forests of Pakistan. In: Lieth H, Al Masoom A (eds) Towards the rational use of high salinity tolerant plants, vol 1. Kluwer Academic Publishers, The Netherlands, pp 89–95. https://doi.org/10.1007/978-94-011-1858-3_8

Rabanal HR, Beuschel GK (1978) The mangroves and related coastal fishery resources in the United Arab Emirates. FAO, Rome

Rahman MM, Chongling Y, Islam KS, Haoliang L (2009) A brief review on pollution and ecotoxicological effects on Sundarbans mangrove ecosystem in Bangladesh. Int J Environ Eng 1(4):369–383. https://doi.org/10.1504/IJEE.2009.027982

Rajkaran A, Adams JB (2012) The effect of harvesting on mangrove forest structure and the use of matrix modelling to determine sustainable harvesting practices in South Africa. In: Garcia J, Diez Casero J (eds) Sustainable forest management-current research. InTech, pp 339–358

Ramasubramanian R, Gnanappazham L, Ravishankar T, Navamuniyammal M (2006) Mangroves of Godavari—analysis through remote sensing approach. Wetl Ecol Manag 14:29–37. https://doi.org/10.1007/s11273-005-2175-x

Ramessur R (2002) Anthropogenic-driven changes with focus on the coastal zone of Mauritius, south-western Indian Ocean. Reg Environ Chang 3(1):99–106. https://doi.org/10.1007/s10113-002-0045-0

Rampadarath S, Bandhoa K, Puchooa D, Jeewon R, Bal S (2018) Metatranscriptomics analysis of mangroves habitats around Mauritius. World J Microbiol Biotechnol 34:59. https://doi.org/10.1007/s11274-018-2442-7

Rasolofo MV (1997) Use of mangroves by traditional fishermen in Madagascar. Mangrove Salt Marsh 1(4):243–253. https://doi.org/10.1023/A:1009923022474

Rasool F (1996) Trafficking of mangrove seedlings. In: Letter to Editor, Daily Dawn, Karachi

Rattanachot E, Stankovic M, Tuntiprapas P, Prempree S, Prathep A (2020) Monitoring of seagrass along southern Andaman coast of Thailand. Ecol Res 35(5):773–779. https://doi.org/10.1111/1440-1703.12123

Ravishankar T, Navamuniyammal M, Gnanappazham L, Nayak SS, Mahapatra GC et al (2004) Atlas of Mangrove Wetlands of India (Part - 3 Orissa) MS Swaminathan Research Foundation, Chennai

Raw JL, Riddin T, Wasserman J, Lehman TWK, Bornman TG, Adams JB (2020) Salt marsh elevation and responses to future sea-level rise in the Knysna Estuary, South Africa. Afr J Aquat Sci 45(1–2):49–64. https://doi.org/10.2989/16085914.2019.1662763

Raw JL, Adams JB, Bornman TG, Riddin T, Vanderklift MA (2021) Vulnerability to sea-level rise and the potential for restoration to enhance blue carbon storage in salt marshes of an urban estuary. Estuar Coast Shelf Sci 107495. https://doi.org/10.1016/j.ecss.2021.107495

Reddy CS, Pattanaik C, Muthy MSR (2007) Assessment and monitoring of mangroves of Bhitarkanika wildlife sanctuary, using remote sensing and GIS. Curr Sci 92:1409–1415. https://www.jstor.org/stable/24097770

Ren C, Wang Z, Zhang Y, Zhang B, Chen L, Xi Y et al (2019) Rapid expansion of coastal aquaculture ponds in China from Landsat observations during 1984–2016. Int J Appl Earth Observ Geoinf 82:101902. https://doi.org/10.1016/j.jag.2019.101902

Riani E, Djuwita I, Budiharsono S, Purbayanto A, Asmus H (2012) Challenging for seagrass management in Indonesia. J Coast Dev 15(3):234–242

Richards DR, Friess DA (2016) Rates and drivers of mangrove deforestation in Southeast Asia, 2000–2012. Proc Natl Acad Sci 113(2):344–349. https://doi.org/10.1073/pnas.1510272113

Rogers CS (2019) Immediate effects of hurricanes on a diverse coral/mangrove ecosystem in the US Virgin Islands and the potential for recovery. Diversity 11(8):130. https://doi.org/10.3390/d11080130

Rogers K (2021) Accommodation space as a framework for assessing the response of mangroves to relative sea-level rise. Singap J Trop Geogr 42(2):163–183. https://doi.org/10.1111/sjtg.12357

Romañach SS, DeAngelis DL, Koh HL, Li Y, Teh SY, Barizan RSR, Zhai L (2018) Conservation and restoration of mangroves: global status, perspectives, and prognosis. Ocean Coast Manag 154:72–82. https://doi.org/10.1016/j.ocecoaman.2018.01.009

Rouphael T, Turak E, Brodie J (1998) Seagrasses and mangroves of Yemen's Red Sea. Ecosystems of the red sea coast of Yemen. Protection of marine ecosystems of the red sea coast of Yemen. Hassell & Assoc, AMSAT and UNOPS, pp 41–49

Ruwa RK (1996) Intertidal wetlands. In: McClanahan TR, Young TP (eds) East African ecosystems and their conservation. Oxford University Press, Oxford, pp 101–127

Sainsbury NC, Genner MJ, Saville GR, Pinnegar JK, O'Neill CK, Simpson SD, Turner RA (2018) Changing storminess and global capture fisheries. Nat Clim Chang 8(8):655–659. https://doi.org/10.1038/s41558-018-0206-x

Saintilan N, Khan NS, Ashe E, Kelleway JJ, Rogers K, Woodroffe CD, Horton BP (2020) Thresholds of mangrove survival under rapid sea level rise. Science 368(6495):1118–1121

Saket M, Matusse R (1994) Study for the determination of the rate of deforestation of the mangrove vegetation in Mozambique. FAO/PNUD/MOZ/92/013: 9, DNFFB

Saleem A, Nileysha A (2003) Characteristics, status and need for conservation of mangrove ecosystems in the Republic of Maldives, Indian Ocean. J Nat Sci Found Sri Lanka 31(1–2):201. https://doi.org/10.4038/jnsfsr.v31i1-2.3033

Sánchez-Núñez DA, Bernal G, Pineda JEM (2019) The relative role of mangroves on wave erosion mitigation and sediment properties. Estuar Coast 42(8):2124–2138. https://doi.org/10.1007/s12237-019-00628-9

Sarmin NS, Hasmadi IM, Pakhriazad HZ, Khairil WA (2016a) The DPSIR framework for causes analysis of mangrove deforestation in Johor, Malaysia. Environ Nanotechnol Monit Manag 6:214–218. https://doi.org/10.1016/j.enmm.2016.11.002

Sarmin NS, Mohd Hasmadi I, Pakhriazad HJ, Khairil WA, Iqbal PAS (2016b) Land cover dynamics of Sungai Pulai Mangrove forest using remote sensing and GIS-preliminary results. J Eng Appl Sci 11(3):441–445

Saw AA, Kanzaki M (2015) Local livelihoods and encroachment into a mangrove forest reserve: a case study of the Wunbaik reserved mangrove forest, Myanmar. Proc Environ Sci 28:483–492. https://doi.org/10.1016/j.proenv.2015.07.058

Scales IR, Friess DA (2019) Patterns of mangrove forest disturbance and biomass removal due to small-scale harvesting in southwestern Madagascar. Wetl Ecol Manag 27(5):609–625. https://doi.org/10.1007/s11273-019-09680-5

Scales IR, Friess DA, Glass L, Ravaoarinorotsihoarana L (2018) Rural livelihoods and mangrove degradation in south-west Madagascar: lime production as an emerging threat. Oryx 52(4):641–645. https://doi.org/10.1017/S0030605316001630

Schile LM, Kauffman JB, Crooks S, Fourqurean JW, Glavan J, Megonigal JP (2017) Limits on carbon sequestration in arid blue carbon ecosystems. Ecol Appl 27(3):859–874. https://doi.org/10.1002/eap.1489

Schile-Beers L, Megonigal JP, Kauffman JB, Crooks S, Fourqurean JW, Campbell J et al (2018) Carbon sequestration in arid blue carbon ecosystems: a case study from the United Arab Emirates. In: Windham-Myers L, Crooks S, Troxler TG (eds) A blue carbon primer. CRC Press, pp 327–339. https://doi.org/10.1201/9780429435362

Schuerch M, Spencer T, Temmerman S, Kirwan ML, Wolff C, Lincke D et al (2018) Future response of global coastal wetlands to sea-level rise. Nature 561(7722):231–234. https://doi.org/10.1038/s41586-018-0476-5

Seeruttun LD, Raghbor P, Appadoo C (2021) First assessment of anthropogenic marine debris in mangrove forests of Mauritius, a small oceanic island. Mar Pollut Bull 164:112019. https://doi.org/10.1016/j.marpolbul.2021.112019

Sejati AW, Buchori I, Kurniawati S, Brana YC, Fariha TI (2020) Quantifying the impact of industrialization on blue carbon storage in the coastal area of Metropolitan Semarang, Indonesia. Appl Geogr 124:102319. https://doi.org/10.1016/j.apgeog.2020.102319

Selvam V (2021) An assessment of mangrove management during the colonial and post-colonial periods. Curr Sci 120(5):766–771

Semesi AK (1992) Developing management plans for the mangrove forest reserves of mainland Tanzania. In: Jaccarini V, Martens E (eds) The ecology of mangrove and related ecosystems. Springer, Dordrecht, pp 1–10. https://doi.org/10.1007/978-94-017-3288-8_1

Sengupta D, Chen R, Meadows ME (2018) Building beyond land: an overview of coastal land reclamation in 16 global megacities. Appl Geogr 90:229–238. https://doi.org/10.1016/j.apgeog.2017.12.015

Senterre B, Henriette E, Labrosse J-C, Pinchart J, Senaratne N (2015) Rapid biodiversity assessment of Anse Royale and North-East Point wetlands (Mahé, Seychelles). Consultancy Report, Government of Seychelles, United Nations Development Programme, Victoria, Seychelles, p 62

Shahbudin S, Zuhairi A, Kamaruzzaman BY (2012) Impact of coastal development on mangrove cover in Kilim river, Langkawi Island, Malaysia. J Forest Res 23(2):185–190. https://doi.org/10.1007/s11676-011-0218-0

Sharifinia M, Daliri M, Kamrani E (2019) Estuaries and coastal zones in the northern Persian Gulf (Iran). In: Wolanski E, Day JW, Elliott M, Ramachandran R (eds) Coasts and estuaries, Elsevier, pp 57–68. https://doi.org/10.1016/B978-0-12-814003-1.00004-6

Sivakumar K, Rilwan A, Priyanka K, Salah M, Kathiresan K (2018) Mangroves of the atolls of the Maldives, rich among the atoll groups of the Indian Ocean. ISME/GLOMIS Electron J 16(3):11–18

Soe-Htun U, Maung A, Mon S, Tun TZ, Hteik A, Lwin ML, Murray-Jones S (2018) Seagrass surveys in the southern Rakhine coastal region, Myanmar: biodiversity, distribution and coverage. J Aquac Mar Biol 7(2):103–110

Soe-Htun U, Maung A, Mon S, Ha ST, Aung ST, Lwin AM, Lunn Z (2015) Seagrass conservation and monitoring in myanmar: the biodiversity, distribution, and coverage of seagrasses in the Tanintharyi and Rakhine. Fauna & Flora International, TCP Report no. 26, Bay of Bengal Large Marine Ecosystem Project

Sougueh MA, Jonasson JP (2018) Sampling strategy and management recommendation for Spanish mackerel and other important fish stock in Djibouti. United Nations University Fisheries Training Program, Iceland [final project]

Sridhar R, Thangaradjou T, Kannan L, Astalakshmi S (2010) Assessment of coastal bio-resources of the Palk Bay, India, using IRS-LISS-III data. J India Soc Remote Sens 38(3):565–575. https://doi.org/10.1007/s12524-010-0040-8

Stankovic M, Hayashizaki KI, Tuntiprapas P, Rattanachot E, Prathep A (2021) Two decades of seagrass area change: organic carbon sources and stock. Mar Pollut Bull 163:111913. https://doi.org/10.1016/j.marpolbul.2020.111913

Steinke TD (1999) Mangroves in South African estuaries. Cambridge University Press, In Estuaries of South Africa, pp 119–140

Stiller D, Ottinger M, Leinenkugel P (2019) Spatio-temporal patterns of coastal aquaculture derived from Sentinel-1 time series data and the full Landsat archive. Remote Sens 11(14):1707. https://doi.org/10.3390/rs11141707

Świerczewski D, Malenovský I, Stroiński A (2017) Haloflata gen. nov. —a new genus from salt marshes in Socotra island (Hemiptera: Fulgoromorpha: Flatidae). In: Annales Zoologici. Museum and institute of zoology, Polish Academy of Sciences, vol 67(2), pp 261–278. https://doi.org/10.3161/00034541ANZ2017.67.2.007

Thampanya U, Vermaat JE, Sinsakul S, Panapitukkul N (2006) Coastal erosion and mangrove progradation of Southern Thailand. Estuar Coast Shelf Sci 68(1–2):75–85. https://doi.org/10.1016/j.ecss.2006.01.011

Thangaradjou T, Bhatt JR (2018) Status of seagrass ecosystems in India. Ocean Coast Manag 159:7–15. https://doi.org/10.1016/j.ocecoaman.2017.11.025

Thangaradjou T, Sridhar R, Senthilkumar S, Kannan S (2008) Seagrass resource assessment in the Mandapam coast of the Gulf of Mannar Biosphere Reserve, India. Appl Ecol Environ Res 6(1):139–146

Thangaradjou T, Nobi EP, Dilipan E, Sivakumar K, Kannan L (2009) Seagrasses in India under threat. Seagrass Watch 39:20–21

Thu-Rein AC, Murray-Jones S (2019) Seagrass surveys in ShweThaung Yan coastal areas, the southern part of Rakhine Coastal Region, Myanmar: biodiversity, coverage and biomass. J Aquac Mar Biol 8(3):105–112

Twomey AJ, Saunders MI, Callaghan DP, Bouma TJ, Han Q, O'Brien KR (2021) Lateral sediment erosion with and without the non-dense root-mat forming seagrass Enhalusacoroides. Estuar Coast Shelf Sci 253:107316. https://doi.org/10.1016/j.ecss.2021.107316

Udagedara S, Dahanayaka DDGL (2020) Current status and checklist of seagrass in Sri Lanka. Int J Aquat Biol 8(5):317–326

Udagedara S, Fernando D, Perera N, Tanna A, Bown R (2017) A first record of Halodule pinifolia Miki den Hartog, and new locality of nationally endangered *Halophila beccarii* Asch, from the eastern coast of Sri Lanka. Int J Aquat Biol 5(5):328–335

Ueda S, Go CSU, Yoshioka T, Yoshida N, Wada E, Miyajima T et al (2000) Dynamics of dissolved O_2, CO_2, CH_4, and N_2O in a tropical coastal swamp in southern Thailand. Biogeochemistry 49(3):191–215. https://doi.org/10.1023/A:1006393214432

Uku J, Daudi L, Alati V, Nzioka A, Muthama C (2021) The status of seagrass beds in the coastal county of Lamu, Kenya. Aquat Ecosyst Health Manag 24(1):35–42. https://doi.org/10.14321/aehm.024.01.07

Unsworth RK, Ambo-Rappe R, Jones BL, La Nafie YA, Irawan A, Hernawan UE et al (2018) Indonesia's globally significant seagrass meadows are under widespread threat. Sci Total Environ 634:279–286. https://doi.org/10.1016/j.scitotenv.2018.03.315

Vanderklift MA, Gorman D, Steven AD (2019) Blue carbon in the Indian Ocean: a review and research agenda. J India Ocean Reg 15(2):129–138. https://doi.org/10.1080/19480881.2019.1625209

Veettil BK, Pereira SFR, Quang NX (2018) Rapidly diminishing mangrove forests in Myanmar (Burma): a review. Hydrobiologia 822(1):19–35. https://doi.org/10.1007/s10750-018-3673-1

Viswanathan C, Purvaja R, Jeevamani JJJ, Samuel VD, Sankar R, Abhilash KR et al (2020) Salt marsh vegetation in India: Species composition, distribution, zonation pattern and conservation implications. Estuar Coast Shelf Sci 242:106792. https://doi.org/10.1016/j.ecss.2020.106792

Wang C, Myint SW (2016) Environmental concerns of deforestation in Myanmar 2001–2010. Remote Sens 8(9):728. https://doi.org/10.3390/rs8090728

Wang Y, Bonynge G, Nugranad J, Traber M, Ngusaru A, Tobey J et al (2003) Remote sensing of mangrove change along the Tanzania coast. Mar Geod 26(1–2):35–48. https://doi.org/10.1080/01490410306708

Webb EL, Jachowski NR, Phelps J, Friess DA, Than MM, Ziegler AD (2014) Deforestation in the Ayeyarwady Delta and the conservation implications of an internationally-engaged Myanmar. Glob Environ Chang 24:321–333. https://doi.org/10.1016/j.gloenvcha.2013.10.007

Witsen DB (2012) Mangrove Restoration and Management in Djibouti. Masters thesis. Virginia Polytechnic Institute and State University

WWF-Pakistan (2006) Vegetation assessment. Preliminary ecological assessment, Indus for all programme

Xie D, Pan C, Wu X, Gao S, Wang ZB (2017) Local human activities overwhelm decreased sediment supply from the Changjiang River: Continued rapid accumulation in the Hangzhou Bay-Qiantang Estuary system. Mar Geol 392:66–77. https://doi.org/10.1016/j.margeo.2017.08.013

Xu S, Qiao Y, Xu S, Yue S, Zhang Y, Liu M et al (2021) Diversity, distribution and conservation of seagrass in coastal waters of the Liaodong Peninsula, North Yellow Sea, northern China: implications for seagrass conservation. Mar Pollut Bull 167:112261. https://doi.org/10.1016/j.marpolbul.2021.112261

Yaakub SM, Lim RL, Lim WL, Todd PA (2013) The diversity and distribution of seagrass in Singapore. Nat Singap 6:105–111

Yaakub SM, McKenzie LJ, Erftemeijer PL, Bouma T, Todd PA (2014) Courage under fire: seagrass persistence adjacent to a highly urbanised city–state. Mar Pollut Bull 83(2):417–424. https://doi.org/10.1016/j.marpolbul.2014.01.012

Yagoup MM, Kolan GR (2006) Monitoring coastal zone land use and land cover changes of Abu Dhabi using remote sensing. J India Soc Remote Sens 34(1):57–68. https://doi.org/10.1007/BF02990747

Yang SC, Riddin T, Adams JB, Shih SS (2014) Predicting the spatial distribution of mangroves in a South African estuary in response to sea level rise, substrate elevation change and a sea storm event. J Coast Conserv 18(4):459–469. https://doi.org/10.1007/s11852-014-0331-2
Yang S, Lim RL, Sheue CR, Yong JW (2011) The current status of mangrove forests in Singapore. In: Proceedings of nature society, Singapore's conference on 'nature conservation for a sustainable Singapore' 16th Oct, pp 99–120
Yee ATK, Ang WF, Teo S, Liew SC, Tan HTW (2010) The present extent of mangrove forests in Singapore. Nat Singap 3(3):139–145
Yessoufou K, Stoffberg GH (2016) Biogeography, threats and phylogenetic structure of mangrove forest globally and in South Africa: a review. S Afr J Bot 107:114–120. https://doi.org/10.1016/j.sajb.2015.11.002
Yuan B, Sun J, Lin B, Zhang F (2020) Long-term morphodynamics of a large estuary subject to decreasing sediment supply and sea level rise. Glob Planet Chang 191:103212. https://doi.org/10.1016/j.gloplacha.2020.103212
Yuvaniyama A, Arunin S (1993) Growth of three halophytic grasses on salt-affected soil in northeast Thailand. In: ACIAR proceedings. Australian Centre for International Agricultural Research, pp 32–32
Zanardi-Lamardo E, Mitra S, Vieira-Campos AA, Cabral CB, Yogui GT, Sarkar SK et al (2019) Distribution and sources of organic contaminants in surface sediments of Hooghly river estuary and Sundarban mangrove, eastern coast of India. Mar Pollut Bull 146:39–49. https://doi.org/10.1016/j.marpolbul.2019.05.043
Zedler JB, Bonin CL, Larkin DJ, Varty A (2018) Salt marshes. In: Encyclopedia of ecology. Elsevier, pp 614–622.
Zöckler C, Aung C (2019) The mangroves of myanmar. In: Gul B, Böer B, Khan M, Clüsener-Godt M, Hameed A (eds) Sabkha ecosystems, Springer, Cham, pp 253–268. https://doi.org/10.1007/978-3-030-04417-6_16

Chapter 10
Blue Carbon Ecosystems of the Indian Ocean: Conservation and Management

Abhra Chanda

Abstract The blue carbon ecosystems like mangroves, seagrasses, and salt marshes are facing severe threats throughout the globe due to several natural and anthropogenic factors. The persistence of such imminent threats has led to a dire need for conservation and management plans for these crucial marine habitats. The present chapter reviewed the overall conservation scenario of the blue carbon habitats in the periphery of the Indian Ocean. Preservation, restoration, rehabilitation, afforestation, and controlled felling maintaining forestry rules are some of the well-defined protocols that are implemented by several Indian Ocean countries to safeguard the blue carbon habitats. However, the involvement of local stakeholders, especially the communities exclusively relying on the services of these ecosystems should be included within the management framework. Several studies carried out in different countries unequivocally indicated that stringent law enforcement from the top authority toward the marginalized society residing in and around the blue carbon ecosystems often does not come up with the desired level of conservation. Instead, a consultation with the local community on a small-scale basis, understanding their needs and views, and accordingly cascading the region-specific conservation protocols to the higher officials in charge of the execution could lead to better outcomes. Poverty alleviation and proportionate access to ecosystem services should be given priority to effectively conserve these marine habitats. In several small island nations, where these ecosystems occupy small areas, spreading awareness of their benefits and roles is a top priority. In many western Indian Ocean countries, baseline studies are required in the first place to comment anything on the conservation status and needs. Overall, this chapter highlighted and discussed the present scenario of conservation status in the Indian Ocean countries, the shortfalls observed by the scientific community, and the endeavors that need to be undertaken soon.

Keywords Mangrove · Seagrass · Salt marsh · Restoration · Rehabilitation · Preservation · Community-participation · Local stakeholder · Payment for ecosystem services · Law enforcement

---

A. Chanda (✉)
School of Oceanographic Studies, Jadavpur University, 188 Raja S. C. Mullick Road, Kolkata, West Bengal 700084, India
e-mail: abhrachanda.sos.rs@jadavpuruniversity.in

A. Chanda et al. (eds.), *Blue Carbon Dynamics of the Indian Ocean*,
https://doi.org/10.1007/978-3-030-96558-7_10

10.1 Introduction

The blue carbon ecosystems provide a plethora of ecosystem services (Lau 2013; Himes-Cornell et al. 2018). The marginalized sections of societies that thrive in and around the mangroves, seagrass, and salt marshes, benefit from these ecosystems (Vierros 2017; Okafor-Yarwood et al. 2020). These ecosystems provide livelihood and offer protection from several natural calamities (Quevedo et al. 2021a; 2021b). Several authors argue that blue carbon ecosystems can act as a tool for poverty alleviation if managed and conserved properly (Hejnowicz et al. 2015). The climate change mitigation potential of these ecosystems is also unparalleled (Lovelock and Duarte 2019). These ecosystems are capable of solving several social and geophysical problems. However, despite having such potential, most anthropogenic developmental activities in the coastlines occur at the cost of sacrificing these ecosystems (Howard et al. 2017; Lovelock et al. 2017; Alongi 2020; Sejati et al. 2020). Lovelock and Duarte (2019) mentioned that the very topic of blue carbon needs an interdisciplinary approach to apprehend the holistic virtues of these ecosystems. Conservation and management of these ecosystems have become the topmost priority for coastal managers. The scientific community, stakeholders, policy managers, and conservationists should join hands to effectively formulate solutions that can reap the optimum benefits from these ecosystems toward humankind and the environment.

The nature of ecosystem services provided by the blue carbon ecosystems and the threats vary across space and time. A unanimous conservation strategy for the entire global blue carbon ecosystem might not be appropriate, as the nuances of threats, management options, and social involvement vary between countries and regions within the same country. Thus, region-specific planning is perhaps the best way to safeguard these crucial ecosystems at the local level. The social dependence on these habitats, their ecosystem services, natural and anthropogenic threats, and conservation opportunities remain intertwined. Conservation becomes a priority in places where these habitats face extensive threats. The blue carbon habitats in the periphery of the Indian Ocean are under severe pressure, and proper characterization of the blue carbon strength is missing altogether for many countries. Thus, it becomes crucial to assess the livelihood dependence, conservation, and management strategies opted in the rim of the Indian Ocean across the different countries to shape this book. The present chapter tried to highlight the crucial points in this regard in a country-wise discussion.

10.2 Conservation and Management Scenarios in the Eastern Indian Ocean Countries

10.2.1 India

Among the conventional blue carbon ecosystems, mangroves have perhaps received the most attention from the scientific communities of India. Unlike the submerged aquatic vegetations like seagrass and salt marshes, Indian Forest Governance considers the mangroves as forested regions. In this regard, mangroves have a due advantage over the other two blue carbon ecosystems. DasGupta and shaw (2013) underpinned that mangrove management in India dates back to ancient times. However, its orientation has significantly changed over time. A step-by-step decentralization in the mangrove conservation protocols followed by India in the post-independence era led to good results. DasGupta and shaw (2013) stated that previously the management strategies focused on preventive measures. However, the present day is full of examples where ameliorative steps have been taken to safeguard these habitats. In the last four decades, the government has succeeded to a large extent in including the local communities in mangrove protection measures. Maintaining the delicate balance of using forest products and conserving them remains a challenge in various places of India. However, recent surveys indicate that mangrove covers have increased in the country's periphery. The state forest departments of India collectively recognized fourteen mangrove forests for management through the formation of a National Mangrove Committee (NMC) (Balasubramanian and Ajmal Khan 2002). Out of 31 marine and coastal protected areas in India, 26 cover mangrove forests (Rajagopalan 2011). This statistic shows the importance provided to mangroves by the national-level policymakers. Besides conservation, mangrove restoration endeavors are also present in India, especially in Indian Sundarbans (Vyas and Sengupta 2012).

India shelters substantial seagrass cover; however, their conservation and management are still in their infancy in India. The scientific community paid more attention to exploring the biotechnological potential of the seagrass rather than looking for avenues to reduce the concomitant threats associated with this ecosystem (Thangaradjou and Bhatt 2018). Seagrass meadows provide habitats to the endemic dugongs and green turtles, and the Indian Government recognized their significance. However, direct orders to safeguard the seagrass meadows were absent in the Indian context. The coastal regulation zone notification for 2011 corroborated the seagrass beds as ecologically sensitive habitats and ushered their direct conservation in India. The collation of traditional knowledge in seagrass conservation is also given due importance (Newmaster et al. 2011). Salt marshes are yet to receive their due attention in the Indian periphery. Several studies aimed at characterizing the spatial cover and species diversity (Banerjee et al. 2017); however, research focusing on conservation strategies of this ecosystem is scarce (Viswanathan et al. 2020).

10.2.2 Bangladesh

Bangladesh shelters the world's largest mangrove forest, Sundarbans. Salt marshes and patches of seagrass are also available in the coastal periphery of the country. However, several threats are omnipresent throughout the country's blue carbon repository. Ahmed et al. (2017a) indicated that shrimp farming in the mangrove environment that significantly contributes to the national economy of Bangladesh leaves deleterious effects on the mangrove stands. Lately, deforestation of mangroves to grow the aquaculture sector has been a practice in several countries, and Bangladesh is not an exception. Ahmed et al. (2017a) envisaged the necessity of switching over to Integrated Multi-Trophic Aquaculture techniques and restoring degraded mangrove habitats. They also opined the possibility of shifting the land-based aquaculture sector to open coastal water that can significantly reduce the burden on the mangrove stands of Bangladesh. They argued that the translocation of the shrimp culture industry to open waters would significantly enhance the blue carbon sequestration potential of the Bangladesh mangroves. Islam and Wahab (2005) recognized this issue of aqua-farming in Bangladesh long back. As a measure to safeguard the mangroves of Bangladesh, the concept of forest protection has been pinned in some of the articles of the Bangladesh constitution (Iftekhar 2008). The Bangladesh Forest Research Institute along with several other governing bodies like Bangladesh Forest Industries Development Corporation have been entrusted with several responsibilities to manage and protect the mangrove forests (Iftekhar 2008). Roy (2016) characterized the willingness of the local communities in Bangladesh to protect the mangroves. A section of this community admitted that poor economic conditions compel several people to trespass and degrade the forest habitat. However, given the proper right to harvest and proprietorship, they are more than willing to safeguard the mangroves for their sake. Roy et al. (2013) also concluded that granting state forest ownership to the local stakeholders can truly implement forest conservation in reality. Rahman et al. (2010) summed up the issues in this regard and called for an advanced silviculture system, rigorous scientific exploration, and novel conservation and restoration measures. Bangladeshi scholars have reported the existence of seagrass and salt marshes in their country (Patro et al. 2017, and the references therein). According to the Coastal Zone Policy (2005) of Bangladesh, seagrasses get a bit of protection; however, stringent implementation is still a challenge. Pieces of literature are too few to comment on the conservation status of Bangladesh salt marshes.

10.2.3 Myanmar

Myanmar has witnessed a catastrophic degradation of mangrove habitat and an overall reduction in the spatial extent of mangrove cover in the last few decades. However, Camacho et al. (2020) indicated that slowly but steadily mangrove conservation and rehabilitation are taking place in Myanmar. One of the promising endeavors in this regard is constituting the Taw HtwinGyi Community Forestry by the Myeik Forest Department to safeguard the mangroves in and around the Myeik Township. They inferred that the success of mangrove restoration and conservation depends on effective addressing of the linkages between the economic activities vis-à-vis the rehabilitation options. The involvement of local people at the grass-root-level is utterly important to maximize the potential of mangrove conservation. Camacho et al. (2020) further advised that traditional knowledge of the local stakeholders should be given priority during species selection for afforestation programs. Prescott et al. (2017) argued that the socio-political backdrop of Myanmar is changing. The authoritarian and centralized economy of Myanmar is gradually shifting toward a decentralized and liberal one. Prescott et al. (2017) feared that such transition might lead to more destruction of forests to reap the short-term benefits from infrastructural development. However, such a switch-over also provides an opportunity to shape the conservation and management protocols, provided good governance prevails at the national level. Aung et al. (2013) emphasized the need for a holistic understanding of the post-cyclone recovery dynamics of the mangroves of Myanmar to effectively conserve this habitat from the natural threats that always lurk over them. From Myanmar's point of view, sustainable rice cultivation without affecting the mangrove cover of the country is perhaps the most significant issue that requires the attention of policy managers (Friess et al. 2020). Unlike the mangroves, research on seagrass is sparse that does not allow the policy managers to characterize the threats to this habitat (Jones et al. 2018). However, some of the seagrass-rich sites of Myanmar fall within the purview of the country's marine protected areas. Soe-Htun et al. (2015) called for more stringent conservation measures to safeguard the fisheries sector and seagrass-endemic species in the coastline of Myanmar. The knowledge of Myanmar's salt marshes is too meager to comment on their conservation need and implementation (Murray et al. 2020).

10.2.4 Thailand

Like Bangladesh, the fundamental threat to the Thai mangroves is the growing shrimp cultivation at the cost of the mangrove habitat destruction. Sathirathai and Barbier (2001) observed that coastal aquaculture at the cost of mangrove stands provides a short-term benefit to many, and hence, gets the government's nod. However, a substantial portion of the Thai coastal communities gets a plethora of direct and indirect benefits from the mangrove stands. Sathirathai and Barbier (2001) documented the willingness of these communities to protect the Thai mangrove stands, and this leads to a kind of tassel between the two groups in favor and opposition to shrimp culture. They finally inferred that shrimp cultivation by irreversibly degrading the mangrove habitat is not a viable option in the long run. Treephan et al. (2019) reported ecofriendly community participation to safeguard the mangroves through managing eco-tourism in the Ban Hua Thang region of Thailand. They observed that a local group of people keen to save the mangroves effectively carries out mangrove-based eco-tourism. This group successfully developed a model where all the stakeholders are free to contribute to boosting the income of the community prioritizing the interest of mangrove conservation. They consider the mangroves as their pride and natural heritage. Kongkeaw et al. (2019) also lauded the community-based-mangrove-conservation approach in Thailand. They reported that such measures have successfully conserved several mangroves stands on the coastline of Thailand. The support provided by non-governmental organizations to the community members narrowed down the gap between them and the government, which in turn, led to the success of conserving mangroves in Thailand. Thompson (2018) argued that mangrove restoration would succeed to a greater extent if the government encourages the initiatives from the local communities rather than forcefully enact environmental conservation policies from the top level. In other words, Thompson (2018) advocates the bottom-up approach as it is far more effective than the mere top-down dissemination of authority without consulting the views of the local stakeholders. Thompson and Friess (2019) indicated that the local communities are quite interested in payments for ecosystem services but not for global services like climate change mitigation. They are more willing to get returns against local ecosystem services. However, they also pointed out that contradiction between the government and the local stakeholders is the prime hindrance that bars the effective implementation of mangrove restoration. However, Rattanachot et al. (2018) observed that government intervention worked well toward conserving the seagrass meadows of Koh Tha Rai in Thailand. The implementation of proper management protocol enhanced the spatial cover and blue carbon content in the seagrass meadows of Koh Tha Rai. However, like many other countries, information on the conservation and management status of Thai marshes is still scarce.

10.2.5 Singapore

Singapore witnessed substantial destruction of mangrove cover in the past few decades and is in dire need of protection to safeguard the existing stands. Lately, Velde et al. (2019) studied the stakeholder discourses on mangrove conservation and management near the urbanized setup of Singapore. They observed contradictions among the stakeholders; however, they reached a consensus regarding no further mangrove deterioration. Velde et al. (2019) observed that the local stakeholders want preferably the total mangrove area under the national marine protected area. They also want consistent propaganda of the mangrove ecosystem's cultural services to enhance the awareness related to mangrove conservation. Friess (2017) mentioned the scope of rehabilitation in urban as well as rural regions of Singapore. However, space constraints are major issues in urban fringe mangrove restoration. Ramos and Hesp (2018) reported that mangrove rehabilitation showed substantial success in the man-made estuarine channels of Singapore. The Sungei Api-Api project exhibited a promising potential to rehabilitate mangroves in urban fringes. Ramos and Hesp (2018) envisaged the utility of such endeavors in restoring mangroves in similar setups. Ellison et al. (2020) mentioned that places like Singapore can act as perfect macrocosms for experimenting on restoration and rehabilitation of mangroves. Results obtained from such experiments can be utilized in other similar setups and can be eye-openers for policy managers and conservationists in other parts of the world. Many of the seagrass sites of Singapore have been enlisted under protected areas; however, the outcomes of such measures are yet to be validated (McKenzie et al. 2016). Wainwright et al. (2019) characterized the differences in the seagrass-associated fungal communities in the Singapore coastline and Malay Peninsula regions. They inferred that these fungal communities can be potential indicators of site suitability for seagrass restoration. Yaakub et al. (2014) suggested that the seagrass communities of the Singapore coastline in urban adjacence exhibited significant persistence in the face of pollution and disturbances. This observation indicates that minimal restoration and conservation can lead to fruitful outcomes. However, literature is scarce on salt marshes to comment on their threat or conservation status in Singapore.

10.2.6 Indonesia

Indonesian coastline has a rich marine biodiversity that is of global significance. This country hosts a substantial share of the world's total mangrove area. The seagrass and salt marsh abundance are also prominent in the Indonesian coastal periphery. Sidik et al. (2018) indicated that close to 22% of the country's total mangrove cover comes under the umbrella of protected areas. Such initiative from the Indonesian government is a huge step toward conserving this ecosystem. However, Sidik et al. (2018) also advocated the need to increase the spatial extent of mangroves in Indonesia. He also cautioned that enhancing blue carbon sink strength should not be brought to the forefront to achieve this agenda, rather, the needs of the local communities should be given the foremost priority. Deliberate and proactive local stakeholder participation is the only way out to solve this problem. Sugandini et al. (2017) characterized the general behavior of a local community of Yogyakarta, Indonesia, who were engaged in mangrove conservation. They attributed the environmental attitude of the stakeholders and the traditional environmental knowledge as the key parameters that govern the success of mangrove conservation at the grass-root-level. Contrary to these optimistic findings, Dharmawan et al. (2016) noted that weak research sector and local level politics are the main hindrances toward successful mangrove conservation. They emphasized that a strong amalgamation of neat scientific research at the grassroots level of traditional knowledge can lead to fruitful conservation protocols. The conflicts of interest that arose between the local stakeholders and the science sector often do not fruitfully resolve, which leads to ambiguities and dissatisfaction among the local communities. Idrus et al. (2019), while working in the East Lombok region of Indonesia, stressed local conservation endeavors to best manage the mangrove stands from further deterioration. Almost two decades ago, Armitage

(2002) thrust on the necessity of challenging the existing government formulated conservation enforcement prototypes and looking out for real-time solutions keeping in mind the economic interests of the local stakeholders and their proactive participation. Besides mangroves, Indonesia also shelters seagrass beds in the coastal periphery. Lukman et al. (2021) observed that local awareness among the fisher community and the government officials is the key driver to achieve seagrass conservation while working in the Berau Regency, Indonesia. They observed that the fisher community of this region is mainly keen on utilizing the provisional services of seagrasses. Awareness of holistic knowledge of resource utilization is of utmost necessity to implement seagrass conservation. Unsworth et al. (2018) indicated that the seagrasses of Indonesia are of global significance and under severe threat. They emphasized incentive-type schemes of conservation to effectively manage seagrasses. Despite several endeavors in understanding the suitable mechanisms for seagrass and mangrove cover, studies on salt marshes remain poorly constrained.

10.2.7 Malaysia

Malaysian mangroves have become equally susceptible like in many other countries due to the ongoing coastal development and changes in land use and land cover. Matang mangrove forest of Malaysia is perhaps one of the well-managed forests of the entire world; however, Chong (2006) indicated that the conservation protocol adopted in the different states of Malaysia varies from each other, and their implementation depends on the sectoral differences. The conflicts and the efficiency of the state governments in employing the center-controlled mangrove management guidelines play a pivotal role in this regard. It is undeniable that Malaysia presented exemplary mangrove management through the way they have managed the Matang mangrove forest. This forest was managed scientifically since the onset of the twentieth century and has been under the lens of management and conservation techniques for almost one hundred and twenty years. Goessens et al. (2014) rationalized the scientific forest felling technique applied in the Matang mangrove forest to supply mangrove timber most sustainably. The thirty-year felling cycle and rigorous monitoring of tree thinning and diameter were crucial in decision-making on which tree to fall. Latiff and Faridah-Hanum (2014) mentioned the model of mangrove management implied in Matang could be a stand-alone example of its kind and can be practiced in other mangroves of the world. They stressed meeting the forestry objectives to effectively implement mangrove conservation. Goh (2016) indicated that the top-down enforcement of central government strategies to the grassroots level does not facilitate proper conservation. On the contrary, disseminating the views of the local stakeholders to the top-most authority of the regional planning level can bring out significant improvements. This study also highlighted the importance of stringent environmental impact assessment concerning any action related to forest management. Jusoff and Taha (2008) indicated that mangrove forest management protocols should be regularly updated to generate consistent conservation over time without any sort of dilution.

Regular inputs from the latest research and development projects strengthen the confidence level of the management protocols. Like mangroves, several studies identified the key aspects that should be prioritized to safeguard and conserve the seagrass beds of this country. Bujang et al. (2006) indicated that seagrass conservation endeavors do not restrict to safeguarding the seagrass habitats but also to look after the areas or processes taking place in the upper reaches that can potentially degrade the marine habitats. The resources of seagrass should be under the scanner and their conservation should be prioritized. Bujang et al. (2016) indicated that the seagrasses of Malaysia are under severe threat from several natural and anthropogenic factors; however, proper planning is in dire need to effectively conserve this unique underwater vegetated ecosystem. Hossain et al. (2019) suggested that the various forms of coastal reclamation activities potentially degrade the seagrass habitats and therefore, require special attention to protect the seagrasses from the site-specific and fluvial disturbances caused due to such activities. Information on Malaysian salt marshes is still meager to comment anything on their conservation status.

10.3 Conservation and Management Scenarios in the Western Indian Ocean Countries of Asia

10.3.1 Yemen

The conservation status in the western (Asian) Indian Ocean countries is comparatively poorer than that observed in the eastern sector. Yemen has a substantial mangrove cover in its coastal periphery. Nagi and Abubakr (2013) mentioned an absence of full proof mangrove management plan in Yemen. They advocated that the Environmental Protection Authority of Yemen should immediately strive to furnish effective management options to safeguard the mangroves and other crucial marine ecosystems. They suggested that environment-friendly development schemes and imposing stringent buffer areas in the landward margin of the mangroves along with discouraging coastal infrastructure enhancement in the mangrove forests are a few of the essential pathways to conserve this habitat. Damme and Banfield (2011) indicated that the mangroves of the Socotra Island off the Yemeni shoreline are on the verge of extinction if not conserved with due attention. They also stressed local traditional knowledge to take up as an efficient tool to conserve and safeguard the marine biodiversity of this UNESCO world heritage site. Lately, Veettil et al. (2020) indicated that the Socotra Island underwent significant changes in vegetation cover and land-use-land-cover. They mentioned the several threats lurking on the marine biodiversity of this island; however, nothing is explicitly discussed with regards to the mangrove conservation measures undertaken. According to the latest report of the Ministry of Water and Environment Protection Authority (2017) of Yemen, the central government is trying to enhance the marine protected area cover to safeguard the crucial blue carbon habitats like mangroves and seagrasses.

10.3.2 Oman

The Omani coastline has witnessed drastic changes in the past few decades due to several anthropogenic activities related to urbanization, tourism, and infrastructural developments. Choudri et al. (2016) indicated that the government of Oman has long foreseen these changes and the effects they incur on the marine habitats. The central government of Oman has prepared an in-depth coastal management protocol that includes the integrated coastal zone management guidelines to safeguard the interest of the blue carbon ecosystems. However, the aspect of local community participation requires more attention. They further mentioned that technical research has increased in both quality and quantity; however, the dissemination of knowledge to the local mass to involve them in the conservation process is yet to be done on a nationwide scale. Studies on seagrass and salt marshes in this regard are not found in recent times.

10.3.3 UAE

UAE has fringe mangroves and dense patches in its coastal periphery. Lately, Samara et al. (2020) characterized the anthropogenic disturbance levels in the mangroves of Khor al Beida, UAE. They found minimal anthropogenic intervention. Despite that, the authors suggested an immediate necessity of conservation to maintain the pristine nature of these mangrove patches. Elmahdy et al. (2020) observed that mangrove cover has increased substantially in the last two decades in several pockets of northern UAE. This observation indicates that the mangroves of UAE are under no threats as such from the anthropogenic intervention observed in several countries of the world. Despite such optimistic findings, the government should strive to protect these crucial marine habitats from any future possibility of degradation. Howari et al. (2009) mentioned that the local population of UAE is much aware of the goodness of mangrove preservation. Widespread afforestation program and shutting down of shipyards facilitated mangrove area expansion in the northwestern parts of UAE. Howari et al. (2009) also indicated that during the time frame of 1994 to 2003, the seagrass beds have also increased in area in this region.

10.3.4 Iran

Iranian coastlines have substantial mangrove cover. Ghasemi et al. (2010) mentioned that lack of awareness among the coastal community on mangroves and their utility is the biggest threat and the same time the biggest challenge toward mangrove conservation. Petrosian et al. (2016) examined the suitability of ex-situ mangrove conservation in the Iranian coastal belt. They observed that the fluctuation of temperature, tidal amplitude, and land proneness are the essential most factors that determine the success rate of such conservation programs. Pour et al. (2018) suggested that conservation protocols should indirectly look for enhancing financial assets for the local communities, which in turn, would automatically reduce the environmental reliance of these people and safeguard the mangroves. The conservation status of Iranian seagrass and salt marshes was hard to trace from the existing literature.

10.3.5 Pakistan

Pakistan has a long coastline and deltaic regimes that shelter substantial mangrove cover. Mukhtar and Hannan (2012) detailed the several natural and anthropogenic threats that the mangroves of the Indus delta, Gawatar bay, Sonmiani, and Kalmat regions of Pakistan. They mentioned that several mangrove conservation endeavors have been undertaken by the Pakistan Government to safeguard these habitats. Mukhtar and Hannan (2012) professed the necessity of large-scale mangrove afforestation in these regions to enhance the ecosystem service potential that can benefit the local communities. This enhancement can act as an incentive at the grass-roots level and facilitate the fruitful management of mangroves. Lately, Gilani et al. (2021) assessed the changes in mangrove cover between 1990 and 2020 in five prominent sites of Pakistan. They observed a net increase in mangrove area in all the five sites that sum up to an increase of almost 1000 km^2. Gilani et al. (2021) attributed the stringent conservation measures behind this success. They further indicated that such optimistic results of conservation would further encourage the conservationists of Pakistan to look for future directives of mangrove management. Rasool et al. (2002) mentioned twenty years ago that the local communities strived their level best to restore the natural species abundance in the Sonmiani coast of Balochistan, Pakistan. Rafique (2018) mentioned that time and again the Pakistan Government and non-governmental sectors have developed several conservation policies and mangrove management protocols to combat the existing natural and anthropogenic threats on the mangrove stands. All these policies tried to involve the local community and raise awareness on mangrove conservation. Memon (2012), while characterizing the mangrove restoration endeavors of Pakistan, stressed integrating disaster risk management to better safeguard the mangroves and restore their habitats. Siddiqui et al. (2008) emphasized the need for more marine protected areas along the coastline of Pakistan to conserve the mangrove habitats with an enhanced confidence level.

10.4 Conservation and Management Scenarios in the Western Indian Ocean Countries of Africa

10.4.1 South Africa

South Africa marks the subtropical limit of mangrove extent in the southern hemisphere. Adams et al. (2004) indicated that tree harvesting for timber is the principal threat to the mangroves of South Africa. According to their estimation, only 6% of the country's mangroves fall under marine protected areas. They inferred that lack of official protection principally led to indiscriminate tree-cutting. Naidoo (2016) also pinpointed that lack of protection is the fundamental problem. The author further suggested that acquiring land from landholders and putting those pieces of land under protected areas is extremely crucial to safeguard marine habitats. Hoppe-Speer et al. (2015) tabulated the conservation status of the mangrove species in the prominent estuarine patches of South Africa. They observed that browsing by cattle is the most significant threat and in this regard, most of the sites are unprotected. Quisthoudt et al. (2013) predicted that due to the ongoing climate change, the latitudinal extent of mangroves is expected to increase down south. However, they are most likely going to colonize the present salt marsh habitats. Thus, conserving both the ecosystems (mangroves and salt marshes) requires a holistic understanding. Quisthoudt et al. (2013) further mentioned that two of the critical management action plans would be to ensure the mangrove plant's seed movement to the southern estuaries and to keep enough intertidal space for the natural rehabilitation of mangroves. Lately, Phair et al. (2021) suggested that the keystone seagrass species of South Africa, *Zostera capensis*, require special conservation measures and insisted on using genomic data in this regard. Adams (2020) indicated that salt marshes are crucial habitats that maintain wide biodiversity in the South African coastal periphery. The authors also opined that the natural expansion of mangroves is a crucial factor that needs to be looked upon apart from the other climate-driven threats.

10.4.2 Mozambique

The deltas and coastlines of Mozambique shelter substantial mangrove cover, especially in the estuarine stretches of the central part. Barbosa et al. (2001) mentioned that the Forests and Land Legislation Act of Mozambique deals with the conservation protocol of mangrove management and they strive to involve the local stakeholders. Back in 2001, they mentioned that a national mangrove management plan was being initiated. However, even in 2020, Charrua et al. (2020) observed that the mangrove management picture is not clear in Mozambique. They stressed managing the climatic hazards and the anthropogenic impacts to uphold the ecosystem services of mangroves to the impoverished section of the coastal communities. Chevallier (2013) professed on balancing infrastructural development and enhancing the ecosystem's resilience to the natural and anthropogenic changes to safeguard the mangroves of Mozambique. The authors also suggested utilizing the ecosystem services provided by the marine ecosystems to the level best to alleviate poverty and enhance the financial security of the marginalized population. The seagrass beds of Mozambique support the crucial fisheries sector of this country. Samoilys et al. (2019) asked for sound regional planning and characterizing the spatiotemporal variability of the fish dynamics to effectively this crucial habitat. They argued that spreading awareness on the benefits of seagrass conservation can truly help maintain an unorganized small-scale fisheries sector. Information on salt marshes is still scarce to comment on the conservation requirement and status of the same.

10.4.3 Tanzania

Like many other East African countries, Tanzanian coastlines are rich in mangrove biodiversity. Wang et al. (2003) observed that the decade of 1990–2000 did not witness any significant change in the mangrove cover of Tanzania. Such an observation indicates less anthropogenic intervention and acts as evidence of the implementation of existing management protocols. According to Wang et al. (2003), Tanzania Mangrove Management Project functioned well to safeguard the mangroves of this region. Beymer-Farris and Bassett (2012), however, argued that centralized schemes like REDD + instigated a revolt at the grassroots community level because when such schemes are enforced from the top, often the ground-based needs and the views of the local stakeholders are ignored. The authors feared that in the utter absence of a bottom-up approach, such schemes can hardly manifest the desired outcomes from such planning and execution. In the same line, Mangora (2011) pointed out that the nationalization of mangroves throughout Tanzania did not bring about the desired improvement from the perspective of mangrove degradation. The author instead discussed the advantages of tactically nurturing the proclivity of the local stakeholders to care for the mangrove forests to reach the ideal scenario of conservation. Semesi (1992) indicated similar requirements to be suitable for mangrove management and conservation in the Tanzanian coastline. Ochieng and Erftemeijer (2003) indicated that though some protected areas and legislative acts protect the seagrasses of Tanzania in pen and paper, the ground-level reality often speaks otherwise due to the loopholes in law enforcement. This problem is widespread in many developing countries worldwide. Gullström et al. (2021) indicated that apart from the South African territory, the knowledge on salt marsh extent and threats is purely based on anecdotal information. More studies are required to address this issue for the entire East African coastline.

10.4.4 Kenya

Kenyan keenness on marine conservation is worthy of mention among all the African nations. The Kenyan government and wildlife service has been very proactive to restore and rehabilitate the marine habitats of this country (Ochieng and Erftemeijer 2003). Huxham et al. (2015) exemplified that devising climate-compatible development schemes in consultation with local stakeholders can lead to fruitful conservation of coastal ecosystems like mangroves. Owuor et al. (2019) recorded a community's willingness to pay to ensure and in return for continued ecosystem services. This observation indicates that the local communities are well aware of the utilities that can be expected on successful conservation of mangroves. Dahdouh-Guebas et al. (2000) mentioned the problem of wood harvesting and poaching; however, the scenario has improved much after two decades of practicing and enforcing conservation. On the contrary, Lang'at and Kairo (2013) indicated that the national authorities responsible for mangrove conservation and management in Kenya lack adequate resource to implement the protocols and safety measures in the ground level. Thus, stricter law enforcements are necessary even in the present date to tighten the conservation measures. Ochieng and Erftemeijer (2003) mentioned that some parts of the seagrass beds come under the jurisdiction of the marine protected area and are therefore in a much better shape. However, beyond the protected areas, the management of seagrass is virtually non-existent. Shilland et al. (2021) mentioned that so far only one project based on payment for ecosystem services has been undertaken including the seagrass beds. However, there is much cause for optimism as the conservation scenario, awareness of local communities, and willingness to preserve mother nature is gradually enhancing from the stakeholder to policy managers of Kenya. The Mikoko Pamoja project executed in Gazi Bay, Kenya, is a stand-alone example in this regard (Githaiga et al. 2017).

10.4.5 Somalia

Somalia is one of the least studied countries of Africa in this regard. Characterization of mangrove cover and its temporal dynamics needs to be studied closely to generate some baseline ideas. Guo et al. (2021) observed a net increase in the mangrove forest area on the coastline of Somalia between 1990 and 2015. Adame et al. (2021) also indicated that human intervention is minimal in the Somali coastline. These observations are good enough to encourage conservation endeavors on a national scale to safeguard the mangroves in arid regions like that of Somalia. Observations on seagrass and salt marshes are too few to comment on their conservation and management statuses.

10.4.6 Djibouti

Research on the blue carbon ecosystems of Djibouti is still in its infancy. Hence, it is difficult to comment on the conservation and management practices for the blue carbon ecosystems. Ahmed et al. (2017b) reported alarming levels of heavy metal and polyaromatic hydrocarbons in the mangrove environments of Djibouti city. Thus, policy managers should encourage more baseline studies to reach a position to formulate clear directives for coastal management.

10.5 Conservation and Management Scenarios in the Indian Ocean Island Countries

10.5.1 Sri Lanka

Sri Lanka is an island nation in the Indian Ocean that is prone to several natural and climatic disasters like tropical cyclones and tsunamis. Sri Lanka witnessed substantial deterioration in mangrove cover in the last four decades. However, Wickramasinghe (2017) indicated that recently the scenario has changed a lot and there has been an increasing focus on mangrove management and conservation throughout the coastal periphery of this country. Restoration and rehabilitation are the only options in many coastal sectors of Sri Lanka, where natural regeneration of mangroves is almost not possible. Wickramasinghe (2017) discussed several encouraging facts that exhibit the degree of community willingness, capacity building, and skill development to safeguard these natural barriers and reap the benefit through their well-being. Kodikara et al. (2017) evaluated the success rate of mangrove restoration throughout the Sri Lankan coastline. They found that only nine out of twenty-three sites showed low to moderate success, and in the rest of the sites mangrove plants did not survive. This observation indicates that a much more holistic study is required to understand the mangrove dynamics and their relationship with the abiotic and geomorphologic factors to enhance the success rate of such restoration endeavors. Seagrasses of Sri Lanka have lately started receiving substantial attention with regards to check listing species diversity (Udagedara and Dahanayaka 2020) and genetic identification (Liu et al. 2020); however, information on conservation endeavors are still hard to scavenge from the existing literature.

10.5.2 Maldives

The Maldives is mainly known for corals and mangroves scarcely received any attention as such, which is extremely unfortunate. Curnick et al. (2019) stated that the Maldivian government built an airport in Kulhudhuffushi Island at the cost of mangrove destruction. Sivakumar et al. (2018) indicated that the Maldives though have a patchy mangrove distribution in the atolls they exhibit significantly rich species diversity. Yet, it is a pity that due to the small area covered, mangroves are most neglected in the Maldives (Stevens and Froman 2019). All the studies indicate that the Maldives desperately need proper intention and conservation protocols to safeguard the existing mangroves, before becoming virtually extinct from the Maldivian coastlines. Peer-reviewed publications on the Maldivian seagrass communities are scarce to date. However, Stevens and Froman (2019) reported a proliferation of seagrass beds due to the cultural eutrophication taking place out of the enhanced tourism activities. Thus, more studies are also required in this regard to develop a holistic understanding of the conservation necessity and status of the blue carbon ecosystems in this tourist's heaven.

10.5.3 Seychelles

Very little is known about the mangroves of Seychelles. However, the Aldabra Atoll in Seychelles is known for its pristine environment and it shelters mangroves and seagrasses. Claudino-Sales (2019) mentioned that the Aldabra Atoll is difficult to access and that is the prime reason behind the minimal anthropogenic footprint in this region. Constance et al. (2021) observed a slight increase in the mangrove areas of this atoll in the last two decades. Observations, in particular, related to the seagrass and mangroves are scarce to come by; however, Clifton et al. (2019) pointed out the disagreements between the governmental and non-governmental sectors that could ruin the conservation intentions in an island nation like Seychelles, where maintaining the marine biodiversity is the prime moto to sustain the economy of the country through consistent tourist flow. More comprehensive studies are required in the future to understand the socio-ecological nitty–gritty related to blue carbon ecosystem management.

10.5.4 Madagascar

Being the fourth largest island in the world, Madagascar shelters substantial mangroves in its coastal periphery. Several threats to the blue carbon ecosystems are imminent in this island nation. However, Rakotomahazo et al. (2019) exhibited that participatory mangrove management initiatives led to fruitful outcomes. They exemplified that involving the local stakeholders in conservation research can lead to results that go beyond copybook research and effectively implement the safeguard measures. This type of initiative enhanced awareness among the local population and it was easy to understand the relationship between the blue carbon habitats and the people who directly benefit from them. Despite such optimistic findings, Scales et al. (2018) cautioned about the new trend of using mangrove wood as fuel to locally produce lime from seashells. This trade is mainly taken up as an income option by the marginalized and impoverished sections of the society, however, this approach has major ramifications on the existing mangrove cover of several regions on the coastline of Madagascar. Scales et al. (2018) further professed that understanding the socio-ecological needs and devising methods of poverty alleviation should be prioritized to corroborate mangrove conservation at the grassroots level. Rakotomahazo et al. (2021) in this regard, observed that the local communities in and around the mangroves of Madagascar whole-heartedly accepted the scheme of payment for ecosystem services. They showed an active interest in pursuing this protocol to conserve the mangroves for their betterment. The authors mentioned that the needs and perseverance of the communities should be meticulously judged to reap the benefits of such schemes on the practical ground. Unlike the mangroves, pieces of literature on seagrass conservation status in Madagascar are hard to trace to date.

10.5.5 Comoros

Anecdotal information suggests that only the Moheli Marine Park hosts a small mangrove cover in Comoros. However, very little is known about the threats and the conservation status. The Moheli Marine Park along with a few other regions exhibits patchy seagrass beds; however, peer-reviewed information is still poorly constrained.

10.5.6 Mauritius

Mauritius, a small island nation, shelters less than 1.5 km^2 of mangroves. Seeruttun et al. (2021) indicated that anthropogenic debris is widespread in the mangrove patches of Mauritius that exhibit the need for their conservation. However, Abib and Appadoo (2021) carried out a socio-ecological survey and observed that the local communities are least bothered about mangroves and not at all willing to pay to conserve this ecosystem. Such observations indicate a serious lack of awareness that needs to be addressed by the competent authorities concerned. Daby (2003) mentioned the same negligence level toward seagrasses that are easily compromised for tourism-related activities. The observations of Hammond et al. (2015) also indicate a dire need for mass awareness that is lacking in many small island nations or regions where the blue carbon ecosystems occupy small areas.

References

Abib S, Appadoo C (2021) Local people and mangroves: ecosystem perception and valuation on the southwest coast of Mauritius. West Indian Ocean J Mar Sci 20(1):11–19. https://doi.org/10.4314/wiojms.v20i1.2

Adame MF, Reef R, Santini NS, Najera E, Turschwell MP, Hayes MA et al (2021) Mangroves in arid regions: ecology, threats, and opportunities. Estuar Coast Shelf Sci 248:106796. https://doi.org/10.1016/j.ecss.2020.106796

Adams JB (2020) Saltmarsh at the tip of Africa: patterns, processes, and changes in response to climate change. Estuar Coast Shelf Sci 237:106650. https://doi.org/10.1016/j.ecss.2020.106650

Adams JB, Colloty BM, Bate GC (2004) The distribution and state of mangroves along the coast of Transkei, Eastern Cape Province. South Africa. Wetl Ecol Manag 12(5):531–541. https://doi.org/10.1007/s11273-005-5165-0

Ahmed MM, Doumenq P, Awaleh MO, Syakti AD, Asia L, Chiron S (2017) Levels and sources of heavy metals and PAHs in sediment of Djibouti-city (Republic of Djibouti). Mar Pollut Bull 120(1–2):340–346. https://doi.org/10.1016/j.marpolbul.2017.05.055

Ahmed N, Cheung WW, Thompson S, Glaser M (2017) Solutions to blue carbon emissions: Shrimp cultivation, mangrove deforestation, and climate change in coastal Bangladesh. Mar Pol 82:68–75. https://doi.org/10.1016/j.marpol.2017.05.007

Alongi DM (2020) Global significance of mangrove blue carbon in climate change mitigation. Sci 2(3):67. https://doi.org/10.3390/sci2030067

Armitage D (2002) Socio-institutional dynamics and the political ecology of mangrove forest conservation in Central Sulawesi, Indonesia. Glob Environ Chang 12(3):203–217. https://doi.org/10.1016/S0959-3780(02)00023-7

Aung TT, Mochida Y, Than MM (2013) Prediction of recovery pathways of cyclone-disturbed mangroves in the mega delta of Myanmar. For Ecol Manag 293:103–113. https://doi.org/10.1016/j.foreco.2012.12.034

Balasubramanian T, Ajmal Khan S (2002) Mangroves of India, state of the art report. Envis Publication Series

Banerjee K, Sappal SM, Ramachandran P, Ramesh R (2017) Saltmarsh: ecologically important, yet least studied blue carbon ecosystems in India. J Clim Chang 3(2):59–72. https://doi.org/10.3233/JCC-170014

Barbosa FM, Cuambe CC, Bandeira SO (2001) Status and distribution of mangroves in Mozambique. S Afr J Bot 67(3):393–398. https://doi.org/10.1016/S0254-6299(15)31155-8

Beymer-Farris BA, Bassett TJ (2012) The REDD menace: resurgent protectionism in Tanzania's mangrove forests. Glob Environ Chang 22(2):332–341. https://doi.org/10.1016/j.gloenvcha.2011.11.006

Bujang JS, Zakaria MH, Arshad A (2006) Distribution and significance of seagrass ecosystems in Malaysia. Aquat Ecosyst Health Manag 9(2):203–214. https://doi.org/10.1080/14634980600705576

Bujang JS, Zakaria MH, Short FT (2016) Seagrass in Malaysia: issues and challenges ahead. In: Finlayson CM et al (eds) The wetland book, the wetland book II: distribution, description and conservation. Springer, Dordrecht, pp 1–9. https://doi.org/10.1007/978-94-007-6173-5_268-1

Camacho LD, Gevaña DT, Sabino LL, Ruzol CD, Garcia JE, Camacho ACD et al (2020) Sustainable mangrove rehabilitation: lessons and insights from community-based management in the Philippines and Myanmar. APN Sci Bull 10(1). https://doi.org/10.30852/sb.2020.983

Charrua AB, Bandeira SO, Catarino S, Cabral P, Romeiras MM (2020) Assessment of the vulnerability of coastal mangrove ecosystems in Mozambique. Ocean Coast Manag 189:105145. https://doi.org/10.1016/j.ocecoaman.2020.105145

Chevallier R (2013) Balancing development and coastal conservation: mangroves in Mozambique. Research report 14 on Governance of Africa's Resources Programme, SAIIA

Chong VC (2006) Sustainable utilization and management of mangrove ecosystems of Malaysia. Aquat Ecosyst Health Manag 9(2):249–260. https://doi.org/10.1080/14634980600717084

Choudri BS, Baawain M, Ahmed M (2016) An overview of coastal and marine resources and their management in Sultanate of Oman. J Environ Manag Tour 7(1):21–32
Claudino-Sales V (2019) Aldabra atoll, Seychelles. In: Coastal world heritage sites. Coastal Research Library, vol 28. Springer, Dordrecht, pp 265–270. https://doi.org/10.1007/978-94-024-1528-5_39
Clifton J, Osman EO, Suggett DJ, Smith DJ (2019) Resolving conservation and development tensions in a small island state: a governance analysis of Curieuse Marine National Park, Seychelles. Mar Pol 103617. https://doi.org/10.1016/j.marpol.2019.103617
Coastal Zone Policy (2005) Ministry of water resources, Government of the People's Republic of Bangladesh. http://nda.erd.gov.bd/files/1/Publications/Sectoral%20Policies%20and%20Plans/Costal-Zone-Policy-2005.pdf. Accessed 7 June 2016
Constance A, Haverkamp PJ, Bunbury N, Schaepman-Strub G (2021) Extent change of protected mangrove forest and its relation to wave power exposure on Aldabra Atoll. Glob Ecol Conserv 27:e01564. https://doi.org/10.1016/j.gecco.2021.e01564
Curnick D, Pettorelli N, Amir A, Balke T, Barbier E, Crooks S et al (2019) The value of small mangrove patches. Science 363(6424):239. https://doi.org/10.1126/science.aaw0809
Daby D (2003) Effects of seagrass bed removal for tourism purposes in a Mauritian bay. Environ Pollut 125(3):313–324. https://doi.org/10.1016/S0269-7491(03)00125-8
Dahdouh-Guebas F, Mathenge C, Kairo JG, Koedam N (2000) Utilization of mangrove wood products around Mida Creek (Kenya) amongst subsistence and commercial users. Econ Bot 54(4):513–527. https://doi.org/10.1007/BF02866549
Damme KV, Banfield L (2011) Past and present human impacts on the biodiversity of Socotra Island (Yemen): implications for future conservation. Zool Middle East 54(3):31–88. https://doi.org/10.1080/09397140.2011.10648899
DasGupta R, Shaw R (2013) Changing perspectives of mangrove management in India–an analytical overview. Ocean Coast Manag 80:107–118. https://doi.org/10.1016/j.ocecoaman.2013.04.010
Dharmawan B, Böcher M, Krott M (2016) The failure of the mangrove conservation plan in Indonesia: weak research and an ignorance of grassroots politics. Ocean Coastal Manag 130:250–259. https://doi.org/10.1016/j.ocecoaman.2016.06.019
Ellison AM, Felson AJ, Friess DA (2020) Mangrove rehabilitation and restoration as experimental adaptive management. Fron Mar Sci 7:327. https://doi.org/10.3389/fmars.2020.00327
Elmahdy SI, Ali TA, Mohamed MM, Howari FM, Abouleish M, Simonet D (2020) Spatiotemporal mapping and monitoring of mangrove forests changes from 1990 to 2019 in the Northern Emirates, UAE using random forest, kernel logistic regression and naive Bayes tree models. Front Environ Sci 8:102. https://doi.org/10.3389/fenvs.2020.00102
Friess DA (2017) Mangrove rehabilitation along urban coastlines: a Singapore case study. Reg Stud Mar Sci 16:279–289. https://doi.org/10.1016/j.rsma.2017.09.013
Friess DA, Yando ES, Abuchahla GM, Adams JB, Cannicci S, Canty SW et al (2020) Mangroves give cause for conservation optimism, for now. Curr Biol 30(4):R153–R154. https://doi.org/10.1016/j.cub.2019.12.054
Ghasemi S, Zakaria M, Abdul-Hamid H, Yusof E, Danehkar A, Rajpar MN (2010) A review of mangrove value and conservation strategy by local communities in Hormozgan province. Iran. J Am Sci 6(10):329–338
Gilani H, Naz HI, Arshad M, Nazim K, Akram U, Abrar A, Asif M (2021) Evaluating mangrove conservation and sustainability through spatiotemporal (1990–2020) mangrove cover change analysis in Pakistan. Estuar Coast Shelf Sci 249:107128. https://doi.org/10.1016/j.ecss.2020.107128
Githaiga MN, Kairo JG, Gilpin L, Huxham M (2017) Carbon storage in the seagrass meadows of Gazi Bay, Kenya. PloS One 12(5):e0177001. https://doi.org/10.1371/journal.pone.0177001
Goessens A, Satyanarayana B, Van der Stocken T, Quispe Zuniga M, Mohd-Lokman H, Sulong I, Dahdouh-Guebas F (2014) Is Matang Mangrove Forest in Malaysia sustainably rejuvenating after more than a century of conservation and harvesting management? PloS One 9(8):e105069. https://doi.org/10.1371/journal.pone.0105069

Goh HC (2016) Assessing mangrove conservation efforts in Iskandar Malaysia. Malaysia Sustainable Cities Program, Working Paper Series, pp 1–33

Gullström M, Dahl M, Lindén O, Vorhies F, Forsberg S, Ismail RO, Björk M (2021) Coastal blue carbon stocks in Tanzania and Mozambique: Support for climate adaptation and mitigation actions. IUCN, Gland, Switzerland, p 80

Guo Y, Liao J, Shen G (2021) Mapping large-scale mangroves along the maritime silk road from 1990 to 2015 using a novel deep learning model and Landsat data. Remote Sens 13(2):245. https://doi.org/10.3390/rs13020245

Hammond DS, Gond V, Baider C, Florens FBV, Persand S, Laurance SGW (2015) Threats to environmentally sensitive areas from peri-urban expansion in Mauritius. Environ Conserv 42(3):256–267. https://doi.org/10.1017/S0376892914000411

Hejnowicz AP, Kennedy H, Rudd MA, Huxham MR (2015) Harnessing the climate mitigation, conservation, and poverty alleviation potential of seagrasses: prospects for developing blue carbon initiatives and payment for ecosystem service programmes. Front Mar Sci 2:32. https://doi.org/10.3389/fmars.2015.00032

Himes-Cornell A, Pendleton L, Atiyah P (2018) Valuing ecosystem services from blue forests: a systematic review of the valuation of salt marshes, seagrass beds, and mangrove forests. Ecosyst Serv 30:36–48. https://doi.org/10.1016/j.ecoser.2018.01.006

Hoppe-Speer SC, Adams JB, Bailey D (2015) Present state of mangrove forests along the Eastern Cape coast. South Africa. Wetl Ecol Manag 23(3):371–383. https://doi.org/10.1007/s11273-014-9387-x

Hossain MS, Hashim M, Bujang JS, Zakaria MH, Muslim AM (2019) Assessment of the impact of coastal reclamation activities on seagrass meadows in Sungai Pulai estuary, Malaysia, using Landsat data (1994–2017). Int J Remote Sens 40(9):3571–3605. https://doi.org/10.1080/01431161.2018.1547931

Howard J, McLeod E, Thomas S, Eastwood E, Fox M, Wenzel L, Pidgeon E (2017) The potential to integrate blue carbon into MPA design and management. Aquat Conserv Mar Freshw Ecosyst 27:100–115. https://doi.org/10.1002/aqc.2809

Howari FM, Jordan BR, Bouhouche N, Wyllie-Echeverria S (2009) Field and remote-sensing assessment of mangrove forests and seagrass beds in the northwestern part of the United Arab Emirates. J Coast Res 25(1):48–56. https://doi.org/10.2112/07-0867.1

Huxham M, Emerton L, Kairo J, Munyi F, Abdirizak H, Muriuki T et al (2015) Applying climate compatible development and economic valuation to coastal management: a case study of Kenya's mangrove forests. J Environ Manag 157:168–181. https://doi.org/10.1016/j.jenvman.2015.04.018

Idrus AA, Syukur A, Zulkifli L (2019). The livelihoods of local communities: Evidence success of mangrove conservation on the coast of East Lombok Indonesia. In: AIP conference proceedings, AIP Publishing LLC, 2199(1), pp 050010. https://doi.org/10.1063/1.5141308

Iftekhar MS (2008) An overview of mangrove management strategies in three South Asian countries: Bangladesh, India, and Sri Lanka. Int for Rev 10(1):38–51. https://doi.org/10.1505/ifor.10.1.38

Islam MS, Wahab MA (2005) A review on the present status and management of mangrove wetland habitat resources in Bangladesh with emphasis on mangrove fisheries and aquaculture. In: Segers H, Martens K (eds) Aquatic biodiversity II. Developments in hydrobiology, vol 180, Springer, Dordrecht, pp 165–190. https://doi.org/10.1007/1-4020-4111-X_19

Jones BL, Cullen-Unsworth LC, Howard R, Unsworth RK (2018) Complex yet fauna-deficient seagrass ecosystems at risk in southern Myanmar. Bot Mar 61(3):193–203. https://doi.org/10.1515/bot-2017-0082

Jusoff K, Taha D (2008) Managing sustainable mangrove forests in Peninsular Malaysia. J Sustain Dev 1(1):88–96

Kodikara KAS, Mukherjee N, Jayatissa LP, Dahdouh-Guebas F, Koedam N (2017) Have mangrove restoration projects worked? An in-depth study in Sri Lanka. Restor Ecol 25(5):705–716. https://doi.org/10.1111/rec.12492

Kongkeaw C, Kittitornkool J, Vandergeest P, Kittiwatanawong K (2019) Explaining success in community-based mangrove management: Four coastal communities along the Andaman Sea, Thailand. Ocean Coast Manag 178:104822. https://doi.org/10.1016/j.ocecoaman.2019.104822
Lang'at JK, Kairo JG (2013) Conservation and management of mangrove forests in Kenya. Kenya Marine and Fisheries Research Institute Technical Report. http://hdl.handle.net/1834/6989
Latiff A, Faridah-Hanum I (2014) Mangrove ecosystem of Malaysia: status, challenges and management strategies. In: Faridah-Hanum I, Latiff A, Hakeem K, Ozturk M (eds) Mangrove ecosystems of Asia. Springer, New York, NY, pp 1–22. https://doi.org/10.1007/978-1-4614-8582-7_1
Lau WW (2013) Beyond carbon: Conceptualizing payments for ecosystem services in blue forests on carbon and other marine and coastal ecosystem services. Ocean Coast Manag 83:5–14. https://doi.org/10.1016/j.ocecoaman.2012.03.011
Liu SYV, Kumara TP, Hsu CH (2020) Genetic identification and hybridization in the seagrass genus Halophila (hydrocharitaceae) in Sri Lankan waters. PeerJ 8:e10027. https://doi.org/10.7717/peerj.10027
Lovelock CE, Atwood T, Baldock J, Duarte CM, Hickey S, Lavery PS, Masque P, Macreadie PI, Ricart AM, Serrano O, Steven A (2017) Assessing the risk of carbon dioxide emissions from blue carbon ecosystems. Front Ecol Environ 15(5):257–265. https://doi.org/10.1002/fee.1491
Lovelock CE, Duarte CM (2019) Dimensions of blue carbon and emerging perspectives. Biol Lett 15(3):20180781. https://doi.org/10.1098/rsbl.2018.0781
Lukman KM, Uchiyama Y, Quevedo JMD, Kohsaka R (2021) Local awareness as an instrument for management and conservation of seagrass ecosystem: case of Berau Regency, Indonesia. Ocean Coast Manag 203:105451. https://doi.org/10.1016/j.ocecoaman.2020.105451
Mangora MM (2011) Poverty and institutional management stand-off: a restoration and conservation dilemma for mangrove forests of Tanzania. Wetl Ecol Manag 19(6):533–543. https://doi.org/10.1007/s11273-011-9234-2
McKenzie LJ, Yaakub SM, Tan R, Seymour J, Yoshida RL (2016) Seagrass habitats of Singapore: environmental drivers and key processes. Raffles Bull Zool 34:60–77. http://zoobank.org/urn:lsid:zoobank.org:pub:7F125428-8891-48FC-B315-8D13981B200D
Memon SH (2012) An overview of mangrove restoration efforts in Pakistan. In: Proceedings and a call for action from an MFF regional colloquium 30–31, Sharing lessons on mangrove restoration, August 2012, Mamallapuram, India, pp 51–63
Ministry of Water and Environment Environment Protection Authority (2017) National Biodiversity Strategy and Action Plan II achieving a resilient, productive, and sustainable socio- ecosystem by 2050. The Republic of Yemen Report.
Mukhtar I, Hannan A (2012) Constrains on mangrove forests and conservation projects in Pakistan. J Coast Conserv 16(1):51–62. https://doi.org/10.1007/s11852-011-0168-x
Murray NJ, Keith DA, Tizard R, Duncan A, Htut WT, Hlaing N et al (2020) Threatened ecosystems of Myanmar. An IUCN red list of ecosystems assessment. https://library.wcs.org/doi/ctl/view/mid/33065/pubid/PUB25374.aspx. Accessed on 16 Oct 2021
Nagi HM, Abubakr MM (2013) Threats status to the mangrove ecosystem along the coastal zone of Yemen. J King Abdulaziz Univ 24(1):101. https://doi.org/10.4197/Mar.24-1.8
Naidoo G (2016) The mangroves of South Africa: an ecophysiological review. S Afr J Bot 107:101–113. https://doi.org/10.1016/j.sajb.2016.04.014
Newmaster AF, Berg KJ, Ragupathy S, Palanisamy M, Sambandan K, Newmaster SG (2011) Local knowledge and conservation of seagrasses in the Tamil Nadu State of India. J Ethnobiol Ethnomed 7(1):1–17. https://doi.org/10.1186/1746-4269-7-37
Ochieng CA, Erftemeijer PLA (2003) The seagrasses of Kenya and Tanzania. In Green EP et al (ed) World atlas of seagrasses, pp 82–92
Okafor-Yarwood I, Kadagi NI, Miranda NA, Uku J, Elegbede IO, Adewumi IJ (2020) The blue economy–cultural livelihood–ecosystem conservation triangle: the African experience. Front Mar Sci 7:586. https://doi.org/10.3389/fmars.2020.00586

Owuor MA, Mulwa R, Otieno P, Icely J, Newton A (2019) Valuing mangrove biodiversity and ecosystem services: a deliberative choice experiment in Mida Creek, Kenya. Ecosyst Serv 40:101040. https://doi.org/10.1016/j.ecoser.2019.101040

Patro S, Krishnan P, Deepak Samuel V, Purvaja R, Ramesh R (2017) Seagrass and salt marsh ecosystems in South Asia: an overview of diversity, distribution, threats and conservation status. In: Prusty B, Chandra R, Azeez P (eds) Wetland Science. Springer, New Delhi, pp 87–104. https://doi.org/10.1007/978-81-322-3715-0_5

Petrosian H, Kar AD, Ashrafi S, Feghhi J (2016) Investigating environmental factors for locating mangrove ex-situ conservation zones using GIS spatial techniques and the logistic regression algorithm in mangrove forests in Iran. Pol J Environ Stud 25(5):2097–2106. https://doi.org/10.15244/pjoes/62640

Phair NL, Nielsen ES, von der Heyden S (2021) Applying genomic data to seagrass conservation. Biodivers Conserv 30(7):2079–2096. https://doi.org/10.1007/s10531-021-02184-w

Pour MD, Barati AA, Azadi H, Scheffran J (2018) Revealing the role of livelihood assets in livelihood strategies: towards enhancing conservation and livelihood development in the Hara Biosphere Reserve, Iran. Ecol Indic 94:336–347. https://doi.org/10.1016/j.ecolind.2018.05.074

Prescott GW, Sutherland WJ, Aguirre D, Baird M, Bowman V, Brunner J et al (2017) Political transition and emergent forest-conservation issues in Myanmar. Conserv Biol 31(6):1257–1270. https://doi.org/10.1111/cobi.13021

Quevedo JMD, Uchiyama Y, Kohsaka R (2020) Perceptions of local communities on mangrove forests, their services, and management: Implications for Eco-DRR and blue carbon management for Eastern Samar, Philippines. J Forest Res 25(1):1–11. https://doi.org/10.1080/13416979.2019.1696441

Quevedo JMD, Uchiyama Y, Kohsaka R (2021) Local perceptions of blue carbon ecosystem infrastructures in Panay Island. Philippines. Coast Eng J 63(3):227–247. https://doi.org/10.1080/21664250.2021.1888558

Quevedo JMD, Uchiyama Y, Muhamad Lukman K, Kohsaka R (2021) How blue carbon ecosystems are perceived by local communities in the coral triangle: comparative and empirical examinations in the Philippines and Indonesia. Sustainability 13(1):127. https://doi.org/10.3390/su13010127

Quisthoudt K, Adams J, Rajkaran A, Dahdouh-Guebas F, Koedam N, Randin CF (2013) Disentangling the effects of global climate and regional land-use change on the current and future distribution of mangroves in South Africa. Biodivers Conserv 22(6):1369–1390. https://doi.org/10.1007/s10531-013-0478-4

Rafique M (2018) A review on the status, ecological importance, vulnerabilities, and conservation strategies for the Mangrove ecosystems of Pakistan. Pak J Bot 50(4):1645–1659

Rahman MM, Rahman MM, Islam KS (2010) The causes of deterioration of Sundarban mangrove forest ecosystem of Bangladesh: conservation and sustainable management issues. Aquac Aquar Conserv Legis 3(2):77–90. http://hdl.handle.net/10535/6481

Rajagopalan R (2011) Marine Protected Areas of India. In Sanders JS, Gréboval D, Hjort A (eds) Marine protected areas, country case studies on policy, governance and institutional issues. Food and agriculture organization of the United Nations, Rome, Itlay, pp 33–49

Rakotomahazo C, Ravaoarinorotsihoarana LA, Randrianandrasaziky D, Glass L, Gough C, Todinanahary GGB, Gardner CJ (2019) Participatory planning of a community-based payments for ecosystem services initiative in Madagascar's mangroves. Ocean Coast Manag 175:43–52. https://doi.org/10.1016/j.ocecoaman.2019.03.014

Rakotomahazo C, Razanoelisoa J, Ranivoarivelo NL, Todinanahary GGB, Ranaivoson E, Remanevy ME et al (2021) Community perceptions of a payment for ecosystem services project in southwest Madagascar: a preliminary study. Land 10(6):597. https://doi.org/10.3390/land10060597

Ramos S, Hesp PA (2018) Mangrove establishment in an artificially constructed estuarine channel, Sungei Api-Api, Singapore. In: Makowski C, Finkl C (eds) Threats to Mangrove forests. Coastal Research Library, vol 25. Springer, Cham, pp 419–427. https://doi.org/10.1007/978-3-319-73016-5_19

Rasool F, Tunio S, Hasnain S, Ahmad E (2002) Mangrove conservation along the coast of Sonmiani, Balochistan. Pakistan. Trees 16(2):213–217. https://doi.org/10.1007/s00468-001-0151-5
Rattanachot E, Stankovic M, Aongsara S, Prathep A (2018) Ten years of conservation efforts enhance seagrass cover and carbon storage in Thailand. Bot Mar 61(5):441–451. https://doi.org/10.1515/bot-2017-0110
Roy AKD (2016) Local community attitudes towards mangrove forest conservation: lessons from Bangladesh. Mar Pol 74:186–194. https://doi.org/10.1016/j.marpol.2016.09.021
Roy AKD, Alam K, Gow J (2013) Community perceptions of state forest ownership and management: a case study of the Sundarbans Mangrove Forest in Bangladesh. J Environ Manag 117:141–149. https://doi.org/10.1016/j.jenvman.2012.12.004
Samara F, Solovieva N, Ghalayini T, Nasrallah ZA, Saburova M (2020) Assessment of the environmental status of the mangrove ecosystem in the United Arab Emirates. Water 12(6):1623. https://doi.org/10.3390/w12061623
Samoilys MA, Osuka K, Mussa J, Rosendo S, Riddell M, Diade M et al (2019) An integrated assessment of coastal fisheries in Mozambique for conservation planning. Ocean Coast Manag 182:104924. https://doi.org/10.1016/j.ocecoaman.2019.104924
Sathirathai S, Barbier EB (2001) Valuing mangrove conservation in southern Thailand. Contemp Econ Pol 19(2):109–122. https://doi.org/10.1111/j.1465-7287.2001.tb00054.x
Scales IR, Friess DA, Glass L, Ravaoarinorotsihoarana L (2018) Rural livelihoods and mangrove degradation in south-west Madagascar: lime production as an emerging threat. Oryx 52(4):641–645. https://doi.org/10.1017/S0030605316001630
Seeruttun LD, Raghbor P, Appadoo C (2021) First assessment of anthropogenic marine debris in mangrove forests of Mauritius, a small oceanic island. Mar Pollut Bull 164:112019. https://doi.org/10.1016/j.marpolbul.2021.112019
Sejati AW, Buchori I, Kurniawati S, Brana YC, Fariha TI (2020) Quantifying the impact of industrialization on blue carbon storage in the coastal area of Metropolitan Semarang, Indonesia. Appl Geogr 124:102319. https://doi.org/10.1016/j.apgeog.2020.102319
Semesi AK (1992) Developing management plans for the mangrove forest reserves of mainland Tanzania. In: Jaccarini V, Martens E (eds) The ecology of mangrove and related ecosystems, Springer, Dordrecht, pp 1–10. https://doi.org/10.1007/978-94-017-3288-8_1
Shilland R, Grimsditch G, Ahmed M, Bandeira S, Kennedy H, Potouroglou M, Huxham M (2021) A question of standards: Adapting carbon and other PES markets to work for community seagrass conservation. Mar Pol 129:104574. https://doi.org/10.1016/j.marpol.2021.104574
Siddiqui PJ, Farooq S, Shafique S, Farooqi Z (2008) Conservation and management of biodiversity in Pakistan through the establishment of marine protected areas. Ocean Coast Manag 51(5):377–382. https://doi.org/10.1016/j.ocecoaman.2008.01.006
Sidik F, Supriyanto B, Krisnawati H, Muttaqin MZ (2018) Mangrove conservation for climate change mitigation in Indonesia. Wiley Interdisciplinary Reviews: Clim Chang 9(5):e529. https://doi.org/10.1002/wcc.529
Sivakumar K, Rilwan A, Priyanka K, Salah M, Kathiresan K (2018) Mangroves of the atolls of the Maldives, rich among the atoll groups of the Indian Ocean. ISME/GLOMIS Electron J 16(3):11–18
Soe-Htun U, Maung A, Mon S, Ha ST, Aung ST, Lwin AM, Lunn Z (2015) Seagrass Conservation and monitoring in Myanmar: the biodiversity, distribution, and coverage of seagrasses in the Tanintharyi and Rakhine. Bay of Bengal Large Marine ecosystem project report
Stevens GM, Froman N (2019) The Maldives archipelago. In Sheppard C (ed) World seas: an environmental evaluation, Academic Press, pp 211–236. https://doi.org/10.1016/B978-0-08-100853-9.00010-5
Sugandini D, Rahatmawati I, Arundati R (2017) Environmental attitude on the adoption decision mangrove conservation: an empirical study on communities in special region of Yogyakarta, Indonesia. Rev Integr Bus Econ Res 7(1). https://ssrn.com/abstract=3063374
Thangaradjou T, Bhatt JR (2018) Status of seagrass ecosystems in India. Ocean Coast Manag 159:7–15. https://doi.org/10.1016/j.ocecoaman.2017.11.025

Thompson BS (2018) The political ecology of mangrove forest restoration in Thailand: institutional arrangements and power dynamics. Land Use Pol 78:503–514. https://doi.org/10.1016/j.landusepol.2018.07.016

Thompson BS, Friess DA (2019) Stakeholder preferences for payments for ecosystem services (PES) versus other environmental management approaches for mangrove forests. J Environ Manag 233:636–648. https://doi.org/10.1016/j.jenvman.2018.12.032

Treephan P, Visuthismajarn P, Isaramalai SA (2019) A model of participatory community-based ecotourism and mangrove forest conservation in Ban Hua Thang. Thailand. Afr J Hosp Tour Leis 8(5):1–8

Udagedara S, Dahanayaka DDGL (2020) Current status and checklist of seagrass in Sri Lanka. Int J Aquat Biol 8(5):317–326. https://doi.org/10.22034/ijab.v8i5.619

Unsworth RK, Ambo-Rappe R, Jones BL, La Nafie YA, Irawan A, Hernawan UE et al (2018) Indonesia's globally significant seagrass meadows are under widespread threat. Sci Total Environ 634:279–286. https://doi.org/10.1016/j.scitotenv.2018.03.315

Veettil BK, Van DD, Quang NX (2020) Biodiversity and nature conservation in island ecosystems: spatiotemporal changes in Socotra archipelago (Yemen). J Coast Conserv 24(5):1–9. https://doi.org/10.1007/s11852-020-00774-4

Velde KV, Hugé J, Friess DA, Koedam N, Dahdouh-Guebas F (2019) Stakeholder discourses on urban mangrove conservation and management. Ocean Coast Manag 178:104810. https://doi.org/10.1016/j.ocecoaman.2019.05.012

Vierros M (2017) Communities and blue carbon: the role of traditional management systems in providing benefits for carbon storage, biodiversity conservation, and livelihoods. Clim Chang 140(1):89–100. https://doi.org/10.1007/s10584-013-0920-3

Viswanathan C, Purvaja R, Jeevamani JJJ, Samuel VD, Sankar R, Abhilash KR et al (2020) Salt marsh vegetation in India: Species composition, distribution, zonation pattern and conservation implications. Estuar Coast Shelf Sci 242:106792. https://doi.org/10.1016/j.ecss.2020.106792

Vyas P, Sengupta K (2012) Mangrove conservation and restoration in the Indian Sundarbans. In: Sharing lessons on mangrove restoration. Proceedings and a Call for Action from an MFF regional colloquium, pp 93–101

Wainwright BJ, Zahn GL, Zushi J, Lee NLY, Ooi JLS, Lee JN, Huang D (2019) Seagrass-associated fungal communities show distance decay of similarity that has implications for seagrass management and restoration. Ecol Evolut 9(19):11288–11297. https://doi.org/10.1002/ece3.5631

Wang Y, Bonynge G, Nugranad J, Traber M, Ngusaru A, Tobey J et al (2003) Remote sensing of mangrove change along the Tanzania coast. Mar Geod 26(1–2):35–48. https://doi.org/10.1080/01490410306708

Wickramasinghe D (2017) Regreening the coast: community-based mangrove conservation and restoration in Sri Lanka. In: DasGupta R, Shaw R (eds) Participatory mangrove management in a changing climate. Disaster risk reduction (methods, approaches, and practices). Springer, Tokyo, pp 161–171. https://doi.org/10.1007/978-4-431-56481-2_10

Yaakub SM, McKenzie LJ, Erftemeijer PL, Bouma T, Todd PA (2014) Courage under fire: seagrass persistence adjacent to a highly urbanised city–state. Mar Pollut Bull 83(2):417–424. https://doi.org/10.1016/j.marpolbul.2014.01.012

The manufacturer's authorised representative in the EU is Springer Nature Customer Service Centre GmbH, Europaplatz 3, 69115 Heidelberg, Germany. If you have any concerns regarding our products, please contact ProductSafety@springernature.com

Printed and bound by CPI Group (UK) Ltd, Croydon, CR0 4YY
28/09/2026
02224093-0009